CONTEMPORARY

Business

Mathematics

for Colleges

James E. Deitz, Ed.D.
President Emeritus, Heald Colleges

James L. Southam, Ph.D.
San Francisco State University

13th EDITION

THOMSON
SOUTH-WESTERN

Australia · Canada · Mexico · Singapore · Spain · United Kingdom · United States

THOMSON

SOUTH-WESTERN

Contemporary Business Mathematics for Colleges, 13/e

James E. Deitz, Ed.D. and James L. Southam, Ph.D

Vice President/Team Director:
Mike Roche

Publisher:
Mike Mercier

Senior Developmental Editor:
Susanna C. Smart

Marketing Managers:
Julie Lindsay and Mignon Tucker

Production Editor:
Margaret M. Bril

Manufacturing Coordinator:
Sandee Milewski

Production House:
Pre-Press Company

Printer:
Courier
Kendallville, Indiana 46755

Design Project Manager:
Michelle Kunkler

Cover & Internal Designer:
Grannan Graphic Design
Cincinnati, Ohio

Cover Image:
©Artville, Inc.

Photography Manager:
Deanna Ettinger

Photo Researchers:
Darren Wright and Sam Marshall

Library of Congress Cataloging-in-Publication Data

Deitz, James E.
 Contemporary business mathemat-
ics for colleges / James E. Deitz, James
L. Southam.—13e [ed.].
 p. cm.
 Includes index.
 ISBN 0-324-04299-X (student
ed.)(softcover)
 1. Business mathematics. I. Title:
Business mathematics. II. Southam,
James L. III. Title.

HF5691.D43 2002
650'.01'513—dc21

 2001057682

To the Student

During its previous twelve editions, *Contemporary Business Mathematics for Colleges* has sold more copies than any other business math textbook. The goal of the thirteenth edition is to make a successful book even better. This edition focuses on practical, real-world business mathematics, problems, and step-by-step solutions to help you solve these problems. The authors have done this with an eye toward contemporary business issues and the needs of today's business students, through new features that make learning easier and more enjoyable. *Contemporary Business Mathematics for Colleges* presents the basic principles of mathematics and immediately applies them in a series of practical business problems. The thirteenth edition is designed to provide a balance among conceptual understanding, skill development, and business applications.

Why Is Business Mathematics Important?

In the business world, everyone (employees and managers alike) needs knowledge of and skill in business mathematics. While computers and calculators are used for many calculations, it is important to understand the concepts behind mechanical computations. The purpose of a business mathematics course is to increase your math knowledge and skills as they apply to many aspects of business, and to help make you a more valuable player in the business arena.

Content and New Organization

COVERAGE. In addition to updating mathematical coverage from the previous edition, the authors have updated and expanded technology content. These practical additions of timely topics will help you better understand and prepare for positions in today's technology-oriented workplaces.

ORGANIZATION. The organization of the thirteenth edition has been revised to reflect the need for expanded coverage of some topics, and decreased coverage of others. Instead of two chapters on decimals, there is now one chapter on basic mathematical functions with decimals, titled *Decimals* (Chapter 4), and the two chapters on fractions have also been combined into one, *Fractions* (Chapter 5). The chapter on percents has been moved up as Chapter 6, *Percents in Business*, for more logical sequencing.

Key Features

Before you begin your journey through the thirteenth edition of *Contemporary Business Mathematics for Colleges,* take our guided tour through the special features that will make your journey a valuable learning experience.

Internet Chapter Openers

Feature. The two-page Internet chapter openers have been retained from the previous edition, with many being updated. These consist of screens from the Internet sites of real companies, most of which you may recognize, along with their Internet addresses.

Benefit. These company screens were chosen either to show relationships between the chapter coverage and the companies selected, or to lead you to sites that describe related career opportunities.

Learning Objectives

Feature. In addition, Learning Objectives and chapter-opening vignettes provide a preview of what the chapter holds, and introductory questions have been added. The Learning Objectives are repeated in the margins where first discussed, again in the Bottom Line feature, and also are identified for each end-of-chapter assignment.

Benefit. These enable you to easily find and focus on mastery of the chapter objectives.

NEW! CyberConnections

Feature. These new Internet features come in two sizes: mini features and full-sized. The mini-CyberConnections are interspersed throughout the chapters while full-sized CyberConnections—one or two per chapter— round out the end of each chapter. Most include a problem that is answered at the end of the text.

Benefit. These features enhance the text coverage with real-world math-related issues to show how mathematics is woven through our daily lives. Interesting facts, news stories, and Web sites are highlighted, and all are derived from the Internet.

NEW! Internet Margin Notes

Feature. New Internet margin notes are included in each chapter, and all are available on the text Web site as links for quick access.
Benefit. Internet addresses are provided at relevant points throughout to direct you to Web sites that will offer more help with chapter topics, or provide added information showing the relationships between business and mathematics.

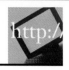

www.aspm.org/
Visit the Web site of the American Society for Payroll Management and read its newsletters on payroll and tax issues. This site also provides many links to other payroll information Web sites.

Concept Checks

Feature. Throughout the book, Concept Checks at the ends of chapter sections allow you to test your mastery of the topics you just studied.
Benefit. These self-checks let you know immediately if you need more practice, instead of having to wait until assignments are turned in or exams are taken.

> **CONCEPT CHECK 6.4**
>
> A company had sales of $144,000 this year and $120,000 last year (*B*). Find both the amount of increase (*P*) and the rate of increase (*R*).
>
> The amount of increase is
> $P = \$144,000 - \$120,000 = \$24,000$
>
> The Rate of increase is
> $R = P \div B = \$24,000 \div \$120,000 = 0.20 = 20\%$
>
> COMPLETE ASSIGNMENTS 6.2 AND 6.3.

NEW! Visit the Web

Feature. New to this edition is a list of Web sites relevant to the chapter material.
Benefit. You may visit these sites for additional information about the topics being covered in the chapter, for more depth, or simply to see how chapter topics are applied in the real world.

> **Visit the Web**
>
> **http://www.tax.org/defaultf.htm**—Articles and information on all types of taxes are available here.
>
> **http://www.taxpenalty.com/index.htm**—Time Value Software's tax penalty help site offers more tax information.
>
> **http://www.tax.state.ny.us/Statistics/Policy-Special/Sourcebook00/Sourcebook00_Contents.htm**—Total state taxes for each state are available here.
>
> **http://www.ncsl.org/programs/fiscal/tctelcom.htm** and **http://www.caltax.org/ecommerce.htm**—Visit this site for information on Internet taxation.

The Bottom Line

Feature. Bottom Line features, placed before the chapter assignments, are another method of review before beginning chapter assignments. The Learning Objectives are reviewed, an example problem is given, and answers are provided at the end of the feature.
Benefit. This new self-test helps you review the chapter material. If you had difficulty completing a Concept Check, this is another opportunity to see how well you have mastered the skills covered in the chapter.

> **The Bottom Line**
>
> *Summary of chapter learning objectives:*
>
Learning Objective	Summary	Example
> | **10.1** | Compute sales taxes, using rate tables and percents | 1. The Farley family lives in a state in which the sales tax rate is 8%. When they purchased a dining room table and chairs regularly priced at $1,400, they were given a discount of 15%. Shipping charges were $50. What was the total cost to the Farleys? |
> | | | 2. Lucy Carter lives in a state in which the state tax on gasoline is $0.18 a gallon. Federal tax is $0.19 a gallon. If she purchased an average of 8 gallons per week during the 52-week year, how much did she pay in state and federal taxes combined? |
> | **10.2** | Compute assessed valuations and property taxes based on assessed valuation | 3. The Nguyen family lives in a town in which the assessed valuation on property is 65% of market value. The Parker family lives in a town in which the assessed valuation on property is 80% of market value. Each home has a market value of $162,000. How much is the assessed valuation of each home? |
> | **10.3** | Compute tax rates in percents and mills | 4. The town of Belton has a total assessed valuation of $850,000,000. For the coming year the city must raise $11,730,000 for operating expenses. a. What will be the tax rate? b. What will the semiannual taxes be on a home with an assessed valuation of $135,000? |
> | | | 5. a. Convert 400 mills to its dollar equivalent. b. Convert $0.90 to mills. |

Self-Check | Review Problems for Chapter 8

1 Alex Muñoz is paid $15 per hour for the first 40 hours and one and one-half times his regular rate for all hours worked over 40 per week.
 a. Determine Alex's gross pay for the week if he works 45 hours.
 b. Calculate the amount to be deducted for Social Security and Medicare taxes for the week.
 c. Determine the amount to be withheld for federal income tax, using percentage method, if Alex is single and claims one withholding allowance.
 d. What is Alex's net pay for the week, assuming that his only payroll deductions are for Social Security, Medicare, and federal income tax?

2 Determine the amount to be withheld for federal income tax for each of the following, using both the percentage and the wage-bracket methods.
 a. A married employee, claiming 2 allowances, has weekly gross pay of $650.
 b. A single employee, with 1 allowance, has weekly gross pay of $525.

3 Calculate the employer's payroll taxes for each of the first 3 months of the year for three employees who are paid as follows:

Albertson, K.	$3,000 per month
Becket, W.	$4,000 per month
Jones, C.	$2,100 per month

Include FUTA (0.8%), SUTA (5.4%), Social Security (6.2%), and Medicare (1.45%) taxes. Be sure to consider the maximum taxable for unemployment taxes ($7,000) per employee.

4 Determine the taxes to be reported on the quarterly 941 form for an employer who paid total gross wages of $62,000 and withheld $7,800 for federal income tax.

Social Security	_____
Medicare	_____
Federal income tax	_____
Total	_____

5 Determine the amount to be withheld from the current period's gross pay of $6,500 for Social Security and Medicare for an employee whose cumulative wages were $78,500, not including pay for the current period. Use the rates and taxable maximum given in the chapter.

6 Employees of Xper Co. are paid at their regular rate for the first 40 hours, at one and one-half times their regular rate for hours worked between 40 and 48, and double their regular rate for all hours worked over 48, per week. Calculate each employee's gross pay for the week.
 John Kowalski, regular rate $12.16, worked 47 hours
 Martha Madison, regular rate $9.50, worked 50 hours
 Joy Weston, regular rate $10.80, worked 42 hours

NEW! Self-Check Review Problems

Feature. These new features at the end of each chapter come just before the chapter assignments.

Benefit. These problems provide yet another opportunity for you to test yourself before completing the end-of-chapter assignments. Answers are provided at the end of the text.

VIDEO

Percent of Increase or Decrease

NEW! Video Icons

Feature. Video icons are placed where appropriate throughout the text.

Benefit. These icons direct you to the video clips available on the CD-ROM that comes with the text. The videos highlight the use of mathematics in business and also walk through a number of problems from several chapters, a help whether you are a visual or audio learner.

Other Important Features

The thirteenth edition once again contains many of the features that were useful in making the text a student-friendly book in the previous edition.

- **Step-by-step format for solving word problems.** Chapter 2 presents the step-by-step illustration for solving word problems, and this process is carried out throughout the text, more than in the previous edition. By utilizing this format, you can increase your ability to correctly solve word problems.

- **Illustrated step-by-step solutions.** Short, concise text presentations of topics are followed by examples that show the step-by-step solutions. You will learn mathematical concepts by immediately applying practical solutions to common business problems.

- **Directed assignments.** At specific points within each chapter, you are directed to appropriate end-of-chapter assignments. This allows you to work end-of-chapter problems before completing the entire chapter.

- **End-of-chapter assignments.** Revised and expanded, the assignments continue to serve as a gauge of your understanding of the chapter materials. Each assignment is tied to the chapter's learning objectives.

STEPS to Figure the Amount of Federal Income Tax Withholding, Using the Percentage Method

1. Determine the employee's gross earnings.
2. Multiply the appropriate (weekly/monthly) "one withholding allowance" amount (from the Withholding Table) by the number of allowances the employee claims.
3. Subtract that amount from the employee's gross earnings.
4. From the appropriate (weekly/monthly and single/married) Percentage Method table, subtract the "of excess over" figure to get the amount subject to the tax.
5. Multiply the amount from Step 4 by the appropriate percentage from the Percentage Method table.
6. If required, add the base tax amount (if any) shown next to the percentage from the Percentage Method Table. (For example, see Table 1, WEEKLY Payroll Period, Married, the second line of the table: $125.50 plus 28% of excess over $960.)

Assignment 6.4: Allocation of Overhead

5 LEARNING OBJECTIVES

Name _____

Date _____ Score _____

A (20 points) Complete the square feet, percent, and distribution columns below. Round percents to the nearest whole number. (1 point for each correct answer in column 1; 2 points for each correct answer in columns 2 and 3)

1. Carla Viberti owns restaurants in four districts of town: (a) Charleston, (b) Empire, (c) Englewood, and (d) Milner Crest. She buys all the food through a central office, which she maintains at the Charleston restaurant. Monthly expenses for the office are distributed among the four restaurants, based on the floor space of each. In the following table, complete the distribution table for monthly expenses of $12,000.

Store	Space Occupied	Square Feet	Percent of Total	Distribution of Expense
(a) Charleston	30 ft × 70 ft	_____	_____	_____
(b) Empire	40 ft × 45 ft	_____	_____	_____
(c) Englewood	40 ft × 60 ft	_____	_____	_____
(d) Milner Crest	30 ft × 40 ft	_____	_____	_____
Total		7,500	100%	$12,000

Score for A (20)

B (16 points) Complete the percent and distribution columns in the following table. Before computing the distribution, round each percent to the nearest whole number. (2 points for each correct answer)

2. Diane Kingsley owns a temporary services company. She employs four types of employees whom she places into temporary positions: (a) accountants, (b) secretaries, (c) food service people, and (d) hotel service people. Diane rents office space for $4,200 per month. She distributes the rent among the four labor groups, according to the number of people from each group placed into jobs. Calculate the percents and the resulting distributions.

	Number of Employees	Percent of Total	Distribution of Rent
(a) Accountants	18	_____	_____
(b) Secretaries	42	_____	_____
(c) Food Service	36	_____	_____
(d) Hotel Service	24	_____	_____
Total	120	100%	$4,200

Score for B (16)

Suggestions for Using This Book

1. Read the text, study the step-by-step illustrations and the clear examples carefully, then work the assignments.
2. Read the instructions carefully for each assignment before solving the problems. Master the technique for solving word problems early, as presented in Chapter 2.
3. Do your own work. You can learn only by doing the calculations yourself. Ask your instructor for help if you have difficulty in understanding what you are asked to do, or how to do it.
4. Before working a problem, try to estimate your answer. The early chapters present methods for doing this.
5. Try to use shortcuts in your calculations. Shortcuts are presented in several chapters.
6. Write figures clearly through regular practice so you will not make errors as a result of mistaking one number for another.
7. Align figures in columns to avoid errors in addition or subtraction.
8. Remember that space is provided on the assignment sheets to compute most problems. You should show each step in the solution so that in case of error, your instructor can help you locate the cause of the difficulty.
9. Work the Concept Checks, The Bottom Line problems, and the Self-Check Review problems. Also turn to the Software Tutorial, the Online Quizzes, and the Practice Notebook for more practice in solving problems. These features and supplements give you a comprehensive review of the problems in each chapter, before you get to the assignments.
10. Record your scores for each assignment on the Progress Record at the end of the book.

Supplements to the Text

Contemporary Business Mathematics for Colleges comes with an integrated learning package.

A Robust Web Site

The text Web site at http://deitz.swcollege.com provides an even more robust array of learning resources than previously. It now includes Online Quizzes, the *Business Math Handbook*, all Internet links from the text, and more.

Business Math Handbook

The *Business Math Handbook,* by Keith H. Wieland of the University of Cincinnati–Clermont College, contains useful materials you will want to use as reference long after your business math courses are over. This is available only at the text Web site at http://deitz.swcollege.com

The *Practice Notebook for Business Mathematics*

A new supplement, this spiral-bound notebook gives you another resource for practicing problems, with ample space for working problems, and allows you to erase errors without erasing problems. Answers are provided at the end of the notebook. Prepared by Ann Aron, Keith E. Weidkamp, and Leland E. Mansuetti.

Online Quizzes

New to this edition, Online Quizzes prepared by Keith E. Weidkamp, of Sierra College, and Leland E. Mansuetti are available on the text Web site at http://deitz.swcollege.com. These quizzes provide

answer feedback, so you know why the correct answer is correct.

Tutorial Software

Even more problems are available to help you master the mathematical skills you need. The software prepared by Keith E. Weidkamp, of Sierra College, and Leland E. Mansuetti, provides you with an opportunity to work math problems on the computer, giving you mathematics and computer practice at the same time. The Tutorial Software is available in Windows on both the CD-ROM and the text Web site.

Spreadsheet Software

Spreadsheet templates are another computer supplement to give you practice with both mathematics and spreadsheet software. The Excel Spreadsheet software, prepared by Adele Stock of Normandale Community College, is available on the CD-ROM and the text Web site.

Topic Review Video

This video covers 12 major mathematics concepts (including fractions, percents, discounts and markups, and creating graphs) and applies them to a series of practical business problems. These professionally-produced, brief video segments have been digitized and placed on the CD-ROM for easier access, and will serve as a useful learning tool.

Student Resource CD-ROM

The Student Resource CD-ROM is packaged with every new text and includes the Software Tutorial, Excel Spreadsheets, *Business Math Handbook,* and digitized Topic Review Videos.

Acknowledgements

The authors would like to thank the many instructors who have reviewed and assisted with revisions of the many editions of this book; it is your suggestions and comments that have helped make this text what it is today.

Linda Johnson
Northern Illinois University

Fran March
Chattanooga State Technical
Community College

Cheryl Macon
Butler County Community College

Steven C. Teeter
Utah Valley State College

Jimmy Anderson
College of the Albemarle

Dale Dean
Athens Area Technical College

Kenneth J. Larson
LDS Business College

Zona J. Elkins
Blue Ridge Community
College

Alan Moggio
Illinois Central College

Elizabeth R. King
Heald Business College

Russ Nail
Hernando Community College

Allan L. Sheets
Indiana Business College

Nellie U. Edmundson
Miami Dade Community College

Kay Finlay
Indiana Vocational Technical College

William B. Harrison
DeVry Institute of Technology

Ann Aron
Aims Community College

Estelle Kochis
Suffolk County Community College

Paul Martin
Aims Community College

William D. Foster
Fontbonne College

John Falls
North Central State College

Charles Trester
Northeast Wisconsin Technical College

LaVerne Vertrees
St. Louis Community College

Queen Young
DeKalb Technical Institute

Dianne F. Hendrickson
Becker College

Karen O'Rourke
Lane Community College

Carol Perry
Marshall University Community and
Technical College

John Northrup
Bismarck State College

We would also like to thank Ann Aron, Keith Wieland, Leland Mansuetti, Keith Weid-kamp, and Adele Stock for their assistance with the supplements. In addition, we thank the staff at South-Western/Thomson Learning who worked to make this new edition the best business mathematics text possible: Marge Bril, our production editor, Mike Mercier, our acquisitions editor, Julie Lindsay, our marketing manager, and especially Susan Smart, our developmental editor, whose heart and soul and innovations have greatly improved the last two editions.

James E. Deitz
James L. Southam

Brief Contents

Contents

CONTEMPORARY

Business

Mathematics

for Colleges

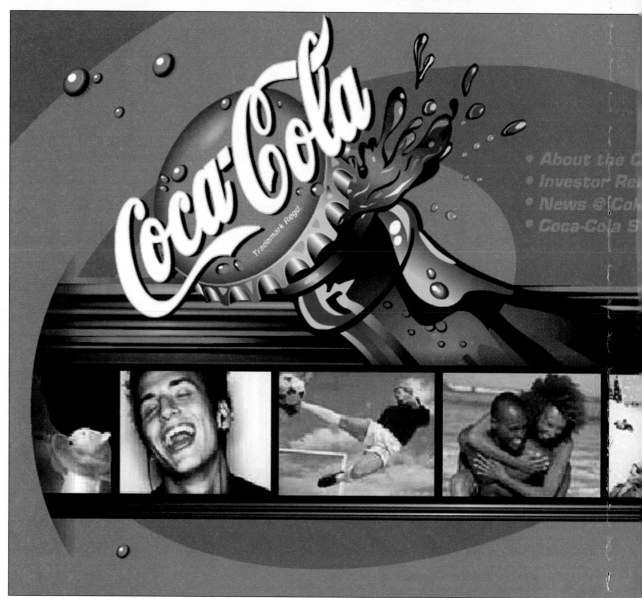

Although the Coca-Cola formula remains the world's best-kept secret, the company's name and products are famous throughout the world. Coca-Cola products are sold in more than 170 countries.

What are the employment opportunities at Coca-Cola? What is the company's mission? What is the current price for a share of Coca-Cola stock? Coke is worth a visit, at www.cocacola.com.

Fundamental Processes

1

LEARNING OBJECTIVES

By studying this chapter and completing all assignments you will be able to:

1
LEARNING OBJECTIVES
Use shortcuts and simplifications to perform the fundamental process of addition rapidly and accurately.

2
LEARNING OBJECTIVES
Use shortcuts and simplifications to perform subtraction rapidly and accurately.

3
LEARNING OBJECTIVES
Use simplifications to perform the fundamental process of multiplication.

4
LEARNING OBJECTIVES
Use shortcuts and simplifications to perform division rapidly and accurately.

5
LEARNING OBJECTIVES
Estimate answers before performing operations.

To perfect a skill, you must practice the basics until you are proficient.

With math in business, you practice the fundamentals of addition, subtraction, multiplication, and division over and over with many applications.

1. Why should you estimate answers for problems before you finalize your exact answers?
2. If subtraction is the opposite of addition, what is the opposite of multiplication?
3. Why should you learn shortcuts for adding columns of figures?

Addition

About half of all computations used in business involve addition. The more skilled you become in adding, the more rapidly you will get accurate answers. Addition is the process of finding the **sum** (total) of two or more **addends** (any of a set of numbers to be added).

Number Combinations

Certain aids can help you add more accurately and rapidly. One of the most helpful is to combine any two numbers that total 10. The following combinations total 10. Practice the combinations until you can identify them instantly.

1	2	3	4	5	9	8	7	6	5
9	8	7	6	5	1	2	3	4	5

When these combinations are found sequentially in any column of numbers, you should add them as 10. In example A, you might add the numbers in the right-hand (ones) column by saying, as you count down the column, "9 plus 4 is 13, plus 6 is 19, plus 3 is 22, plus 7 is 29, plus 8 is 37" (or simply "13, 19, 22, 29, 37"). By using the combinations of 10, however, you can simply add down the column by saying "9 plus 10 is 19, plus 10 is 29, plus 8 is 37" (or "9, 19, 29, 37").

The number 3 is carried over to the top of the next column and written in a small figure above the number 7 in the center column. The combinations of 10 are used in adding the center (tens) column by simply saying "10, 20, 30."

In adding the left-hand (hundreds) column, you carry over the number 3 from the center column total. You can simply say "8, 18, 28, 32."

EXAMPLE A

```
      ³   ³
    5  7  9
    4  2  4
    6  8  6
    9  0  3
    1  5  7
    4  5  8
  3,2  0  7
```

Also, learn to recognize the combinations of three numbers that total 10.

1	1	1	1	2	2	2	3
1	2	3	4	2	3	4	3
8	7	6	5	6	5	4	4

The numbers in each combination may appear in any order. For example, the numbers 2, 3, and 5 may appear in any of six arrangements.

2	3	2	3	5	5
3	2	5	5	2	3
5	5	3	2	3	2

When these numbers totaling 10 appear in sequence in a column, you should combine them and add them as 10. In example B, you might add the numbers in the ones column as you add down the column, "10, 18, 28, 38, 41." Write the number 4, which is

carried over as a small figure above the 1 in the tens column. Then use the combinations of 10 in adding the tens column by saying "5, 15, 25, 35, 43."

EXAMPLE B

$$
\begin{array}{r}
4 \\
(5) \quad 1\ \boxed{7} \\
\boxed{6}\ \boxed{2} \\
(15)\ \boxed{4}\ \boxed{1}\ (10) \\
\boxed{2}\ 8\ (18) \\
2\ \boxed{5} \\
(25)\ \boxed{6}\ \boxed{5}\ (28) \\
\boxed{5}\ \boxed{4} \\
2\ 4 \\
(35)\ \boxed{3}\ \boxed{2}\ (38) \\
(43)\ \underline{\ 8\ \ 3\ }\ (41) \\
4\ \ 3\ \ 1
\end{array}
$$

Repeated Digits

When you're adding a column in which many of the digits are the same, it is often quicker to count the number of repeated digits and then multiply the digit by that number. In example C, the ones column totals 33: 10 + 10 + 13. The tens column shows five 4s, equaling 20: 5 × 4 = 20. The 3 that was carried over and the 5 are then added to the 20 for a total of 28 in the tens column. The total for the problem is 283.

EXAMPLE C

$$
\begin{array}{r}
3 \\
\boxed{4}\ \boxed{1} \\
\boxed{4}\ \boxed{9}\ (10) \\
(5\times4)\ \boxed{4}\ \boxed{8} \\
\boxed{4}\ \boxed{2}\ (10) \\
\boxed{4}\ \boxed{7} \\
\underline{\ 5\ \boxed{6}\ }\ (13) \\
2\ \ 8\ \ 3\ (33)
\end{array}
$$

www.ricksmath.com

Visit this web site to brush up on addition and subtraction through many pratice problems.

Adding from Left to Right
(Columns of Two-Digit Numbers)

When adding columns of two-digit numbers, you can easily count by tens and add the ones column.

EXAMPLE D

Count:

12	12
24	22, 32 + 4 = 36
51	46, 56, 66, 76, 86 +1 = 87
43	97, 107, 117, 127 + 3 = 130
32	140, 150, 160 + 2 = 162
162	

This technique can be used to add four-digit columns in the same manner.

Count left columns *second*

33	
$43, 53 + 1 = 54$	
$64, 74, 84, 94, 104, 114 + 2 = 116$	
$126, 136 + 7 = 143$	

```
  33.28
  21.32
  62.28
  27.33
   1.21
 143.00
 144.21
```

Count right columns *first*

28
$38, 48, 58 + 2 = 60$
$70, 80 + 8 = 88$
$98, 108, 118 + 3 = 121$

Checking Addition

You should always check the accuracy of your addition. To do so, add the columns again in the opposite direction—that is, if you added down, add up for the check.

Horizontal Addition

When using business records, you may need to add numbers horizontally, or across the line, to save the time of recopying the numbers vertically. You may check several horizontal additions by adding the columns vertically and then adding these totals horizontally. This method is called **cross-checking.** The sums obtained by adding the totals horizontally and vertically should be the same.

282	+	346	+	723	+	409	+	716	=	2,476
113	+	806	+	629	+	916	+	620	=	3,084
240	+	318	+	718	+	312	+	309	=	1,897
716	+	501	+	423	+	716	+	114	=	2,470
872	+	417	+	909	+	704	+	472	=	3,374
2,223	+	2,388	+	3,402	+	3,057	+	2,231	=	13,301

CyberConnection

Stocks in the Dow Mon, 08 Jan 2001, 3:17pm EST

Company	Price per Share	Change	% Change	Volume of Shares Traded
Alcoa Inc.	33.5625	0.4375	1.32	5,992,267
American Express Co.	48.8750	−2.6875	−5.21	4,934,224
AT&T Corp.	19.7500	−0.4375	−2.17	8,354,111
Boeing Co.	59.0000	0.2500	0.43	2,365,000
Caterpillar Inc.	45.6250	−0.2500	−0.54	142,160

Source: http://www.bloomberg.com/markets/dow30.html. Copyright © 2001 by bloomberg.com; used with permission.

1. If you owned one share of each stock listed, what would be the total value of your stock ownership?

2. What is the total volume traded on the stock exchange for these five companies?

CONCEPT CHECK 1.1

Add horizontally and vertically; compare horizontal and vertical totals to verify accuracy. Use combinations to simplify addition.

2 4	+	7 6	+	6 3	=	163	(4 + 6)
3 6	+	2 4	+	2 5	=	85	(6 + 4)
2 7	+	4 3	+	1 2	=	82	(7 + 3)
8 7	+	1 4 3	+	1 0 0	=	330	

(Note horizontal combinations)

COMPLETE ASSIGNMENT 1.1.

Subtraction

Subtraction is the process of finding the difference between the **minuend** (number from which subtraction is being made) and the **subtrahend** (number being subtracted); the result is the **difference.** When the subtrahend is greater than the minuend, the result is a negative difference. In business, a negative difference may be called a **credit balance.** A credit balance is frequently shown in parentheses.

LEARNING OBJECTIVES

Use shortcuts and simplifications to perform subtraction rapidly and accurately.

EXAMPLE G

Positive Differences			Negative Difference (Credit Balance)
$32.22	$18.88	Minuend	$12.00
−22.22	−3.63	−Subtrahend	−13.50
$10.00	$15.25	Difference	($ 1.50)

Checking Subtraction

To check subtraction, use addition. If 209 is subtracted from 317, the difference is 108. You can check this result by adding the difference (108) to the subtrahend (209). The sum is 317. You can use the same procedure to check subtraction with a negative difference (credit balance).

EXAMPLE H

Subtract:	Check:
317	108
−209	+209
108	317

EXAMPLE I

Subtract:	Check:
$21.10	($ 3.40)
−24.50	+24.50
($ 3.40)	$21.10

Horizontal Subtraction

When using certain business forms, you may have to subtract numbers horizontally. You can check a number of horizontal subtractions by adding the columns vertically and then subtracting these totals horizontally. This answer should equal the total of the differences in the column at the right.

© PHOTODISC, INC.

Minuend		Subtrahend		Difference
$ 120	−	$ 20	=	$100
283	−	10	=	273
440	−	110	=	330
$ 269	−	$149	=	$120
$1,112	−	$289	=	$823

Subtraction by Changing Numbers

Sometimes you can simplify a subtraction problem, especially one that will involve *borrowing*, by altering it. If you add the same number to both the minuend and the subtrahend, the difference will remain the same. This will frequently make the answer immediately apparent.

Original Problem	Add:	Changed Problem	
45	+2	47	Minuend
−28	+2	−30	−Subtrahend
17		17	Difference

 CONCEPT CHECK 1.2

Subtract:	Check:	Subtract horizontally: Check by comparing totals	Add 3 to minuend and subtrahend
276	142	27 − 13 = 14	46 + 3 49
−134	+134	24 − 11 = 13	−27 + 3 30
		36 − 10 = 26	19 19
142	276	87 − 34 = 53	

COMPLETE ASSIGNMENT 1.2.

Multiplication

LEARNING OBJECTIVES

Use simplifications to perform the fundamental process of multiplication.

Multiplication, stated simply, is repeated addition. When two numbers (called **factors**) are multiplied, one number is repeated as many times as there are units in the other. The factor that is multiplied is called the **multiplicand.** The factor that indicates how many times to multiply is the **multiplier.** The result is the **product.**

CyberConnection

Source: http://www.figurethis.org/index40.htm. Copyright © 2001 Widmeyer Communications. Used with permission.

Which car would go the most highway miles on one tank of gas? Which car would go the most city miles on one tank of gas?

Car	Highway Mileage (mpg)	Tank Capacity (gal)	Approximate Distance (mi)
Century	29	17.5	508
Metro	49	10.3	505
Navigator	*17*	*30*	*510*

Car	City Mileage (mpg)	Tank Capacity (gal)	Approximate Distance (mi)
Century	20	17.5	350
Metro	*44*	*10.3*	*453*
Navigator	13	30	390

STEPS to Multiply Two Numbers

1. Make the smaller factor the multiplier.
2. Multiply from right to left.
3. Add the products to get the final product.

EXAMPLE L

	456	(multiplicand)
STEP 1	×237	(multiplier)
STEP 2	3 192	(product)
STEP 2	13 680	(product)
STEP 2	91 200	(product)
STEP 3	108,072	(final product)

In other words:

$$7 \times 456 = 3{,}192$$
$$30 \times 456 = 13{,}680$$
$$200 \times 456 = 91{,}200$$
$$237 \times 456 = 108{,}072$$

Checking Multiplication

The best method of checking multiplication is to divide the product by the multiplier to obtain the multiplicand. Example M shows the relationship between multiplication and division.

EXAMPLE M

Multiplicand	22	\longrightarrow	22
Multiplier	×6	\longrightarrow	6)132
Product	132	\longrightarrow	

Multiplying Numbers Ending in Zero

To multiply a number by 10, simply add a zero to the end of the number. To multiply a number by 100, add two zeros to the end: $10 \times 46 = 460$; $7{,}689 \times 100 = 768{,}900$.

STEPS to Multiply Numbers with Zeros

1. Make the multiplier the factor with the smaller number of digits after ignoring zeros at the right-hand side of the number.
2. Ignore the right-hand zeros and multiply the remaining numbers.
3. Insert the zeros ignored in Step 2 to the right-hand side of the product.

EXAMPLE N

STEP 1	370 × 200:	Make 2 the multiplier. Ignored:
	37	(1 zero)
	×2	(2 zeros)
STEP 2	74	(3 zeros)
STEP 3	74 000 = 74,000	

EXAMPLE O

STEP 1	1,200 × 160,800:	Make 12 the multiplier. Ignored:
	1,608	(2 zeros)
	×12	(2 zeros)
	3 216	
	16 08	
STEP 2	19,296	(4 zeros)
STEP 3	19,296 0000 = 192,960,000	

Multiplying When the Multiplier Contains Zero Not on the End

Often a zero appears in the center of the multiplier rather than at the end. To multiply 42,674 by 401, first multiply the multiplicand by 1 and write down the product. Then multiply by 4 (which is really 400) and write the result two places, instead of one, to the left. In other words, one extra place is left for each zero in the multiplier.

EXAMPLE P

```
    42,674
  ×   401
    42 674
 17 069 6        (2 places)
 17,112,274
```

Whenever more than one zero appears in the multiplier, the multiplication process is similar. To multiply 33,222 by 2,004, as in Example Q, first multiply 33,222 by 4. Then multiply 33,222 by 2, writing the answer three places to the left. Remember, extra places must be left for the two zeros (1 place + 2 extra places = 3 places).

EXAMPLE Q

```
    33,222
  × 2,004
   132 888
 66 444       (3 places)
 66,576,888
```

Multiplying the Product of Two Factors

Sometimes in business you will need to multiply two factors and then multiply the product of those factors by a third factor. As shown in example R, you begin by multiplying the first two factors and then multiply that product by the third factor.

EXAMPLE R

$21 \times 30 \times 12 = 7{,}560$

$$
\begin{array}{r}
21 \\
\times 30 \\
\hline
630
\end{array}
\qquad
\begin{array}{r}
630 \\
\times 12 \\
\hline
1\,260 \\
6\,30 \\
\hline
7{,}560
\end{array}
$$

Multiplying by 25

A shortcut for multiplying by 25 is to multiply by 100 (increase by two zeros) and divide by 4.

EXAMPLE S

321×25

$32{,}100 \div 4 = 8{,}025$

EXAMPLE T

828×25

$82{,}800 \div 4 = 20{,}700$

Multiplying by 50

A shortcut for multiplying by 50 is to multiply by 100 (increase by two zeros) and divide by 2.

EXAMPLE U

732×50

$73{,}200 \div 2 = 36{,}600$

CONCEPT CHECK 1.3

To multiply by 10, add one zero to the end of the number:

$36 \times 10 = 360$

To multiply by 100, add two zeros to the end of the number:

$36 \times 100 = 3{,}600$

Multiply:

$$
\begin{array}{r}
214 \longrightarrow \text{multiplicand} \\
102 \longrightarrow \text{multiplier} \\
428 \\
21\ 4 \qquad \text{(two places)} \\
\hline
21{,}828 \longrightarrow \text{product}
\end{array}
$$

Check:

$$
\begin{array}{r}
214 \\
102\overline{)21{,}828}
\end{array}
$$

COMPLETE ASSIGNMENT 1.3.

Division

Division is the process of finding how many times one number (the **divisor**) is contained in another (the **dividend**). The result is called the **quotient.** If anything remains after the division is completed, it is called the **remainder.** In example V, $47 \div 2 = 23$ (with 1 left over), 47 is the dividend, 2 is the divisor, 1 is the remainder, and 23 with a remainder of (1) is the quotient.

LEARNING OBJECTIVES

Use shortcuts and simplifications to perform division rapidly and accurately.

EXAMPLE V

```
   23 (1)
2)47
   4
  ─
   7
   6
  ─
   1
```

EXAMPLE W

```
        174          STEP 3
164)28,536           STEPS 1 & 2
   16 4              STEP 4
   ────
   12 13             STEP 5
   11 48
   ────
      656            STEP 6
      656
      ───
        0
```

When the partial dividend is smaller than the divisor, a zero must be placed in the quotient above that digit. This process is continued until the partial dividend is at least as large as the divisor. Then continue the long division steps, as shown in example X.

EXAMPLE X

```
    20,108
34)683,672
   68
   ──
    3 6
    3 4
    ───
      272
      272
      ───
        0
```

Checking Division

To check division, simply multiply the quotient by the divisor and add any remainder to the product. The result will equal the original dividend. (Examples Y and Z provide checks for examples W and X.)

EXAMPLE Y

$$\begin{array}{r} 174 \\ \times 164 \\ \hline 696 \\ 10\,44 \\ 17\,4 \\ \hline 28{,}536 \end{array}$$

EXAMPLE Z

$$\begin{array}{r} 20{,}108 \\ \times 34 \\ \hline 80\,432 \\ 603\,24 \\ \hline 683{,}672 \end{array}$$

Note: Division is the reverse process of multiplication.

Dividing by 10

To divide by 10, drop the digit at the extreme right of the dividend; the dropped digit will be the remainder.

EXAMPLE AA

$790 \div 10 = 79$ (0 remainder)

EXAMPLE BB

$3{,}652 \div 10 = 365$ (2 remainder)

Dividing by 100

To divide by 100, drop the two right-hand digits of the dividend—they will be the remainder.

EXAMPLE CC

$81{,}400 \div 100 = 814$ (0 remainder)

EXAMPLE DD

$257{,}948 \div 100 = 2{,}579$ (48 remainder)

Dividing When the Divisor and Dividend End with Zeros

When a divisor and dividend both end with zeros, a division shortcut is to delete the ending zeros common to both and then divide.

EXAMPLE EE

Both divisor and dividend end with zeros	Zeros common to divisor and dividend have been dropped	Answer
$8{,}400 \div 200$	$84 \div 2$	42
$46{,}000 \div 2{,}300$	$460 \div 23$	20
$42{,}000 \div 100$	$420 \div 1$	420
$20{,}000{,}000 \div 4{,}000$	$20{,}000 \div 4$	5,000
$2{,}760 \div 270$	$276 \div 27$	10 (6 remainder)
$3{,}200 \div 1{,}000$	$32 \div 10$	3 (2 remainder)

www.std.com/anpn/

The Adult Numeracy Network provides math problems and practice designed for adults, as well as links to other useful sites.

CONCEPT CHECK 1.4

Divide:

$$
\begin{array}{r}
21 \quad \text{quotient}\\
\text{divisor} \longrightarrow 32\overline{)683} \longleftarrow \text{dividend}\\
\underline{64}\\
43\\
\underline{32}\\
11 \longleftarrow \text{remainder}
\end{array}
$$

Dividing by 10: $860 \div 10 = 86$
 $863 \div 10 = 86$ (3 remainder)

Dividing by 100: $19,300 \div 100 = 193$
 $19,346 \div 100 = 193$ (46 remainder)

Check:

divisor × quotient + remainder = dividend

$$32 \quad \times \quad 21 \quad + \quad 11 \quad = \quad 683$$

COMPLETE ASSIGNMENT 1.4.

Estimating

5
LEARNING OBJECTIVES

Estimate answers before performing operations.

Estimating When Multiplying

Is estimating important? *Yes*, it is! In using a calculator to make computations, you may possibly omit keystrokes, accidentally repeat keystrokes, or incorrectly shift/omit decimal points. There is a great deal of difference between three times $14.87 and three times $1,487. When working with calculations in any manner—such as entering items into a spreadsheet, a cash register, or a calculator—you should always have a mental estimate of the final product.

Mentally estimating an answer provides a good method for checking whether your product is a reasonable answer.

VIDEO
Estimating Answers

STEPS to Estimate a Multiplication Answer

1. Round both the multiplicand and multiplier to the nearest 10 for two-digit numbers, the nearest 100 for three-digit numbers, the nearest 1,000 for four-digit numbers, etc.
2. Drop the zeros to the right of the nonzero numbers.
3. Mentally multiply the nonzero numbers to determine the base product.
4. Reinsert *all* zeros dropped in Step 2.

EXAMPLE FF

| | | | | Reinsert Zeros | |
Problem	Round to	Drop Zeros	Base Product	Estimated Answer	Real Answer
68 × 21	70 × 20	7 × 2	14	1,400	1,428
693 × 1,957	700 × 2,000	7 × 2	14	1,400,000	1,356,201
7,869 × 43,242	8,000 × 40,000	8 × 4	32	320,000,000	340,271,298
9 × 511,739	9 × 500,000	9 × 5	45	4,500,000	4,605,651
891 × 39 × 104	900 × 40 × 100	9 × 4 × 1	36	3,600,000	3,613,896

Estimating When Dividing

Before doing long division problems, estimate a whole-number answer. The process of mentally estimating whole-number answers helps to avoid major and embarrassing errors.

STEPS **to Estimate a Long Division Answer**

1. Round both the divisor and dividend to the nearest 10 for two-digit numbers, the nearest 100 for three-digit numbers, the nearest 1,000 for four-digit numbers, etc.
2. Drop the number of zeros common to both.
3. Mentally divide the remaining divisor into the remaining dividend.

EXAMPLE GG

Problem	Round to	Drop Zeros	Estimated Answer	Real Answer
77 ÷ 39	80 ÷ 40	8 ÷ 4	2	1.97
196 ÷ 63	200 ÷ 60	20 ÷ 6	3*	3.11*
2,891 ÷ 114	3,000 ÷ 100	30 ÷ 1	30	25.36
592 ÷ 29	600 ÷ 30	60 ÷ 3	20	20.41
18,476 ÷ 384	20,000 ÷ 400	200 ÷ 4	50	48.11
917 ÷ 186	900 ÷ 200	9 ÷ 2	4*	4.93*
21,716,412 ÷ 40,796	20,000,000 ÷ 40,000	2,000 ÷ 4	500	532.32
99,624 ÷ 476	100,000 ÷ 500	1,000 ÷ 5	200	209.29
29,200 ÷ 316	30,000 ÷ 300	300 ÷ 3	100	92.41

CONCEPT CHECK 1.5

ESTIMATING MULTIPLICATION ANSWERS

Problem	Round to	Drop Zeros	Base Product	Reinsert Zeros Estimated Answer	Real Answer
47 × 31	50 × 30	5 × 3	15	1,500	1,457
498 × 221	500 × 200	5 × 2	10	100,000	110,058

ESTIMATING DIVISION ANSWERS

Problem	Round to	Drop Zeros	Estimated Answer	Real Answer
88 ÷ 29	90 ÷ 30	9 ÷ 3	3	3.03
9,811 ÷ 394	10,000 ÷ 400	100 ÷ 4	25	24.90

COMPLETE ASSIGNMENT 1.5.

*Because 20 ÷ 6 and 9 ÷ 2 would result in a remainder, we can reasonably assume that the real number will be *larger*.

The Universal Product Code (UPC) was the first bar code symbology widely adopted. Its birth is usually set at April 3, 1973, when the grocery industry formally established UPC as the standard bar code symbology for product marking. Foreign interest in UPC led to the adoption of the (EAN) code format, similar to UPC, in December 1976.

There are now five versions of UPC, two versions of EAN, and a Japanese Article Numbering (JAN) code identical to one of the EAN versions. By 2005, all U.S. retailers will have to be able to scan all EAN/UPC article numbers (8-, 12-, 13-, and 14-digit).

Chapter Terms for Review

addend	dividend	multiplicand	remainder
credit balance	divisor	multiplier	subtrahend
cross-checking	factors	product	sum
difference	minuend	quotient	

Visit the Web

Visit these sites for information on business mathematics topics in other countries:

http://www.doc.mmu.ac.uk/courses/course-bsc-business-mathematics.html

http://www.puk.ac.za/bmi/

http://www3.oup.co.uk/imaman/

http://www.lse.ac.uk/undergraduate/courses/business/default.htm

http://www.doc.mmu.ac.uk/COURSES/BM/page2.html

CYBERCONNECTION

Peet's brews up an IPO

Coffee roaster expects to trade late in January; BTI Telecom pulls issue

January 5, 2001: 2:11 p.m. ET

NEW YORK (CNNfn) - Peet's Coffee & Tea Inc., a specialty coffee roaster, set the pricing date Friday of its planned IPO for later this month.

Emeryville, Calif.-based Peet's Coffee plans to offer 3.3 million shares at $10 to $14 each via lead underwriters WR Hambrecht & Co. The issue is expected to begin trading the week of Jan. 22.

IPOs ground to a halt in late 2000 and Peet's is one of the first companies to brave the waters this year. The 34-year-old Peet's has developed a strong following along the West Coast, especially in the San Francisco Bay area. Peet's is targeting the specialty coffee market, competing against such established companies as Starbucks Corp. (SBUX: Research, Estimates).

Peet's sells coffee in four states and derives much of its revenue from the sale of whole bean coffee. For the nine months ended Oct. 1, Peet's lost $2.7 million on revenue of $59.4 million.

Peet's Coffee & Tea, Inc. went public and sold 3.3 million shares of its stock. You had $2,000 to spend on stocks and wanted to get in on this deal. If the stock was selling for $12 per share, how many did you buy?

Source: http://cnnfn.cnn.com/2001/01/05/deals/ipos, January 5, 2001. Copyright © 2001 by Cable News Network; used with permission.

The Bottom Line

Summary of chapter learning objectives:

Learning Objective	Summary	Example
1.1	Use shortcuts and simplifications to perform the fundamental process of addition rapidly and accurately	Add the following, using the technique indicated. **Number combinations** / **Repeated digits** / **Counting by tens** 1. 8 2. 18 3. 52 4. 23 2 62 58 41 3 43 57 37 2 27 52 56 +5 +80 +51 +42 Add and then check by adding both vertically and horizontally. 5. $22 + 54 + 63 + 37 =$ _____ $27 + 82 + 44 + 19 =$ _____ $83 + 39 + 72 + 12 =$ _____ $91 + 71 + 21 + 84 =$ _____ __ + __ + __ + __ = _____
1.2	Use shortcuts and simplifications to perform subtraction rapidly and accurately	Subtract the following and then check by addition. 6. 228 ____ 7. 335 ____ −134 +134 −217 +217 Subtract horizontally and check. Subtract by changing numbers. 8. $245 - 130 =$ _____ 9. 53 $432 - 212 =$ _____ −18 $381 - 270 =$ _____ $183 - 111 =$ _____ __ − __ = _____
1.3	Use simplifications to perform the fundamental process of multiplication	Multiply. **Multiplying by numbers ending in zero** 10. 227 11. 437 12. 879 ×143 ×100 ×10 *32,461* *43,700* *8,790* **Multiplying by 25** **Multiplying by 50** 13. 354 14. 846 ×25 ×50 *8950*

Summary of chapter learning objectives:

Learning Objective	Summary	Example
1.4	Use shortcuts and simplifications to perform division rapidly and accurately	Divide and check the answer by multiplication. 15. $27)\overline{1,512}$ $\begin{array}{r} 56 \\ \\ 27 \\ \times\ 56 \\ \hline 1,512 \end{array}$ **Dividing by numbers ending in 0** 16. $8,430 \div 10 =$ _843_ 17. $127,400 \div 100 =$ _1274_ *Tip: Zeros can be Dropped and not Added back in the Answer.* **Dividing when both divisor and dividend end with zeros** 18. $7,400 \div 200 =$ _____ 19. $53,200 \div 400 =$ _____ 20. $140,000 \div 2,000 =$ _____
1.5	Estimate answers before performing operations	Estimate these multiplication answers. Show your rounding, dropping of zeros with base product, estimated answer, and real answer.

Problem	Round to	Dropped Zeros and Base Product	Estimated Answer	Real Answer
21. 47×31	50 X 30	5 X 3	1500	1457
22. 498×221	500 X 200	5 X 2	100,000	~~109,258~~ 110,058

Estimate these division answers. Show your rounding, dropping of zeros, estimated answer, and real answer.

Problem	Round to	Drop Zeros	Estimated Answer	Real Answer
23. $88 \div 29$	90 ÷ 30	_____	~~3~~	3.03
24. $9,811 \div 394$	10,000 ÷ 400	_____	~~25~~	24.90

Self-Check | Review Problems for Chapter 1

1 $8 + 9 + 3 + 12 + 6 =$ _____

2 $32 + 47 + 36 + 12 =$ _____
$17 + 22 + 17 + 11 =$ _____
$14 + 98 + 47 + 81 =$ _____
$77 + 62 + 21 + 44 =$ _____
_____ _____ _____ _____ $=$ _____

3 9,078
 $-6,382$

4 $717 \div 14 =$ _____

5 $98 \times 13 =$ _____

6 $789 \div 36 =$ _____

7 $842 \times 200 =$ _____

8 $974 \div 12 =$ _____

9 $27\overline{)876}$

10 Give an estimated answer: 78×31

11 Give estimated answer: $396 \div 79$

12 $800 \div 25 =$ _____

13 $4,000,000 \div 400 =$ _____

14 $\$370 - \$148 =$ _____
$\$422 - \$109 =$ _____
$\$982 - \$777 =$ _____
$\$\underline{\quad} - \$\underline{\quad} =$ _____

15 1,472
 $\times 28$

16 $704 \times 1,002 =$ _____

17 $704 \div 25 =$ _____

18 $16,000 \div 25 =$ _____

19 Estimate an answer: $899 \div 32$

20 Estimate an answer: $4,078 \times 394$

Assignment 1.1: **Addition**

Name _____

Date _____ Score _____

A **(10 points) Add the following. Where possible, use combinations of ten. (1 point for each correct answer)**

1. 14	**2.** 41	**3.** 19	**4.** 34	**5.** 97	**6.** 50	**7.** 72	**8.** 82	**9.** 38	**10.** 94
56	29	54	33	44	54	99	43	39	71
25	17	14	43	33	54	99	47	22	56
53	36	81	37	76	47	89	93	45	55
52	44	28	36	32	59	47	58	47	84
35	15	11	34	72	54	63	34	25	31
32	56	43	32	34	55	40	22	13	96
48	62	51	38	76	55	62	46	29	76
32	66	76	32	27	35	68	73	79	62

Score for A (10) _____

B **(10 points) Add the following. (1 point for each correct answer)**

11. 209	**12.** 782	**13.** 127	**14.** 920	**15.** 347	**16.** 852	**17.** 251	**18.** 885	**19.** 275	**20.** 479
486	280	145	751	399	428	271	115	342	413
225	438	665	359	354	112	244	316	342	180
219	473	818	822	334	238	234	584	898	418
684	655	682	807	192	959	589	736	505	415

Score for B (10) _____

C **(10 points) Add the following. (1 point for each correct answer)**

21. 248.28	**22.** 201.22	**23.** 234.81	**24.** 238.69	**25.** 586.32
820.14	513.14	371.60	982.30	222.48
306.80	250.54	271.37	376.48	348.86
521.98	2,647.55	408.55	728.90	237.64

26. 703.91	**27.** 126.92	**28.** 442.71	**29.** 535.13	**30.** 333.42
422.38	32.15	71.93	44.78	444.48
721.05	873.19	416.90	208.17	222.25
446.21	872.52	236.19	6,481.29	666.25

Score for C (10) _____

Assignment 1.1 *Continued*

D **(10 points)** Add the following. Use the count-by-tens-and-add-the-ones method. (1 point for each correct answer)

31. 10.76	**32.** 20.43	**33.** 33.79	**34.** 45.86	**35.** 33.27	**36.** 11.43	**37.** 88.71	**38.** 94.32	**39.** 55.93	**40.** 22.79
31.43	82.76	42.56	22.18	98.21	27.43	56.32	74.23	10.70	43.28
88.33	30.42	12.70	33.81	90.01	11.51	83.70	21.44	30.46	12.48
33.08	64.22	21.20	10.04	11.33	21.48	44.12	63.01	47.05	53.20
12.33	56.03	22.19	80.31	33.04	11.80	23.51	34.20	80.11	30.22

Score for D (10)

E **(30 points)** Business Application. The following is the first part of a weekly sales summary—the Weekly Sales Report for the computer department. Complete the totals, both horizontal and vertical, and verify your addition by comparing the vertical and horizontal grand totals. (2 points for each column/row; 4 points for grand total)

DEPARTMENT SALES REPORT
Week of December 11–17, 20XX

Department: COMPUTERS

SALESPERSON	SUN	MON	TUE	WED	THU	FRI	SAT	TOTAL
Whalen	3,443	—	—	8,643	3,176	7,885	9,378	
Tsao	—	8,772	—	9,483	7,339	8,113	9,771	
Culver	8,722	2,443	3,114	5,729	6,193	—	—	
Hernandez	6,117	8,783	—	—	5,685	9,473	11,492	
Ingake	—	3,114	8,492	7,652	3,994	14,119	12,378	
Greenberg	—	—	5,141	2,739	8,941	2,836	10,242	
Total								

Score for E (30)

F **(30 points)** Business Application. The following is the second part of the weekly sales summary—the Consolidated Sales Report for the entire store. Fill in the figures from the Department Sales Report and complete the totals, both horizontal and vertical. Verify your addition by comparing the horizontal and vertical grand totals. (2 points for each column/row; 2 points for grand total)

STORE SALES REPORT
Week of December 11–17, 20XX

DEPARTMENT	SUN	MON	TUE	WED	THU	FRI	SAT	TOTAL
Home Audio	3,465	1,147	1,523	2,403	1,773	2,873	3,432	
Auto Audio	1,278	1,785	1,713	2,117	2,563	3,499	9,971	
Video/TV	15,230	12,377	10,429	9,384	8,773	11,245	13,486	
Computers								
Telecomm	849	722	531	733	1,012	1,239	1,375	
Games	882	248	379	287	415	978	1,015	
Repairs	732	892	384	658	981	1,043	1,774	
Total								

Score for F (30)

Assignment 1.2: Subtraction

Name

Date Score

A **(18 points) Subtract the following. (One point for each correct answer)**

1. 77	**2.** 50	**3.** 72	**4.** 63	**5.** 53	**6.** 38	**7.** 92	**8.** 83	**9.** 60
−16	−23	−25	−29	−48	−49	−16	−65	−10

10. 39	**11.** 40	**12.** 13	**13.** 73	**14.** 37	**15.** 68	**16.** 99	**17.** 57	**18.** 91
−36	−31	−26	−14	−22	−39	−27	−43	−69

Score for A (18)

B **(12 points) Subtract the following. Then check your subtraction by adding the subtrahend and the difference and comparing your total to the minuend. (Two points for each correct answer)**

19. 584	**20.** 963	**21.** 92	**22.** 714	**23.** 616	**24.** 7,116
−173	−874	−491	−30	−333	−3,007

Score for B (12)

C **(6 points) Subtract the following. (1 point for each correct answer)**

25. $97.17	**26.** $15.67	**27.** $71.69	**28.** $43.21	**29.** $80.41	**30.** $90.61
−23.19	−0.88	−10.87	−47.18	−41.80	−20.50

Score for C (6)

D **(9 points) Subtract the following. (One and one-half points for each correct answer)**

31. $8,042.88	**32.** $964.38	**33.** $9,011.09	**34.** $7,430.29	**35.** $3,385.03	**36.** $1,029.27
−3,400.07	−201.83	−795.08	−2,597.73	−233.42	−89.27

Score for D (9)

E **(15 points) Sometimes a double subtraction is necessary. The following problems are of this type. (3 points for each correct final answer)**

37. $7,672.18	**38.** $11,739.93	**39.** $734.12	**40.** $745.89	**41.** $1,837,042.03
−564.27	−3,142.18	−672.18	−250.15	−6,218.18
−124.13	−1,694.25	−13.14	−224.13	−39,917.16

Score for E (15)

Assignment 1.2 *Continued*

F **(20 points) Business Application. In many cases, multiple subtractions each are required to complete a business transaction. (1 point for each intermediate answer; 2 points for each final answer)**

WINTER CATALOG CLEARANCE SALE ON SOFTWARE AND GAMES
10% REDUCTIONS ON CATALOG ORDERS
10% PREFERRED CUSTOMER DISCOUNTS
MAIL-IN REBATE OFFERS

Item	Football Pro	Nemesis	Harpoon 97	Mind	Formula 1
List price	$43.95	$45.70	$42.25	$49.95	$53.75
Less 10% catalog rate	−4.40	−4.57	−4.23	−5.00	−5.38
Less 10% preferred customer rate	−3.96	−4.11	−3.80	−4.50	−4.84
Mail-in rebate	−7.50	−6.25	−7.50	−6.75	−5.75
Your price					

Score for F (20)

G **(20 points) Business Application. Maintaining a budget involves both addition and subtraction. Keeping a budget sometimes involves a continuous record of cash income and expenses. Study the example and then complete the balances. (2 points for each balance)**

Date 2/1/98	To	Subtract Expenses	Add Income	Balance
				$1,475.38
2/2/98	Salary income		$700.00	2,175.38
2/3/98	Hinson Real Estate	$550.00		1,625.38
2/5/98	PG&E	23.22		
2/6/98	Pacific Bell	18.76		
2/6/98	Macy's	43.22		
2/10/98	Chevron	15.75		
2/16/98	Salary income		$700.00	
2/17/98	Fitness USA	25.00		
2/18/98	John Simms, D.D.S.	30.00		
2/23/98	Prudential Insurance	17.73		
2/25/98	Visa	85.42		
2/27/98	General Motors Finance	257.87		

Score for G (20)

Assignment 1.3: **Multiplication**

Name _____

Date _____ Score _____

A **(12 points) Multiply the following. (One-half point for each correct answer)**

1. $2 \times 12 =$ _____ **2.** $8 \times 16 =$ _____ **3.** $13 \times 40 =$ _____ **4.** $12 \times 36 =$ _____

5. $9 \times 10 =$ _____ **6.** $5 \times 15 =$ _____ **7.** $15 \times 16 =$ _____ **8.** $70 \times 6 =$ _____

9. $8 \times 9 =$ _____ **10.** $6 \times 12 =$ _____ **11.** $12 \times 12 =$ _____ **12.** $50 \times 8 =$ _____

13. $6 \times 8 =$ _____ **14.** $8 \times 12 =$ _____ **15.** $4 \times 20 =$ _____ **16.** $56 \times 80 =$ _____

17. $6 \times 6 =$ _____ **18.** $7 \times 22 =$ _____ **19.** $8 \times 11 =$ _____ **20.** $11 \times 800 =$ _____

21. $2 \times 14 =$ _____ **22.** $9 \times 22 =$ _____ **23.** $8 \times 17 =$ _____ **24.** $31 \times 50 =$ _____

Score for A (12)

B **(24 points) Find the products. (2 points for each correct answer)**

25. 1,728 × 42	**26.** 3,026 × 372	**27.** 38,246 × 8,297	**28.** 5,017 × 201	**29.** 3,600 × 300	**30.** 7,179 × 71
31. 8,222 × 509	**32.** 67,406 × 3,006	**33.** 1,236 × 444	**34.** 27,000 × 420	**35.** 8,125 × 279	**36.** 4,017 × 375

Score for B (24)

C **(10 points) Multiply by using shortcuts. (2 points for each correct answer)**

37. 3,684 × 50	**38.** 4,999 × 50	**39.** 6,642 × 25	**40.** 3,212 × 50	**41.** 1,376 × 25

Score for C (10)

D **(18 points) Multiply the three factors. (2 points for each final product)**

42. $17 \times 22 \times 31 =$ _____ **43.** $47 \times 16 \times 70 =$ _____ **44.** $33 \times 33 \times 33 =$ _____

45. $14 \times 100 \times 7 =$ _____ **46.** $915 \times 40 \times 20 =$ _____ **47.** $10 \times 10 \times 10 =$ _____

48. $30 \times 30 \times 30 =$ _____ **49.** $1,116 \times 3 \times 18 =$ _____ **50.** $2,000 \times 7 \times 4 =$ _____

Score for D (18)

Assignment 1.3 *Continued*

E (12 points) Complete the five multiplication problems and then add the five products. (1 point for each correct answer)

51. $12 \times 12.00 =$ _____

52. $27 \times 8.16 =$ _____

53. $104 \times 3.52 =$ _____

54. $6 \times 92.92 =$ _____

55. $55 \times 32.50 =$ _____

56. Total $=$ _____

57. $21 \times 7 \times 16 =$ _____

58. $13 \times 101 \times 22 =$ _____

59. $33 \times 14 \times 7 =$ _____

60. $99 \times 11 \times 100 =$ _____

61. $3 \times 88 \times 100 =$ _____

62. Total $=$ _____

Score for E (12)

F (24 points) Business Application. Complete the merchandise inventory TOTAL column. (1 point for each correct total; 8 points for correct grand total)

MERCHANDISE INVENTORY
JUNE 30, 20xx

Stock Number	Description	Price	# in Stock	Total
G473-2	Blood & Magic	$39.99	58	
G763-4	Circle of Blood	$39.99	172	
G865-A	Time Lapse	$49.95	98	
G2238-1	Nemesis	$34.99	225	
G873-2	Harpoon 97	$42.75	88	
S876-3	Word Perfect Suite Upgrade	$98.77	178	
S4433	Uninstaller 4	$32.59	85	
S887-32	MagnaRAM 97	$45.79	110	
S4536	Netscape Navigator	$38.79	100	
S1322	Norton Utilities 95 2.0	$67.85	68	
S458-2	Quicken 6.0	$27.75	205	
S5382	MS Visual J++	$95.69	80	
E5673-E	Typing Tutor 7	$26.59	108	
E82-18	Corel ChemLab	$52.49	25	
E2442	Encarta 97	$45.29	307	
E3578-1	Perfect Spanish	$44.79	80	
			TOTAL	

Score for F (24)

Assignment 1.4: **Division**

Name _____

Date _____ Score _____

A (10 points) Divide the following problems mentally. (One-half point for each correct quotient)

1. $72 \div 6 =$ _____
2. $90 \div 6 =$ _____
3. $96 \div 8 =$ _____
4. $110 \div 5 =$ _____
5. $126 \div 3 =$ _____
6. $154 \div 7 =$ _____
7. $88 \div 22 =$ _____
8. $144 \div 12 =$ _____
9. $169 \div 13 =$ _____
10. $135 \div 9 =$ _____
11. $990 \div 33 =$ _____
12. $187 \div 11 =$ _____
13. $156 \div 12 =$ _____
14. $900 \div 15 =$ _____
15. $336 \div 12 =$ _____
16. $84 \div 12 =$ _____
17. $104 \div 2 =$ _____
18. $300 \div 20 =$ _____
19. $119 \div 7 =$ _____
20. $225 \div 15 =$ _____

Score for A (10)

B (10 points) Divide by shortcut methods. Express remainders in parentheses. (1 point for each correct answer)

21. $3,450 \div 487 =$ _____
22. $12,800 \div 100 =$ _____
23. $58,700 \div 100 =$ _____
24. $2,200 \div 100 =$ _____
25. $7,800 \div 20 =$ _____
26. $6,450 \div 320 =$ _____
27. $9,005 \div 100 =$ _____
28. $387 \div 10 =$ _____
29. $38,500 \div 1,000 =$ _____
30. $3,250,000 \div 10,000 =$ _____

Score for B (10)

C (50 points) Divide. Show the remainder in parentheses after the whole number in the quotient. (2 points for each correct answer)

31. $21\overline{)478}$
32. $13\overline{)2,795}$
33. $23\overline{)14,076}$
34. $7\overline{)4,919}$

35. $36\overline{)6,436}$
36. $23\overline{)478}$
37. $271\overline{)50,001}$
38. $33\overline{)97,382}$

39. $926\overline{)926,007}$
40. $77\overline{)12,770}$
41. $506\overline{)10,238}$
42. $9\overline{)818,173}$

43. $700\overline{)362,497}$
44. $111\overline{)34,173}$
45. $88\overline{)97,817}$
46. $13\overline{)\$67,209}$

47. $6\overline{)\$13.20}$
48. $54\overline{)78,540}$
49. $51\overline{)100}$
50. $26\overline{)111,013}$

51. $66\overline{)73,428}$
52. $1,014\overline{)20,016}$
53. $66\overline{)17,209}$
54. $65\overline{)372,000}$

55. $29\overline{)58,004,316}$

Score for C (50)

Assignment 1.4 *Continued*

D **(10 points) Divide and check the following problems. (2 points for each correct answer)**

56. $22\overline{)1364}$ **57.** $31\overline{)1395}$ **58.** $92\overline{)7284}$ **59.** $21\overline{)2214}$ **60.** $31\overline{)642}$

Check:

\times _____ \times _____ \times _____ \times _____ \times _____

$=$ _____ $=$ _____ $=$ _____ $=$ _____ $=$ _____

Score for D (10)

E **(20 points) Business Applications. As an estimator for a printing company, you must estimate the paper costs for printing jobs. Paper is priced by the ream, which is 500 pages. Compute the paper costs of the jobs below. (1 point for each correct computation)**

No. of Booklets	No. of Pages	Total Pages	Reams of Paper	Cost per Ream	Total Paper Cost
250	66			$2.00	
120	150			$4.25	
75	220			$4.83	
110	250			$3.75	
25	280			$3.15	
30	250			$4.10	
		Total reams		Total paper cost	

Score for E (20)

Assignment 1.5: **Estimating**

Name _____

Date _____ Score _____

A (60 points) Estimate an answer for each of the following problems. Show your rounding, dropping of zeros with base product, and final estimate. (1 point for each correct answer)

Problem	Rounding to	Dropped Zeros and Base Product	Estimated Answer
1. 1,095 × 427	_____	_____	_____
2. 78,221 × 6,099	_____	_____	_____
3. 34,007 × 80	_____	_____	_____
4. 56 × 1,528	_____	_____	_____
5. 18 × 2,855 × 93	_____	_____	_____
6. 20 × 17 × 19	_____	_____	_____
7. 2,997 × 13	_____	_____	_____
8. 41 × 19 × 3	_____	_____	_____
9. 212 × 101 × 99	_____	_____	_____
10. 23 × 10,322	_____	_____	_____
11. 777 × 777	_____	_____	_____
12. 29,301 × 21	_____	_____	_____
13. 72,111 × 108	_____	_____	_____
14. 13 × 100 × 6	_____	_____	_____
15. 99 × 99 × 99	_____	_____	_____
16. 28 × 42	_____	_____	_____
17. 111 × 39	_____	_____	_____
18. 7 × 99	_____	_____	_____
19. 204 × 17	_____	_____	_____
20. 11 × 12 × 13	_____	_____	_____

Score for A (60)

Assignment 1.5 *Continued*

B **(20 points) Estimate an answer for each of the following problems. Show your rounding, dropping of zeros with base product, estimated answer, and real answer. (1 point for each correct answer)**

Problem	Round to	Dropped Zeros and Base Product	Estimated Answer	Real Answer
21. 883 × 294	_____	_____	_____	_____
22. 42,100 × 412	_____	_____	_____	_____
23. 19,965 × 492	_____	_____	_____	_____
24. 89 × 33	_____	_____	_____	_____
25. 983 × 204	_____	_____	_____	_____

Score for B (20)

C **(20 points) Estimate an answer for each of the following division problems. Show your rounding, dropping of zeros, estimated answer, and real answer. Round to two decimal places. (1 point for each correct answer)**

Problem	Round to	Drop Zeros	Estimated Answer	Real Answer
26. 123 ÷ 41	_____	_____	_____	_____
27. 612 ÷ 12	_____	_____	_____	_____
28. 4,836 ÷ 78	_____	_____	_____	_____
29. 19,760 ÷ 95	_____	_____	_____	_____
30. 21,508 ÷ 1,132	_____	_____	_____	_____

Score for C (20)

Notes

Dell.com | Accessories | About dell | Contact | Search | Support.Dell.com | Orde

United States

Main | Direct Access | Media | Investor Relations | Dell

▷ Careers Home
▷ Search For a Job
▷ Submit Your Resume
▷ Recruiting Events

▷ Working at Dell
▷ Benefits
▷ Living in Nashville
▷ Other Dell Locations

Dell-Nashville Career

Job Opportunities

- Sales
- Administrative Services
- Marketing
- Information Technology
- Legal
- Product Development
- Facilities

- Techr Supp
- Procu
- Huma Resou
- Manuf
- Corpo Comm
- Finan
- Custo Servic

Be your best at DFS with a Career at Dell Financial Services!

Living in Nashville

Spotlight on Na

Nashville's Grand Ole Opry is the longest continuously-running live radio program in the world. It has broadcast every

A replica Parthenon ancient G Athens, G Nashville' Park.

Source: http://dellapp.us.dell.com/careers/nashville/. Copyright © 2000 by Dell Computer Corporation; used with permission.

Dell was founded in 1984 by Michael Dell, who believed that by selling computers directly to customers, the company could best understand and meet their needs. Dell is now the world's No. 1 computer systems company, and is enhancing and broadening the fundamental competitive advantages of the direct model by applying the efficiencies of the Internet to its entire business. Dell led commercial migration to the Internet, launching www.dell.com in 1994 and adding e-commerce capability in 1996. The company's Web site, which runs entirely on Dell PowerEdge servers, receives 500 million page requests per quarter at 86 country sites in 21 languages, dialects, and 40 currencies. Dell customers range from the largest global corporations to home and small business customers.

Visit the Dell home page at http://www.dell.com for information on the company, its products, business alliances, services, and career opportunities.

ge

t

ng

tions

lle

nous
ding in
ands in
nial

Word Problems and Equations

2

LEARNING OBJECTIVES

By studying this chapter and completing all assignments you will be able to:

 1 LEARNING OBJECTIVES Use mental computations in simple addition, subtraction, multiplication, and division.

 2 LEARNING OBJECTIVES Use a systematic approach to solve word problems involving basic math processes.

 3 LEARNING OBJECTIVES Apply formulas to solve rate, time, and distance problems.

 4 LEARNING OBJECTIVES Solve simple numeric equations.

 5 LEARNING OBJECTIVES Recognize numeric relationships in a series.

 6 LEARNING OBJECTIVES Do quick mental calculations through a process of rounding numbers.

In business, most problems to be solved are word problems. For example, "We need to ship 100 pianos to Chicago. How many trucks do we need, and how much will it cost?" Your job is to find the numbers and then apply the fundamental processes of addition, subtraction, multiplication, or division.

1. In determining square yards of carpeting needed, which mathematical process would you most likely use?
2. When you're working with equations, why does it help to know that both sides of an equation are equal?

Mental Computations

Use mental computations in simple addition, subtraction, multiplication, and division.

Simple computations need to be made frequently in business. Practicing easy computation drills will improve your speed and accuracy in using the four fundamental math processes.

In example A you should be able to obtain the ten answers without using pencil, paper, or an electronic calculator. Mentally compute each problem quickly. Each computation is done from left to right. In these problems, addition, subtraction, multiplication, and division are done in the sequence in which they appear.

EXAMPLE A

$7 + 3 + 8 + 4$	$=$	22
$27 - 2 - 5 + 8 + 2$	$=$	30
$60 \div 2 \div 3 \div 5$	$=$	2
$3 + 4 + 2 + 10$	$=$	19
$3 \times 4 \times 2 \times 10$	$=$	240
$28 \div 4 \times 5 \times 2 - 7 \div 9$	$=$	7
$26 \div 2 + 2 \times 2 \times 2 \div 6 + 10 =$		20
$180 \times 2 \div 6 - 20 \div 8 \times 5$	$=$	25
$100 \times 5 - 20 - 80 - 40 \div 6 =$		60
$4,000 \div 2 + 100 \div 7 - 299$	$=$	1

CONCEPT CHECK 2.1

Practice computations until you can do them mentally without extra copying or writing. Use the simplification techniques in chapter 1 whenever possible: number combinations, repeated digits, counting by tens and adding ones, subtraction by changing numbers, multiplying numbers ending in zeros, and dividing numbers ending in zeros. Do the following computations mentally.

	7	+	3	+	6	+	6	+	6	\times	20	\div	10	+	31	$= 87$
Think:			10				+ (3 × 6 = 18)		28	(2 × 28) + 0	560	(560) − 0	56	(66, 76, 86, 87)		= 87

	78	−	29	+	7	+	7	+	7	\times	40	$=$	2,800
Think:	(79 − 30 = 49)			+ (3 × 7 = 21)		70			(7 × 4) + 00 = 2,800				

Solving Word Problems

Use a systematic approach to solve word problems involving basic math processes.

You might have little difficulty with computations expressed in numbers only. In example B you would quickly answer 350.

EXAMPLE B

$(15 + 15 + 10) \times 10 - 50 = 350$

However, you might not answer $350 as quickly when the business problem in example C appears, even though it uses the same numeric elements as example B.

EXAMPLE C

A company orders carpeting for three offices measuring 15 square yards, 15 square yards, and 10 square yards, respectively. A carpet dealer sells the carpet for $10 a square yard and gives a $50 discount when the sale is for three or more offices. How much would the company pay to have the three offices carpeted?

15 sq yd + 15 sq yd + 10 sq yd = 40 sq yd
40 sq yd × $10 = $400 gross price
$400 − $50 discount = $350 net price

V I D E O

W o r d P r o b l e m s

Business problems involving computations simply require addition, subtraction, multiplication, and division.

STEPS to Solve Word Problems

Read the entire problem carefully and then:
1. Determine what is being requested.
2. Determine the processes to be used to solve the problem.

We use these steps to solve the word problem in example C.

STEP 1 What is requested: How much money would the company pay?

STEP 2 The process to be used:
Add square yards in the 3 offices: 15 + 15 + 10 = 40.
Multiply the $10 per square yard cost by total square yards: 40 × $10 = $400.
Subtract the $50 discount: $400 − $50 = $350.

Some word problems will involve all four fundamental processes: addition, subtraction, multiplication, and division.

EXAMPLE D

Phoebe Elias owns half of a small bakery. Last week she baked 6 cakes on Monday, 9 on Tuesday, 11 on Wednesday, 8 on Thursday, and 6 on Friday. She sold all cakes for $9 each. It cost Phoebe $5 to make each cake; the rest was her profit on each cake. Phoebe split her profit evenly with her partner. How much did her partner receive from last week's cakes?

www.educationplanet. com

This Web site offers more practice with word problems.

STEP 1 What is requested: How much money did Phoebe's partner receive?

STEP 2 The process to be used:
Add the cakes baked: 6 + 9 + 11 + 8 + 6 = 40.
Subtract the cost from the sales price: $9 − $5 = $4 profit per cake.
Multiply the $4 profit per cake by the number of cakes sold: 40 × $4 = $160.
Divide the total profit by 2: $160 ÷ 2 = $80 received by the partner.

CyberConnection

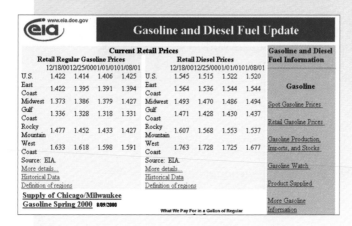

A flower shop in a small town in Alabama has two vans for deliveries. One van averages 120 miles per day, and the other van averages 50 miles per day. The vans use regular gas, and each van gets 17 miles per gallon. How much does gas cost per week for both vans, based on the Department of Energy retail prices?

Source: http://www.eia.doe.gov/.

CONCEPT CHECK 2.2

Summary of steps for solving word problems:

1. Determine what is being requested.
2. Determine the processes to be used to solve the problem.

Problem: Maria wants to upholster three chairs. Two chairs will require 4 yards of material each; the third will require 3 yards. One material costs $32 per yard; the other is $24 per yard. What is the difference between the costs of the two materials for upholstering the chairs?

STEP 1 What is requested: Difference in cost between the two materials

STEP 2 The process to be used:
Add amount of material needed: 4 yd + 4 yd + 3 yd = 11 yd.
Cost of material for three chairs, first material: 11 yd × $32 per yd = $352.
Cost of material for three chairs, second material: 11 yd × $24 per yd = $264.
Difference in cost between the two materials: $352 − $264 = $88 difference in cost.

Solving Rate, Time, and Distance Problems

LEARNING OBJECTIVES

Apply formulas to solve rate, time, and distance problems.

In some business word problems, you must determine how much is done in a given amount of time at a specific speed. These rate, time, and distance problems are solved with a simple formula: Rate (speed) × Time = Distance (amount done). Given any two factors, find the third.

Rate × Time = Distance
Distance ÷ Time = Rate
Distance ÷ Rate = Time

EXAMPLE E

Jan traveled at 35 miles per hour for 5 hours. How far did Jan travel?
35 mph × 5 hr = 175 mi
(Rate × Time = Distance)

EXAMPLE F

Jan traveled 175 miles in 5 hours. How fast was Jan traveling?
175 mi ÷ 5 hr = 35 mph
(Distance ÷ Time = Rate)

EXAMPLE G

At 35 miles per hour, how long would it take Jan to travel a total of 175 miles?
175 mi ÷ 35 mph = 5 hr
(Distance ÷ Rate = Time)

EXAMPLE H

Jan and Ahmed start traveling toward each other from 300 miles apart. Jan is traveling at 35 miles per hour; Ahmed is traveling at 40 miles per hour. How much time will elapse before they meet?
Distance = 300 mi
Total rate = 35 mph (Jan) + 40 mph (Ahmed) = 75 mph
300 mi ÷ 75 mph = 4 hr
(Distance ÷ Rate = Time)

EXAMPLE I

Jan and Ahmed start traveling toward each other from 300 miles apart. Jan is traveling at 35 miles per hour; Ahmed is traveling at 40 miles per hour. How much distance will Jan travel before they meet?
Total rate = 35 mph (Jan) + 40 mph (Ahmed) = 75 mph
Time = 300 mi ÷ 75 mph = 4 hr
Jan's distance = 35 mph (Jan's Rate) × 4 hr (Time) = 140 mi

EXAMPLE J

Mary needs to type a term paper that will be 30 pages long. Each page contains about 200 words. If Mary can type 40 words per minute, how many minutes will it take her to complete the paper?
Choose a formula: We know distance (amount done) and speed (rate). Therefore we choose the formula for time.
Distance (amount done) ÷ Rate (speed) = Time
30 pages × 200 words = 6,000 words ÷ 40 wpm = 150 min

EXAMPLE K

Flora also had a paper to type, but hers was 9,000 words in length. She was able to type it in 150 minutes. How fast did she type?
Choose a formula: We know distance (amount done) and time. Therefore we choose the formula for rate.
Distance (amount done) ÷ Time = Rate (speed)
9,000 words ÷ 150 min = 60 wpm

EXAMPLE L

It is approximately 400 miles from San Francisco to Los Angeles. Roy's friends tell him that he can make the trip in 6 hours if he averages 60 miles per hour. Is this true?

Choose a formula: We know the rate and the time, so we choose the formula for distance.

Rate (speed) × Time = Distance (amount done)

60 mph × 6 hr = 360 mi

Can he get there in 6 hours? *No.*

 CONCEPT CHECK 2.3

The basic formula:

a. Rate (speed) × Time = Distance (amount done)

 If you know any *two* factors, you can find the *third*.

b. Distance (amount done) ÷ Time = Rate (speed)

c. Distance (amount done) ÷ Rate (speed) = Time

Apply the appropriate formula to answer the following question: A machine that produces tortillas at the Baja Restaurant can produce 200 tortillas per hour, or 1,600 tortillas in an 8-hour day. A new machine can produce 3,000 tortillas in 6 hours. How many more tortillas per hour can the new machine produce than the old one?

Distance (amount done) ÷ Time = Rate

1,600 tortillas ÷ 8 hr = 200 per hr

3,000 tortillas ÷ 6 hr = 500 per hr

Difference: 500 − 200 = 300 more tortillas per hr

Solving Simple Numeric Equations

Solve simple numeric equations.

A **numeric sentence** in which both sides of an equals sign contain calculations is called an **equation.** For example, five plus five equals twelve minus two (5 + 5 = 12 − 2) is an equation, as is seven minus one equals thirty divided by five (7 − 1 = 30 ÷ 5).

For an equation to be true, the numbers on the left of the equals sign must always compute to the same answer as the numbers on the right of the equals sign. Moving a number from one side of the equation to the other changes its sign. A plus sign will change to minus; a minus sign will change to plus. A multiplication sign will change to division; a division sign will change to multiplication.

EXAMPLE M **Addition—Subtraction**

6 + 4 + 5 = 17 − 2

Change the + 5 and the − 2: Change only the − 2:

6 + 4 + 2 = 17 − 5 6 + 4 + 5 + 2 = 17

Check: 6 + 4 + 2 = 12 Change only the + 5:

 17 − 5 = 12 · 6 + 4 = 17 − 2 − 5

 6 + 4 = 10 and 17 − 2 − 5 = 10

Multiplication—Division

$3 \times 8 = 48 \div 2$

Change the \times 8 and \div 2: Change only the \div 2:

$3 \times 2 = 48 \div 8$ $3 \times 8 \times 2 = 48$

Check: $3 \times 2 = 6$ Change only the \times 8:

 $48 \div 8 = 6$ $3 = 48 \div 2 \div 8$

A numeric equation may be incomplete, with one factor missing, but provide enough information to be completed.

EXAMPLE O

$6 + 2 = 5 + ?$

$6 + 2 = 8$ so $5 + ? = 8$

Therefore $? = 3$

Or change a number

$6 + 2 - ? = 5$

Therefore $? = 3$

EXAMPLE P

$15 - 3 = 2 + ?$

$15 - 3 = 12$ so $2 + ? = 12$

Therefore $? = 10$

Or change a number

$15 - 3 - ? = 7$

Therefore $? = 5$

EXAMPLE Q

$7 + 3 + 6 = 4 + 4 + ?$

$7 + 3 + 6 = 16$ so $4 + 4 + ? = 16$

Therefore $? = 8$

Or change a number

$7 + 3 + 6 - ? = 4 + 4$

Therefore $? = 8$

EXAMPLE R

$20 \div 5 = 2 \times ?$

$20 \div 5 = 4$ so $2 \times ? = 4$

Therefore $? = 2$

Or change a number

$20 \div 5 \div ? = 2$

Therefore $? = 2$

In business, numeric sentences with equations frequently compare items. Note the following examples.

EXAMPLE S

4 items at $0.50 each = 10 items at ? each

4 items at $0.50 each = $2.00

10 items at ? each = $2.00

$2.00 \div 10$ items $= \$0.20$

Therefore $? = \$0.20$

Or change a number

$4 \times 0.50 \div ? = 10$

Therefore $? = 0.20$

EXAMPLE T

6 tickets at $5 each = 15 tickets at ? each

6 tickets at $5 each = $30

15 tickets at ? each = $30

$30 \div 15$ tickets $= \$2$

Therefore $? = \$2$

Or change a number

$6 \times 5 \div ? = 15$

Therefore $? = \$2$

EXAMPLE U

A company had sales of $25,000 and $20,000 for January and February of last year, respectively. If January sales this year were $30,000, what is the amount needed for February in order to equal last year's sales for the two months?

January LY $25,000 + February LY $20,000 = $45,000

January $30,000 + February (?) $= \$45,000$

$\$45,000 - \$30,000 = \$15,000$

Therefore $?$ $= \$15,000$

CONCEPT CHECK 2.4

Both sides of a true equation are equal. Each side may contain calculations.

$8 + 4 = 14 - 2$

$3 \times 6 = 36 \div 2$

A number may be moved from one side of an equation to the other by reversing its sign.

$8 = 6 + 2$	$8 - 2 = 6$	$7 + 3 = 10$	$7 = 10 - 3$
$12 = 4 \times 3$	$12 \div 3 = 4$	$24 \div 12 = 2$	$24 = 2 \times 12$

Numerical Relationships in a Series

LEARNING OBJECTIVES

Recognize numeric relationships in a series.

Relationships in a series of numbers may be found by comparing the first three or four terms in a series and then extrapolating the numbers that would most logically come next. For example, examining the series 320, 160, 80, 40 indicates that each term is found by dividing the preceding number by 2. The next two numbers in the series would logically be 20 and 10—that is, $40 \div 2 = 20$ and $20 \div 2 = 10$.

Examining the series 7, 14, 21, 28 suggests the addition of 7 to each preceding number. The next two numbers in this series would logically be 35 and 42 ($28 + 7 = 35$ and $35 + 7 = 42$).

In the series 5, 15, 35, 75, 155, seeing a relationship is difficult; however, a relationship does exist. Each number results from multiplying the preceding number by 2 and then adding 5. In this series, the next number would logically be 315 ($155 \times 2 + 5 = 315$).

Recognizing numeric and series relationships can be important in analyzing, communicating, and computing numbers. These relationship series are also used frequently in initial employment tests.

CONCEPT CHECK 2.5

In studying relationships in a numeric series, look for patterns. Patterns most commonly fall into categories:

Addition	2, 7, 12, 17, 22, 27	(+ 5, or 32)
Alternating addition/subtraction	12, 24, 18, 30, 24, 36, 30	(+ 12, − 6, or 42, 36)
Subtraction	39, 32, 25, 18, 11, 4	(− 7, or − 3)
Alternating subtraction/addition	64, 59, 61, 56, 58, 53, 55	(− 5, + 2, or 50, 52)
Multiplication	4, 12, 36, 108, 324, 972	(× 3, or 2,916)
Division	384, 192, 96, 48, 24	(÷ 2, or 12)

You can also devise patterns such as multiplication with addition or subtraction, division with addition or subtraction, and many other combinations.

Making Quick Calculations by Rounding Numbers

Quick calculations are beneficial when working in business situations. *Rounding* odd and difficult-to-compute amounts to even whole numbers that are easier to compute is a technique often used in business. By rounding, you will be able to get quick and accurate answers without having to write out the computations.

Do quick mental calculations through a process of rounding numbers.

EXAMPLE V

How much would 5 items at $2.99 each cost?
To make this computation easily, think "$2.99 is $0.01 less than $3.00." Then think "5 times $3 equals $15." Finally, think "$15.00 less $0.05 (5 × $0.01) is $14.95," which is the correct answer.

EXAMPLE W

The total cost of 3 equally priced items is $119.85. How much does one of these items cost?
To figure out this problem easily, think "$119.85 is $0.15 less than $120.00." Then think "$120 divided by 3 = $40, and $40.00 less $0.05 ($0.15 ÷ 3) is $39.95," the correct answer.

EXAMPLE X

At 19 miles per gallon, how many miles would a car go on 16 gallons of gas?
To figure out this problem easily, think "19 is just 1 mile less than 20." Then think "16 times 20 = 320, and 320 minus 16 (16 × 1) is 304," the correct answer.

 CONCEPT CHECK 2.6

You may have noticed that making quick calculations is quite similar to making estimations, which you did in chapter 1. In fact, quick calculation is only an additional step. After estimating an answer, you determine the degree to which the estimated, or rounded, answer differs from the actual answer by mentally correcting for the amount of the estimation or rounding.

COMPLETE ASSIGNMENTS 2.1 AND 2.2

This page, at http://www.edhelper.com/math.htm, provides worksheets for business math—payroll problems, depreciation problems, interest and stock worksheets, and more. The study of business mathematics is a good base from which to go on to careers in accounting, retailing, computer software development, finance (financial advisor or analyst), management, market research, statistics, education, and many others.

Source: http://www.edhelper.com/businessmath.htm#S1.
Copyright © 2001 by edHelper.com.

Chapter Terms for Review

equation numeric sentence

Visit the Web

http://www.loanvalue.com/loanvalue/fincalc.htm—Find a few business math calculators at this Web site.

http://tier2.census.gov/zbp/—This site provides business data by zip codes or by county.

http://www.toolkit.cch.com/tools/tools.asp—The Business Owner's ToolKit is a useful site for small business information.

CYBERCONNECTION

Nobody happy with California power rate increase

'We're not sure we can keep the lights on for our customers'

January 4, 2001
Web posted at: 7:46 p.m. EST (0046 GMT)

SAN FRANCISCO (CNN) — Amid protests by both consumer advocates and state power companies, the California Public Utilities Commission approved rate hikes Thursday on businesses and residents.

At the same time, the Justice Department reviewed a request for a criminal investigation into the state's energy woes.

The cost of electricity and natural gas for millions of Californians will rise between 7 percent to 15 percent under a temporary rate surcharge approved Thursday by the state Public Utilities Commission.

The 90-day increase is expected to raise the average California homeowner's monthly energy bill from $60 to $65.40. . . .

How much of an increase?

The surcharge — unveiled Wednesday and approved a day later by PUC members meeting in San Francisco — is about one-third the amount sought by PG&E and SoCal Edison, which serve more than three-fourths of California's population.

- The 1-cent-per-kilowatt surcharge plan results in rate hikes of approximately:
- 9 percent for residential customers
- 7 percent for small businesses
- 12 percent for medium-sized businesses
- 15 percent for large businesses and industrial customers

All money generated by the rate hikes would be put into an account controlled by the PUC until an independent audit of PG&E and SoCal Edison is completed. If the audit determines the companies do not need the money, it would be refunded to customers. . . .

Source: http://www.cnn.com/. Copyright © 2001 by Cable News Network; used with permission.

Monthly CG&E bills chill the blood

Record lows, higher rates collide

By Mike Boyer and Patrick Stack
The Cincinnati Enquirer

Saturday, January 06, 2001

Tristate residents who heat their homes with natural gas are reeling after getting their Cincinnati Gas & Electric bills for December, a month that was 41 percent colder than usual.

And the utility is just starting to mail out bills for January, which will reflect a 20 percent rate increase — coming on top of an average 33 percent rate hike in December.

The higher rates coincided with the nation's coldest November-December period on record. The two-month period averaged 33.8 degrees, compared to the old record of 34.2 set in 1898, according to records kept by the National Climatic Data Center in Asheville, N.C.

"I was shocked," said Kris Mullaney, 35, whose most recent bill to heat her 80-year-old Westwood home was $270 (up from $110 the month before). "I knew they (rates) were going up, but that first bill was sticker shock," she said.

"I think it's a little high," said Mary E. Jenkins, 84, of Reading. Ms. Jenkins' latest gas bill was $291.32, up from $112.88 the month before.

This month's rate hike will increase the bill of a typical Cincinnati customer using 25,000 cubic feet of gas by about $93 over what that person paid in January last year, a 60 percent hike. . . .

Source: http://enquirer.com/. Copyright © 1995–2000 by *The Cincinnati Enquirer*. Used with permission from *The Cincinnati Enquirer*/Mike Boyle and Patrick Stack.

These articles—one from CNN and the other from The *Cincinnati Enquirer*—report on electricity and natural gas price hikes during the winter of 2000–2001.

Were these costs higher or lower in the West than in the Midwest? What was the total percentage rate increase in Cincinnati?

The Bottom Line

Summary of chapter learning objectives:

Learning Objective	Summary	Example
2.1	Use mental computations in simple addition, subtraction, multiplication, and division	Use mental computations. 1. Add: $4 + 3 + 8 + 11 + 9 + 2 + 3 =$ _____ 2. Add by combining numbers: $4 + 6 + 8 + 8 + 8 + 30 + 10 =$ _____ 3. Subtract: $84 - 7 - 12 - 23 =$ _____ 4. Subtract and add: $9 + 4 - 2 - 8 + 4 =$ _____ 5. Multiply and divide: $28 + 4 \times 2 \div 4 + 14 =$ _____ 6. Multiply and divide: $18 \div 3 + 10 - 5 \times 3 =$ _____
2.2	Use a systematic approach to solve word problems involving basic math processes	Use the two-step process to solve the word problem. 7. Martha is preparing to make two dresses. One will require 3 yards of material; the other will require 4 yards of material. The material for the first dress costs $12.00 per yard; the material for the second costs $15.00 per yard. Buttons and trimming will cost $8.00 for each dress. What will be the total cost? Determine what is being requested. Determine the processes to be used to solve the problem. Answer: *112* _____
2.3	Apply formulas to solve rate, time, and distance problems	8. At an average rate of 50 miles per hour, how long would it take to drive 650 miles? *13hrs* 9. At an average rate of 60 miles per hour, how far could you drive in 6 hours? *360 miles* 10. If you drove 70 miles per hour and covered 280 miles, how much time did it take? *4hrs*
2.4	Solve simple numeric equations	11. $7 + 8 - 2 = 5 + 9 - 1$ Change the 8 to the opposite side and test the equation. _____ 12. $5 \times 12 = 120 \div 2$ Change the 12 to the opposite side and test the equation. _____ 13. $44 \div 11 = 13 - 9$ Change the 9 to the opposite side and test the equation. _____ 14. 18 items at $3.00 each = 6 items at ? ? = _____

Summary of chapter learning objectives:

Learning Objective	Summary	Example
2.5	Recognize numeric relationships in a series	Determine the relationships in the following problems and insert the next two numbers. 15. 4, 7, 6, 9, 8, 11, _10_ , _____ Pattern: +3 −1 16. 12, 48, 24, 96, 48, _____ , _____ Pattern: _____
2.6	Do quick mental calculations through a process of rounding numbers	17. What is the cost of 8 items at $3.99 each? 18. At 59 miles per hour, how far would a car go in 20 hours?

Answers: 1. 40 2. 74 3. 42 4. 7 5. 30 6. 33 7. $112 8. 13 hr 9. 360 mi 10. 4 hr
11. 7 − 2 = 5 + 9 − 1 − 8 12. 5 ÷ 120 = 120 ÷ 2 ÷ 12 13. 44 ÷ 11 + 9 = 13 14. $9.00 15. (+ 3, − 1) 10, 13
16. (× 4, ÷ 2) 192, 96 17. (8 × $4.00) = $32.00 − 0.08 = $31.92 18. (60 × 20) = 1200 − 20 = 1,180 mi

Self-Check | Review Problems for Chapter 2

1 Add: $7 + 9 + 4 + 8 + 2 =$ _____

2 Subtract: $70 - 7 - 4 - 8 - 3 - 6 =$ _____

3 Multiply: $4 \times 2 \times 3 \times 2 \times 2 =$ _____

4 Divide: $120 \div 2 \div 3 \div 5 \div 2 =$ _____

5 A corporation had 4 monthly earnings of $12,493, $6,007, $3,028, and $9,728. What was its total earnings in the 4 months? _____

6 If the corporation in question 5 had 12-month earnings of $74,500, how much did it earn in the last 8 months of the year? _____

7 If a tour bus gets 7 miles per gallon of gas and used 61 gallons in a week, how many miles did it travel in the week? _____

8 An employer earned $4,000. Half the earnings went into an employee bonus pool. The pool was split among 5 employees. How much did each employee receive? _____

9 A delivery firm bought 21 gallons of gas on Monday, 15 on Tuesday, 24 on Wednesday, 34 on Thursday, and 11 on Friday. If gas cost $2.15 per gallon, how much did the delivery firm pay for the week's gas? _____

10 A store owner planned to give away $1,200 at Christmas. The owner gave $150 to each of 5 full-time employees and $50 to each of 4 part-time employees. The remainder was given to a local charity. How much did the charity receive? _____

11 How long would it take to travel 1,265 miles at 55 miles per hour? _____

12 Bob and Mary start traveling toward each other from 1,330 miles apart. Bob is traveling at 30 miles per hour, Mary at 40 miles per hour. How many hours elapse before they meet? _____

13 Bob and Mary start traveling toward each other from 960 miles apart. Bob is traveling at 25 miles per hour, Mary at 55 miles per hour. How many hours elapse before they meet? _____

14 $41 - 6 = 27 +$ _____

15 $72 + 72 = 300 -$ _____

16 $10 \times 3 = 90 \div$ _____

17 Four items at $9 each $=$ _____ items at $12 each

18 What is the next number in the series 3, 7, 8, 12, ? _____

19 What is the next number in the series 5, 20, 10, 40, ? _____

20 To find the price of 7 items at $1.99 you would think: 7 times $ _____ less 7 times $ _____ $= $13.93

Assignment 2.1: Word Problems, Equations, and Series

LEARNING OBJECTIVES

Name _____

Date _____ Score _____

A **(20 points) Do the steps in the order in which they occur. Do not use an electronic calculator. (1 point for each correct answer)**

1. $14 + 5 + 3 + 4 =$ _____
2. $6 \times 6 - 4 \div 8 \times 2 =$ _____
3. $17 - 3 - 2 - 5 =$ _____
4. $14 \div 2 \times 5 \times 2 + 5 =$ _____
5. $80 \div 4 \div 2 \div 5 =$ _____
6. $9 \times 2 + 2 \times 6 - 20 \div 4 =$ _____
7. $3 \times 2 \times 5 \times 3 =$ _____
8. $(4 - 3) \times 5 \times 5 \times 5 - 3 =$ _____
9. $25 \div 5 \times 3 + 1 + 11 + 2 - 6 =$ _____
10. $(12 + 12 + 12 + 14) \div 5 \times 3 + 3 =$ _____
11. $100 \times 5 - 50 \div 9 + 5 \div 11 \times 3 =$ _____
12. $(36 \div 3 \div 4 + 10 + 5 - 3) \times 3 =$ _____
13. $(15 \div 3 \times 2 + 8 - 3 + 12) \div 3 =$ _____
14. $(10 \times 8) + (20 \times 3 \div 6 \div 5) + 4 =$ _____
15. $9 \div 3 \times 7 + 4 + 5 \times 4 - 6 =$ _____
16. $680 \div 2 \div 2 + 10 \div 6 \times 2 + 6 =$ _____
17. $32 \times 2 \div 8 \times 100 + 200 \div 4 + 1 =$ _____
18. $12 + 10 + 3 + 26 + 29 \div 4 \times 3 =$ _____
19. $1,000 \times 4 \times 2 - 5,000 \div 6 =$ _____
20. $3 + 4 + 5 + 6 + 7 \div 5 \times 800 =$ _____

Score for A (20)

B **(10 points) Do these problems without using scratch paper or an electronic calculator. (2 points for each correct answer)**

21. How much would you pay for six gallons of gasoline selling at $2.05 per gallon? _____
22. How many items would you have if you had three books, seven cards, and twenty-one pencils? _____
23. If six people divided three pizzas so that each person got one piece, how many slices would each pizza have? _____
24. How much would you have if you received $7.00 from one person, $23.00 from a second, $10.50 from a third, and $4.00 from a fourth? _____
25. If 27 people were divided into three equal groups and each group added two additional members, how many members would be in each group? _____

Score for B (10)

C **(10 points) Do the steps in the order in which they occur. Do these problems without using scratch paper or an electronic calculator. (1 point for each correct answer)**

26. 12 items at $3 each plus $2 tax = _____
27. 15 watches at $30 each less a $50 discount = _____
28. 3 lamps at $22 each plus 7 bulbs at $2 each = _____
29. 100 belts at $4 each less discounts of $60 and $30 = _____
30. 3 dozen scissors at $11.20 per dozen plus a $4 shipping charge = _____
31. 8 pounds of pears at $3 per pound plus 50¢ per pound for packaging = _____

32. $38 sale price plus $3 tax less a $12 discount plus a $3 delivery charge = _____

33. 6 bath towels at $8 each and 4 hand towels at $3 each plus $2.50 tax = _____

34. 4 dozen brushes at $25 per dozen plus $5 tax plus $7 shipping charge = _____

35. 2 shirts at $20 each, 4 ties at $5 each, and 7 pairs of socks at $2 each = _____

Score for C (10)

D **(20 points) Complete the following equations by supplying the missing items. (1 point for each correct answer)**

36. $21 + 3 =$ _____ $+ 8$

37. $13 +$ _____ $= 7 + 26$

38. _____ $+ 4 = 4 + 12$

39. $400 = 17 - 2 +$ _____

40. $22 - 9 =$ _____ $- 6$

41. $36 -$ _____ $= 17 + 8$

42. $9 + 17 - 3 = 4 \times$ _____ $- 5$

43. $160 \div 4 + 2 = 7 \times 7 -$ _____

44. $13 - 11 \times$ _____ $= 8 \times 8 + 16$

45. _____ $\times 3 \times 3 = 9 \div 3 \times 9$

46. $4 \times 20 =$ _____ $76 + 4$

47. _____ $\div 2 = 9 - 1$

48. $64 \div 32 = 900 \div$ _____

49. $15 - 9 - 2 = 25 -$ _____

50. _____ $+ 6 = 43 - 12$

51. $(7 \times 8) - 6 =$ _____

52. $15 \times 2 \times 2 =$ _____

53. $13 \times$ _____ $= 77 - 12$

54. _____ $\times 9 = 81 - 9$

55. $4 \times$ _____ $= 10 \times 6$

Score for D (20)

E **(40 points) In each of the following problems, a definite relationship exists among the numbers in each series. Extend each series two items by following the correct process. (8 points for each problem; 2 points for each correct line)**

56. Extend each series below through addition.

 a. 4, 8, 12, 16, _____

 c. 2, 4, 7, 11, 13, _____

 b. 1, 4, 5, 8, _____

 d. 5, 10, 20, 35, 55, _____

57. Extend each series below through subtraction.

 a. 50, 45, 40, 35, _____

 c. 100, 90, 81, 73, _____

 b. 50, 45, 43, 38, _____

 d. 610, 600, 580, 550, _____

58. Extend each series below through multiplication.

 a. 4, 8, 16, 32, _____

 c. 2, 4, 20, 40, _____

 b. 5; 25; 125; _____

 d. 2, 4, 12, 48, _____

59. Extend each series below through division.

 a. 15,625; 3,125; 625; 125; _____

 c. 10,000, 2,000, 1,000, 200, _____

 b. 729, 243, 81, 27, _____

 d. 10,000, 5,000, 1,000, 500, _____

60. Extend each series below through combinations of the four processes above.

 a. 72, 75, 69, 72, _____

 c. 6, 9, 18, 21, 42, _____

 b. 200, 100, 300, 150, _____

 d. 240, 120, 600, 300, 1,500, _____

Score for E (40)

Assignment 2.2: Word Problems, Formulas, and Equations

Name _____

Date _____ Score _____

A **(40 points) Solve the following word problems. (5 points for each correct answer)**

1. A store regularly sold 2 cans of soup for $0.64. It advertised a special sale of 6 cans for $1.56. A customer bought 12 cans at the sale. How much did the customer save over the regular price? __3.84__

2. A sales representative's car gets 18 miles to a gallon of gas. It was driven 120 miles each day for 30 days. Gas cost an average of $1.27 per gallon. What was the sales representative's total 30-day cost for gas? __254__

3. A store clerk sold a customer a ruler for $0.67, three pencils for $0.29 each, notebook paper for $0.99, and an eraser for $0.35 and was given $10.00 in payment. How much change did the clerk give the customer from the $10.00? (All prices include tax.) __7.12__

4. A college student worked at a local store for $6.00 per hour, as his class schedule permitted. The student worked 3 hours each Monday, Tuesday, Wednesday, and Thursday. He also worked 2 hours each Friday and 8 hours each Saturday. How many weeks did the student have to work to earn $792 for a new bicycle? __6 weeks__

5. A box, a crate, and a trunk weigh a total of 370 pounds. The crate weighs 160 pounds. The trunk weighs 4 pounds more than the box. What does the box weigh? _____

6. A hotel has 12 floors. Each floor has 20 single-person rooms and 40 two-person rooms. What is the total guest capacity of the hotel? _____

7. A department store offers its customers socks for $1.50 per pair or $15.00 per dozen. If two customers buy 1 dozen together and each pays half the cost, how much will each customer save by paying the quantity price? _____

8. Supply Clerk A ordered 5 staplers for $27.50 total and 2 large boxes of staples for $1.75 each. Supply Clerk B ordered a box of computer disks for $8.50 and a box of computer paper for $39.95. How much more did Clerk B spend than Clerk A? (All prices include tax.) _____

Score for A (40)

Assignment 2.2 *Continued*

B **(10 points) Solve the following time, rate, distance problems. (5 points for each correct answer)**

9. Wendy leaves Chicago to travel the 2,000 miles to Los Angeles, driving at a speed of 55 miles per hour. Mark leaves Los Angeles to travel the same 2,000-mile route to Chicago, driving at a speed of 45 miles per hour. How many miles will Mark have traveled when they meet? _____

10. Car A traveled to a destination 450 miles away at 50 miles per hour. Car B traveled to a destination 550 miles away at 55 miles per hour. How much longer did Car B travel than Car A? _____

Score for B (10)

C **(40 points) Solve each of the problems without writing any computations on paper and without using a calculator or a computer. (2 points for each correct answer)**

11. 5 items at $1.99 = _____ 12. 2 items at $7.98 = _____

13. 4 items at $19.98 = _____ 14. 2 items at $49.96 = _____

15. 15 items at $0.99 = _____ 16. 10 items at $9.99 = _____

17. 6 items at $3.95 = _____ 18. 5 items at $1.02 = _____

19. 19 items at $40 = _____ 20. 3 items at $19.99 = _____

21. 20 items at $40.05 = _____ 22. 30 items at $1.99 = _____

23. 20 items at $39.98 = _____ 24. 2 items at $5.99 = _____

25. 48 items at $5 = _____ 26. 5 items at $1.97 = _____

27. 7 items at $7.97 = _____ 28. 2 items at $99.98 = _____

29. 30 items at $2.98 = _____ 30. 99 items at $1.90 = _____

Score for C (40)

D **(10 points) In each of the equations below, rewrite the equation by moving the last number on each side of the equals sign to the other side and making appropriate sign changes so that the equation is still true. (Example: Given $13 + 7 + 2 = 10 + 12$: Answer $13 + 7 - 12 = 10 - 2$) (1 point for each correct equation)**

31. $6 + 4 + 5 = 17 - 2$ 32. $6 \times 2 \div 3 = 8 \div 4 \times 2$

33. $9 - 3 - 3 = 2 + 1$ 34. $8 \div 2 \times 4 = 24 \div 3 \times 2$

35. $20 + 1 - 7 = 16 - 2$ 36. $3 \times 3 \times 3 = 18 \div 2 \times 3$

37. $12 + 3 - 5 = 7 + 3$ 38. $7 \times 4 \div 2 = 28 \times 2 \div 4$

39. $64 - 32 - 16 = 8 + 8$ 40. $63 \div 7 \times 2 = 3 \times 2 \times 3$

Score for D (10)

Notes

Company Information

The Burger King Corporation Press Kit

Press Kit — Click here to download

- The Background of Burger King Corporation
- Facts about Burger King Corporation
- Facts for the 2000's
- Burger King® Timeline History
- Burger King® International Locations
- Burger King® Breakfast Fun Facts
- Latin America Corporate Facts

Home · What's Hot · Big Kids · Community · Nutrition · Locations · Company

Press Releases

Burger King Corporation press releases are available

Burger King is an international food service company with branches in many countries. Its home page states: "We're proud to say that we serve individuals, not billions."

As one of the largest food service companies in the United States, Burger King offers many job opportunities, from food server or preparer to shift leader, assistant manager, or manager. Office positions in the Miami headquarters and at regional offices require a thorough grounding in mathematics and accounting.

Burger King's home page stresses feedback on customer satisfaction as part of the company's quality control program. Visit Burger King at http://www.burgerking.com.

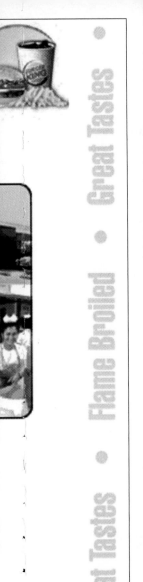

Weights and Measures

3

LEARNING OBJECTIVES

By studying this chapter and completing all assignments you will be able to:

1 Convert weights and measures from the given unit to a smaller or larger unit on the same scale.

2 Add, subtract, multiply, and divide, using U.S. customary units of weight and measure.

3 Use math processes to solve business problems involving U.S. customary units of weight and measure.

The standard kilogram weight is kept inside three glass domes in an underground chamber by the International Bureau of Weights and Measures outside Paris, France. Why three domes? Because a superficial scratch or abrasion or a particle of dust or microscopic air pollution can change the accuracy of the weight.

1. When might you need to know the number of ounces in a pound? Ounces in a pint?
2. In what type of job would it be helpful to know the number of pounds in a ton?
3. In what type of job would it be helpful to know the number of square feet in a square yard?
4. In what occupation would it be helpful for you to know the number of gallons in a cubic foot?

Converting Weights and Measures

Many areas of business rely on **U.S. customary units of weight and measure,** which represent standards for expressing weight, liquid measure, length (distance), area, and volume.

Figure 3-1: U.S. Customary Units of Weight and Measure

Distance
12 inches (in.) = 1 foot (ft)
3 feet (ft) = 1 yard (yd)
5,280 feet (ft) = 1 mile (mi)
1,760 yards (yd) = 1 mile (mi)

Liquid
8 fluid ounces (fl oz) = 1 cup (c)
2 cups (c) = 1 pint (pt)
2 pints (pt) = 1 quart (qt)
4 quarts (qt) = 1 gallon (gal)

Weights
16 ounces (oz) = 1 pound (lb)
2,000 pounds (lb) = 1 ton (t)

Volume
1,728 cubic inches (cu in.) = 1 cubic foot (cu ft)
27 cubic feet (cu ft) = 1 cubic yard (cu yd)

Area
144 square inches (sq in.) = 1 square foot (sq ft)
9 square feet (sq ft) = 1 square yard (sq yd)
43,560 square feet (sq ft) = 1 acre (a)
640 acres (a) = 1 square mile (sq mi)

To convert from smaller units (inches, for example) to larger units (feet, for example), *divide.* To convert from larger units (gallons, for example) to smaller units (quarts, for example), *multiply.*

Convert the following:

From (smaller) in. to (larger) ft	300 in. = ? ft	300 ÷ 12 = 25 ft
From (smaller) oz to (larger) lb	80 oz = ? lb	80 ÷ 16 = 5 lb
From (larger) gal to (smaller) qt	20 gal = ? qt	20 × 4 = 80 qt
From (larger) mi to (smaller) ft	2 mi = ? ft	2 × 5,280 = 10,560 ft
From (larger) cu yd to (smaller) cu ft	5 cu yd = ? cu ft	5 × 27 = 135 cu ft
From (smaller) cu ft to (larger) cu yd	162 cu ft = ? cu yd	162 ÷ 27 = 6 cu yd
From (smaller) sq in. to (larger) sq ft	432 sq in. = ? sq ft	432 ÷ 144 = 3 sq ft

EXAMPLE A

Convert the following distance units into the requested distance units.

Problem	Information Needed	Computation	Answer
9 yd = ? ft	1 yd = 3 ft	3 × 9	9 yd = 27 ft
14 ft = ? in.	1 ft = 12 in.	14 × 12	14 ft = 168 in.
4 yd = ? in.	1 yd = 3 ft and 1 ft = 12 in.	4 × 3 × 12	4 yd = 144 in.
3 mi = ? ft	1 mi = 5,280 ft	3 × 5,280	3 mi = 15,840 ft
14,080 yd = ? mi	1 mi = 1,760 yd	14,080 ÷ 1,760	14,080 yd = 8 mi
31,680 ft = ? mi	1 mi = 5,280 ft	31,680 ÷ 5,280	31,680 ft = 6 mi

EXAMPLE B

Convert the following liquid units into the requested liquid units.

Problem	Information Needed	Computation	Answer
12 gal = ? qt	4 qt = 1 gal	12 × 4	12 gal = 48 qt
24 oz = ? c	8 oz = 1 c	24 ÷ 8	24 oz = 3 c
10 qt = ? pt	1 qt = 2 pt	10 × 2	10 qt = 20 pt
4 qt = ? c	1 qt = 2 pt and		
	1 pt = 2 c	4 × 2 × 2	4 qt = 16 c

EXAMPLE C

Convert the following weight units into the requested weight units.

Problem	Information Needed	Computation	Answer
80 oz = ? lb	1 lb = 16 oz	80 ÷ 16	80 oz = 5 lb
1 t = ? oz	2,000 lb = 1 t and		
	1 lb = 16 oz	1 × 2,000 × 16	1 t = 32,000 oz

EXAMPLE D

Convert the following volume units into the requested volume units.

Problem	Information Needed	Computation	Answer
3 cu yd = ? cu ft	1 cu yd = 27 cu ft	3 × 27	3 cu yd = 81 cu ft
2 cu ft = ? cu in.	1 cu ft = 1,728 cu in.	2 × 1,728	2 cu ft = 3,456 cu in.
216 cu ft = ? cu yd	1 cu yd = 27 cu ft	216 ÷ 27	216 cu ft = 8 cu yd

EXAMPLE E

Convert the following area units into the requested area units.

Problem	Information Needed	Computation	Answer
4 sq ft = ? sq in.	1 sq ft = 144 sq in.	4 × 144	4 sq ft = 576 sq in.
864 sq in. = ? sq ft	1 sq ft = 144 sq in.	864 ÷ 144	864 sq in. = 6 sq ft
6 sq yd = ? sq ft	1 sq yd = 9 sq ft	6 × 9	6 sq yd = 54 sq ft
130,680 sq ft = ? a	1 a = 43,560 sq ft	130,680 ÷ 43,560	130,680 sq ft = 3 a
3,200 a = ? sq mi	1 sq mi = 640 a	3,200 ÷ 640	3,200 a = 5 sq mi

 CONCEPT CHECK 3.1

Complete the following conversions.

	Original	Process		Solution
Converting Distance Measures				
Change inches to feet	48 in.	48 ÷ 12	=	4 ft
Feet to inches	5 ft	5 × 12	=	60 in.
Feet to yards	15 ft	15 ÷ 3	=	5 yd
Yards to feet	6 yd	6 × 3	=	18 ft
Miles to feet	2 mi	2 × 5,280	=	10,560 ft
Feet to miles	5,280 ft	5,280 ÷ 5,280	=	1 mi
Miles to yards	4 mi	4 × 1,760	=	7,040 yd
Yards to miles	10,560 yd	10,560 ÷ 1,760	=	6 mi

www.
sciencemadesimple.
com/conversions.html

You should know how to make conversions mathematically when you don't have access to a calculator or to a computer. But sometimes you can use the conversion calculator at this Web site.

	Original	Process		Solution
Converting Liquid Measures				
Ounces to cups	24 fl oz	24 ÷ 8	=	3 c
Cups to ounces	6 c	6 × 8	=	48 fl oz
Cups to pints	8 c	8 ÷ 2	=	4 pt
Pints to cups	5 pt	5 × 2	=	10 c
Pints to quarts	10 pt	10 ÷ 2	=	5 qt
Quarts to pints	8 qt	8 × 2	=	16 pt
Quarts to gallons	12 qt	12 ÷ 4	=	3 gal
Gallons to quarts	3 gal	3 × 4	=	12 qt
Converting Weights				
Ounces to pounds	48 oz	48 ÷ 16	=	3 lb
Pounds to ounces	5 lb	5 × 16	=	80 oz
Converting Volume Measures				
Cubic inches to cubic feet	5,184 cu in.	5,184 ÷ 1,728	=	3 cu ft
Cubic feet to cubic inches	5 cu ft	5 × 1,728	=	8,640 cu in.
Cubic feet to cubic yards	135 cu ft	135 ÷ 27	=	5 cu yd
Cubic yards to cubic feet	50 cu yd	50 × 27	=	1,350 cu ft
Converting Area Measures				
Square inches to square feet	1,584 sq in.	1,584 ÷ 144	=	11 sq ft
Square feet to square inches	24 sq ft	24 × 144	=	3,456 sq in.
Square feet to square yards	180 sq ft	180 ÷ 9	=	20 sq yd
Square yards to square feet	30 sq yd	30 × 9	=	270 sq ft

Performing Math Operations with U.S. Customary Units of Weight and Measure

2 LEARNING OBJECTIVES

Add, subtract, multiply, and divide, using U.S. customary units of weight and measure.

Adding Weights and Measures

When you're adding weights and measures, first arrange units in a right-to-left sequence of smallest to largest. In expressing a final answer, convert smaller units to larger units whenever possible—for example, ounces to pounds, quarts to gallons, and feet to yards.

EXAMPLE F

Add and convert, where possible, the following distance measurements. Refer to Figure 3-1 as necessary.

```
 7 yd   2 ft  11 in.
+2 yd   1 ft   6 in.
```

7 yd	2 ft	11 in.
+2 yd	1 ft	6 in.
9 yd	3 ft	17 in.

(+1) ⟵ (−12)

| 9 yd | 4 ft | 5 in. |

(+1) ⟵ (−3)

| 10 yd | 1 ft | 5 in. |

EXAMPLE G

Add and convert, where possible, the following liquid measurements. Refer to Figure 3-1 as necessary.

2 gal	3 qt	1 pt
3 gal	2 qt	1 pt
+4 gal		1 pt

2 gal	3 qt	1 pt
3 gal	2 qt	1 pt
+4 gal	___	1 pt
9 gal	5 qt	3 pt
	(+1) ⟵ (−2)	
9 gal	6 qt	1 pt
(+1) ⟵ (−4)		
10 gal	2 qt	1 pt

Subtracting Weights and Measures

When you're subtracting weights and measures, first arrange units in a right-to-left sequence of smallest to largest. When you're subtracting weights and measures, "borrow" quantities needed from the next higher unit. In expressing your final answer, convert smaller units into larger units whenever possible.

EXAMPLE H

Subtract the following distance measurements.

13 yd	2 ft	5 in.
−3 yd	2 ft	9 in.

13 yd	2 ft	5 in.
	1 (2 − 1) ⟶	17 (5 + 12)
12 (13 − 1) ⟶	4 (1 + 3)	
−3 yd	2 ft	9 in.
9 yd	2 ft	8 in.

EXAMPLE I

Subtract the following liquid measurements.

6 gal	2 qt	1 pt
−2 gal	3 qt	

6 gal	2 qt	1 pt
5 (6 − 1) ⟶	6 (2 + 4)	
−2 gal	3 qt	
3 gal	3 qt	1 pt

EXAMPLE J

Subtract the following area measurements.

28 sq yd	3 sq ft
−15 sq yd	8 sq ft

28 sq yd	3 sq ft
27 (28 − 1) ⟶	12 (3 + 9)
−15 sq yd	8 sq ft
12 sq yd	4 sq ft

Multiplying Weights and Measures

In business you will multiply and divide weights and measures most often to compute the square feet, square yards, or cubic measurements of a given area and to compute prices for various quantities and measurements of merchandise.

EXAMPLE K

How many square feet are there in an office 12 feet wide and 15 feet long?
12 ft × 15 ft = 180 sq ft

EXAMPLE L

How many square yards are there in an exhibit area 30 yards wide and 66 yards long?
30 yd × 66 yd = 1,980 sq yd

To find a total cubic measurement, multiply length times width times height.

EXAMPLE M

How many cubic feet are there in a warehouse that measures 100 feet by 60 feet by 12 feet?
100 ft × 60 ft × 12 ft = 72,000 cu ft

EXAMPLE N

How many cubic yards are there in a room measuring 21 feet long by 18 feet wide by 9 feet high?
21 ft × 18 ft × 9 ft = 3,402 cu ft
3,402 cu ft ÷ 27 cu ft per cu yd = 126 cu yd

Dividing Weights and Measures

EXAMPLE O

A store owner wants to divide her building of 30,000 square feet into 6 equal department spaces. How many square feet will each department have?
30,000 sq ft ÷ 6 = 5,000 sq ft per department

EXAMPLE P

A contractor wants to excavate a pit containing 720 cubic yards of dirt. If he has 6 customers who will purchase the dirt and he wants to allocate an equal amount to each customer, how much will each receive?
720 cu yd ÷ 6 = 120 cu yd each

CONCEPT CHECK 3.2

Adding and Converting Weights and Measures

5 yd	2 ft	9 in.
+4 yd	2 ft	11 in.
9 yd	4 ft	20 in.
	5 (4 + 1) ⟵	8 (20 − 12)
10 (9 + 1) ⟵ 2 (5 − 3)		
10 yd	2 ft	8 in.

Subtracting and Converting Weights and Measures

10 gal	3 qt	1 pt
	2 (3 − 1) ⟶	3 (1 + 2)
−3 gal	2 qt	2 pt
7 gal	0 qt	1 pt

Multiplying and Converting Weights and Measures

How many cubic feet are there in a storage cabinet measuring 8 feet wide by 6 feet deep by 9 feet high?

8 ft × 6 ft × 9 ft = 432 cu ft

Dividing and Converting Weights and Measures

A contractor is excavating for two swimming pools in the same neighborhood. The first is 15 feet wide, 30 feet long, and 9 feet deep.

The second is 18 feet wide, 30 feet long, and 6 feet deep.

His largest truck will haul 5 cubic yards on each trip. How many trips will it require to haul away the dirt?

1st pool = 15 ft × 30 ft × 9 ft = 4,050 sq ft ÷ 27 = 150 cu yd
2nd pool = 18 ft × 30 ft × 6 ft = 3,240 sq ft ÷ 27 = 120 cu yd
Total 270 cu yd
Number of trips = 270 cu yd ÷ 5 cu yd per trip = 54 trips

Business Applications

Businesses constantly use weight and measure computations in transactions. They use gross weight to determine the price of commodities sold by ounces, pounds, tons, and so on. They use square yards to price carpet installations. They use cubic yards to determine the cost of removing dirt for a basement or swimming pool. They use gallons to compute the cost per mile of gasoline or other fuels used to operate cars and trucks.

LEARNING OBJECTIVES

Use math processes to solve business problems involving U.S. customary units of weight and measure.

EXAMPLE Q

What is the total dollar value for the following sales of a group of spices selling at $9.60 per pound?

Sale A = 2 lb 4 oz
Sale B = 2 lb 10 oz
Sale C = 1 lb 2 oz

Add:	2 lb	4 oz
	2 lb	10 oz
	1 lb	2 oz
	5 lb	16 oz

Convert: 5 lb 16 oz = 6 lb
Multiply: 6 lb × $9.60 = $57.60 total value

EXAMPLE R

A liquid spray sells for $6.40 per gallon.

1. How much will 1 quart cost?
 Divide: $6.40 per gal ÷ 4 qt per gal = $1.60 per qt
2. How much will 1 pint cost?
 Divide: $1.60 per qt ÷ 2 pt per qt = $0.80 per pt
3. How much will 3 gallons, 3 quarts, and 1 pint cost?
 Multiply and add: 3 gal × $6.40 = $19.20
 3 qt × $1.60 = 4.80
 1 pt × $0.80 = .80
 Total Cost $24.80

© PHOTODISC, INC.

A homeowner wants to build a swimming pool requiring dirt removal to a depth of 9 feet, for an area 15 feet wide and 45 feet long. At a removal cost of $40 per cubic yard, how much will the homeowner pay?

Multiply: 45 ft × 15 ft × 9 ft = 6,075 cu ft
Divide: 6,075 cu ft ÷ 27 cu ft per cu yd = 225 cu yd
Multiply: 225 cu yd × $40 = $9,000

How much will it cost to carpet a room that measures 15 feet by 21 feet if the installed carpet costs $22 per square yard?

21 ft × 15 ft = 315 sq ft
315 sq ft ÷ 9 sq ft per sq yd = 35 sq yd
35 sq yd × $22 per sq yd = $770

A business plans to recarpet its office, which is 27 feet wide by 45 feet long. Supplier A charges $30 per square yard installed; Supplier B charges $26 per square yard installed. How much will the business save by using Supplier B?

27 ft × 45 ft = 1,215 sq ft
1,215 sq ft ÷ 9 sq ft per sq yd = 135 sq yd
Supplier A: 135 sq yd × $30 per sq yd = $4,050
Supplier B: 135 sq yd × $26 per sq yd = $3,510
Supplier A − Supplier B: $4,050 − $3,510 = $540 savings
or
$30 − $26 = $4 × 135 sq yd = $540

Land sells for $2 per square foot. How much will 3 acres cost?

1 a = 43,560 sq ft
43,560 sq ft per a × 3 a = 130,680 sq ft
130,680 sq ft × $2 per sq ft = $261,360

www.webcrawler.com/
education/reference/
weights_and_
measures/
conversions_and_calcs/
financial_calculators/

Weights and measures, calculators, and conversions can be used through the Webcrawler Web site.

CONCEPT CHECK 3.3

Review steps in solving word problems:

1. Determine what is being requested.
2. Determine the processes to be used to solve the problem.

A homeowner wants to carpet a room that measures 18 feet by 24 feet. The carpet will cost $27 per square yard, installed. In addition to the carpet, he will put a baseboard around all four sides of the room, less one 6-foot open archway. The cost of the baseboard is $6 per yard. What will be the total cost?

What is requested: Total cost of carpeting and baseboard
The process to be used:
Multiply: 18 ft × 24 ft = 432 sq ft
Divide: 432 sq ft ÷ 9 = 48 sq yd of carpet needed
Multiply: 48 sq yd carpet × $27 per sq yd = $1,296, cost of carpet
Add: 18 ft + 18 ft + 24 ft + 24 = 84 ft around room
Subtract: 84 ft − 6 ft (open archway) = 78 ft
Divide: 78 ft ÷ 3 = 26 yd of baseboard needed
Multiply: 26 yd of baseboard × $6 per yd = $156
Add: Cost of carpet $1,296
 Cost of baseboard $ 156
 $1,452, total cost of project

A manufacturer of computer chips packs them into cartons, allowing exactly one cubic inch (1 in. × 1 in. × 1 in.) per chip. She has boxes that measure 2 feet by 3 feet by 4 feet. How many packed chips can she ship in 8 boxes?

What is requested: Number of cubic inches in 8 boxes
The process to be used:
Multiply: 2 ft × 3 ft × 4 ft = 24 cu ft per box
Multiply: 24 cu ft per box × 8 boxes = 192 cu ft in all boxes
Multiply: 192 cu ft × 1,728 cu in. per cu ft = 331,776 cu in., or 331,776 chips

COMPLETE ASSIGNMENTS 3.1 AND 3.2.

Chapter Terms for Review

U.S. customary units of weight and measure

Visit the Web

http://www.google.com—This search engine can do a search for "calculating weights and measures."

http://muextension.missouri.edu/xplor/agguides/crops/g04020.htm—This site contains conversions for many weights and measures, including those for crops.

http://www.lib.usf.edu/virtual/browse/index.php?s=weights—Reference tools can be found at this site at the University of South Florida.

http://www.butlercountyohio.org/auditor/wm_equivalent.cfm—Butler County, Ohio, has this Web site about weights and measures in the county, and includes tables of international weights and measures.

Here are some other departments of weights and measures to visit.

http://datcp.state.wi.us/core/consumerinfo/weights_measures—Wisconsin

http://www.weights.az.gov/—Arizona

http://www.state.wv.us/labor/weights/faq.html—West Virginia

CYBERCONNECTION

http://wminfo.htm

Department Secretary:
David L. Volk

Joe Hjermstad, Director

Mailing Address:
Weights & Measures
Program
118 West Capitol
Pierre, SD 57501

Phone: (605) 773-3697
FAX: (605) 773-6631

SOUTH DAKOTA DEPARTMENT OF COMMERCE AND REGULATION

Weights & Measures Program

Welcome

Mission Statement:

The Weights and Measures Program is responsible for checking the accuracy of all commercial weighing and measuring equipment in South Dakota, including gas pumps, scales, fuel meters, LPG meters and large capacity scales.

- About Weights and Measures
- Metrology Laboratory
- South Dakota Service Agency Registration
- Complaint Filing
- Fees
- Labeling and Packaging
- Installation of Weighing and Measuring Devices
- Other Helpful Web Sites NIST Office of Weights and Measures

This is the Web site of the South Dakota State department of Commerce and Regulation, Weights & Measures Program. This program is responsible for ensuring that commercial scales, gas pumps, meters, and other measuring devices are accurate. Find out what else it does.

1. If you bought a box of 12 inkjet print cartridges, and each cartridge contained 2 ounces, how many ounces would you have? How many pints would you have?

2. Suppose that, unbeknownst to you, the statement that each cartridge contained 2 ounces was false because the ink company's scale was wrong, and each cartridge actually contained only 1.5 ounces. If you used one 2-ounce cartridge per month in the past, how long would the new box of 1.5-ounce cartridges last?

Source: http://www.state.sd.us/dcr/wm/.

The Bottom Line

Summary of chapter learning objectives:

Learning Objective	Summary	Example
3.1	Convert weights and measures from the given unit to a smaller or larger unit on the same scale	Convert the following weights and measures to the requested units. 1. 2,448 in. = _____ ft 2. 56 fl oz = _____ c 3. 240 oz = _____ lb 4. 8,640 cu in. = _____ cu ft 5. 8,640 sq in. = _____ sq ft 6. 21 yd = _____ ft 7. 4 pt = _____ c 8. 8 t = _____ lb 9. 5 cu yd = _____ cu ft 10. 12 sq yd = _____ sq ft 11. 36,960 ft = _____ mi 12. 8 qt = _____ pt 13. 130,680 sq ft = _____ a
3.2	Add, subtract, multiply, and divide, using U.S. customary units of weight and measure	14. 5 yd 3 ft 9 in. 17. 9 yd 3 ft 7 in. +1yd 2 ft 4 in. −2 yd 4 ft 6 in. 15. 5 gal 3 qt 1 pt 18. 3 gal 3 qt 1 pt +2 gal 2 qt 1 pt −1 gal 1 qt 2 pt 16. 30 sq yd 8 sq ft 19. 42 sq yd 3 sq ft +12 sq yd 7 sq ft −17 sq yd 4 sq ft 20. How many square yards are there in a classroom that is 30 feet wide and 42 feet long? _____ 21. How many cubic yards are there in a room 18 feet by 12 feet by 9 ft? _____ 22. A contractor has 36 cubic yards of gravel that he wants to distribute equally among 4 work sites. How many cubic yards of gravel will go to each work site?
3.3	Use math processes to solve business problems involving U.S. customary units of weight and measure	23. How much will it cost to carpet a room that measures 27 feet by 21 feet if the carpet costs $28.00 per square yard and an extra 6 square yards are needed for matching? _____

Answers: 1. 204 ft 2. 7 c 3. 15 lb 4. 5 cu ft 5. 60 sq ft 6. 63 ft 7. 8 c 8. 16,000 lb 9. 135 cu ft 10. 108 sq ft 11. 7 mi 12. 16 pt 13. 3 a 14. 8 yd 1 in. 15. 8 gal 2 qt 16. 43 sq yd 6 sq ft 17. 6 yd 2 ft 1 in. 18. 2 gal 1 qt 1 pt 19. 24 sq yd 8 sq ft 20. 140 sq yd 21. 72 cu yd 22. 9 cu yd 23. $1,932

Self-Check | Review Problems for Chapter 3

1 Add: 4 yd 2 ft 11 in.
 2 yd 2 ft 4 in.

2 Add: 7 gal 3 qt 1 pt
 2 gal 1 qt 1 pt

3 Add: 4 sq yd 3 sq ft 100 sq in.
 2 sq yd 8 sq ft 22 sq in.

4 Subtract: 17 yd 2 ft 7 in.
 7 yd 2 ft 9 in.

5 Subtract: 9 gal 2 qt 1 pt
 4 gal 1 qt 1 pt

6 A pick-up truck could carry $1\frac{1}{2}$ tons. Its load contained 750 pounds of dirt. How much additional dirt could be hauled in one load? _____

7 A pick-up truck could carry $1\frac{1}{2}$ tons. How many 50 pound boxes could the truck carry? _____

8 Five front-line Viking players weighed 278, 260, 240, 275, and 290, respectively. Five front-line Rams players weighed 310, 255, 258, 247, and 244, respectively. Which line had the higher front-line weight?

9 A cleaning additive requires 2 ounces of cleaner fluid be added to every 3 quarts of water. If you wanted to make 6 gallons of cleaning fluid, how many ounces of cleaning additive would be required? _____

10 A building site requires 2 feet of soil be removed from an area 90 feet by 90 feet. How many cubic yards of soil must be removed? _____

11 A pet store has an aquarium measuring 60 inches long and 30 inches wide. If the aquarium is filled to a depth of 24 inches, compute the cubic feet of water the aquarium contains. _____

12 The director of a recreation center could purchase a large quantity of rope less expensively than several small quantities. The recreation center needed 60 individual jumping ropes of 5 feet each and 18 jumping ropes of 15 feet each. Compute the number of yards of rope the recreation center needed. _____

13 The sandbox at a day care center is 6 feet long, 4 feet wide, and 18 inches high. If the box is filled to capacity, compute the cubic feet of sand it will hold.

14 A rancher has 160 acres of grazing land. Each cow needs 21,780 square feet for grazing. Compute the number of cows he can keep. _____

15 A painter is planning to paint all four walls and the ceiling of a hotel lobby. The lobby is 90 feet long, 45 feet wide, and 15 feet high. If a gallon of paint will cover 180 square feet, compute the number of gallons of paint the painter will use. _____

Assignment 3.1: **Weights and Measures**

Name _____

Date _____ Score _____

A **(6 points) Use Figure 3-1 to convert the following weights and measures into the requested units. (1 point for each correct answer)**

1. 6 miles = _____ feet

2. 192 inches = _____ feet

3. 48 yards = _____ inches

4. 5 square feet = _____ square inches

5. 150 gallons = _____ pints

6. 720 square feet = _____ square yards

Score for A (6)

B **(66 points) Add, subtract, multiply, and divide the following weights and measures. (6 points for each correct answer)**

7. 4 yd 2 ft 7 in.
 18 yd 1 ft 4 in.
 +2 yd 2 ft 7 in.

8. 7 yd 1 ft 8 in.
 2 yd 6 in.
 +4 yd 2 ft 11 in.

9. 3 lb 13 oz
 12 lb 4 oz
 +24 lb 1 oz

10. 3 t 1,100 lb
 +4 t 1,000 lb

11. 2 gal 3 qt 1 pt
 +6 gal 2 qt 1 pt

12. 3 sq yd 7 sq ft
 +20 sq yd 6 sq ft

13. Undeveloped land in Ogden, Utah is offered for sale at $2.50 per square foot. How much will 5 acres cost?

14. If 8 acres of land sell for $696,960, what is its cost per square foot? _____

15. Cheese sells for $7 per pound. What is the total cost of three packages that weigh 2 lb 11 oz; 1 lb 6 oz, and 4 lb 15 oz, respectively? _____

16. Which costs more: 7 gallons of gasoline priced at $1.62 per gallon or 9 quarts of milk priced at $1.39 per quart? _____

17. Twenty acres of land sold for $5,227,200. What was the selling price per square foot? _____

Score for B (66)

Assignment 3.1 *Continued*

C **(28 points) Use Figure 3-1 to do the following problems. In the final answers, convert smaller units to larger units whenever possible. (4 points for each correct answer)**

18. If an air conditioner changed 36 cu yd of air per minute, how many minutes would it take to change the air in a room measuring 18 feet long, 12 feet wide, and 9 feet high? _____

19. Liquid soap sells for $38.40 per gallon. How much will it cost to fill 64 dispensers if each dispenser holds 4 fluid ounces? _____

20. If a cargo of rice weighs 44 tons, what would be its value if rice were priced at $0.05 per pound? _____

21. A toy manufacturer packages tiny cars in boxes 4 inches long, 3 inches wide, and 2 inches high. How many cubic feet of storage space would be required to store 18,000 boxed cars? _____

22. If a liquid lawn spray sells for $12.80 per gallon, how much will a merchant pay for 2 gallons, 3 quarts, and 1 pint? _____

23. The Astoria Park swimming pool is 90 feet long and 30 feet wide and averages 6 feet in depth. The water flows through the pipe that fills the pool at the rate of 18 cubic feet per minute. How long will be required to fill the Astoria Park pool? _____

24. An office has 30 workers in a room that is 60 feet in length by 24 feet in width and 12 feet high. What is the number of cubic feet of air available for each person in the room? _____

Score for C (28)

Assignment 3.2: Business Applications

Name _____

Date _____ Score _____

A **(100 points) Solve the following word problems and business applications involving weights and measures. (10 points for each correct answer)**

1. A room is 30 feet long and 24 feet wide. How much will it cost to cover the floor with carpet costing $18 per square yard (9 square feet) if 4 extra square yards are purchased for matching? _____

2. A company packs its fish in cans. Each can weighs 1 ounce empty and is filled with 12 ounces of fish. An empty carton weighs 8 ounces. Two dozen cans are packed in a carton. A customer orders 5 cartons. How many pounds does the shipment weigh? _____

3. A state is building a freeway. It needs a piece of property that is 1,936 feet long and 180 feet wide. If the price of land is $972 per acre, how much will the state pay? _____

4. Carpet costs $31 per square yard. Carpet padding costs $6 per square yard. Carpet installation charges are $3 per square yard. How much will it cost to cover two areas with carpet and pad if area A measures 20 feet by 30 feet and area B measures 30 feet by 40 feet? _____

5. Room A measures 15 feet by 15 feet and is being carpeted with carpet that costs $22 per square yard installed. Room B measures 12 feet by 21 feet and is being carpeted with carpet that costs $20 per square yard installed. Which room will be cheaper to carpet? _____

6. In an effort to improve his general health, the president of Farrell Machine Company has begun exercising regularly. One Sunday he walks at an average pace of three miles an hour. He starts his walk at 3:30 and returns at 6:00. During that time he stops to talk to a friend for 12 minutes. He also stops to jump rope for 15 minutes and spends 3 minutes giving directions to a stranger. How far does he walk?

7. In developing a machine model for a client, an engineer uses 9 inches of black wire, 6 inches of green wire, 21 inches of white wire, and 12 inches of yellow wire. How many feet of wire does she use?

8. A lifeguard buys rope to tie up boats at the Community Sailing Center. He needs 12 feet for each of three boats, 14 feet for each of two boats, and 17 feet for each of four boats. How many yards of rope does he need? _____

9. The captain of a commercial cruise ship walks around the deck (a distance of 1,056 feet) 15 times each day for 8 days. How many miles does she walk during that time? _____

10. A florist shop owner waters the plants in his greenhouse once a week. He puts three pints of water on each of the 16 plants in row A and two quarts of water on each of the 14 plants in row B. How many gallons of water does he use? _____

Score for A (100)

Notes

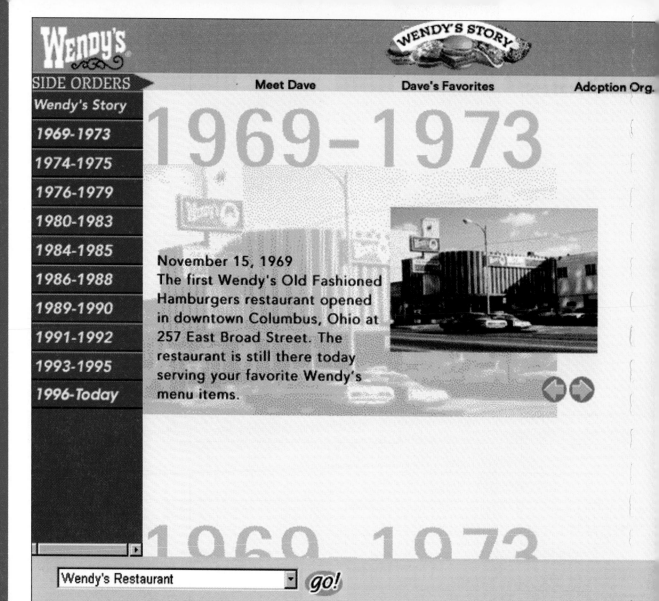

Source: http://www.wendys.com/wendys_story/wendys_story.html. Copyright © 2001 by Wendy's International; used with permission.

In November 1969, Dave Thomas opened the first Wendy's restaurant in Columbus, Ohio, and in 1972, the first Wendy's franchise opened in Indianapolis. In 5 years, total restaurant sales were nearly $25 million, and in 1976 the chain began to sell shares. In 1995, Wendy's merged with Tim Hortons, the largest coffee and fresh-baked goods restaurant chain in Canada. Together, the chains boast more than 7,000 locations and combined annual sales of more than $7 billion.

You may recognize Dave Thomas from the Wendy's television commercials. But did you know that in 1993, 45 years after dropping out of high school, Thomas returned to receive an honorary GED? Dave also wrote two books about his experiences, and he donates the proceeds to the Foundation for Adoption.

Visit Wendy's site at http://www.wendys.com for more information on the organization's history, community activities, foundations, companies, and employment.

Decimals

4

LEARNING OBJECTIVES

By studying this chapter and completing all assignments you will be able to:

 1 Read decimal numbers.

 2 Round decimal numbers.

3 Add two or more decimal numbers.

4 Subtract one decimal number from another.

 5 Multiply two decimal numbers.

 6 Divide one decimal number by another decimal number.

 7 Multiply and divide by decimal numbers that end with zeros.

 8 Approximate products and quotients.

You can see examples of the use of decimals in many everyday situations, especially on electronic digital displays in retail stores, on gas pumps, and in electronic banking.

 1. How are calculations with decimal numbers similar to and different from calculations with whole numbers?

Fractions Versus Decimal Numbers

McDonald's restaurant sells a hamburger sandwich called the Quarter Pounder. The sandwich is named for the amount of meat: one-quarter pound of ground beef. McDonald's—or anyone—can describe the same amount of meat in four different ways: 4 ounces, $\frac{1}{4}$ pound, 0.25 pound, or 25% of a pound. To express less than one pound, McDonald's could use smaller units, fractions, decimals, or percentages.

All four expressions are useful, but which one is best? It depends on what you're doing: whether you're buying or selling, whether you're speaking or writing, whether you're just estimating or making accurate financial records, or whether you're working with large volumes of something cheap or small quantities of something very expensive. For McDonald's, a Four Ouncer wouldn't sell as well as a Quarter Pounder, but Bloomingdale's sells perfume by the (fluid) ounce rather than by the gallon, quart, pint, or even cup.

Historically, partial quantities were first expressed with smaller units of measure. You have already reviewed the most common U.S. units of measure in chapter 3. Verbal expressions such as "half of a candy bar" or "a third of the pizza" are so common that children learn them before they can even read. We review fractions with you in chapter 5. With the development of inexpensive handheld electronic calculators, most calculations are now performed using decimal numbers. We review decimals with you here in chapter 4. Percents are a combination of decimal numbers and a few common fractions. Percents are as easy to calculate as decimal numbers and also allow relatively simple verbal expressions like fractions do. We review percents with you in chapter 6.

Decimal Numbers and Electronic Displays

© PHOTODISC, INC.

A customer in a delicatessen might ask for a fourth of a pound or maybe four ounces of salami. However, the electronic scale in the delicatessen will display 0.25 or 0.250 or even 0.2500. To indicate one fourth as a fraction, we write $\frac{1}{4}$; three fourths is $\frac{3}{4}$. In the U.S. monetary system, a *quarter* is the actual name of the coin worth twenty-five cents; three quarters are worth seventy-five cents. But when we write monetary amounts, we use either whole numbers or decimals instead of fractions. We write either 25¢ and 75¢ or $0.25 and $0.75. We do not normally write $\frac{1}{4}$ or $\frac{3}{4}$.

For most business transactions and record keeping, decimal numbers are much better than fractions. Calculations in decimal numbers are both easy and accurate. The development of modern electronic digit displays, such as that on your calculator, has also helped decrease the importance of fractions. The photo at left shows the electronic digital display on a gasoline pump. Service stations use gasoline pumps to measure gasoline volumes accurately to three decimal places. The pump in the photo displays a purchase of 12.661 gallons of gasoline. The number 12.661 is sometimes called a **mixed decimal.** The 12 is the whole number part of the number; the 661 is the **pure decimal** part. The period that separates the 12 from the 661 is the **decimal point.**

CyberConnection

Dewey Decimal System

Many libraries continue to use the Dewey Decimal Classification (DDC), which was invented in 1876 by a young American librarian, Melvil Dewey. He decided to divide the field of human knowledge into ten major classes, to subdivide these classes into ten divisions, each division into ten sections, and so on. Physics for example is subdivided as follows:

530 Physics **534 Sound** **537 Electricity**
531 Solid mechanics **535 Light** **538 Magnetism**
532 Fluid mechanics **536 Heat** **539 Modern physics**
533 Gas mechanics

With the use of a decimal point, and extra digits after the point, each of these sections can be further subdivided to cover increasingly narrower subject areas. For example, for Electricity the subdivisions include:

537 **Electricity** **537.5342** **Long waves**
537.5 **Electronics** **537.5343** **Short waves**
537.534 **Radio waves**

The longer the classification number, the more specific is the topic. The shorter the number, the more general it is. So, on the shelves you will find that the more general books come first in the sequence, and as you move along the shelves you will find the books becoming more specialized.

Visit the Web sites http://sulacco.library.ucg.ie/misc/dewey.html and http://www.lib.duke.edu/libguide/fi_books_dd.htm for more information. Go to your library and see if it uses the Dewey decimal system or some other system.

Many calculators today permit you to change the number of decimal places displayed. When you purchase a new calculator, it may be preset to display exactly two decimal places. Divide 1 by 8 on your calculator. The correct answer should be 0.125. If the calculator displays 0.12 or 0.13, it is preset to display two decimal places. Read the instruction manual of your calculator to see if your calculator permits you to change the number of decimal places. Also, note that your calculator displayed a zero (0) to the left of the decimal point. We follow that convention in this book.

Reading Decimal Numbers

Reading decimal numbers is just like reading whole numbers: Each place represents a different value. Starting at the decimal point and reading to the left, the places represent ones, tens, hundreds, and so on; starting at the decimal point and reading to the right, the places represent tenths, hundredths, thousandths, and so on.

To read the pure decimal part of a number, read the numbers as if they were not a decimal; then, attach the name of the *place value* where the last digit on the right

LEARNING OBJECTIVES 1

Read decimal numbers.

Figure 4-1: Number System on Both Sides of the Decimal Point

4 ones
9 tens
1 hundred
7 thousands
0 ten thousands
6 hundred thousands

3 tenths
5 hundredths
8 thousandths
2 ten-thousandths
4 hundred-thousandths

6 0 7, 1 9 4.3 5 8 2 4

appears. In the photo on page 72, the last digit is in the thousandths place. The decimal point is read as the word *and*.

Figure 4-1 illustrates the number system on both sides of the decimal point for the number 607,194.35824. The place value for each digit is given.

EXAMPLE A

The decimal part of the number in Figure 4-1 is .35824. With the digits in the figure, you can write several decimal numbers, including 0.3, 0.35, 0.008, 0.0082, and 0.35824. These numbers are read as follows.

Decimal	Words
0.3	three tenths
0.35	thirty-five hundredths
0.008	eight thousandths
0.0082	eighty-two ten-thousandths
0.35824	thirty-five thousand eight hundred twenty-four hundred-thousandths

Reading Long Decimal Numbers

The entire number in Figure 4-1—607,194.35824—is read as "six hundred seven thousand one hundred ninety-four and thirty-five thousand eight hundred twenty-four hundred-thousandths." For a long number, reciting it orally is inefficient, and might be confusing to the listener. For such a number, it may be better simply to read the digits and commas, from left to right. The word *point* is used for the decimal point.

EXAMPLE B

Recite orally the number 607,194.35824.

Number	Oral Recitation
607,194.35824	"six zero seven comma one nine four point three five eight two four"

 CONCEPT CHECK 4.1

a. Write 28.074 using words: Twenty-eight and seventy-four thousandths
b. Write thirteen and five hundredths using digits: 13.05

CONCEPT CHECK 4.5

a. Multiply 3.15 × 2.4.

$$
\begin{array}{r}
3.15 \quad \text{(2 places)} \\
\times\ 2.4 \quad \text{(1 place)} \\
\hline
1260 \\
630 \\
\hline
7.560 \quad \text{(3 places)}
\end{array}
$$

May be written as 7.56

b. Multiply 0.315 × 0.24.

$$
\begin{array}{r}
0.315 \quad \text{(3 places)} \\
\times\ 0.24 \quad \text{(2 places)} \\
\hline
1260 \\
630 \\
\hline
0.07560 \quad \text{(5 places; insert 1 zero)}
\end{array}
$$

May be written as 0.0756

Dividing Decimal Numbers

When dividing decimal numbers, remember that a whole number will have its decimal point immediately to the right of the units digit.

To divide one decimal number by another, follow these steps.

LEARNING OBJECTIVES

Divide one decimal number by another decimal number.

STEPS to Divide one Decimal Number by Another

1. Arrange the divisor, dividend, and division bracket ($\overline{)}$) as in whole number long division.
2. Move the decimal point in the divisor to the right until the divisor is a whole number. (You won't have to move it if the divisor is already a whole number.)
3. Move the decimal point in the dividend to the right exactly the same number of decimal places as you did in Step 2. If necessary, add more zeros to the right end of the dividend. (See example L.)
4. Write the decimal point in the quotient directly above the decimal point in the dividend.
5. If necessary, write zeros in the quotient between the decimal point and the first nonzero digit. (See example M.)
6. Divide as you would for whole numbers.

EXAMPLE L

	STEP 1	STEP 2 STEP 3	STEP 4	STEP 6

$$
2.4 \div 0.16 \quad \text{is} \quad 0.16\overline{)2.4} \;=\; 0.16.\overline{)2.40.} \;=\; 16.\overline{)240.} \;=\; 16.\overline{)240.}
$$

$$
\begin{array}{r}
15. \\
16\overline{)240.} \\
\underline{16} \\
80 \\
\underline{80} \\
0
\end{array}
$$

| | STEP 1 | STEPS 2, 3, & 4 | STEPS 5 & 6 |

$$0.056 \div 4 \quad is \quad 4\overline{)0.056} \; = \; 4.\overline{)0.056} \; = \; 4.\overline{)0.056}$$

$$\begin{array}{r} 0.014 \\ 4.\overline{)0.056} \\ -4 \\ \hline 16 \\ -16 \\ \hline 0 \end{array}$$

Recall from chapter 1 that, in long division with two whole numbers, you write a *remainder* when the division doesn't come out evenly. For example, $17 \div 8 = 2$ with a remainder of 1. In division with decimals, you don't write remainders. You simply keep dividing until you have some required number of decimal places. To get the required number of decimal places, you may have to keep adding zeros to the right end of the dividend. (See example N.)

Calculate $17 \div 8$ to three decimal places.

| | STEP 1 | STEP 2, 3, & 4 | STEP 6 |

$$17 \div 8 \quad is \quad 8\overline{)17} \; = \; 8.\overline{)17.} \; = \; 8.\overline{)17.} \; = \; 8.\overline{)17.000}$$

$$\begin{array}{r} 2. \\ 8.\overline{)17.} \\ -16 \\ \hline 1 \end{array} \qquad \begin{array}{r} 2.125 \\ 8.\overline{)17.000} \\ -16 \\ \hline 10 \\ -8 \\ \hline 20 \\ -16 \\ \hline 40 \\ 40 \end{array}$$

✓ **CONCEPT CHECK 4.6**

Divide 1.026 by 15.

| STEPS 1 & 4 | STEPS 5 & 6 |

$$15\overline{)1.026} \; = \; \begin{array}{r} 0.0684 \\ 15\overline{)1.0260} \\ \underline{90} \\ 126 \\ \underline{120} \\ 60 \\ \underline{60} \\ 0 \end{array}$$

Divide 0.007 by 0.4.

| STEPS 1, 2, 3, & 4 | STEPS 5 & 6 |

$$0.4.\overline{)0.0.07} \; = \; \begin{array}{r} 0.0175 \\ 4\overline{)0.0700} \\ \underline{4} \\ 30 \\ \underline{28} \\ 20 \\ \underline{20} \\ 0 \end{array}$$

Using Multipliers and Divisors that End with Zeros

In chapter 1 we showed simple multiplication and division shortcuts when the multiplier or the divisor is a whole number ending in zeros (e.g., 30, 200, or 1000). The same shortcuts may be used with decimal numbers.

LEARNING OBJECTIVES

Multiply and divide decimal numbers that end with zeros.

If the multiplier is 10, 100, 1000, and so on, there is just one step.

Step 1 Move the decimal point in the multiplicand to the *right* the same number of places as the number of zeros in the multiplier. (See example O.) If necessary, add zeros to the *right* end of the multiplicand before multiplying. (See example P.)

EXAMPLE O

0.56 × 10 = 0.5.6 = 5.6
(1 place)

EXAMPLE P

4.73 × 1,000 = 4.730 = 4,730
(3 places)

If the multiplier ends in zeros (e.g., 100 or 1,000) but has a first digit that is not 1, there are two steps.

Step 1 Multiply the multiplicand by the nonzero part of the multiplier.

Step 2 Move the decimal point in the product from step 1 to the *right* the same number of places as the number of zeros in the multiplier.

EXAMPLE Q

Multiply 3.431 by 2,000

Multiply by 2: 3,431 × 2 = 6,862
Move the decimal point three places to the right: 3.431. ⟶ 3,431.

If the divisor is 10, 100, 1000, and so on, there is just one step.

Step 1 Move the decimal point in the dividend to the *left* the same number of places as the number of zeros in the divisor. (See example R.) If necessary, add zeros to the *left* end of the dividend before dividing. (See example S.)

EXAMPLE R

735.1 ÷ 100
735.1 ÷ 100 = 7.35.1 = 7.351
(2 places)

EXAMPLE S

9.64 ÷ 1,000
9.64 ÷ 1,000 = .009.64 = 0.00964
(3 places)

If the divisor ends in zeros (e.g., 100 or 1,000) but has a first digit that is not 1, there are two steps.

Step 1 Divide the dividend by the nonzero part of the divisor.

Step 2 Move the decimal point in the quotient from Step 1 to the *left* the same number of places as the number of zeros in the divisor.

EXAMPLE T

Divide 615.24 by 300

Divide by 3: 6.1524 ÷ 3 = 2.0508.

Move the decimal point two places to the left: 615.24. ⟶ 6.1524

CONCEPT CHECK 4.7

a. Multiply 0.413 by 300.

Move the decimal point
two places to the right:

0.413 ⟶ 41.3
41.3 × 3 = 123.9

b. Divide 4.375 by 10.

Move the decimal point
one place to the left:

4.375 ÷ 10 = .4375 = 0.4375

COMPLETE ASSIGNMENT 4.2

Approximating Products and Quotients

8
LEARNING OBJECTIVES

Approximate products
and quotients.

Business people today use calculators almost exclusively to do important computations. But calculators are perfect only if every single key is pressed correctly. Often you can discover a calculator error by doing some simple mental approximations. The objective is to determine whether the answer is approximately the right size—that is, whether the decimal point is in the correct position. To do so, we round each decimal number to only one nonzero digit and all the rest zeros. Follow these steps.

STEPS to Approximate a Multiplication Problem

1. In each factor, round the first nonzero digit on the left. (How does the digit to its right compare to 5?)
2. Change all the digits to the right of the first nonzero digit to zero.
3. Multiply the two new factors.
4. Place the decimal point in the product.

EXAMPLE U

Approximate 3.764 × 7.1

	STEP 1	STEPS 2 & 3
3.764 ⟶	4.000	4
× 7.1 ⟶	× 7.0	× 7
		28

EXAMPLE V

Approximate 0.089 × 61.18.

	STEP 1	STEPS 2 & 3
0.089 ⟶	0.090	0.09
× 61.18 ⟶	× 60.00	× 60
		5.40

The actual answers are 26.7244 and 5.44502.

In division, the mental arithmetic will be easier if you change the decimal numbers so that the division will end evenly after one step. That will happen if you round the divisor to one nonzero digit and the dividend to two nonzero digits, evenly divisible by the new divisor.

STEPS **to Approximate a Division Problem**

1. Round the divisor to a *single nonzero digit* at the left, followed by all zeros.
2. Round the dividend to a *two-digit number* at the left, followed by all zeros. Select the two-digit number so that it is evenly divisible by the new divisor.
3. Divide the new dividend by the new divisor.
4. Place the decimal point correctly in the quotient.

EXAMPLE W

Approximate $3.764 \div 7.1$.

STEP 1 STEPS 2 STEPS 3 & 4

$7.1\overline{)3.764}$ \longrightarrow $7.0\overline{)3.764}$ \longrightarrow $7.\overline{)3.500}$ \longrightarrow $7.\overline{)3.5}$

$$
\begin{array}{r}
0.5 \\
7.\overline{)3.5} \\
\underline{3\,5} \\
0
\end{array}
$$

EXAMPLE X

Approximate $61.18 \div 0.089$.

STEP 1 STEPS 2 STEPS 3 & 4

$0.089\overline{)61.18}$ \longrightarrow $0.090\overline{)61.18}$ \longrightarrow $0.09\overline{)63.00}$ \longrightarrow

$$
\begin{array}{r}
700. \\
9.\overline{)6300.} \\
\underline{63} \\
0
\end{array}
$$

The actual answers are 0.5301 and 687.4157.

CONCEPT CHECK 4.8

a. Approximate 5.932×0.407.

$5.932 \longrightarrow 6.000$
$0.407 \longrightarrow 0.400$

$$
\begin{array}{r}
0.4 \quad \text{(1 place)} \\
\times 6 \quad \text{(0 places)} \\
\hline
2.4 \quad \text{(1 place)}
\end{array}
$$

Compare with $5.932 \times 0.407 = 2.414324$.

b. Approximate $0.0738 \div 92.65$.
Remember to round off the divisor first.

$92.65 \longrightarrow 90.00$
$0.0738 \longrightarrow 0.0720$

$$
\begin{array}{r}
. \\
90\overline{)0.072}
\end{array}
\longrightarrow
\begin{array}{r}
.0008 \\
90\overline{)0.0720} \\
\underline{720} \\
0
\end{array}
$$

Compare with $0.0738 \div 92.65 = 0.000796546$.

COMPLETE ASSIGNMENT 4.3

Chapter Terms for Review

decimal point

mixed decimal

pure decimal

rounding

rounding off

rounding up

Visit the Web

http://www.aaamath.com/dec.html—If you're rusty on decimals, find some refreshers at All About Decimals.

http://www.mathleague.com/help/decwholeexp/decwholeexp.htm—More decimals help is available at this Web site.

http://pittsford.monroe.edu/jefferson/calfieri/decimals/decimalsmain.html—This Web site is yet another source of help with decimals.

http://forum.swarthmore.edu/dr.math/faq/faq.fractions.html—Frequently asked questions about decimals and fractions are at Ask Dr. Math.

http://www.math.com/school/subject1/lessons/S1U1L2GL.html—This is another excellent resource for decimals brush-up.

CYBERCONNECTION

SEC Orders Securities Markets to Begin Trading in Decimals on July 3, 2000
Decimal Pricing Implementation Plan Due in 45 Days

WASHINGTON, DC, January 28, 2000—The Securities and Exchange Commission today ordered the securities markets to begin quoting securities prices in decimals by July 3, 2000. The Commission order requires the markets to submit a decimals pricing implementation plan within 45 days and requires the options and equities markets to phase in decimal pricing by year end.

During the phase-in period the markets can quote equity securites in price increments of up to 5 cents. The markets must also create and conduct a pilot program during the phase-in period in which a sample of securities will be quoted in one-cent increments.

SEC Chairman Arthur Levitt said, "Decimal trading will benefit investors by allowing greater competition and making it easier to compare prices. It will also bring about consistency with foreign markets, which now trade in decimals. We look forward to working with the industry to ensure that the transition to decimals goes smoothly."

On January 28, 2000, the SEC issued an Order requiring the major U.S. stock exchanges and other market participants to change to decimal pricing in the U. S. securities markets. The Order set out a timetable for the participants to begin some trading in decimals by July 3, 2000, and additional trading by January 3, 2001.

On March 6, 2000, the Nasdaq Stock market, Inc. ("Nasdaq") announced it would not be able to meet the target dates and also expressed concern about overall industry readiness. The Nasdaq requested that the SEC work with the industry and the markets to determine an appropriate time frame that would not impose unnecessary risks on investors.

On April 13, 2000, the SEC issued an Order staying the original deadlines for decimalization. In this Order, the SEC also requested the participants to comment on two alternatives for implementing decimal pricing this year.

The first alternative would begin decimal pricing in all exchange-listed securities on or before September 4, 2000 ("Dual Pricing"). (Congressmen Thomas Bliley, Michael Oxley, and Edward Markey had strongly urged the implementation of decimal pricing on or before September 4, 2000 because of the benefits to investors.) The second alternative envisioned a temporary or "pilot" program to begin decimal pricing in certain exchange-listed securities and options on or before September 4, 2000 ("Decimals Pilot"). Under both alternatives, all stocks would be traded in decimals by March 31, 2001. The New York Stock Exchange (NYSE) began trading in decimals on January 29, 2001.

Discussion Questions: What is the difference between "trading" stocks in decimals and "quoting" stock market prices in decimals? How can the use of decimals in stock market trading help investors?

Source: http://www.sec.gov

CYBERCONNECTION

Subdivisions of class five climbing

In *Mountaineering, Freedom of the Hills*, author Ed Peters explains the subdivisions of class 5:

"The experienced climber, having accomplished or attempted free climbs of varying degrees of difficulty in the YDS class 5 range, gains an understanding of the level of difficulty involved. To the beginner, however, these ratings are simply a set of numbers, understandably, easy if rated 5.0 and impossible if rated 5.13. To provide a slightly better understanding within the class for the beginner the following tongue-in-cheek description is provided:

5.0 to 5.4
There are two hand- and two footholds for every move; the holds become progressively smaller as the number increases.

5.5 to 5.6
The two hand- and two footholds are there, obvious to the experienced, but not necessarily so to the beginner.

5.7
The move is missing one hand- or foothold.

5.8
The move is missing two holds of the four, or missing only one but is very strenuous.

5.9
The move has only one reasonable hold which may be for either a foot or a hand.

5.10
No hand- or footholds. The choices are to pretend a hold is there, pray a lot, or go home.

5.11
After thorough inspection you conclude this move is obviously impossible; however, occasionally someone actually accomplishes it. Since there is nothing for a handhold, grab it with both hands.

5.12
The surface is as smooth as glass and vertical. No one has really ever made this move, although a few claim they have.

5.13
This is identical to 5.12 except it is located under overhanging rock."

Ratings are established on lead; the follower has a somewhat easier climb.

Return to the Yosemite Decimal System index or the Climber.Org resource page.

Latitudes and longitudes, the binary system of computing, clocks, measurements, the stock market, and even the difficulty levels of mountain climbing all require the use of decimals.

Just for fun, visit the Yosemite Decimal System Web site for a set of numeric ratings describing the difficulty of climbs. It is not the only rating system used by climbers, but it is the one preferred by most people who use Climber.Org. You'll find there is also a European Rating System and an Alaskan Rating System.

The Bottom Line

Summary of chapter learning objectives:

Learning Objective	Summary	Example
4.1	Read decimal numbers	1. Write 7.0019, using words. 2. Write fifty-two and eleven thousandths, using digits.
4.2	Round decimal numbers	3. Round 0.406501 to the nearest thousandth (i.e., to three decimal places). 4. Round up 12.25062 to the next hundredth (i.e., to two decimal places).
4.3	Add two or more decimal numbers	5. Add 81.9, 14.372, and 2.06
4.4	Subtract one decimal number from another	6. Subtract 8.3042 from 25.06
4.5	Multiply two decimal numbers	7. Multiply 4.68 × 3.5 _____
4.6	Divide one decimal number by another decimal number	8. Divide 0.076 ÷ 4 _____ 9. Divide 0.046 ÷ 2.5 _____
4.7	Multiply and divide by decimals that end with zeros	10. Multiply 0.049765 × 1,000 _____ 11. Divide 9.462 by 100 _____ 12. Multiply 0.0623 × 30 _____ 13. Divide 84.8 by 200 _____
4.8	Approximate products and quotients	14. Approximate 48.79 × 0.47 _____ 15. Approximate 0.2688 ÷ 0.713 _____

Answers: 1. seven and nineteen ten-thousandths **2.** 52.011 **3.** 0.407 **4.** 12.26 **5.** 98.332 **6.** 16.7558 **7.** 16.38 **8.** 0.019 **9.** 0.0184 **10.** 49.765 **11.** 0.09462 **12.** 1.869 **13.** 0.424 **14.** 25 **15.** 0.4

Self-Check | Review Problems for Chapter 4

1 Write "two hundred six and fourteen thousandths" as a number. _____

2 Write 5,025.3084, using words. _____

3 Round 5.875 feet to the nearest tenth _____ **4** Round $14.333 to the nearest cent _____

5 Add 3.79475 and 739.85 _____ **6** Add 12.45, 0.087, and 5.7 _____

7 Subtract 8.693 from 12.43 _____ **8** Subtract 162.78 from 341.5512 _____

9 Multiply 3.722 by 0.483 (do not round off) _____ **10** Multiply $17.75 by 14.62 (round off to the nearest cent) _____

In problems 11 and 12, divide to three places and round to the nearest hundredth.

11 Divide 45.88 by 14.2 _____ **12** Divide $6.25 by 8.41 _____

In problems 13 and 14, use shortcuts to solve each problem and round to the nearest hundredth.

13 Multiply 74.21 by 100 _____ **14** Divide $1,294.50 by 1,000 _____

In problems 15 and 16, pick the best approximate answers from the possible answers.

15 Multiply 48.98 by 11.2 _____ (a) 0.5 (b) 5 (c) 50 (d) 500 (e) 5,000

16 Divide $6.65 by 8.21 _____ (a) $0.008 (b) $0.08 (c) $0.80 (d) $8.00 (e) $80.0

17 The owner of a chain of very large, upscale bookstores, DeLois McBryde, decided to start selling coffee drinks such as espresso and cappuccino at one of her stores. During the first day, the store had total sales of $3,241.73. Of the total, $284.75 was from coffee drinks. How much of the total was from books and other items? _____

18 Gary Gehlert operates a tennis and golf shop in a resort. Last year he started selling on the Internet as well. He had the following profits last year: Tennis (shop), $48,627.14; Golf (shop), $129.651.12; Tennis (Internet), $8,993.84; and Golf (Internet), $18,745.49. What were the total profits from these sources? _____

19 A landscape contractor, Roger Hillman, needed to plant 350 one-gallon plants and 85 five-gallon plants. Roger uses about 0.5 cubic foot of planting soil for each one-gallon plant and 1.5 cubic feet of soil for each five-gallon plant. How many cubic feet of planting soil will Roger need for all these plants? _____

20 Planting soil is sold by the cubic yard. How many cubic yards of planting soil will Roger Hillman need to do his planting in question 19? (Round the answer to two decimal places.) _____

Assignment 4.1: Addition & Subtraction of Decimal Numbers

LEARNING OBJECTIVES

Name _____

Date _____ Score _____

A **(13 points) Use digits to write each number that is expressed in words. Use words to write each number that is expressed in digits. (1 point for each correct answer)**

1. Six hundred thirteen thousandths _____

2. Nineteen ten-thousandths _____

3. Forty-two hundredths _____

4. Seventy-four and twelve thousandths _____

5. Eight hundred sixty and ninety-eight hundred-thousandths _____

6. Sixty and six ten-thousandths _____

7. 15.73 _____

8. 0.004 _____

9. 492.3 _____

10. 0.0081 _____

11. 58.909 _____

12. 9.016 _____

13. 2,403.07 _____

Score for A (13)

B **(24 points) Round as indicated. (1 point for each correct answer)**

Nearest Tenth

14. 6.3499 gal _____

15. 48.53 mi _____

16. 3.824 qt _____

17. 374.29 yd _____

18. 4.25 ft _____

19. 6.375 in. _____

Nearest Cent

20. $5.426 _____

21. $0.068 _____

22. $942.3749 _____

23. $3.1077 _____

24. $0.0449 _____

25. $72.625 _____

Nearest Thousandth

26. 4.37275 pt _____

27. 0.00549 qt _____

28. 12.6325 oz _____

29. 5.040603 in. _____

30. 8.9989 yd _____

31. 0.100499 lb _____

UP to the *Next* Cent

32. $4.621 _____

33. $0.159 _____

34. $42.525 _____

35. $2.0917 _____

36. $11.4485 _____

37. $0.6545 _____

Score for B (24)

Assignment 4.1 *Continued*

C **(27 points) Write the following numbers in columns, and then add. (3 points for each correct answer)**

38. 3.84, 42.81, 604.927 **41.** 24.78, 71.402, 8.3176 **44.** 337.51, 15.8042, 0.452

39. 0.7323, 4.084, 17.42 **42.** 6.084, 0.005, 48.2007 **45.** 36.7, 208.51, 3.992

40. 15.4, 32.574, 9.51, 74.0822 **43.** 6.4, 3.211, 12.6, 28.59 **46.** 0.592, 1.82, 0.774, 6.5

Score for C (27)

D **(36 points) Subtract the following. (3 points for each correct answer)**

47.	0.829	**50.**	0.7212	**53.**	5.0739	**56.**	4.37
	−0.37		−0.034		−2.843		−1.9055

48.	0.04264	**51.**	32.	**54.**	708.932	**57.**	8.432
	−0.00497		−4.37		−419.058		−5.947

49.	26.04	**52.**	804.07	**55.**	0.395	**58.**	1.0045
	−9.108		−167.1		−0.296		−1.003

Score for D (36)

Assignment 4.2: Multiplication & Division of Decimal Numbers

Name

Date Score

A (32 points) Multiply the following. Round monetary products to the nearest cent. Do not round non-monetary products. (4 points for each correct answer)

1. $14.25	**2.** $24.60	**3.** $410.00	**4.** $57.80
× 56	× 3.5	× 0.706	× 0.25
_____	_____	_____	_____
5. 107.21	**6.** 51.97	**7.** 285.70326	**8.** 916.03
× 0.84	× 0.43	× 0.08	× 0.402
_____	_____	_____	_____

Score for A (32)

B (24 points) Divide the following. Round monetary quotients to the nearest cent. Round nonmonetary quotients to two decimal places. (4 points for each correct answer)

9. 7)$12.95 **10.** 0.36)$6.75 **11.** 1.2)$54.30

_____ _____ _____

Assignment 4.2 *Continued*

12. $1.5\,\overline{)2.5\,7}$ **13.** $0.11\,\overline{)0.6735}$ **14.** $0.09\,\overline{)0.7888}$

_____ _____ _____

Score for B (24)

C **(12 points) Multiply and/or divide by just moving the decimal point or by doing some simple multiplication/division and moving the decimal point. Round monetary answers to the nearest cent. Do not round nonmonetary answers. (1 point for each correct answer)**

15. 0.0425×100 = _____ **21.** $\$72.41 \times 20$ = _____

16. $3.0608 \times 1{,}000$ = _____ **22.** $\$32.25 \times 300$ = _____

17. $0.0972 \times 10{,}000$ = _____ **23.** $\$0.06 \times 8{,}000$ = _____

18. $863 \div 10$ = _____ **24.** $\$2.50 \times 600$ = _____

19. $0.571 \div 100$ = _____ **25.** $\$42.60 \div 20$ = _____

20. $369.512 \div 1{,}000$ = _____ **26.** $\$9{,}216 \div 300$ = _____

Score for C (12)

D **(32 points) For each of the following problems, underline the estimate that is most nearly correct. (2 points for each correct answer)**

		(a)	**(b)**	**(c)**	**(d)**
27.	0.079×0.053	4.0	0.4	0.04	0.004
28.	7.65×0.8499	0.064	0.64	6.4	64
29.	3.8×66.758	0.28	2.8	28	280
30.	0.00867×642	0.054	0.54	5.4	54
31.	0.8388×0.5503	0.0048	0.048	0.48	4.8
32.	0.00389×0.0957	0.04	0.004	0.0004	0.00004
33.	278.25×93.99	27,000	2,700	270	27
34.	0.00058×0.0049	0.000003	0.00003	0.0003	0.003
35.	0.21×5987	120	1,200	12,000	120,000
36.	$7.958 \div 0.0544$	16	160	1,600	16,000
37.	$3.865 \div 873.12$	0.004	0.04	0.4	4
38.	$0.00612 \div 0.00358$	0.15	1.5	15	150
39.	$83.72 \div 0.6186$	0.14	1.4	14	140
40.	$0.5447 \div 0.00585$	900	90	9	0.9
41.	$43.85 \div 6.397$	70	7	0.7	0.07
42.	$0.02655 \div 0.09481$	300	30	3	0.3

Score for D (32)

Assignment 4.3: Decimal Numbers in Business

Name

Date Score

A **(36 points) Business Applications and Critical Thinking. Solve the following. Do not round your final answers. (6 points for each correct answer)**

1. Mike Haynes had 18.5 feet of rope. He cut off a piece 12.75 feet long. How much did he have left?

2. A jeweler had only 2.8 ounces of gold on hand, so she bought 12.5 ounces more to make Christmas items. She used 9.7 ounces for gold rings. How much gold did she have left?

3. Sherry Tanaka reads meters for the electric company. She walked 2.4 miles on Monday; 3.8 miles on Tuesday, 2.9 miles on Wednesday, 3.25 miles on Thursday, and 3.1 miles on Friday. What was her total distance for the week?

4. Four drivers bring their cars in to a service station to buy gasoline. Individually, they buy 10.4, 8.3, 13.6, and 15.9 gallons. How much did they purchase all together?

5. Cho Jewelers had 50.8 ounces of gold on hand. After the firm received 120 ounces and used 117.9 ounces, how much gold was on hand?

6. Brainard Paving Co. delivered 7.3 tons of asphalt. It used 5.2 tons for a parking lot in a shopping center and 1.4 tons for an access road. How many tons of asphalt were left?

 Score for A (36)

Assignment 4.3 *Continued*

7. Bill Wells Hardware sells a large-diameter plas-
tic pipe for $0.08 per foot and copper pipe for
$1.02 per foot. How much will Jesse Cullen save
by using plastic pipe if she needs 250 feet of
pipe? _____

8. Benoit Landscaping sent three truckloads of
topsoil to a job. The soil cost $21.50 per cubic
yard. Two trucks carried 7.25 cubic yards each;
the third carried 6.75 cubic yards. What was the
total cost of all the topsoil? _____

9. Wholesale, 1,000 two-ounce plastic bottles cost
3.5 cents each, and 2,000 four-ounce bottles
cost 4.5 cents each. What is the total cost of all
3,000 bottles? _____

10. Carol Wilson commutes to work. On Monday,
she bought 14.62 gallons of regular gasoline at
$1.279 per gallon. On Thursday, she bought
15.51 gallons at $1.319. How much did she pay
for gasoline that week? _____

11. James Cotton paid $24.60 for 300 feet of nylon
rope. The rope was 0.25 inch in diameter. What
was James's cost per foot for this particular
rope? _____

12. A pizza chef has 24 pounds of flour on hand.
He needs 3.75 pounds of flour for one large
recipe of pizza dough. How many recipes can
he make with the flour on hand? (Round to the
nearest tenth.) _____

13. Paint thinner costs $1.29 per gallon. How many
gallons can a painting contractor buy for $10?
(Round to the nearest tenth.) _____

14. Peggy Sommer earns $17.48 per hour. How
many hours did she work during a partial day
for which her pay was $109.25? _____

Score for B (64)

Notes

Although the use of decimals has taken over fractions in many business areas—the stock market, for example—fractions are still used in many industries, such as the shoe industry. Did you know that shoe sizes are different the world over? A men's $8\frac{1}{2}$ in the U.S. is a 7 in England, a $41\frac{1}{2}$ in the rest of Europe, and a 26 in Japan. You can find a conversion chart at the Dexter Shoe Company Web site at http://www.dextershoe.com/custserv/conversion_chart.asp#.

The Dexter Shoe Company was incorporated in 1957 in Dexter, Maine. Dexter shoes are distributed in 102 countries through department stores, independent footwear retailers, Web sites, and Dexter Factory Stores. Visit their site for information about the company, including career opportunities, an online survey, company history, and more. http://www.dextershoe.com.

Fractions | 5

LEARNING OBJECTIVES

By studying this chapter and completing all assignments you will be able to:

 Change improper fractions and mixed numbers.

 Change fractions to lower and higher terms.

 Add fractions and mixed numbers.

 Subtract fractions and mixed numbers.

 Multiply fractions, mixed numbers, and whole numbers.

 Divide fractions, mixed numbers, and whole numbers.

1. How are fractions similar to and different from decimals?
2. In fractions, do all four of the arithmetic operations (addition, subtraction, multiplication, and division) require common denominators? Explain.
3. In fractions, can you use cancellation in all four of the arithmetic operations (addition, subtraction, multiplication, and division)? Explain.
4. What is the rule for dividing one fraction by another fraction?
5. How can you use a calculator to multiply two fractions?

Fractions are a natural part of all cultures. Very young children who cannot yet read learn simple fractions such as one half and one third when their parents teach them about sharing a candy bar or a pizza. Before the development of inexpensive handheld calculators, fractions were perhaps more important than today because they permitted shortcuts in arithmetic. However, fractions are still important in some industries. And the rules of fractions will always remain important in algebra and higher mathematics.

Vocabulary of Fractions

A restaurant cuts its small pizzas into six pieces. Each piece is "*one sixth*" of the pizza. If you take two pieces of pizza, you have "*two sixths*" of the pizza. With numbers, two sixths is written as $\frac{2}{6}$. The 2 is called the **numerator,** and the six is called the **denominator.** Because the numerator, 2, is smaller than the denominator, 6, we say that $\frac{2}{6}$ is a **proper fraction.** When you buy two small pizzas and each is cut into six pieces, you have twelve pieces, or twelve sixths, or $\frac{12}{6}$. Here the numerator, 12, is greater than or equal to the denominator, 6, so we say that $\frac{12}{6}$ is an **improper fraction.** If you eat one of the twelve slices of pizza, eleven pieces remain, or $\frac{11}{6}$, or one whole pizza and $\frac{5}{6}$ of the other pizza. We can write this result as $1\frac{5}{6}$, which is called a **mixed number.** That is, $1\frac{5}{6}$ is just another way to write $\frac{11}{6}$. Figure 5-1 illustrates these concepts.

Figure 5-1

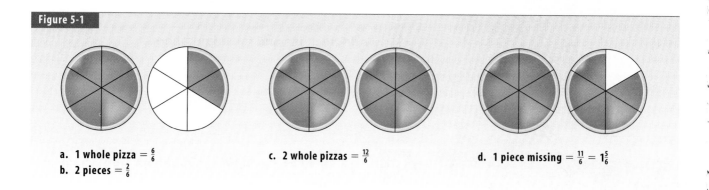

a. **1 whole pizza** $= \frac{6}{6}$
b. **2 pieces** $= \frac{2}{6}$

c. **2 whole pizzas** $= \frac{12}{6}$

d. **1 piece missing** $= \frac{11}{6} = 1\frac{5}{6}$

Changing Improper Fractions and Mixed Numbers

LEARNING OBJECTIVES

Change improper fractions and mixed numbers.

With simple arithmetic, we can change improper fractions to mixed numbers and mixed numbers to improper fractions.

STEPS **to Change an Improper Fraction to a Mixed Number**

1. Divide the numerator by the denominator.
2. Put the quotient in the whole-number position.
3. Put the remainder in the numerator position of the fraction part.
4. Put the original denominator in the denominator position of the fraction part.

EXAMPLE A

Change $\frac{11}{8}$ to a mixed number.

STEP 1 STEPS 2, 3, & 4

$$\frac{11}{8} = 8)\overline{\begin{array}{c} 1\ R3 \\ 11 \end{array}} \qquad \text{Thus,} \qquad \frac{11}{8} = 1\frac{3}{8}$$
$$\begin{array}{c} \underline{8} \\ 3 \end{array}$$

Note: Refer to Point A in Figure 5-2 to see where this mixed number appears on a ruler.

Figure 5-2

$$\frac{11}{8} = 1\frac{3}{8} \qquad\qquad\qquad 3\frac{1}{2} = \frac{7}{2}$$

A B

C

$$\frac{12}{16} = \frac{6}{8} = \frac{3}{4}$$

STEPS to Change a Mixed Number to an Improper Fraction

1. Multiply the denominator of the fraction by the whole number.
2. Add the numerator of the fraction to the product of Step 1. Put the sum in the numerator position of the improper fraction.
3. Put the denominator of the fraction of the mixed number in the denominator position of the improper fraction.

EXAMPLE B

Change $3\frac{1}{2}$ to an improper fraction.

STEP 1 STEPS 2 & 3

$$2 \times 3 = 6 \qquad \text{Thus,} \qquad 3\frac{1}{2} = \frac{6+1}{2} = \frac{7}{2}$$

See Point B in Figure 5-2.

Changing Fractions to Lower and Higher Terms

LEARNING OBJECTIVES

Change fractions to lower and higher terms.

Look at a measuring tape shown in Figure 5-2. Point C marks the distance $\frac{12}{16}$ of an inch, but it could also be written as $\frac{6}{8}$ or $\frac{3}{4}$ of an inch. Thus $\frac{12}{16}$, $\frac{6}{8}$, and $\frac{3}{4}$ are three ways to write the same value. We say that $\frac{6}{8}$ is in **lower terms** and $\frac{12}{16}$ is in **higher terms.** We also say that $\frac{3}{4}$ is in **lowest terms** because it cannot be changed to any lower terms. When we change a fraction to lower terms, we are *reducing* the fraction to lower terms. If we change a mixed number such as $2\frac{12}{16}$ to $2\frac{3}{4}$, we have reduced the mixed number to its lowest terms. When we change a fraction to higher terms, we are *raising* the fraction to higher terms.

STEPS **to Reduce a Fraction to Lowest Terms**

1. Divide both the numerator and the denominator by a common divisor greater than 1 to arrive at a reduced fraction.

2. If necessary, repeat Step 1 until the fraction is in lowest terms.

Rule: If a fraction's numerator and denominator have no common divisor greater than 1, the fraction is already in lowest terms.

EXAMPLE C

Reduce $\frac{12}{16}$ to lowest terms.

$$\frac{12}{16} = \frac{12 \div 2}{16 \div 2} = \frac{6}{8} = \frac{6 \div 2}{8 \div 2} = \frac{3}{4} \quad \text{or} \quad \frac{12}{16} = \frac{12 \div 4}{16 \div 4} = \frac{3}{4}$$

Note that dividing by 4 once is faster than dividing by 2 twice. Always try to use the greatest common divisor that you can find.

www.visualfractions. com/

Visit this Web site for an online tutorial and review of fractions.

STEPS **to Raise a Fraction to Higher Terms**

1. Divide the new denominator by the old denominator. The quotient is the *common multiplier.*

2. Multiply the old numerator by the multiplier.

3. Multiply the old denominator by the multiplier.

VIDEO

Reducing and Raising Fractions

EXAMPLE D

Raise $\frac{3}{4}$ to twenty-fourths.

STEP 1

$$\frac{3}{4} = \frac{?}{24} \qquad 24 \div 4 = 6$$

STEPS 2 & 3

$$\text{So, } \frac{3}{4} = \frac{3 \times 6}{4 \times 6} = \frac{18}{24}$$

Adding Fractions and Mixed Numbers

Fractions and mixed numbers are all numbers—they can be added and subtracted just like whole numbers. However, **before fractions and mixed numbers can be added or subtracted, they must have common denominators.**

When you add fractions and/or mixed numbers, you must first find the **common denominator,** which is the denominator shared by all of the fractions and is the denominator of the fraction part of the answer. The smallest common denominator possible is called the **least common denominator.** However, if the least common denominator is not easily apparent, it may be quicker to use the first common denominator that you can discover and then reduce the answer. The product of all of the denominators will always be a common denominator, but there will often be a smaller common denominator.

LEARNING OBJECTIVES

Add fractions and mixed numbers.

V I D E O

Adding and Subtracting Fractions and Mixed Numbers

STEPS	**to Add Two or More Fractions and/or Mixed Numbers**

1. If necessary, change the fraction parts to fractions with common denominators.
2. Add the numerators to make the numerator of the fraction part of the answer. If there are any whole number parts, add them to make the whole number part of the answer.
3. If necessary, reduce the fraction part to a mixed number in lowest terms and mentally combine any whole number parts to make a final mixed number answer.

EXAMPLE E

Add $1\frac{7}{8}$ and $2\frac{5}{8}$.

The fractions already have a common denominator of 8.

$$
\begin{array}{r}
1\dfrac{7}{8} \\[4pt]
+\,2\dfrac{5}{8} \\
\hline
3\dfrac{12}{8} = 3 + 1\dfrac{4}{8} = 4\dfrac{1}{2}
\end{array}
$$

EXAMPLE F

Add $\frac{5}{6}$ and $\frac{3}{4}$.

A common denominator is $6 \times 4 = 24$.

$$
\begin{array}{r}
\dfrac{5}{6} = \dfrac{5 \times 4}{6 \times 4} = \dfrac{20}{24} \\[6pt]
+\,\dfrac{3}{4} = \dfrac{3 \times 6}{4 \times 6} = \dfrac{18}{24} \\
\hline
\dfrac{38}{24} = 1\dfrac{14}{24} = 1\dfrac{7}{12}
\end{array}
$$

EXAMPLE G

Add $3\frac{3}{8}$, $7\frac{5}{6}$, and $\frac{1}{4}$.

The least common denominator is 24.

STEP 1

$$
\begin{array}{rcl}
3\dfrac{3}{8} &=& 3 \quad \dfrac{9}{24} \\[6pt]
7\dfrac{5}{6} &=& 7 \quad \dfrac{20}{24} \\[6pt]
+\,\dfrac{1}{4} &=& +\ \ \dfrac{6}{24} \\
\hline
\end{array}
$$

STEP 2

$$10 \quad \dfrac{35}{24}$$

STEP 3

$$10 + 1\dfrac{11}{24} = 11\dfrac{11}{24}$$

© PHOTODISC, INC.

Subtracting Fractions and Mixed Numbers

LEARNING OBJECTIVES 4

Subtract fractions and mixed numbers.

The procedure for subtracting one fraction from another is essentially the same as the procedure for adding one fraction to another. When you calculate the difference $3\frac{1}{4} - \frac{3}{4}$, $3\frac{1}{4}$ is called the *minuend* and $\frac{3}{4}$ is called the *subtrahend*, as in the subtraction of whole numbers.

Borrowing 1

Sometimes, as with $3\frac{1}{4} - \frac{3}{4}$, the fraction part of the minuend is smaller than the fraction part of the subtrahend. To make the fraction part of the minuend larger than the fraction part of the subtrahend, you have to "borrow 1" from the whole number part of the minuend. Actually, you're just rewriting the minuend. Remember that $3\frac{1}{4}$ means $3 + \frac{1}{4}$, or the same as $2 + 1 + \frac{1}{4}$, or $2 + \frac{4}{4} + \frac{1}{4}$, or $2\frac{5}{4}$. Figure 5-3 is useful in explaining borrowing.

VIDEO

Adding and Subtracting Fractions and Mixed Numbers

Figure 5-3

a. 3 whole units plus $\frac{1}{4}$ of a unit

b. 2 "whole units" plus $\frac{5}{4}$ of a unit

1. If necessary, change the fractions so that all fractions have a common denominator. The common denominator is the denominator in the fraction part of the answer.

2. If necessary, "borrow 1" from the whole number part of the minuend to make the fraction part of the minuend larger than the fraction part of the subtrahend.

3. Subtract the numerators of the fractions to make the numerator of the fraction part of the answer.

4. If there are any whole number parts, subtract them to make the whole number part of the answer.

5. If necessary, reduce the fraction in the answer to lowest terms.

EXAMPLE H

STEP 3 STEP 5

$$\frac{7}{8} - \frac{5}{8} = \frac{7-5}{8} = \frac{2}{8} = \frac{1}{4}$$

EXAMPLE I

STEP 1 STEP 3

$$\frac{3}{4} - \frac{1}{5} = \frac{3 \times 5}{4 \times 5} - \frac{1 \times 4}{5 \times 4} = \frac{15}{20} - \frac{4}{20} = \frac{11}{20}$$

EXAMPLE J

STEP 1

$$\begin{array}{r} 5\frac{3}{4} = \quad 5\frac{9}{12} \\ - 2\frac{1}{3} = - 2\frac{4}{12} \\ \hline \end{array}$$

STEPS 3 & 4 $3\frac{5}{12}$

EXAMPLE K

STEP 1 STEP 2

$$\begin{array}{r} 4\frac{2}{9} = \quad 4\frac{4}{18} = \quad 3\frac{18}{18} + \frac{4}{18} = \quad 3\frac{22}{18} \\ - 1\frac{5}{6} = - 1\frac{15}{18} = - 1\frac{15}{18} \qquad\qquad = - 1\frac{15}{18} \\ \hline \end{array}$$

STEPS 3 & 4 $2\frac{7}{18}$

CyberConnection

This Web site also offers practice with fractions.

Rule 27: Compound Fractions

A compound fraction is sometimes called a mixed number. Recall that $3\frac{1}{2}$, $4\frac{1}{5}$, and $-8\frac{2}{3}$ are examples of compound fractions.

• To manipulate compound fractions, just convert them to simple fractions and follow rules 1 through 23 for simple fractions.

Example 1: Convert $3\frac{1}{2}$ to a simple fraction.

Answer. $3\frac{1}{2}$ can be written $3 + \frac{1}{2}$. Write 3 as the fraction $\frac{3}{1}$. Now $3 + \frac{1}{2}$ can be written $\frac{3}{1} + \frac{1}{2}$ and

$$\frac{3}{1} + \frac{1}{2} = \frac{3}{1} \times \frac{2}{2} + \frac{1}{2} = \frac{6}{2} + \frac{1}{2} = \frac{7}{2}.$$

• To convert a simple fraction to a compound fraction, the numerator must be larger than the denominator. Separate the whole number first.

Source: http:www.sosmath.com/algebra/fraction/frac1/frac1.html.
Copyright © 2001 MATH MEDICS, LLC; used with permission.

CONCEPT CHECK 5.2

a. Subtract $\frac{5}{6}$ from $\frac{7}{8}$.

The least common denominator is 24.

$$\begin{array}{r} \dfrac{7}{8} = \dfrac{21}{24} \\[2ex] -\dfrac{5}{6} = -\dfrac{20}{24} \\[2ex] \hline \dfrac{1}{24} \end{array}$$

b. Subtract $2\frac{7}{10}$ from $5\frac{4}{15}$.

The least common denominator is 30.

$$\begin{array}{r} 5\dfrac{4}{15} = \ \ 5\dfrac{8}{30} = \ \ 4\dfrac{38}{30} \\[2ex] -2\dfrac{7}{10} = -2\dfrac{21}{30} = -2\dfrac{21}{30} \\[2ex] \hline 2\dfrac{17}{30} \end{array}$$

COMPLETE ASSIGNMENT 5.1.

Multiplying Fractions, Mixed Numbers, and Whole Numbers

LEARNING OBJECTIVES

Multiply fractions, mixed numbers, and whole numbers.

In fractions, multiplication is the simplest operation and division is the next simplest. The reason is that multiplication and division do not require common denominators like addition and subtraction do. Recall that any mixed number can be changed to an improper fraction. Also, a whole number can be written as an improper fraction by writing the number in the numerator with a denominator of 1. For example, the whole-number 5 can be written as the fraction $\frac{5}{1}$.

STEPS to Multiply Fractions, Mixed Numbers, and Whole Numbers

1. If necessary, change any mixed or whole numbers to improper fractions.
2. Multiply all the numerators to get the numerator of the product.
3. Multiply all the denominators to get the denominator of the product.
4. If necessary, write the product as a fraction or mixed number in lowest terms.

VIDEO

Multiplication and Division of Mixed Numbers

EXAMPLE L

| STEP 1 | STEPS 2 & 3 | STEP 4 |

$$1\frac{1}{3} \times \frac{4}{5} = \frac{4}{3} \times \frac{4}{5} = \frac{4 \times 4}{3 \times 5} = \frac{16}{15} = 1\frac{1}{15}$$

EXAMPLE M

| STEPS 2 & 3 | STEP 4 |

$$\frac{2}{3} \times \frac{4}{5} \times \frac{5}{6} = \frac{2 \times 4 \times 5}{3 \times 5 \times 6} = \frac{40}{90} = \frac{4}{9}$$

Note: The word *of* often means *multiply* when it is used with fractions. For example, you know that "$\frac{1}{2}$ *of* 6 bottles" is 3 bottles. And $\frac{1}{2}$ of $6 = \frac{1}{2} \times \frac{6}{1} = \frac{6}{2} = 3$

Canceling Common Factors in Numerators and Denominators

As the last step in example M, we reduced the fraction $\frac{40}{90}$ to its lowest terms, $\frac{4}{9}$. Recall that reducing this fraction means that we divide both the numerator and the denominator by 10. As an option, we can do the division in advance, before doing any multiplication. Examining the three numerators and three denominators we discover that they have common factors of 2 and 5 ($2 \times 5 = 10$). Divide out, or **cancel,** both common factors in the numerators and denominators as shown in example N. This division of the common factors is often called **cancellation.** Canceling common factors is an option; it isn't required to calculate the correct product.

EXAMPLE N

Multiply the three fractions, using cancellation.

$$\frac{2}{3} \times \frac{4}{5} \times \frac{5}{6} = \frac{2}{3} \times \frac{4}{\cancel{5}} \times \frac{\overset{1}{\cancel{5}}}{6} = \frac{2}{3} \times \frac{\overset{2}{\cancel{4}}}{\cancel{5}} \times \frac{\overset{1}{\cancel{5}}}{\cancel{6}} = \frac{2 \times 2 \times 1}{3 \times 1 \times 3} = \frac{4}{9}$$

EXAMPLE O

Multiply the fraction and the whole number, using cancellation

$$8 \times \frac{3}{4} = \frac{8}{1} \times \frac{3}{4} = \frac{\overset{2}{\cancel{8}}}{1} \times \frac{3}{\cancel{4}} = \frac{2 \times 3}{1 \times 1} = \frac{6}{1} \quad or \quad 6$$

EXAMPLE P

Multiply the fraction and the mixed number, using cancellation

STEP 1	STEPS 2 & 3

$$\frac{2}{5} \times 1\frac{3}{4} = \frac{2}{5} \times \frac{7}{4} = \frac{\overset{1}{\cancel{2}}}{5} \times \frac{7}{\underset{2}{\cancel{4}}} = \frac{1 \times 7}{5 \times 2} = \frac{7}{10}$$

CONCEPT CHECK 5.3

Multiply the fraction, whole number, and mixed number, using cancellation.

STEP 1	STEPS 2 & 3	STEP 4

$$\frac{1}{4} \times 3 \times 2\frac{2}{3} = \frac{1}{4} \times \frac{3}{1} \times \frac{8}{3} = \frac{1}{\underset{1}{\cancel{4}}} \times \frac{\overset{1}{\cancel{3}}}{1} \times \frac{\overset{2}{\cancel{8}}}{\underset{1}{\cancel{3}}} = \frac{1 \times 1 \times 2}{1 \times 1 \times 1} = \frac{2}{1} = 2$$

Dividing Fractions, Mixed Numbers, and Whole Numbers

Recall that with whole numbers division is the *inverse* of multiplication. You can check a multiplication problem by division. With fractions, you actually perform a division problem by doing multiplication. That is, you *invert the divisor and multiply*.

Divide fractions, mixed numbers, and whole numbers

STEPS to Divide Fractions, Mixed Numbers, and Whole Numbers

1. If necessary, change the dividend and/or the divisor from mixed or whole numbers to improper fractions.
2. Invert the divisor (i.e., exchange the numerator and denominator).
3. Change the division symbol to a multiplication symbol.
4. Multiply the two factors (canceling where possible, if desired).
5. Write the quotient as a proper fraction or mixed number in lowest terms.

VIDEO

Multiplication and Division of Mixed Numbers

EXAMPLE Q

$$\frac{3}{10} \div \frac{2}{5} = \frac{3}{10} \times \frac{5}{2} = \frac{3}{\overset{}{10}_{2}} \times \frac{\overset{1}{5}}{2} = \frac{3 \times 1}{2 \times 2} = \frac{3}{4}$$

STEPS 2 & 3 · STEP 4

EXAMPLE R

$$6 \div 1\frac{3}{5} = \frac{6}{1} \div \frac{8}{5} = \frac{6}{1} \times \frac{5}{8} = \frac{\overset{3}{6}}{1} \times \frac{5}{\overset{8}{4}} = \frac{3 \times 5}{1 \times 4} = \frac{15}{4} = 3\frac{3}{4}$$

STEP 1 · STEPS 2 & 3 · STEP 4 · STEP 5

www.mathleague.com/
help/fractions/
fractions.htm

More fraction help is available at this Web site.

✓ CONCEPT CHECK 5.4

Divide $3\frac{3}{4}$ by $1\frac{1}{2}$.

Change both mixed numbers to improper fractions: $\frac{15}{4} \div \frac{3}{2}$
Invert the divisor $\frac{3}{2}$ to $\frac{2}{3}$ and multiply:

$$\frac{15}{4} \times \frac{2}{3} = \frac{\overset{5}{15}}{\underset{2}{4}} \times \frac{\overset{1}{2}}{\underset{1}{3}} = \frac{5 \times 1}{2 \times 1} = \frac{5}{2} = 2\frac{1}{2}$$

COMPLETE ASSIGNMENT 5.2.

Decimal Equivalents of Fractions

When you're working without a calculator, some arithmetic problems are easier to solve in fractions and some are easier to solve in decimals. That's especially true in addition and subtraction of fractions because you must always find a common denominator. When you're using a calculator, you can quickly change fractions to decimal numbers and do the arithmetic in decimals. You can also use the calculator to check all the work that you do in fractions.

The fraction bar separating the numerator and denominator can be interpreted as a division symbol. For example, to change $\frac{3}{4}$ to a decimal, divide 3 by 4; the quotient is 0.75. Likewise, $\frac{4}{5} = 4 \div 5 = 0.8$. The numbers 0.75 and 0.8 are called the **decimal equivalents** of $\frac{3}{4}$ and $\frac{4}{5}$, respectively, which are similarly called the **fractional equivalents.**

EXAMPLE S

Compute $\dfrac{3}{4} + \dfrac{4}{5}$, using decimals.

$$\frac{3}{4} = 0.75$$
$$+\frac{4}{5} = \frac{+0.8}{1.55}$$

EXAMPLE T

Compute $\dfrac{3}{4} \times \dfrac{4}{5}$, using decimals.

$$\frac{3}{4} = 0.75$$
$$\times\frac{4}{5} = \frac{+0.8}{0.600}$$

Chapter Terms for Review

cancel
cancellation
common denominator
decimal equivalent
denominator

fractional equivalent
fractions
higher terms
improper fraction
least common denominator

lower terms
lowest terms
mixed number
numerator
proper fraction

Visit the Web

http://pittsford.monroe.edu/jefferson/calfieri/fractions/fractionsmain.html—You can practice fraction problems at this Web site.

http://school.discovery.com/homeworkhelp/webmath—More help with fractions is available at this Discovery.com site.

http://www.mcs.surrey.ac.uk/Personal/R.Knott/Fibonacci/cfINTRO.html—Visit this Web site for a little challenge in fractions.

http://home.xnet.com/~fidler/triton/math/review/mat045/fraction/fract.htm—Test yourself on fractions.

http://forum.swarthmore.edu/library/topics/fractions—Visit the Math Forum for links to more fractions help.

CyberConnection

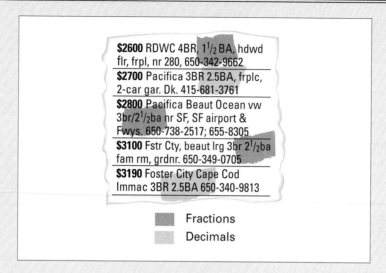

$2600 RDWC 4BR, 1$\frac{1}{2}$ BA, hdwd flr, frpl, nr 280, 650-342-9662

$2700 Pacifica 3BR 2.5BA, frplc, 2-car gar. Dk. 415-681-3761

$2800 Pacifica Beaut Ocean vw 3br/2$\frac{1}{2}$ba nr SF, SF airport & Fwys. 650-738-2517; 655-8305

$3100 Fstr Cty, beaut lrg 3br 2$\frac{1}{2}$ba fam rm, grdnr. 650-349-0705

$3190 Foster City Cape Cod Immac 3BR 2.5BA 650-340-9813

Fractions

Decimals

Fractions are used so often in everyday speech that people aren't even aware that they're speaking "math language." People tell a half-truth, spend a quarter, cut a pizza in eighths, use a third cup of sugar, and a person may be one-sixteenth Italian. They also often *see* fractions in business—sweaters $\frac{1}{2}$ off, coats marked down $\frac{1}{3}$, and stocks up $\frac{1}{4}$.

One of the advertisements above is for a house with $1\frac{1}{2}$ baths. If you had two of these houses, each with $1\frac{1}{2}$ baths, would you have 3 whole baths? Explain.

The Bottom Line

Summary of chapter learning objectives:

Learning Objective	Summary	Example
5.1	Change improper fractions and mixed numbers	1(a). Change $\frac{26}{7}$ to a mixed number. 1(b). Change $3\frac{2}{5}$ to an improper fraction.
5.2	Change fractions to lower and higher terms	2(a). Reduce $\frac{30}{42}$ to lowest terms. 2(b). Raise $\frac{7}{12}$ to sixtieths; that is, $\frac{7}{12} = \frac{?}{60}$.
5.3	Add fractions and mixed numbers	3. Add $\frac{7}{8}, \frac{5}{6},$ and $1\frac{1}{4}$.
5.4	Subtract fractions and mixed numbers	4. Subtract $1\frac{3}{4}$ from $4\frac{2}{5}$.
5.5	Multiply fractions, mixed numbers, and whole numbers	5. Multiply: $\frac{2}{5} \times \frac{3}{7}$
5.6	Divide fractions, mixed numbers, and whole numbers	6. Divide: $1\frac{4}{5} \div \frac{3}{4}$

Answers: 1(a). $3\frac{5}{7}$ 1(b). $\frac{17}{5}$ 2(a). $\frac{5}{7}$ 2(b). $\frac{35}{60}$ 3. $2\frac{23}{24}$ 4. $2\frac{13}{20}$ 5. $\frac{6}{35}$ 6. $2\frac{2}{5}$

Write all answers as proper fractions or mixed numbers in lowest terms.

1 Change $3\frac{5}{6}$ to an improper fraction _____

2 Change $\frac{75}{12}$ to a mixed number _____

3 Reduce $\frac{42}{60}$ to lowest terms _____

4 Raise $\frac{9}{14}$ to 56ths _____

5 Add $\frac{2}{3}$, $\frac{2}{5}$, and $\frac{3}{10}$ _____

6 Add $\frac{7}{8}$ and $1\frac{1}{6}$ _____

7 Add $\frac{3}{4}$, $2\frac{4}{5}$, and 3 _____

8 Subtract $\frac{1}{3}$ from $\frac{3}{5}$ _____

9 Subtract $\frac{8}{9}$ from $2\frac{5}{6}$ _____

10 Subtract $1\frac{5}{9}$ from $4\frac{2}{5}$ _____

11 Multiply $\frac{5}{9}$ by $\frac{6}{25}$ _____

12 Multiply $\frac{9}{16}$ by $1\frac{13}{15}$ _____

13 Multiply $2\frac{1}{10}$, $\frac{8}{15}$, and $2\frac{1}{12}$ _____

14 Divide $\frac{15}{16}$ by $\frac{5}{12}$ _____

15 Divide $1\frac{11}{25}$ by $\frac{24}{35}$ _____

16 Divide $1\frac{5}{7}$ by $1\frac{13}{14}$ _____

17 Jackie Barner decided to use an expensive, but effective, herbicide to kill weeds and brush on a client's land. For one part of the land, she needed $3\frac{2}{3}$ quarts of herbicide; for a second part, she needed $2\frac{1}{4}$ quarts; and for the third part, she needed $1\frac{5}{6}$ quarts. In total, how many quarts of herbicide did Jackie need for this client? _____

18 Cabinetmaker Vic Pettijon needs to make a cabinet door. The cabinet drawing shows an opening $24\frac{3}{16}$ inches wide. Vic wants a space of 1/8 inch on each side of the cabinet door. How wide should he make the door? _____

19 The Grand Hotel just hired Jim Spurgeon as a chef. Jim makes a hot sauce that uses $1\frac{3}{4}$ tablespoons of chili powder, but he needs to increase the recipe by $3\frac{1}{2}$ times. How many tablespoons of chili powder should he use? _____

20 How many pieces of copper $2\frac{5}{16}$ inches long can be cut out of one piece that is $24\frac{1}{2}$ inches long? _____

Assignment 5.1: Addition & Subtraction of Fractions

LEARNING OBJECTIVES

Name _____

Date _____ Score _____

A (12 points) Change the improper fractions to whole numbers or to mixed numbers. Change the mixed numbers to improper fractions. (1 point for each correct answer)

1. $\dfrac{13}{7}$　　2. $\dfrac{32}{10}$　　3. $\dfrac{18}{8}$　　4. $\dfrac{20}{15}$

5. $\dfrac{11}{6}$　　6. $\dfrac{25}{5}$　　7. $3\dfrac{7}{10}$　　8. $2\dfrac{11}{15}$

9. $4\dfrac{5}{8}$　　10. $3\dfrac{1}{4}$　　11. $6\dfrac{3}{5}$　　12. $1\dfrac{7}{12}$

Score for A (12) _____

B (15 points) In problems 13–20, reduce each fraction to lowest terms. In problems 21–27, raise each fraction to higher terms, as indicated. (1 point for each correct answer)

13. $\dfrac{15}{25}$　　14. $\dfrac{8}{24}$　　15. $\dfrac{10}{15}$　　16. $\dfrac{8}{20}$

17. $\dfrac{32}{48}$　　18. $\dfrac{24}{30}$　　19. $\dfrac{42}{56}$　　20. $\dfrac{16}{32}$

21. $\dfrac{5}{6} = \dfrac{}{18}$　　22. $\dfrac{3}{4} = \dfrac{}{20}$　　23. $\dfrac{7}{8} = \dfrac{}{24}$　　24. $\dfrac{5}{12} = \dfrac{}{36}$

25. $\dfrac{11}{12} = \dfrac{}{48}$　　26. $\dfrac{2}{3} = \dfrac{}{12}$　　27. $\dfrac{2}{5} = \dfrac{}{45}$

Score for B (15) _____

C (24 points) Add the following fractions and mixed numbers. Write the answers as fractions or mixed numbers, with fractions in lowest terms. (3 points for each correct answer)

28. $\dfrac{5}{8}$ $+ \dfrac{1}{8}$　　29. $\dfrac{7}{10}$ $+ \dfrac{3}{10}$　　30. $\dfrac{7}{16}$ $+ 2\dfrac{11}{16}$　　31. $1\dfrac{2}{3} =$ $+ 2\dfrac{3}{4} =$

32. $1\dfrac{1}{4} =$ $\dfrac{3}{8} =$ $+ 4\dfrac{11}{12} =$　　33. $4\dfrac{1}{2} =$ $3\dfrac{2}{3} =$ $+ \dfrac{5}{6} =$　　34. $\dfrac{3}{5} =$ $3\dfrac{5}{6} =$ $+ 5\dfrac{1}{3} =$　　35. $5\dfrac{8}{9} =$ $4\dfrac{8}{15} =$ $+ 1\dfrac{1}{5} =$

Score for C (24) _____

Assignment 5.1 *Continued*

D **(24 points) Subtract the following fractions and mixed numbers. Write the answers as proper fractions or mixed numbers, with fractions in lowest terms. (3 points for each correct answer)**

36. $\dfrac{7}{8}$

 $-\dfrac{3}{8}$

37. $3\dfrac{11}{12}$

 $-1\dfrac{1}{12}$

38. $\dfrac{3}{4} =$

 $-\dfrac{7}{16} =$

39. $1\dfrac{3}{4} =$

 $-1\dfrac{2}{3} =$

40. $3\dfrac{2}{3} = \qquad =$

 $-2\dfrac{5}{6} = \qquad =$

41. $4\dfrac{2}{5} = \qquad =$

 $-1\dfrac{3}{4} = \qquad =$

42. $6\dfrac{7}{8} =$

 $-2\dfrac{2}{3} =$

43. $8\dfrac{1}{10} = \qquad =$

 $-4\dfrac{3}{4} = \qquad =$

Score for D (24)

E **(25 points) Business Applications and Critical Thinking. Solve the following. (5 points for each correct answer) Write your answers as fractions or mixed numbers in lowest terms.**

44. Two students were trying to decide which sum is larger: $\frac{1}{2} + \frac{1}{3} + \frac{1}{4}$ or $0.2 + 0.3 + 0.4$. Which sum is larger? _____

45. Olene Erickson specialized in custom painting, but for the first coat she could combine leftover paints when the colors were relatively the same. She had three containers of different shades of white: $3\frac{1}{3}$ gallons; $2\frac{3}{5}$ gallons and $2\frac{1}{2}$ gallons. If Olene combined the contents of all the containers, how much paint did she have? _____

46. Jim Howard has a top board that is $\frac{13}{16}$ inch thick. He wants to use wood screws to attach it to a bottom board. If a wood screw is $1\frac{1}{2}$ inches long, how much of the screw will be left over to go into the bottom board? _____

47. Rick Morrow is planning to attach a plywood panel to the wall with nails that are $1\frac{3}{4}$ inches long. The panel is $\frac{3}{8}$ inch thick. Beneath the panel will be a layer of sheetrock that is $\frac{1}{2}$ inch thick. How many inches of the nail will go into the wood frame that is underneath the sheetrock? _____

48. London Fabric Center sold four pieces of wool fabric to a tailor. The pieces measure $3\frac{2}{3}$ yards, $2\frac{3}{4}$ yards, 3 yards, and $4\frac{1}{2}$ yards. How many yards of wool did the tailor purchase? _____

Score for E (25)

112 | Part 2 Decimals, Fractions, and Percentages

Assignment 5.2: Multiplication and Division of Fractions

Name _____

Date _____ Score _____

A **(32 points) Change whole or mixed numbers to improper fractions and multiply. Cancel if possible. Where the word *of* appears, replace it by the multiplication symbol. Write the answers as mixed numbers or proper fractions in lowest terms. (4 points for each correct answer)**

1. $\dfrac{5}{6} \times \dfrac{4}{15} =$ _____

2. $\dfrac{3}{10} \times \dfrac{4}{7} \times \dfrac{5}{6} =$ _____

3. $\dfrac{3}{4}$ of $\dfrac{5}{6} =$ _____

4. $\dfrac{5}{18} \times \dfrac{4}{9} \times \dfrac{3}{10} =$ _____

5. $1\dfrac{1}{2} \times 1\dfrac{5}{9} =$ _____

6. $\dfrac{3}{8}$ of $10 =$ _____

7. $1\dfrac{7}{8} \times 12 \times \dfrac{3}{10} =$ _____

8. $1\dfrac{1}{3} \times 1\dfrac{7}{8} \times 1\dfrac{4}{5} =$ _____

Score for A (32)

B **(32 points) Change the mixed numbers to improper fractions and divide. Cancel where possible. Write the quotients as mixed numbers or proper fractions in lowest terms. (4 points for each correct answer)**

9. $\dfrac{5}{8} \div \dfrac{3}{4} =$ _____

10. $\dfrac{7}{10} \div \dfrac{8}{15} =$ _____

11. $\dfrac{3}{4} \div \dfrac{5}{8} =$ _____

12. $\dfrac{3}{10} \div 3\dfrac{3}{5} =$ _____

13. $6\dfrac{1}{4} \div 1\dfrac{7}{8} =$ _____

14. $3\dfrac{5}{6} \div 1\dfrac{7}{12} =$ _____

Assignment 5.2 *Continued*

15. $3\frac{1}{3} \div \frac{4}{5} =$ _____

16. $1\frac{3}{4} \div 2\frac{1}{3} =$ _____

Score for B (32)

C **(36 points) Business Applications and Critical Thinking. Use fractions and mixed numbers to solve each of the following. State the answers as whole numbers, proper fractions, or mixed numbers. (6 points for each correct answer)**

17. Last week G. Floyd Construction built a small driveway that required $7\frac{1}{3}$ cubic yards of concrete. This week the company must build another one that is $2\frac{1}{2}$ times larger. How much concrete will be required? _____

18. Joyce Thompson bought eight pieces of copper tubing that were each $6\frac{3}{4}$ inches long. What was the total length of tubing that Joyce bought? (Give the answer in inches.) _____

19. Linda Johanssen had $2\frac{1}{4}$ quarts of liquid fertilizer in a container. Her supervisor asked her to mix $\frac{2}{3}$ of the fertilizer with water and save the remainder. How many quarts of fertilizer did Linda mix with water? _____

20. Wm. Gifford Landscaping needs several pieces of PVC irrigation pipe, each 3 feet 4 inches long. PVC pipe comes in 20-foot lengths. How many pieces can Gifford cut out of one length of pipe? *Hint:* 4 inches equals $\frac{1}{3}$ foot. _____

21. Larry Eickworth has a diesel-powered generator on his ranch. The generator has a tank that holds $3\frac{3}{4}$ gallons of diesel fuel. He stores the diesel fuel in 55-gallon drums (barrels). How many times can Larry refill his generator from one drum of fuel? _____

22. Home builders Bill and John Walter are planning a narrow stairway to an attic. The stairs will each be 2 feet 3 inches long. They will cut the stairs from boards that are 12 feet long. How many whole stairs can they cut from one 12-foot board? *Hint:* 3 inches is $\frac{1}{4}$ foot. _____

Score for C (36)

Notes

My Virtual Agent
Home Planner
Life at Home Newsletter
Promotions
Español
Property Search

If **BUYING** a home were this easy you wouldn't need CENTURY 21-click here

If **SELLING** a home were this easy you wouldn't need CENTURY 21-click here

If **OWNIN** you wouldn't

ABOUT US CAREER OPPORTUNITIES CONTAC US

Source: http://www.century21.com. Courtesy of Century 21 Real Estate Corporation, 2001.

Percents are used for expressing many items in real estate transactions, including commissions (which are commonly computed as a percent of the sales or purchase price), property taxes, commercial rents, and interest.

One of the major national brokers in residential, commercial, and recreational real estate transactions is Century 21. Visit its Web site at www.century21.com for tips on buying or selling a home, financing a real estate purchase, relocation services, and more. Also check the career opportunities offered by this major player in real estate brokerage.

were this easy
JRY 21-click here

SITE
MAP

PRIVACY
POLICY

Percents in Business

6

LEARNING OBJECTIVES

By studying this chapter and completing all assignments you will be able to:

1
LEARNING
OBJECTIVES

Change fractions and decimals to percents.

2
LEARNING
OBJECTIVES

Change percents to decimals.

3
LEARNING
OBJECTIVES

Find Base, Rate, and Percentage.

4
LEARNING
OBJECTIVES

Use percents to measure increase and decrease.

5
LEARNING
OBJECTIVES

Use percents to allocate overhead expenses.

Do you know what profession has one of the lowest percentages of success in the nation? Show business! In the best of times, 93 percent of union actors are out of work. For writers, unemployment hovers at 80 percent year-round.

1. How are calculations with percents similar to or different from calculations with decimals?
2. How are percents used to calculate increases and decreases?
3. What does it mean "to distribute rental expense among three departments based on the percentage of floor space in each department"?

Percents and percentages are used extensively in a wide variety of applications. The National Football League (NFL) computes the completion percentage for each quarterback. The National Basketball Association (NBA) calculates field goal percentages for all the players. When the Wells Fargo Bank loans money, the interest is quoted as a percentage rate. Almost every container of food has a label with nutritional information printed on it, and much of the information is written in percents.

Changing Fractions and Decimals to Percents

LEARNING OBJECTIVES

Change fractions and decimals to percents.

In Chapter 5, we showed that every fraction can be written as a decimal and vice versa. A **percent** is simply a third way to write fractional parts. You learned early in school that when you scored "100%" on an examination, you had answered all the questions correctly. If you scored only 75%, you had scored 75 points out of a perfect 100 points, or $\frac{75}{100}$, or $\frac{3}{4}$. Recall that $\frac{3}{4}$ written as a decimal is 0.75.

Note: The "cent" part of "percent" means 100, from the Latin word "centum." So 75% means "75 per centum," or "75 per 100," or "75 out of 100." Today, in modern French, the word for *100* is "cent" and the words for *percent* are "pour cent." A French person would say "cent pour cent" for 100%, the phrase literally meaning "100 out of 100."

Percents are convenient in written and verbal communication, especially when the equivalent fraction is not common. To some people, quoting a sales tax rate as "five one hundredths" or as "one twentieth" would sound awkward. As a percent, the rate would be "five percent." Even more awkward would be stating a sales tax rate of 8.25%, which is spoken as "eight point two five percent." As a fraction, the rate could be $\frac{8.25}{100}$, or "eight point two five one hundredths." Of course, the fraction could also be written as $\frac{825}{10,000}$ and spoken as "eight hundred twenty five ten-thousandths." As a decimal, the 8.25% tax rate would be written as 0.0825, which is also spoken as "eight hundred twenty five ten-thousandths." In summary, 8.25% both looks better and sounds better than $\frac{8.25}{100}$, $\frac{825}{10,000}$, and 0.0825. Therefore, you need to be able to easily change fractions and decimals to percents.

> **STEPS** **to Change a Fraction or a Decimal to a Percent**
>
> 1. If the number is a fraction or a mixed number, change it to a decimal. (If the decimal doesn't terminate, round it to four places—more if required.)
> 2. In the decimal, move the decimal point two places to the *right* (add zeros if necessary). Note: The decimal may be a mixed decimal (examples F and G) or a whole number (example H).
> 3. Write a percent sign at the *right* end of the new number.

EXAMPLE A		
STEP 1	STEP 2	STEP 3
$\frac{4}{5} = 0.8$ ⟶	0.80. ⟶	80%

EXAMPLE B		
STEP 1	STEP 2	STEP 3
$2\frac{5}{8} = 2.625$ ⟶	2.62.5 ⟶	262.5%

EXAMPLE C

	STEP 2	STEP 3
0.75 ⟶	0.75. ⟶	75%

EXAMPLE D

	STEP 2	STEP 3
0.005 ⟶	0.00.5 ⟶	0.5%

EXAMPLE E

	STEP 2	STEP 3
0.3 ⟶	0.30. ⟶	30%

EXAMPLE F

	STEP 2	STEP 3
2.375 ⟶	2.37.5 ⟶	237.5%

EXAMPLE G

	STEP 2	STEP 3
3.5 ⟶	3.50. ⟶	350%

EXAMPLE H

	STEP 2	STEP 3
4 ⟶	4.00. ⟶	400%

✔ CONCEPT CHECK 6.1

a. Change $2\frac{3}{5}$ to a percent.

$2\frac{3}{5} = 2.6 \longrightarrow 2.60. \longrightarrow 260\%$

b. Change 0.025 to a percent.

$0.025 \longrightarrow 0.02.5 \longrightarrow 002.5 = 2.5\%$

Changing Percents to Decimals

The traditional method for calculating with percents is to convert the percent to a decimal. Most calculators have a key (usually labeled %) that does the conversion for you. For example, if you enter the number 8.25 and press the % key, the calculator will display 0.0825. The calculator has moved the decimal point two places to the left. If you want to change a percent to a decimal without using a calculator, follow the same rule.

LEARNING OBJECTIVES

Change percents to decimals.

STEPS to Change a Percent to a Decimal

1. Remove the percent sign.
2. Move the decimal point two places to the *left* (insert zeros if needed). Remember to write a zero to the left of the decimal point as required.

EXAMPLE I

STEP 1	STEP 2

$35\% \longrightarrow 35 \longrightarrow .35. = 0.35$

EXAMPLE J

STEP 1	STEP 2

$1.5\% \longrightarrow 1.5 \longrightarrow .01.5 = 0.015$

**http://ecep1.louisiana.
edu/ecep/math/j/j.htm**

For more practice with percents, visit this Web site.

EXAMPLE K

STEP 1	STEP 2

$200\% \longrightarrow 200 \longrightarrow 2.00. = 2$

Some percents are written with fractions. The sales tax rate of 8.25% might be written on some charts as $8\frac{1}{4}\%$. Conceivably, you could change this percent to a decimal and write it as $0.08\frac{1}{4}$. However, the more usual method is to change the $\frac{1}{4}$ to a decimal and write the percent as the decimal 0.0825.

STEPS **to Change a Percent Containing a Fraction to a Decimal**

1. Convert the fraction in the percent to its decimal equivalent.
2. Drop the percent sign.
3. Move the decimal point two places to the *left* (insert zeros if needed).

EXAMPLE L

STEP 1	STEP 2	STEP 3

$\frac{1}{2}\% = 0.5\% \longrightarrow 0.5 \longrightarrow .00.5 = 0.005$

EXAMPLE M

STEP 1	STEP 2	STEP 3

$2\frac{3}{4}\% = 2.75\% \longrightarrow 2.75 \longrightarrow .02.75 = 0.0275$

 CONCEPT CHECK 6.2

a. Change 126% to a decimal.

$126\% \longrightarrow 126 \longrightarrow 1.26. = 1.26$

b. Change $\frac{1}{4}\%$ to a decimal.

$\frac{1}{4}\% = 0.25\% \longrightarrow 0.25 \longrightarrow .00.25 = 0.0025$

Finding Base, Rate, and Percentage

LEARNING OBJECTIVES

Find Base, Rate, and Percentage.

Suppose that you have $5 and spend $4 for lunch. Example A showed that the fraction $\frac{4}{5}$ equals 80%. You can say that "you spent 80% of your money ($5) for your lunch ($4)." Without the context of your lunch, you have simply "80% of $5 = $4." In this book we call 80% the **Rate (R),** $5 the **Base (B)** amount, and $4 the **Percentage (P)** amount. The

Base and the Percentage amounts will always have the same units (e.g., dollars, feet, or pounds). The Rate will be the percent. (The word *rate* comes from the word *ratio*—in this case, $\frac{4}{5}$.) It may make sense for you to think of the Base amount as the denominator in the rate (i.e., ratio $= \frac{4}{5}$) because the denominator is the "base" (i.e., bottom) of the fraction.

Note: In practice, the terms percent and percentage are used interchangeably. Sometimes you will see the word *percentage* used to mean a rate and the word *percent* used to mean an amount. You will even see the two words *percentage rate* to mean the rate. In this book, however, we use only the one meaning for each word.

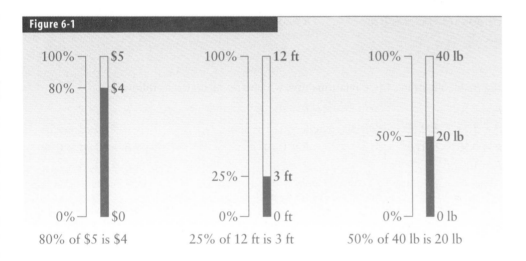

EXAMPLE N	**EXAMPLE O**	**EXAMPLE P**
80% of $5 = $4	25% of 12 ft = 3 ft	50% of 40 lb = 20 lb
80% is the Rate	25% is the Rate	50% is the Rate
$5 is the Base	12 ft is the Base	40 lb is the Base
$4 is the Percentage	3 ft is the Percentage	20 lb is the Percentage

Figure 6-1

80% of $5 is $4 25% of 12 ft is 3 ft 50% of 40 lb is 20 lb

In Figure 6-1 the Rate (or percent) is shown in the left-hand column. The Percentage is represented by the shaded portion of the right-hand column. The Base is represented by the entire right-hand column.

The word **of** often appears in problems that involve percents. Recall from Chapter 5 that normally **of** means **multiply**. We just showed that $80\% = \frac{\$4}{\$5}$. Recall that you can "check" a division problem by multiplication, so we would get 80% × $5 = $4. In words, we said that "80% *of* $5 is $4."

Rule: The number that follows the word *of* is the Base (and is the denominator in the fraction); the number that follows the word *is* is the Percentage amount.

The preceding examples illustrate the basic relationship among the Rate, Base, and Percentage: Rate × Base = Percentage. As a formula, it is written as $R \times B = P$ or as $P = R \times B$.

When you know any two of these three numbers, you can calculate the third by changing the formula:

If you want to find B, the formula becomes $B = P \div R$ or $P \div R = B$.
If you want to find R, the formula becomes $R = P \div B$ or $P \div B = R$.

VIDEO

Base, Rate, Percentage

EXAMPLE Q	EXAMPLE R	EXAMPLE S
Find *P* when *R* = 50% and *B* = 200 qt	Find *R* when *B* = $50 and *P* = $10	Find *B* when *P* = 60 yd and *R* = 75%

Figure 6-2

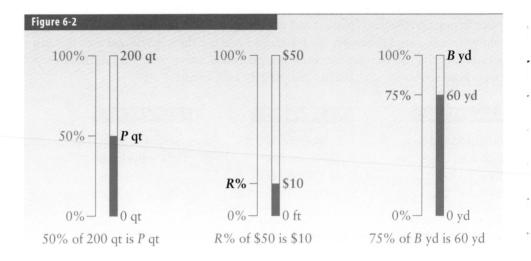

50% of 200 qt is *P* qt *R*% of $50 is $10 75% of *B* yd is 60 yd

Figure 6.2 illustrates these relationships, which are calculated as follows.

$P = R \times B$	$R = P \div B$	$B = P \div R$
$P = 50\% \times 200$ qt	$R = \$10 \div \50	$B = 60$ yd \div 75%
$P = 0.50 \times 200$ qt	$R = 0.20$	$B = 60$ yd \div 0.75
$P = 100$ qt	$R = 20\%$	$B = 80$ yd

 CONCEPT CHECK 6.3

a. Find the Base when the Rate is 15% and the Percentage amount is $60.
 $B = P \div R = \$60 \div 15\% = \$60 \div 0.15 = \$400$

b. Find the Rate when the Base is 96 feet and the Percentage amount 72 feet.
 $R = P \div B = 72$ ft \div 96 ft $= 0.75 = 75\%$

COMPLETE ASSIGNMENT 6.1.

Using Percents in Business

Percent problems occur frequently in business. Examples T and U are typical fundamental applications, in which we solve for the Base (*B*) amount and the Rate (*R*), respectively.

Judith Cruz is a computer programmer. In December, she received a $900 bonus, which equaled 20% of her monthly salary. What was her monthly salary?

P = amount of bonus = $900
R = rate of bonus = 20%
B = monthly salary = ?
As $P \div R = B$,
$P \div R = \$900 \div 20\% = \$900 \div 0.20 = \$4,500$ monthly salary

Last year Express Coffee Stop had total expenses of $250,000. Of that total, $150,000 was the expense for employee salaries. At Express, employee salary expense is what percent of total expenses?

P = employee salaries = $150,000
R = ?
B = total expenses = $250,000
Since $P \div B = R$,
$P \div B = \$150,000 \div \$250,000 = 0.60 = 60\%$

Using Percents to Measure Increase and Decrease

An important application of percents in business is to measure change from one year to the next or from one month to the next. Real estate firms compare the number of homes sold this year with the number of homes sold last year. Read and carefully compare the following four statements about home sales last year and this year.

LEARNING OBJECTIVES

Use percents to measure increase and decrease.

Mateski Realty sold 50% more homes this year than it did last year, when it sold 36 homes.

Alquist & Blenz Real Estate sold 30 more homes this year than last year, which represents 25% more homes this year than last year.

Real estate agent Jane Wu sold 3 fewer homes this year than she did last year, when she sold 12 homes.

Larry Wittmayer, a real estate broker, sold 20 homes last year; this year he sold 26 homes.

The number of homes sold last year is the Base (B) amount (last year is called the *base year*). The change in homes can be reported as a number, which would be the Percentage amount (P), or as a percent, which would be the Rate (R). If any two of the three values are given, the third can be determined using one of the three formulas in this chapter.

Find the number of additional homes (P) that Mateski Realty sold this year.

B = 36 and R = 50%. Since $P = R \times B$,
P = 50% × 36 = 0.50 × 36 = 18 more homes this year

Find the number of homes that Alquist & Blenz Real Estate sold last year (B).

$P = 30$ and $R = 25\%$. Since $B = P \div R$,
$B = 30 \div 25\% = 30 \div 0.25 = 120$ homes sold last year

Find Jane Wu's Rate of decrease (R) from last year's sales.

$P = 3$ and $B = 12$. Since $R = P \div B$,
$R = 3 \div 12 = 0.25 = 25\%$ decrease

To find the percent change when the only numbers reported are the amounts (B) for last year and this year, the first step is to find the **amount of increase** or the **amount of decrease**. P is the difference between the amounts for the two years. Then use $R = P \div B$ to find the **rate of increase** or the **rate of decrease.**

Find Larry Wittmayer's Rate of change (R).
 Larry sold 20 homes last year (B) and 26 this year. The amount of change is

$P = 26 - 20 = 6$ more homes this year.

The Rate of change is

$R = P \div B = 6 \div 20 = 0.30 = 30\%$ increase

VIDEO

Percent of Increase or Decrease

 CONCEPT CHECK 6.4

A company had sales of $144,000 this year and $120,000 last year ($B$). Find both the amount of increase (P) and the rate of increase (R).

The amount of increase is
$P = \$144{,}000 - \$120{,}000 = \$24{,}000$

The Rate of increase is
$R = P \div B = \$24{,}000 \div \$120{,}000 = 0.20 = 20\%$

COMPLETE ASSIGNMENTS 6.2 AND 6.3.

Using Percents to Allocate Overhead Expenses

LEARNING OBJECTIVES

Use percents to allocate overhead expenses.

Many businesses are organized into divisions or departments. Suppose Elnora's is a retailer of sportswear. It has three departments: women's clothes, men's clothes, and children's clothes. Management and owners of Elnora's need to measure the profitability of each department. Elnora's also knows the amounts it paid for the merchandise sold and the salaries of employees in each department. Elnora's can subtract these departmental costs from the departmental revenues.

But what about rent and other general costs such as electricity? These costs that are not directly related to the types of merchandise sold are called **overhead costs**. For example, Elnora's monthly rental expense might be $12,000 for the entire building. How should that single amount be divided among the three departments? Should each department be assigned $\frac{1}{3}$, or $4,000, of the total rent?

Most businesses *allocate*, or distribute, the rent based on a measurement related to the total cost. Rent is a cost of using the building; it could be allocated on the basis of floor space, since each department occupies some of that space.

STEPS **to Allocate an Overhead Cost Based on Total Floor Space (or Some Other Measurement)**

1. Find the total square feet of floor space.
2. Divide the floor space of each department by the total floor space and change to percents.
3. Multiply each percent (in fractional or decimal form) times the total rent (or other overhead value).

EXAMPLE Z

Determine the amount of rent to allocate to the respective departments of Elnora's.

	STEP 1	STEP 2	STEP 3
Department	**Floor Space**	**Percent of Total**	**Distribution of Rent**
Children's	25 ft × 20 ft = 500 sq ft	500 ÷ 2,500 = 20%	$12,000 × 0.20 = $ 2,400
Women's	25 ft × 50 ft = 1,250 sq ft	1,250 ÷ 2,500 = 50%	$12,000 × 0.50 = $ 6,000
Men's	25 ft × 30 ft = 750 sq ft	750 ÷ 2,500 = 30%	$12,000 × 0.30 = $ 3,600
Total	2,500 sq ft	100%	$12,000

This same method is used for many other business expenses, such as utilities, fire insurance, and salaries of office personnel. Examples of other bases that might be used for allocation are number of employees, hours worked, and units produced.

CONCEPT CHECK 6.5

A building maintenance company has two different divisions: commercial and residential. Employees spend 980 hours working in commercial buildings and 420 hours working in residences. The company has a utility expense of $10,000 that it wants to allocate between the two divisions, based on the percent of employee hours used by each division.

Total hours worked: 980 + 420 = 1,400

Commercial: 980 ÷ 1,400 = 0.70, or 70% of employee hours
 70% of $10,000 = 0.70 × $10,000 = $7,000 of office expense

Residential: 420 ÷ 1,400 = 0.30, or 30% of employee hours
 30% of $10,000 = 0.30 × $10,000 = $3,000 of office expense

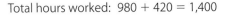

COMPLETE ASSIGNMENT 6.4.

© CORBIS

Chapter Terms for Review

amount of decrease

amount of increase

Base (*B*)

"of"

overhead costs

percent

Percentage (*P*)

Rate (*R*)

rate of decrease

rate of increase

Visit the Web

http://www.aaamath.com/pct/html—For practice on solving percent problems, visit this Web site.

http://pittsford.monroe.edu/jefferson/calfieri/decimals/Percents.html—Take a quiz on percents.

http://www.shodor.org/interactivate/discussions/percents/html—You can read a student/mentor discussion of percents.

http://www.purplemath.com/modules/percents.htm and **http://www.mccc.edu/~kelld/CompFDP.htm**—These are good sites for more percents practice.

CYBERCONNECTION

Golden Rule Mortgage					(800) 991-9922
program	rate	pts	dwn	lip	apr
30-yr. Fixed	6.375	1.63	5	30	6.589
20-yr. Fixed	6.250	1.63	5	30	6.488
15-yr. Fixed	5.875	1.63	5	30	6.297

Source: http://www.nfns.com/cincinnati/. Copyright © 2001. Used with permission from the Cincinnati Enquirer; Courtesy of Golden Rule Mortgage.

As you learned from this chapter, percents are used in many ways in business. In this chart, the interest rate (rate), points (pts), down payment (dwn), and percentage rate (apr) are all in terms of percent. (Lock-in period (lip) is in days.)

Notice the differences between the 30-year, 20-year, and 15-year fixed mortgage interest rates. Visit this site to check the differences between interest rates among several banks and mortgage companies. Find a similar site in your own area and compare the differences and similarities among banks and mortagage companies at this site in your area. You can also visit some bank and mortgage company sites for more information.

CYBERCONNECTION

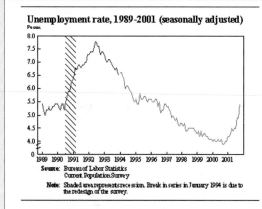

Unemployment rate, 1989-2001 (seasonally adjusted)

Source: Bureau of Labor Statistics
Current Population Survey

Note: Shaded area represents recession. Break in series in January 1994 is due to the redesign of the survey.

For more information, see the latest

- Employment Situation Press release
- Statement of the Commissioner of the Bureau of Labor Statistics to the Joint Economic Committee

For concepts and definitions, see

- How the Government Measures Unemployment
- Technical notes to tables published in Employment and Earnings

Last Modified Date: November 2, 2001

Percents are used often in government reports that would otherwise involve enormous numbers. Do you recall hearing in the news about the unemployment rate? This is always expressed in percents.

Visit http://stats.bls.gov/wh/cpsbref3.htm and find this graph. For more information at this site, click on How the Government Measures Unemployment. Technical notes to tables published in Employment and Earnings contains interesting information on data collection and coverage, concepts and definitions, historical comparisons, sampling, and estimating methods.

Source: http://stats.bls.gov/wh/cpsbref3.htm.

The Bottom Line

Summary of chapter learning objectives:

Learning Objective	Summary	Example
6.1	Change fractions and decimals to percents	1. Change 0.35 to a percent. 2. Change $\frac{5}{8}$ to a percent.
6.2	Change percents to decimals	3. Change 3.75% to a decimal.
6.3	Find Base, Rate, and Percentage	4. Find the Percentage: 15% of 36 $= P$ 5. Find the Rate: R% of 140 $=$ 28 6. Find the Base: 75% of B $=$ 150
6.4	Use percents to measure increase and decrease	7. Increase a $3,000 salary by 30%. 8. From 300 to 240 is a decrease of what percent?
6.5	Use percents to allocate overhead expenses	9. A company has three stores, A, B and C, with 4, 6, and 10 employees, respectively. Based on the number of employees, allocate a $3,000 expense among the stores.

Answers: 1. 35% **2.** 62.5% **3.** 0.0375 **4.** 5.4 **5.** 20% **6.** 200 **7.** $3,900 **8.** 20% **9.** Store A; $600; Store B; $900; Store C; $1,500

Self-Check | Review Problems for Chapter 6

1 Change 0.875 to a percent _____

2 Change 16.1% to a decimal _____

3 Change 0.0025 to a percent _____

4 Change 250% to a decimal _____

5 Change $\frac{3}{5}$ to a percent _____

6 Change 0.05% to a decimal _____

7 15% of 40 = _____

8 150% of 60 = _____

9 25% of _____ = 40

10 100% of _____ = 60

11 _____ % of 40 = 28

12 _____ % of 60 = 150

13 Sales were $10,000 two years ago and increased by 25% last year. How much were sales last year? _____

14 Sales were $12,500 last year and decreased by 25% this year. How much were sales this year? _____

15 Expenses were $2,000 two years ago and $4,000 last year. What was the percent increase last year? _____

16 Expenses were $4,000 last year and $2,000 this year. What was the percent decrease this year? _____

17 Carol Brandon owns a nursery. This year she sold 174 more rose bushes than she did last year. This represents a 15% increase over the previous year. How many rose bushes did Carol's nursery sell last year? _____

18 Sylvester Flake manages Internet sales for a company that started selling its product over the Internet three years ago. Last year, company sales over the Internet were only about $500,000. This year sales were $1,625,000. Calculate the company's percent increase in Internet sales this year.

19 Ron Burch is a bank teller. When he started this morning, his cash drawer had coins worth $114. The coins represented only 1.5% of all the money that Ron had in his cash drawer. What was the total value of all this money?

20 Florence Hasel is an orthopedic surgeon. Last winter, Dr. Hasel performed 40 emergency surgeries. Twenty-six of those surgeries were the result of ski injuries. What percent of Dr. Hasel's emergency surgeries were the result of ski injuries?

Assignment 6.1: Base, Rate, and Percentage

Name _____

Date _____ Score _____

A **(20 points) Change the percents to decimals. Change the nonpercents to percents. (1 point for each correct answer)**

1. 21% = _.21_

2. 100% = _1.00_

3. $3\frac{2}{3}$% = _____

4. 0.375 = _37.5%_

5. 4 = _400%_

6. $33\frac{1}{3}$% = _____

7. 0.12 = _12%_

8. 0.3 = _30.%_

9. $1\frac{1}{4}$ = _1.25. = 125.%_

10. 3.2% = _3.20.%_

11. 324.5% = _3245_

12. 0.0004% = _____

13. 0.52 = _52%_

14. 250% = _25._

15. $0.08\frac{3}{4}$ = _____

16. $\frac{1}{2}$ = _50%_

17. 5.0 = _____

18. 0.000075 = _____

19. 0.1% = _____

20. 1,000% = _____

Score for A (20)

B **(30 points) Solve each of the following problems for the Percentage. (2 points for each correct answer)**

21. 0.375% of 64 =
.00375 × 64 = .24 ✗

22. 0.25% of 1,200 = _____

23. 100% of 13.19 = _____

24. 62.5% of 16 = _10_
.625 × 16

25. 40% of 0.96 = _.384_
.4 × 0.96 =

26. 250% of $24 = _60_
250 × 24 =

27. 25% of $1.12 = _____
.25 × 1.12

28. 120% of $42 = _____
1.20 × 42

29. 2.5% of $200 = _____
.025 × 200 =

30. 50% of $284 = _____
.50 × 284

31. 8% of 20 = _____
.08 × 20

32. 15% of 0.06 = _____
.15 × 0.06 =

33. 187.5% of 32 = _____
✗

34. 1.5% of $96 = _____
✗

35. 0.2% of 560 = _____

Score for B (30)

Assignment 6.1 *Continued*

C **(50 points) Solve each of the following for the Percentage, the Rate, or the Base. Write rates as percents. Round dollars to the nearest cent. (2 points for each correct answer)**

R × b = P

36. $33\frac{1}{3}$% of _____ = $15
 X

37. .25⁹ of $24 = $6 *Rate B - P*

38. 300% of 0.12 = 0.36

39. 100% of 26.9 = 26.9

B = P ÷ R

40. 2.5% of 500 = 12.5 *500*

41. 80% of .6 = $0.48 *X*

42. 1.25% of .44 = 1.8

43. 225% of 60 = 135 *X*

44. .25 of 0.048 = 0.012

(*Places to Right to get %*)

45. 150% of 44 = $66

46. 2.5% of $2,460 = _____

47. 0.75% of _____ = $30

48. 250% of _____ = 14.4

49. 37.5% of _____ = 33

50. 0.025% of $12,800 = _____

51. 120% of _____ = 480

52. 120% of _____ = $39

53. _____ of 5.4 = 2.16

54. _____ of $2,340 = $46,800

55. 15% of $140 = _____

56. 120% of $91 = _____

57. _____ of 75 = 105

58. 175% of _____ = $560

59. 12% of _____ = 3

60. _____ of 2.1 = 0.336

Score for C (50)

Assignment 6.2: Rate of Increase and Rate of Decrease

Name _____

Date _____ Score _____

A **(40 points) Calculate the missing values. ($2\frac{1}{2}$ points for each correct answer)**

1. Increasing the base value of 240 by 25% gives the new value _____ .

2. Decreasing the base value of 140 by 60% gives the new value _____ .

3. Start with 65, increase it by 60%, and end up with _____ .

4. Start with 90, decrease it by 12%, and end up with _____ .

5. Sales were $8,000 last month, but decreased by 3% this month. Sales were _____ this month.

6. Profits were $4,400 last month and increased by 4% this month. Profits were _____ this month.

7. Base value = 271; increase = 100%; new (final) value = _____

8. Base value = 350; decrease = 100%; new (final) value = _____

9. A $12 decrease is 20% of the base value of _____ .

10. An increase of 40 units is 8% of the base value of _____ units.

11. The price increased from $150 to $165; the percent increase was _____ .

12. Production decreased from 5,000 units to 4,000 units; the percent decrease was _____ .

13. $300 is what percent less than $400? _____

14. 320 is what percent greater than 160? _____

15. Sales were $500,000 in June but only $400,000 in July. The rate of decrease was _____ .

16. Profits were $11,000 last month and $10,000 the previous month. The rate of increase was _____ .

Score for A (40)

Assignment 6.2 *Continued*

B (30 points) The following table shows the volumes of various items sold by Thrift's Speed Shop during the past 2 years. Compute the amount of change and the rate of change between this year and last year. Compute the rates to the nearest tenth of a percent. If the amount and rate are increases, write a + in front of them; if they are decreases, enclose them in parentheses (). (1 point for each correct amount; 2 points for each correct rate)

Thrift's Speed Shop
Volume Sold (number of units)

Description of Item	This Year	Last Year	Amount of Change	Rate of Change
17. Batteries	513	439		
18. Brake fluid (pints)	1,784	1,942		
19. Coolant (gallons)	2,045	1,815		
20. Headlight lamps	4,907	4,084		
21. Oil (quarts)	12,458	10,743		
22. Mufflers	439	525		
23. Shock absorbers	895	1,084		
24. Tires, Auto	6,742	5,866		
25. Tires, Truck	2,195	2,803		
26. Wiper blades	1,920	1,842		

Score for B (30)

C (30 points) During May and June, Stewart's Paint Store had sales in the amounts shown in the following table. Compute the amount of change and the rate of change between May and June. Compute the rates of change to the nearest tenth of a percent. If the amount and rate are increases, write a + in front of them; if they are decreases, then enclose them in parentheses (). (1 point for each correct amount; 2 points for each correct rate)

Stewart's Paint Store
Volume Sold (in dollars)

Description of Item	June	May	Amount of Change	Rate of Change
27. Brush, 2" wide	$711.14	$594.67		
28. Brush, 3" wide	564.20	512.51		
29. Brush, 4" wide	229.87	244.27		
30. Drop cloth, 9 × 12	143.50	175.66		
31. Drop cloth, 12 × 15	174.29	151.55		
32. Paint, latex (gal)	35,506.24	37,382.19		
33. Paint, latex (qt)	4,072.35	4,878.96		
34. Paint, oil (gal)	8,308.44	7,864.41		
35. Paint, oil (qt)	3,358.35	4,074.96		
36. Paint scraper	274.10	238.82		

Score for C (30)

Assignment 6.3: Business Applications

Name

Date Score

A **(50 points) Solve the following problems. Round dollar amounts to the nearest cent. Round other amounts to the nearest tenth. Write rates as percents to the nearest tenth of a percent. (5 points for each correct answer)**

1. Fitzwalter Electric shipped 4,500 capacitors in April. Clients eventually returned 6% of the capacitors. How many of the April capacitors were eventually returned? _**270**_

 $P = B \times B$
 $.06 \times 4,500$

2. Stephen D. Fitzwalter, CEO of Fitzwalter Electric, wants the company to reduce the percent of capacitors that customers return. In May, the company shipped 4,000 capacitors, and 200 were eventually returned. What percent of the May shipment was eventually returned? _**5%**_

 $\div B$
 $200 \quad 4,000$

3. By June of the following year, Fitzwalter Electric had reduced the percent of capacitors returned to 4% of the number shipped. If 220 capacitors were returned from that month's shipment, how many had been shipped? _**5500**_

 $B = P \div R$
 $5500 = 220 \quad .04$

4. A food importer, Velures of Europe, imports 60% of its vinegars from France, 30% from Italy, and 10% from Spain. The total value of all the vinegars that it imports is $720,000. What is the value of the vinegars that are *not* imported from France? _**288,000**_

 $B \quad R \times P \quad P = R \times B$
 $6 \quad .040 \qquad .04 \quad 720,000$

5. Next year, Velures is planning to import $440,000 worth of vinegars from France, $200,000 worth of vinegars from Italy, and $160,000 worth of vinegars from Spain. If next year's imports occur as currently being planned, what percent of the total imports will be from France? _____

6. Parker Parka Products, Inc., manufactures only parkas for adults and children. Last year, Parker manufactured all its children's parkas in Asia. Those children's parkas represented 38% of all the Parker production. If the company made a total of 235,000 parkas, how many children's parkas did it produce? _____

7. This year, Parker again plans to manufacture all its children's parkas in Asia, and Parker will expand the children's product line to 40% of the total number of parkas produced. If Parker plans to produce 96,000 children's parkas, how many parkas does the company plan to produce in total? _____

8. Next year, Parker plans to keep the percent of children's parkas at 40% but increase the number of children's parkas produced to 100,000. How many parkas does the company plan to produce for adults? (*Hint:* First you need to calculate the total number of all parkas to be produced next year.) _____

9. Larry Johnson is a single father. He tries to save 15% of his monthly salary for his son's education. In August, Larry's salary was $4,800. How much should he save to meet his objective? _____

10. In September, Larry Johnson got a promotion and a raise. Because his monthly expenses did not increase very much, Larry was able to save more dollars. He saved $1,350, which was 25% of his new salary. How much was Larry's new salary? _____

Assignment 6.3 *Continued*

B (50 points) Solve the following problems. Round dollar amounts to the nearest cent. Round other amounts to the nearest tenth. Write rates as percents to the nearest tenth of a percent. (5 points for each correct answer)

11. Bradley Grayson, a paralegal, will receive a 3% salary increase this month. Hence he will receive $90 more salary this month than he received last month. What was Bradley's salary last month? _____

12. Roberta Coke works in the market research department of a soft drink company. Yesterday Roberta received a raise of $375 per month. Roberta now earns 5% more than she did before the raise. How much does she earn now? _____

13. A farmers' market is held downtown every Saturday. The volume has been increasing by about 4% every week. If the volume was $48,000 this week, what should the volume be next week? _____

14. Marcia Almeida works as a sales analyst for a toy manufacturer. She predicts that toy sales will decrease 6% between May and June. If the amount of the sales decrease is $165,000, what level of sales is she predicting for June? _____

15. Last month, Fred Gerhardt started working as an apprentice machinist. One of his first projects was to reduce the diameter of a metal shaft from 0.160 inch to 0.136 inch. By what percent did he reduce the diameter of the shaft? _____

16. Judy Gregory, a mechanical engineer, was able to increase the efficiency of a manufacturing facility. By doing so, she decreased the cost to manufacture a commercial quality lawnmower by $21, which was 12% of the former cost. What will be the new reduced cost to manufacture the lawnmower? _____

17. Richard Phipps is the purchasing agent for a janitorial service. He orders all the supplies used by his company. Because of new contracts to clean three new office buildings, Richard ordered an additional $6,000 worth of supplies this month. This was a 12% increase from last month. What was the value of the supplies that Richard ordered last month? _____

18. Nancy Yamamoto owns a gift ship that had sales of $65,000 in November. Because of the Christmas holiday season, Nancy predicts that the shop will double its monthly sales in December. What percent increase is Nancy predicting for December over November? _____

19. Suppose that Nancy Yamamoto's gift shop had had sales of $125,000 in November and had doubled its sales in December. What would be the percent increase for December over November? _____

20. Because of Father's Day, Dukes Men's Clothing store had sales of $250,000 in June. Sales decreased by 50% in July. What were Duke's sales in July? _____

Score for B (50)

Assignment 6.4: Allocation of Overhead

Name

Date Score

A **(20 points) Complete the square feet, percent, and distribution columns below. Round percents to the nearest whole number. (1 point for each correct answer in column 1; 2 points for each correct answer in columns 2 and 3)**

1. Carla Viberti owns restaurants in four districts of town: (a) Charleston, (b) Empire, (c) Englewood, and (d) Milner Crest. She buys all the food through a central office, which she maintains at the Charleston restaurant. Monthly expenses for the office are distributed among the four restaurants, based on the floor space of each. In the following table, complete the distribution table for monthly expenses of $12,000.

Store	Space Occupied	Square Feet	Percent of Total	Distribution of Expense
(a) Charleston	30 ft × 70 ft	2100 ft 8 ÷ 7,500	12,000 × .0000 28	N2,000 = 3,360
(b) Empire	40 ft × 45 ft	1800 ft 8 ÷ 7500	12,000 × .0000 24	×2,000 = 2,880
(c) Englewood	40 ft × 60 ft	2400 ft 8 ÷ 7500	12,000 × .0000 32	×12,000 = 3,840
(d) Milner Crest	30 ft × 40 ft	1200 ft 8 ÷ 7500	12,000×.0000 16	16 × 12,000 = 1,920
Total		7,500	100%	$12,000

Score for A (20)

B **(16 points) Complete the percent and distribution columns in the following table. Before computing the distribution, round each percent to the nearest whole number. (2 points for each correct answer)**

2. Diane Kingsley owns a temporary services company. She employs four types of employees whom she places into temporary positions: (a) accountants, (b) secretaries, (c) food service people, and (d) hotel service people. Diane rents office space for $4,200 per month. She distributes the rent among the four labor groups, according to the number of people from each group placed into jobs. Calculate the percents and the resulting distributions.

	Number of Employees	Percent of Total	Distribution of Rent
(a) Accountants	18		
(b) Secretaries	42		
(c) Food Service	36		
(d) Hotel Service	24		
Total	120	100%	$4,200

Score for B (16)

Assignment 6.4 *Continued*

C **(64 points)** The following situations provide practice in distributing monthly overhead expenses at a central office. From the information given in the following table, complete the distributions indicated in problems 3 through 6. Remember: Answers for each problem should sum to the total monthly overhead expense. **(4 points for each correct answer)**

Monthly Overhead Expense		Basis of Distribution	Location				
			East	West	North	South	TOTAL
Rent	$25,000	Square feet	16,800	9,600	14,400	19,200	60,000
Janitorial	12,000	Machine hours worked	14,400	18,000	10,800	28,800	72,000
Utilities	16,000	Units produced	5,700	7,800	10,200	6,300	30,000
Insurance	10,000	Number of employees	90	75	105	30	300

3. Distribute rent based on the number of square feet at each location.

East $\frac{16,800}{60,000}$; .28 West $\frac{9600}{60,000}$; .16 North $\frac{14,400}{60,000}$; .24 South $\frac{19,200}{60,000}$; .32 Check.

.28 X 25,000 .16 X 25,000 .24 X 25,000 .32 X 25,000

$7000 $4,000 6,000 8,000

4. Distribute janitorial expense based on the number of machine hours worked in each location.

East $\frac{14,400}{72,000}$; .20 West $\frac{18,000}{72,000}$; .25 North $\frac{10,800}{72,000}$; .15 South $\frac{28,000}{72,000}$; .40 Check.

.20 X 12,000 .25 X 12,000 .15 X 12,000 .40 X 12,000

$2400 $3000 $1800 $4800

5. Distribute utilities expense based on the units produced at each location.

East $\frac{5,700}{30,000}$; West $\frac{7,800}{30,000}$; North $\frac{10,200}{30,000}$; South $\frac{6,300}{30,000}$ Check.

X 16,000 X 16,000 X 16,000 X 16,000

3,040 + 4 + 5440 + 3,360 = 16,000

6. Distribute insurance expense based on the number of employees at each location.

East _____ ; West _____ ; North _____ ; South _____ Check.

Y 10,000 X 10,000 X 10,000 X 10,000

3,000 2,500 1,000 = 10,000

Score for C (64)

Notes

FDIC

Each depositor insured to $100,000

SITEMAP SEARCH

Federal Deposit Insurance Corpo

| Deposit Insurance | Bank Data | Regulations & Examination | Consumers & Communities | Buying From, Selling to FDIC | News, Eve & FOIA |

Quick Search

Find it
Advanced Search

Bank Data
- Individual Banks
- Banking Industry
 - Analytical
 - Statistical
 - Historical

Bank Data: Individual Banks

Searchable databases allow users to find institutions and their branches in order to determine their status as insured depository institutions, their financial condition and their condition relative to other ins The databases may also contain other financial non-financial information about individual financi institutions as well as certain aggregate financial for comparative use..

Find a Bank, Savings Association Bank Holding Company

Institution Directory

A searchable database for the latest comprehensive financial profile for every FDIC-insured institution, including the most recent quarterly balance sheets, income statements, performance and condition rati Several years' financial data and holding

Once limited to savings, checking, loans, and safe deposit boxes, bank services have changed drastically. Banks now offer VISA and MasterCard cards and automatic teller machines (ATMs), and customers can buy and sell stocks, plan for retirement, exchange foreign currency, and buy insurance there.

Virtually all bank deposits in the United States are insured by the Federal Deposit Insurance Corporation (FDIC). The FDIC is an independent government agency created to stabilize and secure the nation's banking system. It insures most U.S. bank deposits by individuals and companies to a maximum of $100,000 per account.

The FDIC has a highly skilled and diverse workforce. Visit its home page at www.fdic.gov to see the many job opportunities that exist at the FDIC.

Banking

7

LEARNING OBJECTIVES

By studying this chapter and completing all assignments you will be able to:

 1 Understand use of a deposit slip, bank check, and check stubs.

 2 Maintain a checkbook and check register.

 3 Reconcile a bank statement with a checkbook balance.

Most individuals and businesses deal with banks and other financial institutions that offer traditional banking services. Banking customers have accounts to which they make deposits and withdrawals in cash, checks, electronic transfers, and receipts from charge cards and other business transactions.

1. What math procedures will you need most to keep your checkbook in balance? To reconcile your bank book balance with your monthly bank statement?
2. To be sure that you don't overdraw your account, what do you need to know?
3. Do banks charge penalties for writing bad checks? Could writing a bad check damage your credit rating?

Using Deposit Slips and Bank Checks

LEARNING OBJECTIVES

Understand use of a deposit slip, bank check, and check stubs.

Bank customers usually make deposits to their checking accounts by using **deposit slips**. Figure 7-1 shows a typical deposit slip, with spaces to list cash and checks being deposited.

In most businesses, each deposit will include a number of checks. Each check is individually listed on each deposit slip. Deposits are also made electronically. Many employees have their pay electronically transmitted directly from their employers' bank accounts to their individual bank accounts.

A bank **check** is a written order directing the bank to pay a certain sum to a designated party, called the **payee**. Banks normally provide checkbooks to their members. Figures 7-2 and 7-3 show typical bank checks, one with the stub on the left and the other with the stub on the top.

Today, many bank transactions are completed electronically. Funds that are transmitted electronically, primarily via computers, are called **electronic fund transfers (EFTs)**.

Figure 7-1: Deposit Slip

WELLS FARGO BANK

VAN NESS-CALIFORNIA OFFICE 1560 VAN NESS AVENUE SAN FRANCISCO, CA 94109

35-6686
3130

DATE _____ 20 _____

DEPOSITS MAY NOT BE AVAILABLE FOR IMMEDIATE WITHDRAWAL

SIGN HERE FOR LESS CASH IN TELLER'S PRESENCE

HART FURNITURE CO.
1039 BROADWAY
SAN FRANCISCO, CA 94103

USE OTHER SIDE FOR ADDITIONAL LISTING. BE SURE EACH ITEM IS PROPERLY ENDORSED.

CASH	CURRENCY	300	00
	COIN	60	49
LIST CHECKS SINGLY	16–30	250	00
	18–21	125	00
	17–17	216	00
TOTAL FROM OTHER SIDE		209	00
TOTAL		1,160	49
LESS CASH RECEIVED		—	
NET DEPOSIT		1,160	49

PLEASE LIST EACH CHECK SEPARATELY BY BANK NUMBER

CHECKS	DOLLARS					CENTS
1 14–36				7	6	75
2 13–22				1	3	25
3 13–22			1	1	9	00
4						
5						
6						
7						
8						
9						
10						
11						
12						
PLEASE FORWARD TOTAL TO REVERSE SIDE			2	0	9	00

A Seized Opportunity

In 1995, a young man named Patrick Combs received a piece of junk mail for a get-rich-quick scheme that included a document that looked like a check. The "check" was intended to show how much money a person could make by seizing the opportunity. The sample "check" was for $95,093.35, and it included the words "non-negotiable for cash" in the upper right hand corner.

Combs deposited the item to his account through an ATM and was astonished to learn later that his bank had treated it as a real check and credited his account for the full amount. Thus began an odyssey that landed him in the *Wall Street Journal* and on every major news show and talk show. His experience is detailed on his Web site.

After he deposited the check and it was accepted, he asked for a cashier's check from his bank in that amount, and placed the cashier's check in a safe-deposit box. In the meantime, the non-negotiable "check" went through three banks before the problem was discovered. Eventually he returned the cashier's check to the bank when assured that he wouldn't be prosecuted for fraud. Read the details of his experience at his Web site, http://www.dnai.com/~pcombs/$$tablecontents.html.

They include **automatic teller machine (ATM)** transactions by which customers can check their balances, make deposits, and withdraw funds from their accounts without having to wait for the next available bank teller. Computer programs also initiate many electronic fund transfers. These transactions are processed through the Automated Clearing House Association and include direct deposits of payroll checks and Social Security and other government and pension benefit payments.

Figure 7-2: Check with Check Stub on Left

No. 2506	$ 124.35

September 24 20 – –

To *Ace Auto Repair*

For *Delivery truck*

	$	¢
Balance Bro't Fwd	1,332	80
Amount Deposited	1,160	49
Total	2,493	29
Amount This Check	124	35
Balance Car'd Fwd	2,368	94

HART FURNITURE CO.
1039 Broadway
San Francisco, CA 94103

No. 2506

September 24 20 – –

35-6686 / 3130

Pay to the order of *Ace Auto Repair* $ 124.35

One hundred twenty – four and 35/100 DOLLARS

WELLS FARGO BANK
VAN NESS-CALIFORNIA OFFICE 1560 VAN NESS AVENUE SAN FRANCISCO, CA 94109

For *Delivery truck repair*

Robert S. Hart

⑈3130 6686⑈ 2506⑈ 117⑈020⑈8

Figure 7-3: Check with Check Stub on Top

BAL. FOR'D	997	03	DATE 10-1-19--	3500	
DEPOSITS	— 451	04	TO: Men's Wearhouse	NEW BAL. 1,555	08
	— 102	01	FOR: Suit-Slacks	THIS CHECK 300	00
NEW BAL.	1,555	08		BAL. FOR'D 1,255	08

WELLS FARGO BANK 🔒 3500

91-119
1221(1)

VALUED CUSTOMER SINCE 1976

October 1, 20 --

Pay to the order of ___Men's Wearhouse___ $ 300.00

___Three hundred no/100_____ DOLLARS

MARY MAHEW
40 ACELA DR.
TIBURON, CA 94920

For ___Suit-Slacks___ ___Mary Mayhew___

⑈1221011191⑈3500 0255 355521⑈

CONCEPT CHECK 7.1

Fill in the total (as necessary) and balance on each check stub. Carry each balance forward to the next stub.

No. 1 $ 65.00	No. 2 $ 79.00	No. 3 $ 25.00
May 1 20 --	May 4 20 --	May 5 20 --
To Citizens News	To District Utilities	To U.S Postal Service
For Advertising	For Gas & electric	For Stamps

	$	¢		$	¢		$	¢
Balance Bro't Fwd	890	00	Balance Bro't Fwd			Balance Bro't Fwd		
Amount Deposited			Amount Deposited			Amount Deposited	100	00
Total			Total			Total		
Amount This Check	65	00	Amount This Check	79	00	Amount This Check	25	00
Balance Car'd Fwd			Balance Car'd Fwd			Balance Car'd Fwd		

Maintaining Checkbooks and Check Registers

2

LEARNING OBJECTIVES

Maintain a checkbook and check register.

A bank **checkbook** also provides check stubs or a special page on which to record deposits, withdrawals, check numbers, dates, check amounts, other additions and subtractions, and the account balance.

Figure 7-2 shows that check number 2506 was written against the account of Hart Furniture Co. on September 24 to Ace Auto Repair. The check was for $124.35 for repairs to the delivery truck. The stub shows a balance brought forward of $1,332.80, a deposit on September 24 of $1,160.49, the amount of this check ($124.35), and a balance carried forward of $2,368.94.

Summary of chapter learning objectives:

Learning Objective	Summary	Example
7.1	Understand use of a deposit slip, bank check, and check stubs	1. Fill in the New Bal and Bal For'd on each check stub. Carry Bal For'd to the next stub.

#1

Bal For'd $100.00	Date 01/17
Deposit 350.50	To Store
New Bal _____	_____
This Ck 175.09	For Misc
Bal For'd _____	

#2

Bal For'd _____	Date 01/22
Deposit 375.00	To Store
New Bal _____	_____
This Ck 78.88	For Misc
Bal For'd _____	

Learning Objective	Summary	Example
7.2	Maintain a checkbook and check register	2. Fill in the cash balance for each date.

CHECK REGISTER

DATE	CHECK NUMBER	CHECK TO—DEPOSIT INFORMATION	DEPOSIT AMOUNT	CHECK AMOUNT	BALANCE
					$2,077.50
12/11	100	MacDonalds		$ 77.14	
12/12		Monthly Salary Check	$782.17		
12/13	101	R.J. Maples—Rent		$1,000.00	
12/14	102	Macy's		$ 99.97	
12/15	103	True Value		$ 107.16	
12/17		Income from Stocks	$222.24		

Learning Objective	Summary	Example
7.3	Reconcile a bank statement with a checkbook balance	3. Steven Hunt's monthly bank statement balance was $1,238. His checkbook balance was $1,440. He noted the following checks outstanding: #119 for $350 and #125 for $197. He noted a deposit of 1,600 as not recorded by the bank. The bank had charged him $17 for checks and $32 for a bad check he had deposited. The bank had credited his account with an electronic transfer for $900. Reconcile the bank and checkbook balances.

Checkbook balance: $1,440
 Add electronic transfer: _____
 Subtotal _____
 Less bank charges: _____
 _____ _____
Adjusted checkbook balance:
 Bank balance on statement: $1,238
 Add unrecorded deposit: _____
 Subtotal _____
 Less outstanding checks: #119 _____
 #125 _____
Adjusted bank balance _____

Answers: **1.** $450.50; $275.41; $650.41; $571.53 **2.** $2,000.36; $2,782.53; $1,782.53; $1,682.56; $1,575.40; $1,797.64 **3.** $2,291

Self-Check | Review Problems for Chapter 7

① Each of the following items requires an adjustment to either the bank statement balance or the check register balance. Indicate the correct handling of each item by writing the appropriate letter in the blank.

> A = add to bank statement balance
> B = subtract from bank statement balance
> C = add to checkbook balance
> D = subtract from checkbook balance

_____ **(a)** Outstanding check written to the landlord for rent

_____ **(b)** Bank charge for printing checks

_____ **(c)** A deposit made at the end of the period which was not included on the bank statement

_____ **(d)** A customer's check which was returned by the bank for insufficient funds (a bounced check)

_____ **(e)** An error in recording a check in the check register. A check written to Acme Services for $92.20 was recorded in the check register as $95.50

_____ **(f)** Interest on the checking account

_____ **(g)** A bank fee of $20 for the bounced check

_____ **(h)** Bank fees for ATM withdrawals

② The balance in Ferndale Construction Company's check register May 31 was $12,583.40. The bank statement for Ferndale Construction Company listed the following information:

Previous balance (May 1)	$12,620.10
Deposits	16,265.00
Checks and withdrawals	17,805.95
Interest	52.50
Service charges	20.00
Check returned for insufficient funds	150.00
New balance (May 31)	$10,961.65

By comparing the bank statement and the check register, the company's bookkeeper determined that a deposit of $1,850.15 was not included on the statement and that the following checks were outstanding:

No. 602	$ 35.80
No. 610	212.00
No. 612	95.10

While preparing the reconciliation, the company's bookkeeper noted that check number 585, which had been written for $82.50, had been recorded in the check register as $85.50.

Prepare a bank reconciliation statement for Ferndale Construction Company.

Assignment 7.1: Check Register and Check Stubs

Name _____

Date _____ Score _____

A **(20 points) In the following check register, fill in the cash balance resulting from each transaction. (2 points for each correct answer)**

1.

CHECK REGISTER			DEDUCT ALL PER CHECK OR SERVICE CHARGES THAT APPLY			BALANCE
DATE		CHECK NUMBER	CHECKS ISSUED TO OR DEPOSITS RECEIVED FROM	AMOUNT OF CHECK	AMOUNT OF DEPOSIT	**$1,200.00**
Apr	04	842	Capital Mortgage Company	775.00		
	04	–	Deposit weekly cash receipts		3,987.50	
	05	843	U.S. Treasury	1,743.00		
	06	844	State Income Tax	990.00		
	07	845	General Telephone	65.30		
	08	846	Computer Repair Service	72.00		
	12	–	Deposit weekly cash receipts		3,285.75	
	12	847	Midville Utilities	126.42		
	12	848	Central Advertising, Inc.	850.00		
	12	849	Tax Services, Inc.	650.00		
						3201.53

Score for A (20)

B **(15 points) Fill in the new balance (New Bal) and balance forward (BalFor'd) on each check stub, carrying each balance forward to the next stub. (1½ points for each correct answer)**

2.

```
#101
BalFor'd  806.12   Date 6-1    New Bal _____
Deposit   300.00   To  ACE     This Ck  29.30
New Bal _____   For REPAIR  BalFor'd _____
```

5.

```
#104
BalFor'd _____   Date 6-10   New Bal _____
Deposit 2,160.00 To  CHRON      This Ck  83.16
New Bal _____   For  AD      BalFor'd _____
```

3.

```
#102
BalFor'd _____   Date 6-5    New Bal _____
Deposit _____    To  DON     This Ck  312.80
New Bal _____    For NOTE    BalFor'd _____
```

6.

```
#105
BalFor'd _____   Date 6-15   New Bal _____
Deposit  907.16  To  B/A        This Ck  294.28
New Bal _____   For CAR PAYMENT BalFor'd _____
```

4.

```
#103
BalFor'd _____   Date 6-8    New Bal _____
Deposit  862.13   To  NEC       This Ck  740.12
New Bal _____   For COMPUTER BalFor'd _____
```

Score for B (15)

Assignment 7.1 *Continued*

C (20 points) According to the check register of the Boswell Carpet Company, the cash balance on July 1 was $1,650.20. During the month, deposits of $129.67, $681.10, and $385.60 were made. Checks for $98.99, $307.53, $19.56, $212.40, $195.64, and $88.62 were recorded. (15 points for a correct answer in 7; 5 points for a correct answer in 8)

7. What was the cash balance shown in the check register on July 31? _____

8. After entering all the items in the check register, the bookkeeper found that the check recorded as $212.40 was actually written as $224.20. What is the correct cash balance? _____

<div align="right">

Score for C (20)

</div>

D (45 points) The following problems show the deposits and checks that were recorded on a series of check stubs. In each problem, find the bank balance after each deposit or check. (3 points for each correct answer)

9.			10.			11.		
Balance	$4,300	00	Balance	$205	55	Balance	$1,401	36
Check #1	279	10	Check #21	25	00	Deposit	277	00
Balance			Balance			Balance		
Check #2	148	20	Deposit	876	43	Check #31	136	41
Balance			Balance			Balance		
Deposit	976	80	Check #22	188	14	Check #32	364	18
Balance			Balance			Balance		
Check #3	814	00	Check #23	314	82	Deposit	1,124	85
Balance			Balance			Balance		
Check #4	285	17	Check #24	72	38	Check #33	775	10
Balance			Balance			Balance		

<div align="right">

Score for D (45)

</div>

Assignment 7.2: Check Register and Bank Statements

Name

Date Score

A **(40 points) Solve the following problems. (10 points for a correct final balance in 1; 30 points for a correct final answer in 2)**

1. On October 31, the balance of the account of Toyland, Inc., at the Citizens Bank was $922.10. This amount was also the balance on the check register at that time. Company checks written and deposits made during November are shown on the check register. Fill in the cash balance for each transaction.

CHECK REGISTER			DEDUCT ALL PER CHECK OR SERVICE CHARGES THAT APPLY			BALANCE
DATE		CHECK NUMBER	CHECKS ISSUED TO OR DEPOSITS RECEIVED FROM	AMOUNT OF CHECK	AMOUNT OF DEPOSIT	$922.10
Nov	01	551	Muni. Water, Inc. (2 mos)	119.60		
	06	552	Columbia Gas	49.60		
	07	553	Olympia Telephone	74.19		
	07	–	Deposit cash receipts		225.50	
	21	554	City Trash Disposal (3 mos)	112.32		
	21	555	Jack's Janitorial Service	33.33		
	24	556	United Fund	12.00		
	24	557	Guide Dogs for the Blind	67.77		
	26	558	Wilson Insurance	212.00		
	28	559	Security Systems, Inc.	138.00		
	28	–	Deposit cash receipts		94.00	

2. On December 3, Toyland, Inc., whose check register you completed in problem 1, received the following bank statement. Reconcile the balance on the check register at the end of the month with the final balance on the bank statement. In reconciling the bank statement, you can find which of the checks are outstanding by comparing the list of checks on the statement with the register. Interest and a service charge were recorded on the statement.

C B CITIZEN'S BANK

STATEMENT OF ACCOUNT

TOYLAND, INC.
4617 GILMORE ROAD
WHEATLAND, WI 54828-6075

ACCOUNT NUMBER
072 4736
11/30/--
DATE OF STATEMENT

Balance From Previous Statement	Number of Debits	Amount of Checks and Debits	No. of Credits	Amount of Deposits and Credits	Service Charge	Statement Balance
922.10	8	594.81	2	229.70	9.00	547.99

DATE	CHECKS - DEBITS	CHECKS - DEBITS	DEPOSITS - CREDITS	BALANCE
11/03	119.60			802.50
11/05	49.60			752.90
11/09	9.00 SC			743.90
11/09	74.19			669.71
11/09			225.50 ATM	895.21
11/23	112.32	33.33		749.56
11/26	67.77			681.79
11/30	138.00			543.79
11/30			4.20 INT	547.99

PLEASE EXAMINE AND REPORT ANY DISCREPANCIES WITHIN 10 DAYS
DM-Debit Memo OD-Overdraft
ATM-Automated Teller Machine INT-Interest Paid
CM-Credit Memo SC-Service Charge

TOYLAND, INC.
Reconciliation of Bank Statement
November 30

Bank balance on statement
Plus deposit not recorded by bank

Minus outstanding checks:

Checkbook balance
Plus bank interest

Minus service charge

Score for A (40)

Assignment 7.2 *Continued*

B **(60 points) Solve the following problems. (12 points for each correct answer)**

3. Compute the reconciled balance for each of the problems from the information given.

	Bank Statement Balance	Checkbook Balance	Other Information	Reconciled Balance
a.	$ 841.20	$ 858.85	Outstanding checks: $12.45, $25.90	
			Automatic transfer to savings: $50.00	_____
			Automatic charge, safety deposit box: $6.00	
b.	$1,559.39	$1,617.94	Outstanding checks: $84.62, $14.20, $55.00	
			Outstanding deposit: $224.70	_____
			Automatic transfer to savings: $25.00	
			Bank interest credited: $37.33	
c.	$ 893.17	$ 944.73	Outstanding checks: $7.50, $4.18, $62.40	
			Outstanding deposits: $12.32, $120.00	_____
			Bank interest credited: $24.18	
			Charge for printing new checks: $17.50	
d.	$ 984.32	$ 944.39	Outstanding checks: $42.00	_____
			Deposit of $76.89 shown in check register as $78.96	
e.	$ 625.20	$1,127.38	Outstanding checks: $150.00, $37.82	_____
			Outstanding deposit: $440.00	
			Deposit of $250.00 shown twice in check register	

Score for B (60)

Assignment 7.3: Bank Balance
Reconciliation Statements

Name

Date Score

A **(50 points) Using the data provided, prepare a bank reconciliation statement in each of the following problems. Space is provided for your solutions. (25 points for each correct reconciliation)**

1. The balance shown in the bank statement of Central Heating, Inc., on November 30 was $1,132.46. The balance shown on the check register was $896.91. The following checks were outstanding:

No. 148	$26.62	No. 161	$77.17
No. 156	100.16	No. 165	29.10

There was a bank interest credit of $12.00 and a service charge of $9.50 that had not been entered on Central Heating's check register.

2. The June 30 bank statement for Furgison Electric Company shows that a customer's bad check in the amount of $960 was returned and charged against the Furgison Electric Company's account by the bank. This is the first knowledge the company had that one of the checks deposited was not good.

 The balance shown on the Furgison Electric Company's bank statement was $22,367.14. The balance shown on the check register was $24,696.83. The following checks were outstanding:

No. 363	$1,066.20	No. 396	$1,544.14
No. 387	1,972.81	No. 397	772.86

The following items required adjustment on the bank reconciliation statement:

Outstanding deposit:	$3,001.87
Automatic transfer to note payment:	$4,000.00
Bad check returned and charged to Furgison Electric Company's account by the bank:	$ 960.00
Bank interest credit:	$ 276.17

Score for A (50)

Assignment 7.3 Continued

B **(50 points) Using the data provided, prepare a bank reconciliation statement in each of the following problems. Space is provided for your solutions. (25 points for each correct reconciliation)**

3. The balance shown on the July 31 bank statement of Dawson & Sons, Inc., was $17,464.54. The balance shown by the check register was $17,940.44. A deposit of $2,004.35 had not been credited by the bank, and the following checks were outstanding:

| No. 730 | $ 78.60 | No. 753 | 556.25 | No. 761 | 97.73 |
| No. 749 | 2,765.23 | No. 757 | 271.84 | No. 768 | 958.60 |

The following items required adjustment on the bank reconciliation statement:

Charge for printing checks	$ 18.00
Automatic insurance payment charged to depositor's account by the bank	$1,765.00
Check deposited by Dawson & Sons, returned by bank as bad check:	$1,500.00
Interest on bank account credited by the bank	$ 83.20

4. The balance shown on the March 31 bank statement of Melody Music, Inc., was $9,686.56. The balance shown on the check register was $6,855.74. The following checks were outstanding:

| No. 1243 | $ 640.70 | No. 1267 | $ 87.30 | No. 1270 | $143.90 |
| No. 1258 | 1,225.20 | No. 1268 | 221.12 | No. 1272 | 462.70 |

The following items were listed on the bank statement:

Charge made by the bank for safe deposit box	$ 13.86
AA Realty's check incorrectly charged to the Melody Music account by the bank	$ 65.00
Interest on bank account credited by the bank	$128.76

Score for B (50)

Notes

ADP Automatic Data Processing, Inc.

Solutions
and **Services**
that make it easier
to do **business**

We're ADP
The Business Behind Business. Providing paychecks for more than 30 million
worldwide, processing securities transactions for clients in 25 countries, delive
computing solutions to 16,000 auto/truck dealers, and doing over 14 million au
claim estimates annually.

Year-End Central
Information for your
year-end tasks.
 ● Click here

ADP Employer Services
Payroll, HR & Benefits Administration Services in the U.S. including:

- ● Payroll
- ● Human Resources
- ● 401(k)/Retiremen: Services
- ● Expense Management
- ● Time & Labor Management

- ● Tax & Compliance Management
- ● Benefits Administration
- ● Recruitment Services
- ● Screening & Selection Services
- ● Professional Employer Organization (PEO)

Claims Management
Solutions

Helping property/casualty insurers streamline
workflow, increase estimating accuracy and
speed claims processing.

Auto/Truck Dealer
& Manufacturer Systems

Serving auto retailers with technology
solutions that improve efficiency, client
relationships and profitability.

Securitie
& Investo

Helping fi
trades, m
sharehold

Source: http://www.adp.com/. Copyright © 2001 Courtesy of Automatic Data Processing, Inc.

ADP is the largest company in the United States specializing in the processing of payroll records. Its employees produce paychecks and payroll records for more than 18 million wage earners. Its Employer Services serves 300,000 employers with payroll, human resources, tax compliance, and reporting services.

ADP also offers human resources and benefits administration processing, as well as electronic banking and bill payment services for small businesses. Its online *Career Magazine*, featuring articles on job search techniques, career changes, and employee concerns, is available through links from the company's home page at www.adp.com.

Payroll Records

Sidebar

Business Behind Business℠

[GO]

Corporate Overview
- About Us
- Corporate Philosophy
- World Class Service
- Careers
- Investor Information
- Annual Report
- News Releases

Employer Resource Center

Access free information, tools and applications.

ADP Employer Services Worldwide

[Select a country site ▾]

ations.

s providers process tion and distribute ations.

LEARNING OBJECTIVES

By studying this chapter and completing all assignments you will be able to:

 1 Prepare a payroll register.

 2 Compute federal income tax withholding amounts.

 3 Compute Social Security, Medicare, and other withholdings.

4 Complete an employee's earnings record.

 5 Compute an employer's quarterly federal tax return.

 6 Compute an employer's federal and state unemployment tax liability.

Keeping track of time, computing earnings, and computing net pay after taxes and deductions are part of every business. Although much of the processing is done by specialized firms with large computers, the original data are developed by payroll personnel in business, professional, and government offices.

1. Why are businesses required to compute and withhold deductions for federal income tax?
2. What percentage of your salary or wages is withheld for payment into the Social Security Trust Fund?
3. Who earns more per month: A person making $10 an hour or one making $1,800 per month?

Employers must keep payroll records, withhold and pay payroll taxes, and file quarterly and annual reports with state and federal government offices. The payroll records and processes described in this chapter are common to all employers.

Federal taxes paid by all employees include the federal income tax and the two contributions (commonly referred to as taxes) required by the Federal Insurance Contributions Act (FICA): old-age, survivors, and disability insurance, commonly called Social Security; and hospital insurance, commonly called Medicare.

When hiring new employees, employers must verify each employee's eligibility to work in the United States, get the employee's Social Security number, and have the employee complete a **Form W-4**. The W-4 form shown in Figure 8-1 indicates that Kyle Abrum is married and claims four dependents, which constitutes his **withholding allowance**.

Figure 8-1: Form W-4 (2001)

Personal Allowances Worksheet (Keep for your records.)

A Enter "1" for **yourself** if no one else can claim you as a dependent A _/_

B Enter "1" if: { • You are single and have only one job; or
 • You are married, have only one job, and your spouse does not work; or } . . B _____
 • Your wages from a second job or your spouse's wages (or the total of both) are $1,000 or less. }

C Enter "1" for your **spouse**. But, you may choose to enter -0- if you are married and have either a working spouse or more than one job. (Entering -0- may help you avoid having too little tax withheld.) C _/_

D Enter number of **dependents** (other than your spouse or yourself) you will claim on your tax return D _2_

E Enter "1" if you will file as **head of household** on your tax return (see conditions under **Head of household** above) . E _____

F Enter "1" if you have at least $1,500 of **child or dependent care expenses** for which you plan to claim a credit . . F _____
 (**Note:** Do **not** include child support payments. See **Pub. 503**, Child and Dependent Care Expenses, for details.)

G **Child Tax Credit** (including additional child tax credit):
 • If your total income will be between $18,000 and $50,000 ($23,000 and $63,000 if married), enter "1" for each eligible child.
 • If your total income will be between $50,000 and $80,000 ($63,000 and $115,000 if married), enter "1" if you have two
 eligible children, enter "2" if you have three or four eligible children, or enter "3" if you have five or more eligible children. G _____

H Add lines A through G and enter total here. (**Note:** This may be different from the number of exemptions you claim on your tax return.) ▶ H _4_

For accuracy, complete all worksheets that apply. {
 • If you plan to **itemize or claim adjustments to income** and want to reduce your withholding, see the **Deductions and Adjustments Worksheet** on page 2.
 • If you are **single**, have **more than one job** and your combined earnings from all jobs exceed $35,000, **or** if you are **married** and have a **working spouse or more than one job** and the combined earnings from all jobs exceed $60,000, see the **Two-Earner/Two-Job Worksheet** on page 2 to avoid having too little tax withheld.
 • If **neither** of the above situations applies, **stop here** and enter the number from line H on line 5 of Form W-4 below.

- **Cut here and give Form W-4 to your employer. Keep the top part for your records.** - - - - - - - - - - - - - - - - - -

Form W-4
Department of the Treasury
Internal Revenue Service

Employee's Withholding Allowance Certificate

▶ **For Privacy Act and Paperwork Reduction Act Notice, see page 2.**

OMB No. 1545-0010

2001

| 1 Type or print your first name and middle initial Last name | 2 Your social security number | |
|---|---|---|
| *Kyle B.* | *Abrum* | *123 45 6789* |

Home address (number and street or rural route)
4052 Oak Avenue

3 ☐ Single ☒ Married ☐ Married, but withhold at higher Single rate.
Note: If married, but legally separated, or spouse is a nonresident alien, check the Single box.

City or town, state, and ZIP code
Lawton OK 12345

4 If your last name differs from that on your social security card, check here. You must call 1-800-772-1213 for a new card. ▶ ☐

5 Total number of allowances you are claiming (from line **H** above **or** from the applicable worksheet on page 2) **5** *4*

6 Additional amount, if any, you want withheld from each paycheck **6** $

7 I claim exemption from withholding for 2001, and I certify that I meet **both** of the following conditions for exemption:
 • Last year I had a right to a refund of **all** Federal income tax withheld because I had **no** tax liability **and**
 • This year I expect a refund of **all** Federal income tax withheld because I expect to have **no** tax liability.
 If you meet both conditions, write "Exempt" here ▶ **7**

Under penalties of perjury, I certify that I am entitled to the number of withholding allowances claimed on this certificate, or I am entitled to claim exempt status.
Employee's signature
(Form is not valid unless you sign it.) ▶ *Kyle B. Abrum* Date ▶ *5/12/xx*

8 Employer's name and address (Employer: Complete lines 8 and 10 only if sending to the IRS.) 9 Office code (optional) 10 Employer identification number

Cat. No. 10220Q

Preparing a Payroll Register

A **payroll register** is a summary of employee status information, wages earned, payroll deductions, and take-home pay. Whether they do it manually or by computer, all employers maintain some form of payroll register.

A payroll register is prepared for each payroll period. Payroll periods are weekly, bi-weekly, semimonthly, or monthly. Figure 8-2 shows a payroll register for one weekly period ending March 29. The line for Kyle Abrum shows that he is married, claims four withholding allowances, and is paid on an hourly basis at the rate of $11 per hour ($16.50 for overtime hours). For the current week, he worked 40 regular hours and 6 overtime hours, for gross earnings of $539. From his gross pay he had deductions for Social Security ($33.42), Medicare ($7.82), Federal Income Tax ($28.79), Group Medical Insurance ($39), Group Dental Insurance ($12), and Other ($42), totaling $163.03. His net pay was $375.97.

LEARNING OBJECTIVES 1

Prepare a payroll register.

Figure 8.2: Weekly Payroll Register

| NAME | MARITAL STATUS | WITHHOLDING ALLOWANCES | W = WEEKLY H = HOURLY | RATE | HOURS REG | HOURS O/T | GROSS EARN-INGS | DEDUCTIONS SOCIAL SECURITY | MEDI-CARE | FEDERAL INCOME TAX | GROUP MED. INS. | GROUP DENTAL INS. | OTHER | TOTAL DEDUC-TIONS | NET EARNINGS |
|---|---|---|---|---|---|---|---|---|---|---|---|---|---|---|---|
| Abrum, Kyle | M | 4 | H | 11.00 | 40 | 6 | 539.00 | 33.42 | 7.82 | 28.79 | 39.00 | 12.00 | 42.00 | 163.03 | 375.97 |
| Garcia, Fran | S | 2 | W | 680.00 | 40 | — | 680.00 | 42.16 | 9.86 | 79.76 | 18.00 | 9.00 | — | 158.78 | 521.22 |
| Parker, Marie | S | 1 | H | 12.10 | 32 | — | 387.20 | 24.01 | 5.61 | 42.06 | 18.00 | — | — | 89.68 | 297.52 |
| Thomas, Robert | M | 3 | H | 9.40 | 40 | 4 | 432.40 | 26.81 | 6.27 | 21.16 | 39.00 | 12.00 | 13.10 | 118.34 | 314.06 |
| Weber, James | S | 1 | H | 15.80 | 40 | — | 632.00 | 39.18 | 9.16 | 81.93 | 18.00 | 9.00 | — | 157.27 | 474.73 |
| Totals | | | | | | — | 2,670.60 | 165.58 | 38.72 | 253.70 | 132.00 | 42.00 | 55.10 | 687.10 | 1,983.50 |

CONCEPT CHECK 8.1

After completion of the payroll register entries, one way to check on the accuracy of computations is to subtract the Total Deductions column from the Gross Pay total; the difference should equal the total of the Net Earnings column. From the payroll register shown in Figure 8-2, check the accuracy of the column totals:

| | |
|---|---|
| Total of Gross Earnings column | $2,670.60 |
| Less total of Deductions column | 687.10 |
| Total of Net Earnings column | $1,983.50 |

Computing Federal Income Tax Withholding Amounts

LEARNING OBJECTIVES

Compute federal income tax withholding amounts.

The federal income tax is a payroll tax that the employer must withhold from the employee's pay and turn over to the Internal Revenue Service (IRS). The amount of the deduction varies with the amount of earnings, the employee's marital status, and the number of withholding allowances claimed.

The *Employer's Tax Guide*, published annually by the Internal Revenue Service, gives employers two primary methods to figure how much income tax to withhold from their employees. These two methods are the **percentage method** and the **wage-bracket method**.

Figure 8-2 shows that Kyle Abrum's federal income tax withholding amount was $28.79, computed by the percentage method. With the percentage method, a deduction is granted for each withholding allowance claimed, based on a chart in the *Employer's Tax Guide*. The amount for each withholding allowance is provided in a table labeled Income Tax Withholding Percentage Method Table. Figure 8-3 illustrates a recent table. It shows that, for weekly pay, a deduction of $55.77 is allowed for each withholding allowance. (For monthly pay, a deduction of $241.67 is allowed for each withholding allowance.)

Figure 8-3: Percentage Method Amount for One Withholding Allowance

| Payroll Period | One Withholding Allowance |
|---|---|
| Weekly | $55.77 |
| Biweekly | $111.54 |
| Semimonthly | $120.83 |
| Monthly | $241.67 |

After the total withholding allowance is subtracted from an employee's gross earnings, the amount to be withheld is computed by taking a percentage of the difference. The percentage to be used is given by the IRS in the Tables for Percentage Method of Withholding. Figure 8-4 illustrates a recent table for weekly, biweekly, semimonthly, and monthly payroll periods.

http://icarus.stanford.edu/101416

http://128.220.185.18/images/entl/calc.htm

Visit these sites for basic payroll calculations.

STEPS **to Figure the Amount of Federal Income Tax Withholding, Using the Percentage Method**

1. Determine the employee's gross earnings.
2. Multiply the appropriate (weekly/monthly) "one withholding allowance" amount (from the Withholding Table) by the number of allowances the employee claims.
3. Subtract that amount from the employee's gross earnings.
4. From the appropriate (weekly/monthly and single/married) Percentage Method table, subtract the "of excess over" figure to get the amount subject to the tax.
5. Multiply the amount from Step 4 by the appropriate percentage from the Percentage Method table.
6. If required, add the base tax amount (if any) shown next to the percentage from the Percentage Method Table. (For example, see Table 1, WEEKLY Payroll Period, Married, the second line of the table: $125.50 plus 28% of excess over $960.)

Tables for Percentage Method of Withholding
(For Wages Paid in 2001)

TABLE 1—WEEKLY Payroll Period

(a) SINGLE person (including head of household)—

If the amount of wages (after subtracting withholding allowances) is: The amount of income tax to withhold is:

Not over $51 $0

| Over— | But not over— | | of excess over— |
|---|---|---|---|
| $51 | —$552 | 15% | —51 |
| $552 | —$1,196 | $75.15 plus 28% | —552 |
| $1,196 | —$2,662 | $255.47 plus 31% | —1,196 |
| $2,662 | —$5,750 | $709.93 plus 36% | —2,662 |
| $5,750 | | $1,821.61 plus 39.6% | —5,750 |

(b) MARRIED person—

If the amount of wages (after subtracting withholding allowances) is: The amount of income tax to withhold is:

Not over $124 $0

| Over— | But not over— | | of excess over— |
|---|---|---|---|
| $124 | —$960 | 15% | —$124 |
| $960 | —$2,023 | $125.40 plus 28% | —$960 |
| $2,023 | —$3,292 | $423.04 plus 31% | —$2,023 |
| $3,292 | —$5,809 | $816.43 plus 36% | —$3,292 |
| $5,809 | | $1,722.55 plus 39.6% | —$5,809 |

TABLE 2—BIWEEKLY Payroll Period

(a) SINGLE person (including head of household)—

If the amount of wages (after subtracting withholding allowances) is: The amount of income tax to withhold is:

Not over $102 $0

| Over— | But not over— | | of excess over— |
|---|---|---|---|
| $102 | —$1,104 | 15% | —$102 |
| $1,104 | —$2,392 | $150.30 plus 28% | —$1,104 |
| $2,392 | —$5,323 | $510.94 plus 31% | —$2,392 |
| $5,323 | —$11,500 | $1,419.55 plus 36% | —$5,323 |
| $11,500 | | $3,643.27 plus 39.6% | —$11,500 |

(b) MARRIED person—

If the amount of wages (after subtracting withholding allowances) is: The amount of income tax to withhold is:

Not over $248 $0

| Over— | But not over— | | of excess over— |
|---|---|---|---|
| $248 | —$1,919 | 15% | —$248 |
| $1,919 | —$4,046 | $250.65 plus 28% | —$1,919 |
| $4,046 | —$6,585 | $846.21 plus 31% | —$4,046 |
| $6,585 | —$11,617 | $1,633.30 plus 36% | —$6,585 |
| $11,617 | | $3,444.82 plus 39.6% | —$11,617 |

TABLE 3—SEMIMONTHLY Payroll Period

(a) SINGLE person (including head of household)—

If the amount of wages (after subtracting withholding allowances) is: The amount of income tax to withhold is:

Not over $110 $0

| Over— | But not over— | | of excess over— |
|---|---|---|---|
| $110 | —$1,196 | 15% | —$110 |
| $1,196 | —$2,592 | $162.90 plus 28% | —$1,196 |
| $2,592 | —$5,767 | $553.78 plus 31% | —$2,592 |
| $5,767 | —$12,458 | $1,538.03 plus 36% | —$5,767 |
| $12,458 | | $3,946.79 plus 39.6% | —$12,458 |

(b) MARRIED person—

If the amount of wages (after subtracting withholding allowances) is: The amount of income tax to withhold is:

Not over $269 $0

| Over— | But not over— | | of excess over— |
|---|---|---|---|
| $269 | —$2,079 | 15% | —$269 |
| $2,079 | —$4,383 | $271.50 plus 28% | —$2,079 |
| $4,383 | —$7,133 | $916.62 plus 31% | —$4,383 |
| $7,133 | —$12,585 | $1,769.12 plus 36% | —$7,133 |
| $12,585 | | $3,731.84 plus 39.6% | —$12,585 |

TABLE 4—MONTHLY Payroll Period

(a) SINGLE person (including head of household)—

If the amount of wages (after subtracting withholding allowances) is: The amount of income tax to withhold is:

Not over $221 $0

| Over— | But not over— | | of excess over— |
|---|---|---|---|
| $221 | —$2,392 | 15% | —$221 |
| $2,392 | —$5,183 | $325.65 plus 28% | —$2,392 |
| $5,183 | —$11,533 | $1,107.13 plus 31% | —$5,183 |
| $11,533 | —$24,917 | $3,075.63 plus 36% | —$11,533 |
| $24,917 | | $7,893.87 plus 39.6% | —$24,917 |

(b) MARRIED person—

If the amount of wages (after subtracting withholding allowances) is: The amount of income tax to withhold is:

Not over $538 $0

| Over— | But not over— | | of excess over— |
|---|---|---|---|
| $538 | —$4,158 | 15% | —$538 |
| $4,158 | —$8,767 | $543.00 plus 28% | —$4,158 |
| $8,767 | —$14,267 | $1,833.52 plus 31% | —$8,767 |
| $14,267 | —$25,171 | $3,538.52 plus 36% | —$14,267 |
| $25,171 | | $7,463.96 plus 39.6% | —$25,171 |

Using the six steps given, we compute Kyle Abrum's withholding as follows:

| STEP 1 | $539.00 | (gross earnings from payroll register) |
|---|---|---|

| STEP 2 | $ 55.77 | (one withholding allowance) |
|---|---|---|
| | × 4 | (number of withholding allowances claimed) |
| | $223.08 | (total withholding allowance) |

| STEP 3 | $539.00 | (gross earnings) |
|---|---|---|
| | − 223.08 | (total withholding allowance) |
| | $315.92 | (amount subject to withholding) |

| STEP 4 | $315.92 | (amount subject to withholding) |
|---|---|---|
| | − 124.00 | (less $124, "excess amount" in Figure 8-4) |
| | $191.92 | (amount subject to percentage computation) |

| STEP 5 | $191.92 | (amount subject to percentage computation) |
|---|---|---|
| | × 0.15 | (15% computation) |
| | $ 28.79 | (amount of tax withheld) |

STEP 6 The wage range $124–$960 doesn't have a base tax amount and therefore doesn't apply in the case of Kyle Abrum.

CyberConnection

This sample page is from one of the numerous payroll calculating software programs available to businesses.

See technology at work! Symmetry Software creates payroll and related software for business. Its Paycheck-City site contains useful information, links, and tools for tax and salary issues. Visit this Web site and check out the many calculators with which to calculate wages, taxes, and your 401K plan.

The Paycheck Calculator

Calculation based on: Tax Year 2001 ▾ For Arizona ▾

Instructions: Click an item's name for help on contents.

General Information

| | | |
|---|---|---|
| Gross Pay | | Annually ▾ |
| Gross Salary YTD | | (optional) |
| Pay Frequency | Weekly ▾ | |
| Federal Filing Status | Single ▾ | |
| # of Federal Allowances | | |
| Additional Fed. Withholding | | |
| Round Federal Withholding | ○ Yes ◉ No | |

State and Local Information for Arizona

| | |
|---|---|
| % of Federal | 0% Exempt ▾ |
| Additional State W/H | |

Deduction Section

Use 2 ▾ deduction(s) for my paycheck.

| | | |
|---|---|---|
| Deduction #1 Name | | |
| Deduction #1 Amount | | % of gross pay ▾ |
| Ded. #1 Exempt from: | ☐ Federal ☐ FICA ☐ State ☐ Local | |
| Deduction #2 Name | | |
| Deduction #2 Amount | | % of gross pay ▾ |
| Ded. #2 Exempt from: | ☐ Federal ☐ FICA ☐ State ☐ Local | |

Calculate Clear

Source: http://www.paycheckcity.com/images/np1.gif. Copyright © 2001 Courtesy of Symmetry Software.

The second method of figuring the amount of tax to be withheld from an employee's pay, the wage-bracket method, involves use of a series of wage-bracket tables published in the IRS *Employer's Tax Guide*. Figures 8-5 and 8-6 illustrate the tables for single and married persons who are paid on a weekly basis.

Using the tables from Figure 8-6, we see that a married employee earning a weekly wage of between $530 and $540 and claiming four withholding allowances will have $28 withheld. Note that the amount of federal income tax withheld from Kyle Abrum's pay, using the wage-bracket method, is approximately the same as the amount withheld using the percentage method: $28 versus $28.79. Small differences will frequently result because the wage-bracket method uses tables based on $10 divisions and rounded amounts. Over a period of a year, these differences tend to be relatively insignificant and are accepted by the IRS.

CONCEPT CHECK 8.2

Using the percentage method steps given, verify the federal income tax withholding for Fran Garcia as recorded in the payroll register.

STEP 1 $680.00 (gross earnings from payroll register)

STEP 2 $ 55.77 (one withholding allowance)
 × 2 (number of withholding allowances claimed)
 $111.54 (total withholding allowance)

STEP 3 $680.00 (gross earnings)
 − 111.54 (total withholding allowance)
 $568.46 (amount subject to withholding)

STEP 4 $568.46 (amount subject to withholding)
 − 552.00 (less $552; excess amount shown in Figure 8-4)—Table 1(a) WEEKLY/SINGLE
 $ 16.46 (amount subject to percentage computation)

STEP 5 $ 16.46 (amount subject to percentage computation)
 × 0.28 (28% computation)
 $ 4.61 (amount of tax withheld on percentage computation)

STEP 5 $ 4.61 (amount of tax withheld on percentage computation)
 + 75.15 (base tax amount)
 $ 79.76 (total amount of tax withheld)

Use the wage-bracket method to find the federal income tax withholding for Fran Garcia. Then compute the difference between the percentage method and the wage-bracket method.

Percentage method (Step 6) $79.76
Wage-bracket method (Figure 8-5 because she is single) 81.00
Difference $ 1.24

Figure 8-5: Single Persons—Weekly
Payroll Period

SINGLE Persons—WEEKLY Payroll Period

(For Wages Paid in 2001)

| If the wages are— | | And the number of withholding allowances claimed is— | | | | | | | | | | |
|---|---|---|---|---|---|---|---|---|---|---|---|---|
| At least | But less than | 0 | 1 | 2 | 3 | 4 | 5 | 6 | 7 | 8 | 9 | 10 |
| | | The amount of income tax to be withheld is— | | | | | | | | | | |
| $0 | $55 | 0 | 0 | 0 | 0 | 0 | 0 | 0 | 0 | 0 | 0 | 0 |
| 55 | 60 | 1 | 0 | 0 | 0 | 0 | 0 | 0 | 0 | 0 | 0 | 0 |
| 60 | 65 | 2 | 0 | 0 | 0 | 0 | 0 | 0 | 0 | 0 | 0 | 0 |
| 65 | 70 | 2 | 0 | 0 | 0 | 0 | 0 | 0 | 0 | 0 | 0 | 0 |
| 70 | 75 | 3 | 0 | 0 | 0 | 0 | 0 | 0 | 0 | 0 | 0 | 0 |
| 75 | 80 | 4 | 0 | 0 | 0 | 0 | 0 | 0 | 0 | 0 | 0 | 0 |
| 80 | 85 | 5 | 0 | 0 | 0 | 0 | 0 | 0 | 0 | 0 | 0 | 0 |
| 85 | 90 | 5 | 0 | 0 | 0 | 0 | 0 | 0 | 0 | 0 | 0 | 0 |
| 90 | 95 | 6 | 0 | 0 | 0 | 0 | 0 | 0 | 0 | 0 | 0 | 0 |
| 95 | 100 | 7 | 0 | 0 | 0 | 0 | 0 | 0 | 0 | 0 | 0 | 0 |
| 100 | 105 | 8 | 0 | 0 | 0 | 0 | 0 | 0 | 0 | 0 | 0 | 0 |
| 200 | 210 | 23 | 15 | 6 | 0 | 0 | 0 | 0 | 0 | 0 | 0 | 0 |
| 210 | 220 | 25 | 16 | 8 | 0 | 0 | 0 | 0 | 0 | 0 | 0 | 0 |
| 220 | 230 | 26 | 18 | 9 | 1 | 0 | 0 | 0 | 0 | 0 | 0 | 0 |
| 230 | 240 | 28 | 19 | 11 | 3 | 0 | 0 | 0 | 0 | 0 | 0 | 0 |
| 240 | 250 | 29 | 21 | 12 | 4 | 0 | 0 | 0 | 0 | 0 | 0 | 0 |
| 250 | 260 | 31 | 22 | 14 | 6 | 0 | 0 | 0 | 0 | 0 | 0 | 0 |
| 260 | 270 | 32 | 24 | 15 | 7 | 0 | 0 | 0 | 0 | 0 | 0 | 0 |
| 270 | 280 | 34 | 25 | 17 | 9 | 0 | 0 | 0 | 0 | 0 | 0 | 0 |
| 280 | 290 | 35 | 27 | 18 | 10 | 2 | 0 | 0 | 0 | 0 | 0 | 0 |
| 290 | 300 | 37 | 28 | 20 | 12 | 3 | 0 | 0 | 0 | 0 | 0 | 0 |
| 300 | 310 | 38 | 30 | 21 | 13 | 5 | 0 | 0 | 0 | 0 | 0 | 0 |
| 310 | 320 | 40 | 31 | 23 | 15 | 6 | 0 | 0 | 0 | 0 | 0 | 0 |
| 320 | 330 | 41 | 33 | 24 | 16 | 8 | 0 | 0 | 0 | 0 | 0 | 0 |
| 330 | 340 | 43 | 34 | 26 | 18 | 9 | 1 | 0 | 0 | 0 | 0 | 0 |
| 340 | 350 | 44 | 36 | 27 | 19 | 11 | 2 | 0 | 0 | 0 | 0 | 0 |
| 350 | 360 | 46 | 37 | 29 | 21 | 12 | 4 | 0 | 0 | 0 | 0 | 0 |
| 360 | 370 | 47 | 39 | 30 | 22 | 14 | 5 | 0 | 0 | 0 | 0 | 0 |
| 370 | 380 | 49 | 40 | 32 | 24 | 15 | 7 | 0 | 0 | 0 | 0 | 0 |
| 380 | 390 | 50 | 42 | 33 | 25 | 17 | 8 | 0 | 0 | 0 | 0 | 0 |
| 390 | 400 | 52 | 43 | 35 | 27 | 18 | 10 | 1 | 0 | 0 | 0 | 0 |
| 400 | 410 | 53 | 45 | 36 | 28 | 20 | 11 | 3 | 0 | 0 | 0 | 0 |
| 410 | 420 | 55 | 46 | 38 | 30 | 21 | 13 | 4 | 0 | 0 | 0 | 0 |
| 420 | 430 | 56 | 48 | 39 | 31 | 23 | 14 | 6 | 0 | 0 | 0 | 0 |
| 430 | 440 | 58 | 49 | 41 | 33 | 24 | 16 | 7 | 0 | 0 | 0 | 0 |
| 440 | 450 | 59 | 51 | 42 | 34 | 26 | 17 | 9 | 1 | 0 | 0 | 0 |
| 450 | 460 | 61 | 52 | 44 | 36 | 27 | 19 | 10 | 2 | 0 | 0 | 0 |
| 460 | 470 | 62 | 54 | 45 | 37 | 29 | 20 | 12 | 4 | 0 | 0 | 0 |
| 470 | 480 | 64 | 55 | 47 | 39 | 30 | 22 | 13 | 5 | 0 | 0 | 0 |
| 480 | 490 | 65 | 57 | 48 | 40 | 32 | 23 | 15 | 7 | 0 | 0 | 0 |
| 490 | 500 | 67 | 58 | 50 | 42 | 33 | 25 | 16 | 8 | 0 | 0 | 0 |
| 500 | 510 | 68 | 60 | 51 | 43 | 35 | 26 | 18 | 10 | 1 | 0 | 0 |
| 510 | 520 | 70 | 61 | 53 | 45 | 36 | 28 | 19 | 11 | 3 | 0 | 0 |
| 520 | 530 | 71 | 63 | 54 | 46 | 38 | 29 | 21 | 13 | 4 | 0 | 0 |
| 530 | 540 | 73 | 64 | 56 | 48 | 39 | 31 | 22 | 14 | 6 | 0 | 0 |
| 540 | 550 | 74 | 66 | 57 | 49 | 41 | 32 | 24 | 16 | 7 | 0 | 0 |
| 550 | 560 | 76 | 67 | 59 | 51 | 42 | 34 | 25 | 17 | 9 | 0 | 0 |
| 560 | 570 | 79 | 69 | 60 | 52 | 44 | 35 | 27 | 19 | 10 | 2 | 0 |
| 570 | 580 | 82 | 70 | 62 | 54 | 45 | 37 | 28 | 20 | 12 | 3 | 0 |
| 580 | 590 | 84 | 72 | 63 | 55 | 47 | 38 | 30 | 22 | 13 | 5 | 0 |
| 590 | 600 | 87 | 73 | 65 | 57 | 48 | 40 | 31 | 23 | 15 | 6 | 0 |
| 600 | 610 | 90 | 75 | 66 | 58 | 50 | 41 | 33 | 25 | 16 | 8 | 0 |
| 610 | 620 | 93 | 77 | 68 | 60 | 51 | 43 | 34 | 26 | 18 | 9 | 1 |
| 620 | 630 | 96 | 80 | 69 | 61 | 53 | 44 | 36 | 28 | 19 | 11 | 2 |
| 630 | 640 | 98 | 83 | 71 | 63 | 54 | 46 | 37 | 29 | 21 | 12 | 4 |
| 640 | 650 | 101 | 86 | 72 | 64 | 56 | 47 | 39 | 31 | 22 | 14 | 5 |
| 650 | 660 | 104 | 88 | 74 | 66 | 57 | 49 | 40 | 32 | 24 | 15 | 7 |
| 660 | 670 | 107 | 91 | 76 | 67 | 59 | 50 | 42 | 34 | 25 | 17 | 8 |
| 670 | 680 | 110 | 94 | 78 | 69 | 60 | 52 | 43 | 35 | 27 | 18 | 10 |
| 680 | 690 | 112 | 97 | 81 | 70 | 62 | 53 | 45 | 37 | 28 | 20 | 11 |
| 690 | 700 | 115 | 100 | 84 | 72 | 63 | 55 | 46 | 38 | 30 | 21 | 13 |

Figure 8-6: Married Persons—Weekly Payroll Period

MARRIED Persons—WEEKLY Payroll Period

(For Wages Paid in 2001)

| If the wages are— | | And the number of withholding allowances claimed is— | | | | | | | | | | |
|---|---|---|---|---|---|---|---|---|---|---|---|---|
| At least | But less than | 0 | 1 | 2 | 3 | 4 | 5 | 6 | 7 | 8 | 9 | 10 |
| | | The amount of income tax to be withheld is— | | | | | | | | | | |
| $0 | $125 | 0 | 0 | 0 | 0 | 0 | 0 | 0 | 0 | 0 | 0 | 0 |
| 125 | 130 | 1 | 0 | 0 | 0 | 0 | 0 | 0 | 0 | 0 | 0 | 0 |
| 130 | 135 | 1 | 0 | 0 | 0 | 0 | 0 | 0 | 0 | 0 | 0 | 0 |
| 135 | 140 | 2 | 0 | 0 | 0 | 0 | 0 | 0 | 0 | 0 | 0 | 0 |
| 140 | 145 | 3 | 0 | 0 | 0 | 0 | 0 | 0 | 0 | 0 | 0 | 0 |
| 145 | 150 | 4 | 0 | 0 | 0 | 0 | 0 | 0 | 0 | 0 | 0 | 0 |
| 150 | 155 | 4 | 0 | 0 | 0 | 0 | 0 | 0 | 0 | 0 | 0 | 0 |
| 155 | 160 | 5 | 0 | 0 | 0 | 0 | 0 | 0 | 0 | 0 | 0 | 0 |
| 160 | 165 | 6 | 0 | 0 | 0 | 0 | 0 | 0 | 0 | 0 | 0 | 0 |
| 165 | 170 | 7 | 0 | 0 | 0 | 0 | 0 | 0 | 0 | 0 | 0 | 0 |
| 170 | 175 | 7 | 0 | 0 | 0 | 0 | 0 | 0 | 0 | 0 | 0 | 0 |
| 175 | 180 | 8 | 0 | 0 | 0 | 0 | 0 | 0 | 0 | 0 | 0 | 0 |
| 180 | 185 | 9 | 0 | 0 | 0 | 0 | 0 | 0 | 0 | 0 | 0 | 0 |
| 185 | 190 | 10 | 1 | 0 | 0 | 0 | 0 | 0 | 0 | 0 | 0 | 0 |
| 190 | 195 | 10 | 2 | 0 | 0 | 0 | 0 | 0 | 0 | 0 | 0 | 0 |
| 195 | 200 | 11 | 3 | 0 | 0 | 0 | 0 | 0 | 0 | 0 | 0 | 0 |
| 200 | 210 | 12 | 4 | 0 | 0 | 0 | 0 | 0 | 0 | 0 | 0 | 0 |
| 210 | 220 | 14 | 5 | 0 | 0 | 0 | 0 | 0 | 0 | 0 | 0 | 0 |
| 220 | 230 | 15 | 7 | 0 | 0 | 0 | 0 | 0 | 0 | 0 | 0 | 0 |
| 230 | 240 | 17 | 8 | 0 | 0 | 0 | 0 | 0 | 0 | 0 | 0 | 0 |
| 240 | 250 | 18 | 10 | 1 | 0 | 0 | 0 | 0 | 0 | 0 | 0 | 0 |
| 250 | 260 | 20 | 11 | 3 | 0 | 0 | 0 | 0 | 0 | 0 | 0 | 0 |
| 260 | 270 | 21 | 13 | 4 | 0 | 0 | 0 | 0 | 0 | 0 | 0 | 0 |
| 270 | 280 | 23 | 14 | 6 | 0 | 0 | 0 | 0 | 0 | 0 | 0 | 0 |
| 280 | 290 | 24 | 16 | 7 | 0 | 0 | 0 | 0 | 0 | 0 | 0 | 0 |
| 290 | 300 | 26 | 17 | 9 | 1 | 0 | 0 | 0 | 0 | 0 | 0 | 0 |
| 300 | 310 | 27 | 19 | 10 | 2 | 0 | 0 | 0 | 0 | 0 | 0 | 0 |
| 310 | 320 | 29 | 20 | 12 | 4 | 0 | 0 | 0 | 0 | 0 | 0 | 0 |
| 320 | 330 | 30 | 22 | 13 | 5 | 0 | 0 | 0 | 0 | 0 | 0 | 0 |
| 330 | 340 | 32 | 23 | 15 | 7 | 0 | 0 | 0 | 0 | 0 | 0 | 0 |
| 340 | 350 | 33 | 25 | 16 | 8 | 1 | 0 | 0 | 0 | 0 | 0 | 0 |
| 350 | 360 | 35 | 26 | 18 | 10 | 1 | 0 | 0 | 0 | 0 | 0 | 0 |
| 360 | 370 | 36 | 28 | 19 | 11 | 3 | 0 | 0 | 0 | 0 | 0 | 0 |
| 370 | 380 | 38 | 29 | 21 | 13 | 4 | 0 | 0 | 0 | 0 | 0 | 0 |
| 380 | 390 | 39 | 31 | 22 | 14 | 6 | 0 | 0 | 0 | 0 | 0 | 0 |
| 390 | 400 | 41 | 32 | 24 | 16 | 7 | 0 | 0 | 0 | 0 | 0 | 0 |
| 400 | 410 | 42 | 34 | 25 | 17 | 9 | 0 | 0 | 0 | 0 | 0 | 0 |
| 410 | 420 | 44 | 35 | 27 | 19 | 10 | 2 | 0 | 0 | 0 | 0 | 0 |
| 420 | 430 | 45 | 37 | 28 | 20 | 12 | 3 | 0 | 0 | 0 | 0 | 0 |
| 430 | 440 | 47 | 38 | 30 | 22 | 13 | 5 | 0 | 0 | 0 | 0 | 0 |
| 440 | 450 | 48 | 40 | 31 | 23 | 15 | 6 | 0 | 0 | 0 | 0 | 0 |
| 450 | 460 | 50 | 41 | 33 | 25 | 16 | 8 | 0 | 0 | 0 | 0 | 0 |
| 460 | 470 | 51 | 43 | 34 | 26 | 18 | 9 | 1 | 0 | 0 | 0 | 0 |
| 470 | 480 | 53 | 44 | 36 | 28 | 19 | 11 | 2 | 0 | 0 | 0 | 0 |
| 480 | 490 | 54 | 46 | 37 | 29 | 21 | 12 | 4 | 0 | 0 | 0 | 0 |
| 490 | 500 | 56 | 47 | 39 | 31 | 22 | 14 | 5 | 0 | 0 | 0 | 0 |
| 500 | 510 | 57 | 49 | 40 | 32 | 24 | 15 | 7 | 0 | 0 | 0 | 0 |
| 510 | 520 | 59 | 50 | 42 | 34 | 25 | 17 | 8 | 0 | 0 | 0 | 0 |
| 520 | 530 | 60 | 52 | 43 | 35 | 27 | 18 | 10 | 2 | 0 | 0 | 0 |
| 530 | 540 | 62 | 53 | 45 | 37 | 28 | 20 | 11 | 3 | 0 | 0 | 0 |
| 540 | 550 | 63 | 55 | 46 | 38 | 30 | 21 | 13 | 5 | 0 | 0 | 0 |
| 550 | 560 | 65 | 56 | 48 | 40 | 31 | 23 | 14 | 6 | 0 | 0 | 0 |
| 560 | 570 | 66 | 58 | 49 | 41 | 33 | 24 | 16 | 8 | 0 | 0 | 0 |
| 570 | 580 | 68 | 59 | 51 | 43 | 34 | 26 | 17 | 9 | 1 | 0 | 0 |
| 580 | 590 | 69 | 61 | 52 | 44 | 36 | 27 | 19 | 11 | 2 | 0 | 0 |
| 590 | 600 | 71 | 62 | 54 | 46 | 37 | 29 | 20 | 12 | 4 | 0 | 0 |
| 600 | 610 | 72 | 64 | 55 | 47 | 39 | 30 | 22 | 14 | 5 | 0 | 0 |
| 610 | 620 | 74 | 65 | 57 | 49 | 40 | 32 | 23 | 15 | 7 | 0 | 0 |
| 620 | 630 | 75 | 67 | 58 | 50 | 42 | 33 | 25 | 17 | 8 | 0 | 0 |
| 630 | 640 | 77 | 68 | 60 | 52 | 43 | 35 | 26 | 18 | 10 | 1 | 0 |
| 640 | 650 | 78 | 70 | 61 | 53 | 45 | 36 | 28 | 20 | 11 | 3 | 0 |
| 650 | 660 | 80 | 71 | 63 | 55 | 46 | 38 | 29 | 21 | 13 | 4 | 0 |
| 660 | 670 | 81 | 73 | 64 | 56 | 48 | 39 | 31 | 23 | 14 | 6 | 0 |
| 670 | 680 | 83 | 74 | 66 | 58 | 49 | 41 | 32 | 24 | 16 | 7 | 0 |
| 680 | 690 | 84 | 76 | 67 | 59 | 51 | 42 | 34 | 26 | 17 | 9 | 0 |
| 690 | 700 | 86 | 77 | 69 | 61 | 52 | 44 | 35 | 27 | 19 | 10 | 2 |
| 700 | 710 | 87 | 79 | 70 | 62 | 54 | 45 | 37 | 29 | 20 | 12 | 3 |
| 710 | 720 | 89 | 80 | 72 | 64 | 55 | 47 | 38 | 30 | 22 | 13 | 5 |
| 720 | 730 | 90 | 82 | 73 | 65 | 57 | 48 | 40 | 32 | 23 | 15 | 6 |
| 730 | 740 | 92 | 83 | 75 | 67 | 58 | 50 | 41 | 33 | 25 | 16 | 8 |

Computing Social Security, Medicare, and Other Withholdings

3
LEARNING
OBJECTIVES

Compute Social Security, Medicare, and other withholdings.

The **Federal Insurance Contributions Act** (**FICA**) provides for a federal system of old-age, survivors, disability, and hospital insurance. The old-age, survivors, and disability insurance part of FICA is financed by *Social Security tax*. The hospital insurance part of FICA is financed by the *Medicare tax*. These taxes are reported separately and are levied on both the employer and the employee. These taxes have different rates, but only the Social Security tax has a wage base, which is the *maximum* wage that is subject to the tax for the year.

The Social Security tax rate of 6.2% is levied on both the employer and the employee. For 2001, the wage base was $80,400.

The Medicare tax rate of 1.45% is levied on both the employer and the employee. There is no wage base limit for Medicare; all covered wages are subject to the Medicare tax.

Although both rates are subject to change by legislation, they were current when we compiled the payroll register illustrated in this chapter. All amounts are rounded for the nearest cent. The amounts for Kyle Abrum were $33.42 for Social Security and $7.82 for Medicare.

EXAMPLE B

Social Security deduction:

| $539.00 | (gross earnings) |
|---|---|
| × 0.062 | (Social Security rate) |
| $ 33.42 | (Social Security amount) |

EXAMPLE C

Medicare deduction:

| $539.00 | (gross earnings) |
|---|---|
| × .0145 | (Medicare rate) |
| $ 7.82 | (Medicare amount) |

Many employers today provide some form of group medical insurance for their employees. Frequently, the employee is asked to pay a portion of the premium charged for such insurance, based on the number of dependents the employee has named to be insured. For the payroll register shown in Figure 8-2, we assumed the weekly rates for medical and dental plans shown in Figure 8-7.

Figure 8-7: Weekly Medical and Dental Plan Rates

| | Weekly Medical Plan Premium Paid by Employee | Weekly Dental Plan Premium Paid by Employee |
|---|---|---|
| Employee only | $18.00 | $9.00 |
| Employee plus one dependent | $22.00 | $10.00 |
| Employee plus 2 or more dependents | $39.00 | $12.00 |

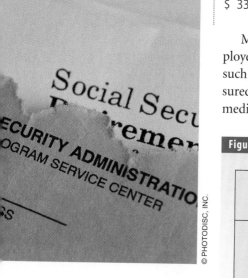

© PHOTODISC, INC.

www.aspm.org/

Visit the Web site of the American Society for Payroll Management and read its newsletters on payroll and tax issues. This site also provides many links to other payroll information Web sites.

The payroll register presented in Figure 8-2 showed that Kyle Abrum subscribed to both the medical and the dental programs. Because of his three dependents, the amounts of his deductions were $39 and $12, respectively.

Frequently, employees will arrange to have special payroll deductions made by the employer to pay union dues, put money into special retirement or savings plans, or make contributions to charitable organizations.

In addition, 42 of the 50 states have some form of state income tax, which normally requires withholding in the same manner as the federal income tax. In such states, state

income tax withholding columns are added to the payroll register and withholdings are made according to wage-bracket or percentage charts established by the state, in the same manner as federal income tax withholdings.

The payroll register illustrated in Figure 8-2 reflects a $42 weekly deduction that Kyle Abrum had requested be made for payment of his union dues (other).

www.aflcio.org/
paywatch/tooclose.htm
The AFL–CIO maintains a Web site that tracks CEO salaries.

CONCEPT CHECK 8.3

Using the format in examples B and C, compute Social Security and Medicare amounts for Fran Garcia, based on her gross weekly earnings of $680.

Social Security deduction:
| | |
|---|---|
| $ 680 | (gross earnings) |
| × 0.062 | (Social Security rate) |
| $42.16 | (Social Security amount) |

Medicare deduction:
| | |
|---|---|
| $ 680 | (gross earnings) |
| × 0.0145 | (Medicare rate) |
| $9.86 | (Medicare amount) |

Completing an Employee's Earnings Record

An employer must submit quarterly and annual reports to the federal government and appropriate state government and pay the amount of taxes withheld from employees' earnings for the period. To obtain the necessary information, most employers keep an **employee's earnings record** for each employee. The employee's earnings record summarizes by quarter the employee's gross earnings, deductions, and net pay.

LEARNING OBJECTIVES

Complete an employee's earnings record.

EXAMPLE D

Figure 8-8: Employee's Earnings Record

Name **Kyle Abrum** Social Security No. **234-12-8765**

Address **4052 Oak Ave.** No. of Allowances **4** Marital Status **Married**

| Period Ending | Total Wages | Cumulative Wages | Deductions | | | | | Net Pay |
|---|---|---|---|---|---|---|---|---|
| | | | Social Security | Medicare | Federal Inc. Tax | Other Deductions | Total | |
| 1/4 | $ 440.00 | $ 440.00 | $ 27.28 | $ 6.38 | $ 13.94 | $ 93.00 | $ 140.60 | $ 299.40 |
| 1/11 | 440.00 | 880.00 | 27.28 | 6.38 | 13.94 | 93.00 | 140.60 | 299.40 |
| 3/29 | 539.00 | 6,952.00 | 33.42 | 7.82 | 28.79 | 93.00 | 163.03 | 375.97 |
| Quarter Totals | $6,952.00 | | $431.20 | $ 99.87 | $365.95 | $908.70 | $1,805.72 | $5,146.28 |

The employee's earnings record presented in Figure 8-8 shows that Kyle Abrum is married, claims four allowances, and for the first quarter of the year earned total wages of $6,952. His net pay was $5,146.28 after first quarter withholdings as follows:

| | |
|---|---|
| Federal income tax withholding | $ 365.95 |
| Social Security withholding | 431.20 |
| Medicare withholding | 99.87 |
| Other deductions | $ 908.70 |
| Total deductions | $1,805.72 |

CONCEPT CHECK 8.4

Assuming that Fran Garcia's weekly earnings and deductions have remained constant for each of the 13 weeks in the first quarter of the year, compute the following totals, which would appear on her employee's earnings record for the first quarter:

| | | |
|---|---|---|
| Total wages | $8,840.00 | ($680.00 × 13) |
| Federal income tax withholding | 1,036.88 | ($79.76 × 13) |
| Social Security withholding | 548.08 | ($42.16 × 13) |
| Medicare withholding | 128.18 | ($9.86 × 13) |
| Group medical insurance deductions | 234.00 | ($18.00 × 13) |
| Group dental insurance deductions | $ 117.00 | ($9.00 × 13) |
| Total deductions | $2,064.14 | |
| Net pay | 6,775.86 | |

Computing an Employer's Quarterly Federal Tax Return

LEARNING OBJECTIVES

Compute an employer's quarterly federal tax return.

Every employer who withholds federal income tax and FICA taxes (Social Security and Medicare) must file a quarterly return, Form 941—**Employer's Quarterly Federal Tax Return**. Figure 8-9 shows the data that the employer must include on Form 941 (the completed form is slightly abbreviated here). The return must be filed with the IRS within one month after the end of the quarter.

The employer obtains Social Security and Medicare amounts by multiplying the taxable wages paid by 12.4% and 2.9%, respectively. These amounts represent the employee's deductions and matching amounts required to be paid by the employer.

EXAMPLE E

For the first quarter of 2001, Yeager Manufacturing paid total wages of $2,132,684.27. The company withheld $372,486.20 for federal income tax. All wages paid were subject to Social Security and Medicare taxes. If during the quarter Yeager had deposited $680,000 toward its taxes due, how much would it be required to send in with its first-quarter Form 941?

| | |
|---|---|
| Gross wages $2,132,684.27 × 12.4% (Social Security) | $264,452.85 |
| Gross wages $2,132,684.27 × 2.9% (Medicare) | 61,847.84 |
| Subtotal | 326,300.69 |
| Income taxes withheld | 372,486.20 |
| Total | 698,786.89 |
| Less deposit | 680,000.00 |
| Balance due | $ 18,786.89 |

Figure 8-9: Form 941 Employer's Quarterly Federal Tax Return (extract)

| | | | | | | |
|---|---|---|---|---|---|---|
| 1 | Number of employees in the pay period that includes March 12th . ▶ | 1 | 5 | | |
| 2 | Total wages and tips, plus other compensation | | | 2 | 60,138 12 |
| 3 | Total income tax withheld from wages, tips, and sick pay | | | 3 | 4,997 45 |
| 4 | Adjustment of withheld income tax for preceding quarters of calendar year | | | 4 | 0 00 |
| 5 | Adjusted total of income tax withheld (line 3 as adjusted by line 4—see instructions) . . . | | | 5 | 4,997 45 |
| 6 | Taxable social security wages | 6a | 60,138 12 | × 12.4% (.124) = | 6b | 7,457 13 |
| | Taxable social security tips | 6c | | × 12.4% (.124) = | 6d | 0 00 |
| 7 | Taxable Medicare wages and tips . . . | 7a | 60,138 12 | × 2.9% (.029) = | 7b | 1,744 01 |
| 8 | Total social security and Medicare taxes (add lines 6b, 6d, and 7b). Check here if wages are not subject to social security and/or Medicare tax ▶ ☐ | | | 8 | 9,201 14 |
| 9 | Adjustment of social security and Medicare taxes (see instructions for required explanation) Sick Pay $ _____ ± Fractions of Cents $ _____ ± Other $ _____ = | | | 9 | 0 00 |
| 10 | Adjusted total of social security and Medicare taxes (line 8 as adjusted by line 9—see instructions) | | | 10 | 9,201 14 |
| 11 | **Total taxes** (add lines 5 and 10) | | | 11 | 14,198 59 |
| 12 | Advance earned income credit (EIC) payments made to employees | | | 12 | 0 00 |
| 13 | Net taxes (subtract line 12 from line 11). **If $2,500 or more, this must equal line 17, column (d) below (or line D of Schedule B (Form 941))** | | | 13 | 14,198 59 |
| 14 | Total deposits for quarter, including overpayment applied from a prior quarter | | | 14 | 14,107 58 |
| 15 | **Balance due** (subtract line 14 from line 13). See instructions | | | 15 | 91 01 |

CONCEPT CHECK 8.5

As displayed in Figure 8-9, the total taxes due the IRS consist of the $4,997.45 in federal income taxes withheld from employees, plus $7,457.13 and $1,744.01 for Social Security and Medicare taxes, respectively, half of which is withheld from employees and half of which is paid by the employer. Although the employer files Form 941 quarterly, the amount of taxes due is usually deposited in a qualified depository (bank) monthly or more often, and it is only the difference between the monthly deposits and the total taxes due that is sent with the Form 941 report.

Computing an Employer's Federal and State Unemployment Tax Liability

6
LEARNING
OBJECTIVES

Compute an employer's federal and state unemployment tax liability.

In the preceding section, you learned that the employer must match the employee's contributions to Social Security and Medicare taxes. In addition, employers must pay two payroll taxes for federal and state unemployment programs.

The **Federal Unemployment Tax Act (FUTA)** requires the employer to pay a 6.2% tax on the first $7,000 paid to each employee to fund the federal unemployment compensation program for those who have lost their jobs. Most states have also passed a **State Unemployment Tax Act (SUTA)**, requiring the employer to pay 5.4% tax on the first $7,000 paid to each employee to fund state programs for the unemployed. This 5.4% state tax is *deductible* from the federal tax payment. Thus, in most cases, employers pay the federal government just 0.8% FUTA tax: 6.2% FUTA − 5.4% SUTA = 0.8% requirement.

EXAMPLE F

During the first quarter, Johnson and Johnson paid wages of $976,550.80. Of this amount, $172,400.60 was paid to employees who had been paid $7,000 earlier in the quarter. What was the employer's liability for FUTA and SUTA taxes, assuming that the state rate was 5.4%?

$976,550.80 − $172,400.60 = $804,150.20 subject to FUTA and SUTA taxes
$804,150.20 × 0.008 = $6,433.20 FUTA tax payment
$804,150.20 × 0.054 = $43,424.11 SUTA tax payment
$6,433.20 + $43,424.11 = $49,857.31

CONCEPT CHECK 8.6

Warner-Lambert Company employed Rojas Perez for 13 weeks during the period January 1 through March 31, 2001. His salary was $1,350 per week. At the end of the quarter, how much in FUTA and SUTA taxes did the company have to pay to the federal and state governments based on Rojas's income?

$1,350 per week × 13 weeks = $17,550 total wage
$7,000 maximum × 0.008 = $56 FUTA tax
$7,000 maximum × 0.054 = $378 SUTA tax
$378 + $56 = $434 total federal and state unemployment taxes

COMPLETE ASSIGNMENTS 8.1 AND 8.2.

Chapter Terms for Review

employee's earnings record
Employer's Quarterly Federal Tax Return
Federal Insurance Contributions Act (FICA)
Federal Unemployment Tax Act (FUTA)
Form W-4

payroll register
percentage method
State Unemployment Tax Act (SUTA)
wage-bracket method
withholding allowance

Visit the Web

http://www.americanpayroll.org/tools.html—American Payroll Association—site of the national association for payroll professionals.

http://www.employersonline.com—Employers Online, Inc.—a major site that brings employers and prospective employees together.

http://www.businesstown.com—Business Town—a strong site for new small business owners with considerable information about payroll items.

http://www.irs.gov—Internal Revenue Service—all encompassing site for Internal Revenue Service.

http://www.ssa.gov—Social Security Administration—all encompassing site for Social Security Administration.

http://www.dol.gov—U.S. Department of Labor—all encompassing site for U.S. Department of Labor.

http://www.eeoc.gov—Equal Employment Opportunity Commission—Full information about rules and regulations concerning Equal Employment Opportunity laws.

http://www.tax.gov—Simplified Tax and Wage Reporting System (STAWRS)—site providing various reporting systems for reporting employment taxes and wages.

CYBERCONNECTION

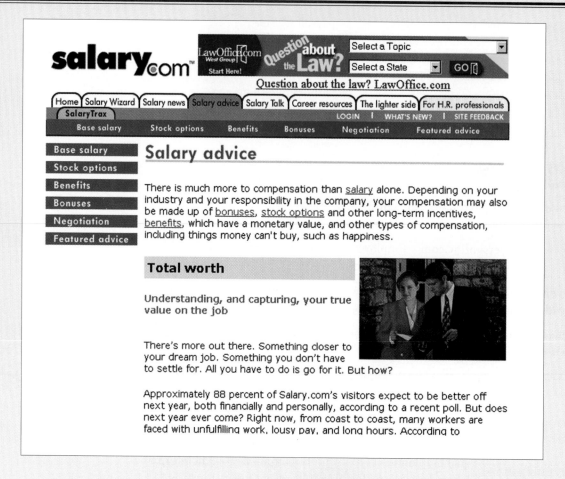

Visit salary.com and click on the Salary Wizard to determine salaries in your area for a specific job you might be interested in. For example, if you live in Fort Wayne, Indiana, and you're interested in an accounting clerk position, follow the directions on the Salary Wizard screen and you'll find that as of January 2001, this job paid a base salary of between $19,606 and $26,803 in that city.

Next, go to the Salary Advice page, shown here, and read about the benefits that can be added to a salary. As an accounting clerk, you would expect to receive at least a benefits package and holiday pay, depending on the size of the company for which you work. Read about benefits and think about how much they might add to your base salary.

Source: http://www.salary.com. Copyright © 2001 Courtesy of Salary.com.

CYBERCONNECTION

Occupational Outlook Handbook

| OOH Home | A-Z Index | Other OOH Info | Order |

Search by occupation: [_____] GO

Administrative Support, Including Clerical Occupations

Search Tips

Payroll and Timekeeping Clerks

Nature of the Work | Employment | Job Outlook

Download a printer-friendly version
(PDF 141K)

Nature of the Work [About this section] ▲ Top

Payroll and timekeeping clerks perform a vital function—ensuring that employees are paid on time and that their paychecks are accurate. If inaccuracies arise, such as monetary errors or incorrect amounts of vacation time, these workers research and correct the records. In addition, they may also perform various other clerical tasks.

The fundamental task of *timekeeping clerks* is distributing and collecting timecards each pay period. They review employee workcharts, timesheets, and timecards, to ensure that information is properly recorded and that records have the signatures of authorizing officials. In companies that bill for the time spent by staff, such as law or accounting firms, timekeeping clerks make sure the hours recorded are charged to the correct job, so clients can be properly billed. These clerks also review computer reports listing timecards that cannot be processed because of errors, and they contact the employee or the employee's supervisor to resolve the problem. In addition, timekeeping clerks are responsible for informing managers and other employees of procedural changes in payroll policies.

Payroll clerks, also called payroll technicians, screen timecards for calculating, coding, or other errors. They compute pay by subtracting allotments, including Federal and State taxes, retirement, insurance, and savings, from gross earnings. Increasingly, computers perform these calculations and alert payroll clerks to problems or errors in the data. In small organizations, or for new employees whose records are not yet entered into a computer system, clerks may perform the necessary calculations manually. In some small offices, clerks or other employees in the accounting department process payroll.

Source: http://stats.bls.gov/oco/ocos149.htm.

The Bottom Line

Summary of chapter learning objectives:

| Learning Objective | Summary | Example |
|---|---|---|
| **8.1** | Prepare a payroll register | Based on the data presented, complete the following payroll register. Fill out the total wages section and then compute the federal income tax, Social Security, Medicare, and other withholdings. Total all columns and check. Use the percentage method. |
| **8.2** | Compute federal income tax withholding amounts | 1. G. Lee is paid $14.20 per hour. He works 40 regular hours and 6 overtime hours during the week ending January 7. He is single and claims one withholding allowance. He takes a weekly medical deduction of $7.

2. E. Berg is paid $13 per hour. He worked 40 regular hours and 8 overtime hours during the week of January 7. He is married and claims 4 withholding allowances. He takes a weekly medical deduction of $15. |
| **8.3** | Compute Social Security, Medicare, and other withholdings | |

| Name | Marital Status | W/H Allow | Total Hours | Regular Earnings | | Overtime Earnings | | | Total Wages | Deductions | | | | | Net Pay |
|---|---|---|---|---|---|---|---|---|---|---|---|---|---|---|---|
| | | | | Rate per Hour | Amt | Hours Worked | Rate per Hour | Amt | | Social Security | Medi-care | Fed. Inc. Tax | Med. Insurance | Total | |
| | | | | | | | | | | | | | | | |
| | | | | | | | | | | | | | | | |
| | | | | | | | | | | | | | | | |

| Learning Objective | Summary | Example |
|---|---|---|
| **8.4** | Complete an employee's earnings record | 3. Complete the earnings record for D. Chan. Use 6.2% for Social Security and 1.45% for Medicare taxes. Use the percentage method for federal income tax withholding. |

Name **D. Chan** Social Security No. **125-11-3296**

Address **7821 Oak Ave.** No. of Allowances **1** Marital Status **Married**

| Period Ending | Total Wages | Cumulative Total | Deductions | | | | | Net Pay |
|---|---|---|---|---|---|---|---|---|
| | | | Social Security | Medicare | Federal Inc. Tax | Other Deductions | Total | |
| 1/31 | $ 3,100 | $ 3,100 | | | | $ 18.00 | | |
| 2/28 | 3,000 | 6,100 | | | | 18.00 | | |
| 3/31 | 3,450 | $9,550 | | | | 18.00 | | |
| Quarter Total | $9,550 | | | | | $ 54.00 | | |

The Bottom Line

Continued

Summary of chapter learning objectives:

| Learning Objective | Summary | Example |
|---|---|---|
| **8.5** | Compute an employer's quarterly federal tax return | 4. The Frazer Company had a total payroll of $279,440 for the first quarter of the year. It withheld $29,700 for federal income tax. It made monthly tax deposits of $24,100. Frazer is now filing its quarterly Form 941. Complete the following to determine the amount of the check that Frazer must send to the IRS for un-deposited taxes due.
a. Social Security tax due for the quarter _____
b. Medicare tax due for the quarter _____
c. Total taxes due for the quarter _____
d. Total deposits for the quarter _____
e. Undeposited taxes due IRS _____ |
| **8.6** | Compute an employer's federal and state unemployment tax liability | 5. Miller Outfitters employed R. Rehnquist for the period from January 1 through March 31, 13 weeks, at a salary of $1,230 per week. At the end of the quarter, how much in FUTA and SUTA taxes are owed to the federal and state governments if the state had a 0.8% FUTA rate and a 5.4% SUTA rate?
a. Total wages
b. FUTA tax
c. SUTA tax
d. Total federal and state unemployment taxes paid |

Answers: 1 and 2. Lee: Reg Earn $568; O/T Earnings $127.80; Total $695.80; Deductions: SS—$43.14; MC—$10.09; FIT—$99.80; MI—$7; Total Deductions $160.03; Net pay $535.71 Berg: Reg Earn $520; O/T Earnings 156.00; Total $676.00; Deductions: SS—$41.91; MC—$9.80; FIT—$49.34; MI—$15. Total Deductions 116.05; Net Pay $559.95 **3.** 1/31: $192.20; $44.95; $348.05; $603.65; $2,496.35 2/28: $186.00; $43.50; $330.05; $577.55; $2,422.45 3/31: $213.90; $50.03; $400.55; $682.48; $2,767.52 Totals: $592.10; $138.48; $1,078.65; $54.00; $1,863.68; $7,686.32 **4. a.** $34,650.56 **b.** 8,103.76 **c.** 72,454.32 **d.** 72,300.00 **e.** $154.32 **5. a.** $15,990.00 **b.** 56.00 **c.** 378.00 **d.** $434.00

Self-Check | Review Problems for Chapter 8

1 Alex Muñoz is paid $15 per hour for the first 40 hours and one and one-half times his regular rate for all hours worked over 40 per week.

a. Determine Alex's gross pay for the week if he works 45 hours.

b. Calculate the amount to be deducted for Social Security and Medicare taxes for the week.

c. Determine the amount to be withheld for federal income tax, using percentage method, if Alex is single and claims one withholding allowance.

d. What is Alex's net pay for the week, assuming that his only payroll deductions are for Social Security, Medicare, and federal income tax?

2 Determine the amount to be withheld for federal income tax for each of the following, using both the percentage and the wage-bracket methods.

a. A married employee, claiming 2 allowances, has weekly gross pay of $650.

b. A single employee, with 1 allowance, has weekly gross pay of $525.

3 Calculate the employer's payroll taxes for each of the first 3 months of the year for three employees who are paid as follows:

| | |
|---|---|
| Albertson, K. | $3,000 per month |
| Becket, W. | $4,000 per month |
| Jones, C. | $2,100 per month |

Include FUTA (0.8%), SUTA (5.4%), Social Security (6.2%), and Medicare (1.45%) taxes. Be sure to consider the maximum taxable for unemployment taxes ($7,000) per employee.

4 Determine the taxes to be reported on the quarterly 941 form for an employer who paid total gross wages of $62,000 and withheld $7,800 for federal income tax.

| | |
|---|---|
| Social Security | _____ |
| Medicare | _____ |
| Federal income tax | _____ |
| Total | _____ |

5 Determine the amount to be withheld from the current period's gross pay of $6,500 for Social Security and Medicare for an employee whose cumulative wages were $78,500, not including pay for the current period. Use the rates and taxable maximum given in the chapter.

6 Employees of Xper Co. are paid at their regular rate for the first 40 hours, at one and one-half times their regular rate for hours worked between 40 and 48, and double their regular rate for all hours worked over 48, per week. Calculate each employee's gross pay for the week.

John Kowalski, regular rate $12.16, worked 47 hours

Martha Madison, regular rate $9.50, worked 50 hours

Joy Weston, regular rate $10.80, worked 42 hours

Assignment 8.1: Payroll Problems

Name _____

Date _____ Score _____

A **(52 points) Complete the payroll. (1 point for each correct answer)**

1. In this company, employees are paid $1\frac{1}{2}$ times their regular rate for overtime hours between 40 and 48 and 2 times their regular rate for overtime hours over 48, per week.

| Name | Total Hours | Regular Rate Per Hour | Regular Earnings | | Time and a Half | | Double Time | | Total Earnings |
|---|---|---|---|---|---|---|---|---|---|
| | | | Hours | Amount | Hours | Amount | Hours | Amount | |
| Avila, Susan | 49 | 9.00 | 40 | | 8 | | 1 | | |
| Carter, Dale | 40 | 8.00 | 40 | | — | | — | | |
| Kula, Mary | 50 | 10.00 | 40 | | 8 | | 2 | | |
| Murphy, Tom | 45 | 9.00 | 40 | | 5 | | — | | |
| Norton, Alice | 40 | 8.80 | 40 | | — | | — | | |
| Payton, Alan | 35 | 8.00 | 35 | | — | | — | | |
| Perry, Lance | 47 | 8.00 | 40 | | 7 | | — | | |
| Polar, Barbara | 41 | 9.00 | 40 | | 1 | | — | | |
| Quinn, Carl | 49 | 8.80 | 40 | | 8 | | 1 | | |
| Reston, Sally | 40 | 8.80 | 40 | | — | | — | | |
| Sacco, Dom | 50 | 9.50 | 40 | | 8 | | 2 | | |
| Warren, Bill | 44 | 10.00 | 40 | | 4 | | — | | |
| TOTALS | | | | | | | | | |

Score for A (52) _____

B **(28 points) Solve the following problems. (7 points for each correct answer)**

2. Joy Agliano is employed at a monthly salary of $3,000. How much is deducted from her monthly salary for FICA taxes (Social Security and Medicare)? _____

3. Coleen Klein is employed by a company that pays her $3,600 a month. She is single and claims one with-holding allowance. What is her net pay after Social Security, Medicare, and federal income tax withholding? Use the percentage method. _____

Assignment 8.1 *Continued*

4. On April 1, the company in Problem 3 changed its pay plan from monthly to weekly and began paying Coleen $830.77 per week. What is her net weekly pay after Social Security, Medicare, and income tax deductions? Use the percentage method. _____

5. William Diggs is married and claims four withholding allowances. His weekly wages are $725. Calculate his Social Security and Medicare deductions and, using the wage-bracket method, his federal income tax withholding. Find his weekly net pay. _____

<div align="right">

Score for B (28)
</div>

C **(20 points) Compute and compare the federal income tax withholding amounts for each of the following individuals using the percentage method and the wage-bracket method. (Follow the steps in Section 8.2 for the percentage method.) (5 points for each correct difference)**

6. Ralph Carson: weekly wages, $320; single; 1 withholding allowance
Percentage method: _____
Wage-bracket method: _____
Difference: _____

7. George Wilson: weekly wages, $445; married; 3 withholding allowances
Percentage method: _____
Wage-bracket method: _____
Difference: _____

8. Mary Suizo: weekly wages, $292; single; 2 withholding allowances
Percentage method: _____
Wage-bracket method: _____
Difference: _____

9. Josephine Creighton: weekly wages, $375; married; 1 withholding allowance
Percentage method: _____
Wage-bracket method: _____
Difference: _____

<div align="right">

Score for C (20)
</div>

Assignment 8.2: Payroll, Earnings Record, Payroll Tax Returns

LEARNING OBJECTIVES

Name _____

Date _____ Score _____

A **(40 points) Solve the following problems. (1 point for each correct answer in the Total Wages column in 1; 2 points for each correct answer in the Net Pay column in 1 and 2)**

1. Complete the following weekly payroll register. Workers receive overtime pay for any time worked in excess of 40 hours per week at the rate of one and one-half their regular rates per hour. The deduction for Social Security is 6.2%, and the deduction for Medicare taxes is 1.45%. Use the wage-bracket method for federal income tax withholding. Be sure to use the correct table based on the marital status of each employee.

| Name | Marital Status | W/H Allow. | Total Hours | Regular Earnings Rate Per Hour | Regular Earnings Amount | Overtime Earnings Hours Worked | Overtime Earnings Rate Per Hour | Overtime Earnings Amount | Total Wages | Deductions Social Security | Deductions Medi-care | Deductions Fed. Inc. Tax | Deductions Med Ins. | Total | Net Pay |
|---|---|---|---|---|---|---|---|---|---|---|---|---|---|---|---|
| Adams, C. | M | 2 | 40 | $10.60 | | — | | | | | | | $ 15.00 | | |
| Brown, W. | S | 1 | 42 | 9.80 | | 2 | | | | | | | 12.00 | | |
| Davids, K. | S | 0 | 32 | 12.40 | | — | | | | | | | 12.00 | | |
| Giles, A. | M | 3 | 40 | 15.00 | | — | | | | | | | 18.00 | | |
| Martin, B. | M | 2 | 48 | 12.50 | | 8 | | | | | | | 18.00 | | |
| Nelson, R. | M | 4 | 43 | 14.60 | | 3 | | | | | | | 18.00 | | |
| Paulson, J. | S | 1 | 40 | 9.80 | | — | | | | | | | 12.00 | | |
| Riley, S. | S | 1 | 42 | 12.00 | | 2 | | | | | | | 12.00 | | |
| Wilson, D. | M | 2 | 40 | 12.50 | | — | | | | | | | 15.00 | | |
| TOTALS | | | | | | | | | | | | | $132.00 | | |

2. The total weekly wages of four employees are listed in the following table. Determine the amount of the deductions and the net pay due to each employee. Use 6.2% for the Social Security and 1.45% for the Medicare tax deductions; use the percentage method for federal income tax withholding. Determine the deductions and totals.

| Name | Marital Status | W/H Allow. | Total Wages | Deductions Social Security | Deductions Medicare | Deductions Federal Income Tax | Total | Net Pay |
|---|---|---|---|---|---|---|---|---|
| Alvarez, George | S | 1 | $ 620.00 | | | | | |
| Coburn, Carol | M | 3 | 480.00 | | | | | |
| Garnett, Mildred | S | 1 | 505.00 | | | | | |
| Kline, Marvin | M | 2 | 700.00 | | | | | |
| TOTALS | | | | | | | | |

Score for A (40)

Assignment 8.2 *Continued*

B (20 points) Solve the following problems. (1 point for each correct weekly answer in the Net Pay column and 2 points for the correct quarter total of that column in 1; 1 point for each correct answer in 2)

3. Complete the employee's earnings record for Michael Carter. Use 6.2% for Social Security and 1.45% for Medicare taxes. Use the percentage method for federal income tax withholding.

Name __Michael Carter__ Social Security No. __125-55-1254__

Address __211 Main St__ No. of Allowances __2__ Marital Status __Married__

| Period Ending | Total Wages | Cumulative Wages | Social Security | Medicare | Federal Inc. Tax | United Fund | Total | Net Pay |
|---|---|---|---|---|---|---|---|---|
| 1/6 | $ 450.60 | $ 450.60 | | | | $ 4.00 | | |
| 1/13 | 412.00 | 862.60 | | | | 4.00 | | |
| 1/20 | 412.00 | 1,274.60 | | | | 4.00 | | |
| 1/27 | 475.50 | 1,750.10 | | | | 4.00 | | |
| 2/3 | 415.20 | 2,165.30 | | | | 4.00 | | |
| 2/10 | 490.25 | 2,655.55 | | | | 4.00 | | |
| 2/17 | 427.50 | 3,083.05 | | | | 4.00 | | |
| 2/24 | 435.90 | 3,518.95 | | | | 4.00 | | |
| 3/3 | 510.00 | 4,028.95 | | | | 4.00 | | |
| 3/10 | 505.60 | 4,534.55 | | | | 4.00 | | |
| 3/17 | 516.00 | 5,050.55 | | | | 4.00 | | |
| 3/24 | 498.50 | 5,549.05 | | | | 4.00 | | |
| 3/31 | 535.80 | 6,084.85 | | | | 4.00 | | |
| Quarter Totals | $6,084.85 | | | | | $ 52.00 | | |

4. The following is a summary of quarterly earnings of a company's employees. Determine the information requested for the employer's quarterly federal tax return.

| Name | Total Wages | Social Security | Medi-care | Fed. Inc. Tax |
|---|---|---|---|---|
| | | Taxes Withheld | | |
| Carter, M. | $ 6,084.85 | $ 377.27 | $ 88.22 | $ 451.42 |
| Davis, L. | 5,368.00 | 332.82 | 77.84 | 437.50 |
| Gordon, J. | 4,266.35 | 264.51 | 61.86 | 398.65 |
| McBride, C. | 7,230.00 | 448.26 | 104.84 | 595.80 |
| Taggert, L. | 6,240.50 | 386.91 | 90.49 | 465.50 |
| Walton, N. | 5,285.92 | 327.73 | 76.65 | 566.00 |
| TOTALS | | | | |

a. Total earnings paid _____

b. Federal income tax withheld _____

c. Total Social Security tax paid _____

d. Total Medicare tax paid _____

e. Total taxes withheld _____

Score for B (20)

Assignment 8.2 *Continued*

C **(40 points) Solve the following problems. (4 points for each correct answer in 5 and 6; 1 point for each correct answer in 7)**

5. The quarterly earnings of the employees of the Alpha Company are listed in the following table. Determine the information needed for the employer's quarterly federal tax return (Form 941).

| Name | Total Wages | Taxes Withheld | | |
|---|---|---|---|---|
| | | Social Security | Medicare | Fed. Inc. Tax |
| Caldwell, Janice | $ 3,420.00 | $ 212.04 | $ 49.59 | $ 423.90 |
| Dorman, J.A. | 3,600.00 | 223.20 | 52.20 | 473.67 |
| Eagie, T.W. | 4,016.50 | 249.04 | 58.24 | 433.33 |
| Fortune, Mark | 3,774.90 | 234.02 | 54.74 | 410.05 |
| Morris, Regina | 3,605.40 | 223.53 | 52.28 | 399.83 |
| Tracy, Joseph | 4,111.60 | 254.92 | 59.62 | 360.17 |
| TOTALS | | | | |

a. Total earnings paid _____

b. Employee's contribution of Social Security tax _____

c. Employee's contribution of Medicare tax _____

d. Federal income tax withheld from wages _____

e. Total taxes _____

6. The Primo Company had a total payroll of $148,600.34 for the first quarter of the current year. It withheld $28,531.27 from the employees for federal income tax during this quarter. The company made the following deposits in a qualified bank depository for the amount of the income and Social Security and Medicare taxes withheld from the employees and for the company's contribution to the FICA tax: $17,050 on February 6; $17,050 on March 4; and $17,050 on April 5. Primo Company's bookkeeper is now filling out Form 941 (quarterly return), which is due by the end of April. Complete the following to determine the amount of the check that the company must send to the IRS for the undeposited taxes due.

a. Total Social Security and Medicare taxes to be paid for quarter _____

b. Total taxes _____

c. Total deposits for quarter (sent to qualified bank depository) _____

d. Undeposited taxes due IRS _____

7. Jordan Mills employed Ruth Liebowitz for the period January 1 through March 31 (13 weeks), at a salary of $1,500 per week. At the end of the first quarter of the year, how much in FUTA and SUTA taxes did the company owe to the federal and state governments if the state had an 0.8% FUTA rate and a 5.4% SUTA rate?

a. Total wages and taxable wages _____ b. FUTA tax _____

c. SUTA tax _____ d. Total federal and state unemployment taxes paid _____

Score for B (20)

The Internal Revenue Service (IRS) is the federal department with which working adults have the most contact—through the income tax return.

The IRS has made efforts to simplify the taxpaying process. Taxpayers can receive more than 100 forms by fax or via the Internet and find answers to questions at the IRS home page at www.irs.ustreas.gov. In addition, the IRS provides pre-recorded telephone information 24 hours a day, 7 days a week and telephone assistance 10 hours each business day.

One of the features of the IRS home page is "The Digital Daily," a daily Internet publication that helps taxpayers find information. Visit the IRS Web site at http://www.irs.gov.

Federal Income Taxes

 9

LEARNING OBJECTIVES

By studying this chapter and completing all assignments you will be able to:

 1 Make the basic computations to determine taxable income for taxpayers who use the standard Form 1040.

 2 Make basic computations to determine taxes owed by taxpayers who use the standard Form 1040.

 3 Make basic computations to determine taxes owed by corporations and other businesses, based on their taxable incomes.

The federal income tax, which began to be collected in the United States in 1914 has become the dominant tax in the U.S. tax system. Today, almost all working people, businesses, and corporations pay taxes and are subject to IRS rules and regulations.

1. How do employers know how much income tax to withhold from an employee's wages?
2. Can you save money by knowing the deductions permitted by the IRS for income tax purposes?
3. Can you claim yourself as a dependent for income tax purposes?
4. How many people can claim the same person as a dependent on their income tax form in any given year?

Personal income taxes provide 48% of all income of the federal government. Social Security and Medicare taxes, which you studied in Chapter 8, provide another 34%. Together these three taxes make up 82% of all federal government income.

Outlays for Social Security, Medicare, and retirement programs constitute 35% of all government expenditures. Payment of interest on government debt represents 12% of all government expenditures.

Figure 9-1 shows the breakdown of federal government income and the allocation of federal government spending.

Determining Taxable Income, Using Standard Form 1040

LEARNING OBJECTIVES 1

Make the basic computations to determine taxable income for taxpayers who use the standard Form 1040.

Form 1040 is the basic form filed by the majority of taxpayers. There are two simplified variations of this form: Form 1040A and Form 1040EZ. The income tax calculation process is illustrated for Form 1040 in Figures 9-2 through 9-7. The label in Figure 9-2 contains spaces for names, address, and Social Security numbers, as well as boxes to check to designate $3 to finance presidential elections.

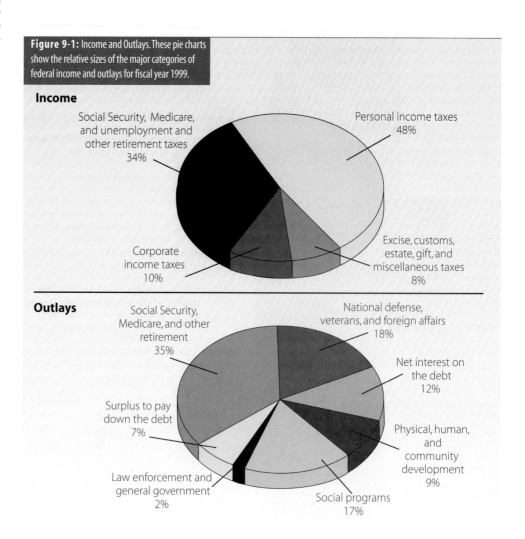

Figure 9-1: Income and Outlays. These pie charts show the relative sizes of the major categories of federal income and outlays for fiscal year 1999.

Income

Social Security, Medicare, and unemployment and other retirement taxes 34%

Personal income taxes 48%

Corporate income taxes 10%

Excise, customs, estate, gift, and miscellaneous taxes 8%

Outlays

Social Security, Medicare, and other retirement 35%

National defense, veterans, and foreign affairs 18%

Net interest on the debt 12%

Physical, human, and community development 9%

Surplus to pay down the debt 7%

Law enforcement and general government 2%

Social programs 17%

A taxpayer's current **filing status** is indicated in the second section of Form 1040, shown in Figure 9-3. Five choices are given. The one selected determines the tax rates the taxpayer uses, as well as many of the taxpayer's deductions.

Personal exemptions, shown in Figure 9-4, are reductions to taxable income for the primary taxpayer and a spouse. One **dependency exemption** is granted for each dependent. Exemptions are phased out for individuals with higher incomes: adjusted gross incomes above $96,700 for a married person filing separately, $161,150 for a head of household, $128,950 for a single person, or $193,400 for a married couple filing jointly. The amount deducted for each exemption is currently $2,800. This amount is usually adjusted for inflation each year.

Taxable income, shown in Figure 9-5, includes wages, salaries, tips, dividends, interest, commissions, back pay, bonuses and awards, refunds of state and local taxes, alimony received, property received for services, severance pay, accrued leave payments, sick pay,

Figure 9-2: Form 1040 Label Section

Figure 9-3: Form 1040 Filing Status Section

Figure 9-4: Form 1040 Exemptions Section

Figure 9-5: Form 1040 Income Section

| | | | | | | | | |
|---|---|---|---|---|---|---|---|---|
| **Income** | 7 | Wages, salaries, tips, etc. Attach Form(s) W-2 | 7 | 66,000 | 00 |
| | 8a | **Taxable** interest. Attach Schedule B if required | 8a | 800 | 00 |
| **Attach Forms W-2 and W-2G here. Also attach Form(s) 1099-R if tax was withheld.** | b | **Tax-exempt** interest. **Do not** include on line 8a . . | 8b | | |
| | 9 | Ordinary dividends. Attach Schedule B if required | 9 | | |
| | 10 | Taxable refunds, credits, or offsets of state and local income taxes (see page 22) . . | 10 | | |
| | 11 | Alimony received | 11 | | |
| | 12 | Business income or (loss). Attach Schedule C or C-EZ | 12 | | |
| | 13 | Capital gain or (loss). Attach Schedule D if required. If not required, check here ▶ ☐ | 13 | | |
| **If you did not get a W-2, see page 21.** | 14 | Other gains or (losses). Attach Form 4797 | 14 | | |
| | 15a | Total IRA distributions . | 15a | | **b** Taxable amount (see page 23) | 15b | | |
| | 16a | Total pensions and annuities | 16a | | **b** Taxable amount (see page 23) | 16b | | |
| **Enclose, but do not attach, any payment. Also, please use Form 1040-V.** | 17 | Rental real estate, royalties, partnerships, S corporations, trusts, etc. Attach Schedule E | 17 | 1,200 | 00 |
| | 18 | Farm income or (loss). Attach Schedule F | 18 | | |
| | 19 | Unemployment compensation | 19 | | |
| | 20a | Social security benefits . | 20a | | **b** Taxable amount (see page 25) | 20b | | |
| | 21 | Other income. List type and amount (see page 25) --------------------------------- | 21 | | |
| | 22 | Add the amounts in the far right column for lines 7 through 21. This is your **total income** ▶ | 22 | 68,000 | 00 |

| | | | | | |
|---|---|---|---|---|---|
| **Adjusted Gross Income** | 23 | IRA deduction (see page 27) | 23 | 4,000 | 00 |
| | 24 | Student loan interest deduction (see page 27) | 24 | | |
| | 25 | Medical savings account deduction. Attach Form 8853 . | 25 | | |
| | 26 | Moving expenses. Attach Form 3903 | 26 | | |
| | 27 | One-half of self-employment tax. Attach Schedule SE . | 27 | | |
| | 28 | Self-employed health insurance deduction (see page 29) | 28 | | |
| | 29 | Self-employed SEP, SIMPLE, and qualified plans . . | 29 | | |
| | 30 | Penalty on early withdrawal of savings | 30 | | |
| | 31a | Alimony paid **b** Recipient's SSN ▶ _____ | 31a | | |
| | 32 | Add lines 23 through 31a | 32 | 4,000 | 00 |
| | 33 | Subtract line 32 from line 22. This is your **adjusted gross income** ▶ | 33 | 64,000 | 00 |

For Disclosure, Privacy Act, and Paperwork Reduction Act Notice, see page 56. Cat. No. 11320B Form **1040** (2000)

unemployment compensation payments, capital gains, and any other income not specifically exempted by statute. Taxable income may include a portion of Social Security payments, IRA distributions, and pensions and annuities. It also includes income from businesses, professions, farming, partnerships, rents, royalties, estates, trusts, and other sources. It does not include income from gifts, inheritances, bequests, interest on tax-exempt state and local municipal bonds, life insurance proceeds at death, workers' compensation benefits, and certain income items for veterans.

The Adjustments to Income section, shown in Figure 9-6, allows the taxpayer to list certain items that are allowed as reductions to the total income. These adjustments include payments by the taxpayer or spouse to an individual retirement account (IRA), student loan interest, payments into a medical savings account, moving expenses, one half of self-employment tax paid, and payments to a retirement plan for the self-employed, penalty on early withdrawal of savings, and alimony paid. **Adjusted gross income** (AGI) is a taxpayer's income after subtraction of adjustments to income from total income. (See lines 32 and 33 of Adjusted Gross Income in Figure 9-6.)

After the adjusted gross income figure is computed, *deductions*—either the standard deduction or itemized deductions—are subtracted in order to figure taxable income (see Figure 9-7). The standard deductions for most taxpayers are shown in Figure 9-8. There are higher standard deductions for individuals who are 65 or over and for individuals who are blind; these are shown in Figure 9-9.

Figure 9-7: Form 1040 Taxable Income and Income Tax Section

Form 1040 (2000) Page **2**

Tax and Credits

| | | | |
|---|---|---|---|
| **34** | Amount from line 33 (adjusted gross income) | **34** | 64,000 00 |
| **35a** | Check if: ☐ **You** were 65 or older, ☐ Blind; ☐ **Spouse** was 65 or older, ☐ Blind. Add the number of boxes checked above and enter the total here . . . ▶ **35a** | | |
| **b** | If you are married filing separately and your spouse itemizes deductions, or you were a dual-status alien, see page 31 and check here ▶ **35b** ☐ | | |
| **36** | Enter your **itemized deductions** from Schedule A, line 28, **or standard deduction** shown on the left. **But** see page 31 to find your standard deduction if you checked any box on line 35a or 35b **or** if someone can claim you as a dependent | **36** | 7,350 00 |
| **37** | Subtract line 36 from line 34 | **37** | 56,650 00 |
| **38** | If line 34 is $96,700 or less, multiply $2,800 by the total number of exemptions claimed on line 6d. If line 34 is over $96,700, see the worksheet on page 32 for the amount to enter . | **38** | 11,200 00 |
| **39** | **Taxable income.** Subtract line 38 from line 37. If line 38 is more than line 37, enter -0- . | **39** | 45,450 00 |
| **40** | **Tax** (see page 32). Check if any tax is from **a** ☐ Form(s) 8814 **b** ☐ Form 4972 . . . | **40** | 5,036 00 |

Figure 9-8: Standard Deduction Chart for Most People

Standard Deduction for Most People

Single:
$4,400

Head of household:
$6,450

Married filing jointly or Qualifying widow(er):
$7,350

Married filing separately:
$3,675

36 Enter your **itemized deductions** from Schedule A, line 28, **or standard deduction** shown on the left. **But** see page 31 to find your standard deduction if you checked any box on line 35a or 35b **or** if someone can claim you as a dependent **36**

Standard Deduction Chart for People Age 65 or Older or Blind—Line 36

If someone can claim you (or your spouse if married filing jointly) as a dependent, use the worksheet below instead.

Enter the number from the box on line 35a of Form 1040 ▶ [] ⚠ CAUTION Do not use the number of exemptions from line 6d.

| IF your filing status is . . . | AND the number in the box above is . . . | THEN your standard deduction is . . . |
|---|---|---|
| Single | 1 | $5,500 |
| | 2 | 6,600 |
| Married filing jointly or Qualifying widow(er) | 1 | $8,200 |
| | 2 | 9,050 |
| | 3 | 9,900 |
| | 4 | 10,750 |
| Married filing separately | 1 | $4,525 |
| | 2 | 5,375 |
| | 3 | 6,225 |
| | 4 | 7,075 |
| Head of household | 1 | $7,550 |
| | 2 | 8,650 |

Some taxpayers choose to itemize deductions rather than use the IRS-approved standard deduction. **Itemized deductions** are deductions allowed for specific payments made by the taxpayer during the tax year. These deductions include charitable contributions, certain interest payments, state and local income and property taxes, a portion of medical and dental expenses, casualty and theft losses, tax preparation fees, and other annually identified deductions. Illustrations, examples, and problems in this book are based on the assumption that all state and local taxes and all donations to charity are deductible.

Computing Taxable Income

Line 39 of Form 1040 shows "taxable income." Taxable income is the amount of income on which the income tax is based. Taxable income for most taxpayers is computed as follows (amounts from the preceding figures):

| | |
|---|---:|
| Total income (income from all sources) (Line 22) | $68,000 |
| Less adjustments to income (reductions of Total Income) (Line 32) | 4,000 |
| Adjusted gross income (Line 33) | 64,000 |
| Less deductions (from Figure 9-8 or 9-9) | 7,350 |
| Less exemptions (Line 6d × $2,800, per Line 38) | 11,200 |
| Taxable income (the amount on which taxes are computed) (Line 39) | $45,450 |

CONCEPT CHECK 9.1

Catherine, a 72-year-old blind widow, had an annual adjusted gross income of $25,000. She filed a return claiming a single exemption and standard deduction. No income tax had been withheld during the year. What was the total income tax due?

| | |
|---|---:|
| Adjusted gross income | $ 25,000 |
| Standard deduction: single, over 65, blind | 6,600 |
| | 18,400 |
| Minus 1 exemption | 2,800 |
| Taxable income | $ 15,600 |
| From Schedule X: | |
| $15,600 × (0.15) | $ 2,340 |

www.irs.treas.gov/
prod/ind_info/webw4/
index.html

The IRS has an exemption calculator that you can use to figure the number of exemptions to include on your Form W-4.

www.irs.treas.gov/
prod/ind_info/
tax_tables/index.html

Check the tax tables to find out how much tax you should have paid during the preceding year.

Determining Taxes Due, Using Standard Form 1040

2
LEARNING
OBJECTIVES

Make basic computations to determine taxes owed by taxpayers who use the standard Form 1040.

Taxes are computed from taxable income (Line 39). **Tax Rate Schedules** (Figure 9-10) show the tax rate for (1) single, (2) married filing joint return (even if only one had income), (3) married filing separate return, (4) head of household, and (5) qualifying widow or widower. The Tax Rate Schedules shown are used for all illustrations, examples, and problems in this book.

The remaining sections of Form 1040 permit listing of special credits, other taxes, and payments, to arrive at the final refund or amount owed, and have spaces for signatures of the taxpayers, and of paid preparers.

Figure 9-10: Tax Rate Schedules

Schedule X—Use if your filing status is **Single**

| If the amount on Form 1040, line 39, is: Over- | But not over- | Enter on Form 1040, line 40 | of the amount over- |
|---|---|---|---|
| $0 | $26,250 | 15% | $0 |
| 26,250 | 63,550 | $3,937.50 + 28% | 26,250 |
| 63,550 | 132,600 | 14,381.50 + 31% | 63,550 |
| 132,600 | 288,350 | 35,787.00 + 36% | 132,600 |
| 288,350 | | 91,857.00 + 39.6% | 288,350 |

Schedule Y-1—Use if your filing status is **Married filing jointly** or **Qualifying widow(er)**

| If the amount on Form 1040, line 39, is: Over- | But not over- | Enter on Form 1040, line 40 | of the amount over- |
|---|---|---|---|
| $0 | $43,850 | 15% | $0 |
| 43,850 | 105,950 | $6,577.50 + 28% | 43,850 |
| 105,950 | 161,450 | 23,965.50 + 31% | 105,950 |
| 161,450 | 288,350 | 41,170.50 + 36% | 161,450 |
| 288,350 | | 86,854.50 + 39.6% | 288,350 |

Schedule Y-2—Use if your filing status is **Married filing separately**

| If the amount on Form 1040, line 39, is: Over- | But not over- | Enter on Form 1040, line 40 | of the amount over- |
|---|---|---|---|
| $0 | $21,925 | 15% | $0 |
| 21,925 | 52,975 | $3,288.75 + 28% | 21,925 |
| 52,975 | 80,725 | 11,982.75 + 31% | 52,975 |
| 80,725 | 144,175 | 20,585.25 + 36% | 80,725 |
| 144,175 | | 43,427.25 + 39.6% | 144,175 |

Schedule Z—Use if your filing status is **Head of household**

| If the amount on Form 1040, line 39, is: Over- | But not over- | Enter on Form 1040, line 40 | of the amount over- |
|---|---|---|---|
| $0 | $35,150 | 15% | $0 |
| 35,150 | 90,800 | $5,272.50 + 28% | 35,150 |
| 90,800 | 147,050 | 20,854.50 + 31% | 90,800 |
| 147,050 | 288,350 | 38,292.00 + 36% | 147,050 |
| 288,350 | | 89,160.00 + 39.6% | 288,350 |

www.irs.treas.gov/ ind_info/tax_trails/ index.html

For an interactive tax quiz, visit this Web site.

© PHOTODISC, INC.

EXAMPLE A

For the Form 1040 illustrated in the text, the tax is computed as follows:

| | |
|---|---|
| Line 39—Taxable income | $45,450 |
| From Schedule Y-1 (married): | |
| Tax on $43,850 | $ 6,577.50 |
| Plus 28% of amount over $43,850 | |
| $45,450 − $43,850 = $1,600 × 0.28 | 448.00 |
| Total tax | $ 7,025.50 rounded to $7,026 |

Filing as head of household, Ken has an adjusted gross income of $110,000. He itemizes the following deductions: $700 to Salvation Army, $900 to his church, $8,200 interest on his mortgage, and $3,300 state taxes. He claims two exemptions. Compute his federal tax. Round to the nearest dollar.

| | | |
|---|---|---|
| Adjusted gross income | $110,000 | |
| Minus itemized deductions | 13,100 | |
| | 96,900 | |
| Minus 2 exemptions | 5,600 | |
| Taxable income | $ 91,300 | |
| From Schedule Z: | | |
| Tax on $90,800 | $ 20,854.50 | |
| Plus 31% (0.31) of excess over $90,800 | | |
| $91,300 − $90,800 = $500 × 0.31 | 155.00 | |
| Total tax | $ 21,009.50 | rounded to $21,010 |

Tax Credits and Net Tax

Credits allowed are subtracted from the tax to calculate the net tax. One of the most common credits is the Child Tax Credit (line 47). Taxpayers with dependent children under age 17 can receive a credit of $500 per qualifying child. The credit phases out at higher income levels.

Figure 9-11 shows that John and Mary Sample received a Child Tax Credit of $500. Look back at Figure 9-4 and note a check mark in the "qualifying child" box for Johnny Sample but not for Maria Sample. This distinction means that the son qualified for the credit because he was under age 17. The daughter qualifies as a dependent for exemption purposes, but no Child Tax Credit is allowed because she is age 17 or older.

Figure 9-11: Form 1040 Credits Section

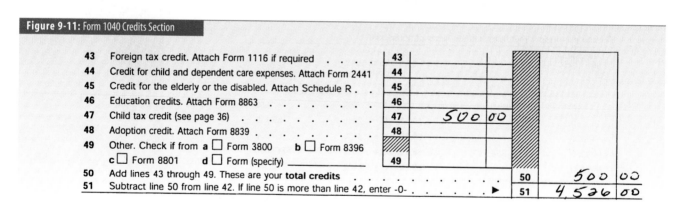

EXAMPLE C

Eric and Audrey Vaughn file a joint return. Their adjusted gross income is $55,650, and they take the standard deduction. They have three children, aged 12, 15, and 17, and claim five exemptions. Compute their net federal income tax after credits.

| | | |
|---|---|---|
| Adjusted gross income | $55,650 | |
| Standard deduction (joint) | 7,350 | |
| | 48,300 | |
| Minus 5 exemptions × $2,800 | 14,000 | |
| Taxable income | $34,300 | |

From Schedule Y-1:

| | |
|---|---:|
| $34,300 × 0.15 = | $ 5,145 |
| Minus child tax credit ($500 × 2) | 1,000 |
| Net tax after credits | $ 4,145 |

CONCEPT CHECK 9.2

Brian and Margaret Lee had wages of $43,200, interest income of $2,400, dividend income of $3,900, and business income of $9,500. They put $2,000 into a deductible IRA. They filed a joint return—claiming three exemptions (Brian, Margaret, and their daughter, aged 5)—and used the standard deduction. During the year, $5,200 in federal income tax had been withheld from their wages. What was the total tax due with the return?

| | |
|---|---:|
| Total income | $59,000 |
| Adjustments to income: IRA deduction | 2,000 |
| Adjusted gross income | 57,000 |
| Standard deduction: married, filing jointly | 7,350 |
| | 49,650 |
| Minus 3 exemptions: 3 × $2,800 | 8,400 |
| Taxable income | $41,250 |
| From Schedule Y-1: | |
| $41,250 × 0.15 = $6,187.50, rounded to | $ 6,188 |
| Minus child tax credit for one child | 500 |
| Net tax after credits | $ 5,688 |
| Minus federal income tax withheld | 5,200 |
| Total tax due | $ 488 |

Determining Taxes for Businesses

For a **sole proprietorship**—a business that has only one owner and is not incorporated—the owner includes the net income or loss from the business on Form 1040 and attaches a substantiating form: Schedule C, Profit or Loss from Business.

For a **partnership**—a business that is not incorporated and has more than one owner—each partner must include his or her share of the income or loss on his or her individual Form 1040.

A **for-profit corporation**—a business that operates for a profit and whose owners are stockholders—must file a special corporate return regardless of the amount of taxable income or loss, even if no tax is due or refundable. The form most commonly used is IRS Form 1120. Most corporations figure their taxes by using a Tax Rate Schedule. Figure 9-12 shows the latest Corporate Tax Rate Schedules used for this book.

LEARNING OBJECTIVES

Make basic computations to determine taxes owed by corporations and other businesses, based on their taxable incomes.

Figure 9-12: Tax Rate Schedule—Corporate

Tax Rate Schedule

If taxable income (line 30,
Form 1120, or line 26, Form
1120-A) on page 1 is:

| Over— | But not over— | Tax is: | Of the amount over— |
|---|---|---|---|
| $0 | $50,000 | 15% | $0 |
| 50,000 | 75,000 | $ 7,500 + 25% | 50,000 |
| 75,000 | 100,000 | 13,750 + 34% | 75,000 |
| 100,000 | 335,000 | 22,250 + 39% | 100,000 |
| 335,000 | 10,000,000 | 113,900 + 34% | 335,000 |
| 10,000,000 | 15,000,000 | 3,400,000 + 35% | 10,000,000 |
| 15,000,000 | 18,333,333 | 5,150,000 + 38% | 15,000,000 |
| 18,333,333 | - - - - - | 35% | 0 |

EXAMPLE D

In 200X, Steel Structures Corp. had taxable income of $72,000. What was the company's total income tax for the year?

| | |
|---|---|
| Tax on $50,000–$75,000 | $ 7,500 |
| Plus $72,000 − $50,000 = $22,000 × 0.25 | 5,500 |
| Total income tax | $13,000 |

EXAMPLE E

In the following year, Steel Structures Corp. had taxable income of $13,546,753. What was the amount of income tax owed? Round to the nearest dollar.

| | |
|---|---|
| Tax on $10,000,000–$15,000,000 | $3,400,000 |
| Plus $13,546,753 − $10,000,000 = $3,546,753 × 0.35 | $1,241,363.55 |
| Total income tax | $4,641,363.55 |
| Rounded to | $4,641,364 |

**www.irs.treas.gov/
bus_info/index.html**

At the IRS Web page for corporate taxes, you'll find answers to FAQs, small-business tax information, employee retirement plans, a business library, and much more.

CONCEPT CHECK 9.3

a. If the Digby Construction Company had taxable income of $22,250,500 last year, what was the amount of income tax owed?

$22,250,500 × 0.35% = $7,787,675

b. The Pearl Diving Corporation had a taxable income of $82,100 last year. What was the amount of its income tax owed?

| | |
|---|---|
| Tax on $75,000–$100,000 | $13,750 |
| Plus $82,100 − $75,000 = $7,100 × 0.34 | $ 2,414 |
| Income tax | $16,164 |

COMPLETE ASSIGNMENTS 9.1 AND 9.2.

Chapter Terms for Review

adjusted gross income (AGI)

child tax credit

filing status

Form 1040

for-profit corporation

itemized deductions

partnership

personal exemptions

sole proprietorship

Tax Rate Schedules

taxable income

Visit the Web

http://www.tax.org/defaultf.htm—Articles and information on all types of taxes are available here.

http://www.taxpenalty.com/index.htm—Time Value Software's tax penalty help site offers more tax information.

http://www.tax.state.ny.us/Statistics/Policy-Special/Sourcebook00/Sourcebook00_Contents.htm—Total state taxes for each state are available here.

http://www.ncsl.org/programs/fiscal/tctelcom.htm and **http://www.caltax.org/ecommerce.htm**—Visit this site for information on Internet taxation.

Among the many additional Web sites on taxes are:

| | |
|---|---|
| **www.irs.ustreas.gov** | **www.taxsys.com** |
| **www.taxweb.com** | **www.irs.gov** |
| **www.hrblock.com/tax** | **www.taxresources.com** |
| **www.ftb.ca.gov/forms** | **www.completetax.com** |
| **www.taxsites.com** | **www.e1040.com** |
| **www.ctj.org** | **www.taxfoundation.org** |
| **www.etaxforms.com** | **www.nite.org** |

CYBERCONNECTION

The Bush Tax Relief Plan
Executive Summary, Washington, D.C.

The President has proposed a bold and fair tax relief plan that will reduce the inequities of the current tax code and help ensure that America remains prosperous. This tax relief plan promotes the values that make the American economy second to none—access to the middle class, family, equal opportunity, and the entrepreneurial spirit. This plan will reduce taxes for everyone who pays income taxes, and it will encourage enterprise by lowering marginal tax rates.

The following is a portion of President Bush's tax relief plan. You can read the full plan at http://www.bushtaxrelief.com/about.asp and compare the tax rate tables.

When President Bush's proposal is fully in place, the typical family with two children will receive at least $1,600 in tax relief. This is real and practical help:

Sixteen hundred dollars will pay the average mortgage for almost two months;

Sixteen hundred dollars will pay for a year's tuition at a community college;

Sixteen hundred dollars will pay the gasoline cost for two cars for a year; and

Sixteen hundred dollars will buy an average family 24 months worth of electric power.

Tax Rates by 2001 Taxable Income*

| Current Code | | | | Bush Plan** | | |
|---|---|---|---|---|---|---|
| **Single** | | | | **Single** | | |
| $0 | $27,050 | 15% | | $0 | $6,000 | 10% |
| $27,050 | $65,550 | 28% | | $6,000 | $27,050 | 15% |
| $65,550 | $136,750 | 31% | | $27,050 | $136,750 | 25% |
| $136,750 | $297,350 | 36% | | $136,750 | — | 33% |
| $297,350 | — | 39.6% | | | | |
| **Head of Household** | | | | **Head of Household** | | |
| $0 | $36,350 | 15% | | $0 | $10,000 | 10% |
| $36,250 | $93,650 | 28% | | $10,000 | $36,250 | 15% |
| $93,650 | $151,650 | 31% | | $36,250 | $151,650 | 25% |
| $151,650 | $297,350 | 36% | | $151,650 | — | 33% |
| $297,360 | — | 39.6% | | | | |
| **Married—Joint Filing** | | | | **Married—Joint Filing** | | |
| $0 | $45,200 | 15% | | $0 | $12,000 | 10% |
| $45,200 | $109,250 | 28% | | $12,000 | $45,200 | 15% |
| $109,250 | $166,500 | 31% | | $45,200 | $166,500 | 25% |
| $166,500 | $297,350 | 36% | | $166,500 | — | 33% |
| $297,350 | — | 39.6% | | | | |

*Taxable income is income less deductions and personal exemptions.
** Rate schedule assumes tax plan is fully phased in.

If you are a head of household with a taxable income of $42,000 a year, what is your tax rate before the plan? What is the rate proposed in the plan? How much will you save with the plan? Calculate some other incomes and taxes and see which income levels will benefit the most.

Source: http://www.bushtaxrelief.com. Copyright © 2001. Courtesy of National Republican Committee.

The Bottom Line

Summary of chapter learning objectives:

| Learning Objective | Summary | Example |
|---|---|---|
| **9.1** | Make the basic computations to determine taxable income for taxpayers who use the standard Form 1040 | 1. Gene Cane is 28 years old and single. He claimed one exemption. In 200X he earned $47,000 in wages and $300 in taxable interest income. He also received dividend income of $375. During the year he invested $1,800 in an individual retirement account. Because of a change of jobs, he also had $1,200 in moving expenses, which qualified as an adjustment to income. He had qualifying deductions of $1,000 in deductible medical bills, $300 in church donations, and $9,600 in interest on the condominium he owned. He also paid $150 in state taxes. He had $2,500 in federal income tax withheld during the year. What was the amount of tax due with his return?

Income:
Less adjustments to income:
Adjusted gross income
Less deductions:

Less exemption:
Taxable income
Tax computation

Less tax withheld during the year
Tax due with return |
| **9.2** | Make basic computations to determine taxes owed by taxpayers who use the standard Form 1040 | 2. Donald and Judy Mason are 72 and 70 years of age, respectively. Judy is blind. They filed as married, filing jointly. Last year they had a total income of $38,000 from investments. They filed a return and claimed the standard deduction. During the year they made quarterly payments of estimated tax in the amount of $2,400. What was the amount of tax due with their return?

Adjusted gross income
Less standard deduction

Less exemptions:
Taxable income
Tax computation

Less payments made during the year on estimated tax
Tax due with return |
| **9.3** | Make basic computations to determine taxes owed by corporations and other businesses, based on their taxable incomes | 3. The SED Corporation earned taxable income of $110,000 in 200X. How much federal income tax did it owe? |

Answers: 1. $2,719 **2.** $975 **3.** $26,150

1. Allison Blake works as a waitress. Last year she earned $15,800 in wages, $8,600 in tips, and $1,500 catering on weekends. She also received $600 interest from her credit union, $800 from a state bond, and an inheritance of $10,000. What was her gross income for federal income tax purposes?

2. Pete and Angel Romero are married and have two children aged 5 and 8. They also support Pete's sister, who lives with them. How much can Pete and Angel subtract from their gross income for exemptions?

3. Jan and Kirsten Bjorg, aged 63 and 66, are married filing a joint tax return. They have itemized deductions totaling $7,900. Should they itemize or use the standard deduction?

4. Eva Jung files as a head of household, has an adjusted gross income of $38,000, claims two exemptions, and uses the standard deduction on her federal return. What is her taxable income?

5. Donald and Katy O'Day are married, filing a joint return, and have taxable income of $65,000. What is the amount of their income tax?

6. The Beckman Corp. had taxable income of $250,000 for the year. Calculate the corporation's federal income tax.

Sales and Property Taxes

10

LEARNING OBJECTIVES

By studying this chapter and completing all assignments you will be able to:

1 Compute sales taxes, using rate tables and percents.

2 Compute assessed valuations and property taxes based on assessed valuation.

3 Compute tax rates in percents and mills.

4 Compute tax payments involving special assessments, prorations, and exemptions.

Sales and property taxes are imposed by states, counties, cities, and special tax districts. Sales taxes are the most common of the state taxes. Property taxes are the most common revenue source for cities and special districts and support most elementary and secondary schools, as well as police and fire departments and libraries.

1. Do people who own houses pay property tax?
2. What is *assessed valuation*?
3. How much sales tax is added to purchases in local area stores?
4. How does a store employee know the amount of sales tax to charge?
5. Is sales tax charged on clothing? Is sales tax charged on a quart of milk?

Most retail businesses collect a sales tax from customers when a sale occurs. The tax money must be turned over to the government. People and companies owning property usually pay taxes on the property's value. In this chapter we explain calculations involving sales and property taxes.

Computing Sales Taxes

Compute sales taxes, using rate tables and percents

A **sales tax** is a government **levy**, or charge, on retail sales of certain goods and services. Most states and many cities and other local government entities levy sales taxes. The state **tax rate**—the percent used to compute the amount of sales tax—currently ranges from 3% to 7%, and city and county rates range from 0.925% to 6%.

Retail sales taxes, which usually are a combination of state and local taxes, are calculated as a single percent of taxable sales. For example, a sale is subject to a state sales tax of 5% and a local sales tax of 1%. The combined rate of 6% is applied to all taxable sales in that locality.

Sales Tax as a Percent of Price

Sales taxes generally are rounded to the nearest cent. For example, sales taxes of 4% and 5% on amounts of up to $1 are charged as shown in Figure 10-1.

Figure 10-1: Sales Taxes

| 4% on Sales of | Tax Due | 5% on Sales of | Tax Due |
|---|---|---|---|
| $0.01 to $0.12 | none | $0.01 to $0.09 | none |
| $0.13 to $0.37 | $0.01 | $0.10 to $0.29 | $0.01 |
| $0.38 to $0.62 | $0.02 | $0.30 to $0.49 | $0.02 |
| $0.63 to $0.87 | $0.03 | $0.50 to $0.69 | $0.03 |
| $0.88 to $1.00 | $0.04 | $0.70 to $0.89 | $0.04 |
| | | $0.90 to $1.00 | $0.05 |

http://thestc.com/ RateCalc.stm

Visit the Sales Tax Clearing House Web site for a sales tax calculator.

STEPS to Compute Sales Tax and Total Sales Amount

1. Multiply the taxable sales amount by the tax rate.
2. Add the sales tax amount to the taxable sales amount to get the total sales amount.

EXAMPLE A

If taxable merchandise of $60.39 is sold in a state with a 5% sales tax, what are the amount of tax and the total amount to be paid?
Amount of tax: $60.39 × 0.05 = $3.019, which rounds to $3.02
Total amount to be paid: $60.39 + $3.02 = $63.41

Most retail stores have cash registers that recognize a code such as the Uniform Product Code (UPC) to determine taxable sales and to calculate the sales tax automatically. The sales receipt usually shows the total taxable sales as a subtotal, the sales tax, and the total sales plus tax. Usually, discounts on a sale are subtracted from the sale price before the tax is figured. Shipping and installation labor charges are generally not taxed.

www.taxadmin.org/ fta/rate/sales.html

Consider the list of sales tax rates for each state.

EXAMPLE B

A customer living in a city with a 6% state sales tax and a 1.5% city sales tax purchased a refrigerator regularly priced at $850. He was given a 10% discount. Delivery charges were $45. What were the amount of tax and the total cost to the buyer?

Discount amount: $850 × 10% = $850 × 0.10 = $85

Price after discount: $850 − $85 = $765, or $850 × 0.90 = $765

Sales tax: $765 × (0.06 + 0.015) = $57.38

Cost to buyer: $765 + $57.38 tax + $45 delivery = $867.38

State laws regarding the items subject to sales tax vary. Most states do not tax groceries; however, most do tax meals served in restaurants. Certain nonfood items also sold in grocery stores (such as laundry detergent) are generally taxed. When nontaxable and taxable items are purchased together, the register usually computes the total price of items purchased and automatically adds the correct amount of tax for each taxable item. The taxable items are clearly marked on the register tape along with the total amount of tax charged.

EXAMPLE C

A customer living in a state in which the tax rate is 4% went to a grocery store and purchased a quart of milk for $1.15, a loaf of bread for $2.79, potatoes for $2.25, and two taxable items—laundry detergent for $8.49 and fabric softener for $5.30. What was her total charge at the checkout counter?

Taxable items: $8.49 + $5.30 = $13.79

Tax: $13.79 × 0.04 = $0.55

Total: $1.15 + $2.79 + $2.25 + $8.49 + $5.30 + $0.55 = $20.53

Sales Tax as an Amount per Unit

All of the states and the District of Columbia levy special taxes on gasoline and cigarettes, usually stated in cents per unit (gallon or pack). State taxes on gasoline vary widely, from $0.075 in Georgia to $0.32 in Connecticut; in addition, the federal tax is currently $0.184 per gallon. State taxes on cigarettes currently range from $0.025 to $1.11 per pack; the federal tax is currently $0.71 per pack.

Excise Tax as an Amount per Unit

An **excise tax** is a tax assessed on each unit. In some states both the excise tax and the general sales tax apply to items such as gasoline, cigarettes, and alcoholic beverages. In such instances, the excise tax may be part of the taxable sales price for general sales tax purposes. For example, in a certain locality gasoline costs $1.20 per gallon, plus state and federal excise taxes of $0.40 and is subject to a general sales tax of 5%. The total price per gallon is $1.68 ($1.20 + $0.40 excise tax + $0.08 general sales tax). The general sales tax is calculated as 6% of $1.60.

CONCEPT CHECK 10.1

In a state in which the combined state and city sales tax rate is 6%, a customer went to a convenience store and purchased the following items: bread, $1.95; ground meat, $6.79; cheese, $4.79; light bulbs, $4.25; and motor oil, $1.79. Only the last two items are taxable. Rounding the tax to the nearest cent, compute the total cost of all items and tax.

Nontaxable items: $1.95 + $6.79 + $4.79 = $13.53
Taxable items: $4.25 + $1.79 = $6.04
Total tax: $6.04 × 0.06 tax rate = $0.36

| | |
|---|---|
| $13.53 | Nontaxable items |
| 6.04 | Taxable items |
| 0.36 | Tax |
| $19.93 | Total |

Computing Assessed Valuations and Property Taxes

2
LEARNING OBJECTIVES

Compute assessed valuations and property taxes based on assessed valuation.

A **property tax** for a business is a tax on real estate or other property, such as machinery, owned by the business. Businesses usually pay property tax bills semiannually. Taxes are based on a value, known as the **assessed valuation**, determined by a representative of the local or state government.

Assessed valuation ordinarily is based on the current **market value** of the property (what the property could be sold for). In many states it is fixed by law at 100%, but it is a fraction of that value in other states. Thus a particular community may use 60% of property values as the basis for tax billing. In most instances, land and buildings are assessed separately.

EXAMPLE D

The Washington family lives in a town in which assessed valuation is 60% of market value. The Bradley family lives in a town in which assessed valuation is 75% of market value. Each home has a market value of $260,000. What is the assessed valuation of each home?

Washington: $260,000 × 0.60 = $156,000
Bradley: $260,000 × 0.75 = $195,000

Assessed valuation often is increased by improvements to the property, such as the addition of an enclosed porch or expensive shrubbery. Ordinary maintenance—a new coat of paint, for instance, or repairs to the roof—isn't justification for an increased assessment.

EXAMPLE E

The Ruiz family and the Walton family live in a town in which assessed valuation is set by law at 80% of market value. They live in identical houses having a market value of $190,000. The Ruiz family added an enclosed deck costing $10,500 and a family room costing $23,000. The Walton family made extensive repairs and repainted the house a new color at a total cost of $12,000. What was the assessed valuation on each home the following year?

Ruiz: $190,000 + $10,500 + $23,000 = $223,500 × 0.8 = $178,800

Walton: $190,000 × 0.8 = $152,000 (repairs and painting are not considered improvements)

✓ CONCEPT CHECK 10.2

a. The Landen family owns a home with a market value of $300,000 in a community that assesses property at 100% of market value. The Jefferson family owns a home with a market value of $400,000 in a community that assesses property at 60% of market value. What is the difference between the actual assessments of the two homes?

 Landen: $300,000 × 1 = $300,000
 Jefferson: $400,000 × 0.6 = $240,000
 Difference = $60,000

b. The Levick family home has a present market value of $200,000 in a community that assesses property at 80% of market value. If they add a family room and an additional bathroom at a cost of $36,000, what will be the new assessed valuation?

 Revised market value: $200,000 + $36,000 = $236,000
 New assessed value: $236,000 × 0.80 = $188,800

Computing Tax Rates in Percents and Mills

Percents

For a city, county, or special district, the tax rate is found by dividing the amount of money the government unit needs to raise by the total assessed valuation of the particular unit.

LEARNING OBJECTIVES

Compute tax rates in percents and mills.

EXAMPLE F

The town of Lakeside has a total assessed valuation of $570,000,000. The amount to be raised by taxation is $9,975,000. What is the tax rate?

 The tax rate is

$9,975,000 ÷ $570,000,000 = 0.0175, or 1.75%

This rate is usually written as 1.75% of value, or $1.75 on each $100 of value.

If a property in Lakeside is assessed for $20,000, what is the tax?

The tax can be found by multiplying the amount by the rate:

$20,000 × 0.0175 = $350

Mills

Tax rates sometimes are expressed in a unit of measure called mills. A **mill** is a tenth of a cent, or $0.001 (one thousandth of a dollar). To convert mills to dollars, divide by 1,000 (move the decimal three places to the left). To convert cents to mills, multiply by 10. Thus a tax rate can be converted from mills to cents or dollars or vice versa by using the following relationships:

| | |
|---|---|
| mills ÷ 10 = cents | 150 mills ÷ 10 = 15¢ |
| mills ÷ 1,000 = dollars | 150 mills ÷ 1,000 = $0.15 |
| cents × 10 = mills | 15¢ × 10 = 150 mills |
| dollars × 1,000 = mills | $0.15 × 1,000 = 150 mills |

Marshall County assesses property at the rate of 160 mills per $100 of assessed value. How much tax would be due on property assessed at $650,000?

$650,000 ÷ 100 = $6,500 to assess millage

160 mills = $0.16

0.16 × $6,500 = $1,040 tax

CONCEPT CHECK 10.3

a. A town has a total assessed valuation of $960,000,000. A total of $12,000,000 must be raised by taxation for the operating expenses of the town. What will be the tax rate?

$12,000,000 ÷ $960,000,000 = 0.0125, or 1.25%

b. Convert $0.57 into mills: 57¢ × 10 = 570 mills, or $0.57 × 1,000 = 570 mills

c. If property in a town is assessed at the rate of 140 mills per $100 of assessed value, how much tax will be due on property assessed at $475,000?

$475,000 ÷ 100 = $4,750 to assess millage

140 mills = $0.14

$4,750 × $0.14 = $665 tax due

Computing Special Assessments, Prorations, and Exemptions

Special assessments can be levied for improvements in a community, such as sewers, roads, or sidewalks. Sometimes the cost is spread over a period of years and added to the annual property tax bill of each property owner.

LEARNING OBJECTIVES

Compute tax payments involving special assessments, prorations and exemptions.

EXAMPLE I

The residents of Jamestown voted to widen their roads and add sidewalks, at a cost of $360 per residence, with the cost to be spread over a 12-year period. The Stevens family had an annual tax bill of $415 before the improvements. If they pay their property taxes semiannually, what will be the amount of their next tax payment?

Annual cost for improvement: $360 ÷ 12 = $30

Annual property tax and improvement payment:

$415 + $30 = $445

Next semiannual tax payment: $445 ÷ 2 = $222.50

Whenever property is sold, it is customary to *prorate*, or distribute, the taxes between seller and buyer as of the date of the settlement.

EXAMPLE J

A home having an annual tax bill of $360 was sold at the end of the seventh month of the taxable year. The seller had already paid the tax for the full year. How much tax was the seller reimbursed on proration of taxes at the time of the sale?

Months prepaid by seller: 12 − 7 = 5

Tax reimbursed by buyer: $360 × $\frac{5}{12}$ = $150

In almost all states, property used exclusively by nonprofit organizations, such as schools, churches, governments, and charities, is exempt from taxation. Some states also allow partial exemptions for veterans and the elderly.

EXAMPLE K

The town of Brookside assesses property at 75% of market value. The tax rate is 1.9%. A church has a total market value of $560,000. How much does the church save each year by being exempt from property taxes?

$560,000 × 0.75 = $420,000 $420,000 × 0.019 = $7,980 saved

EXAMPLE L

A veteran living in Tucson receives a partial exemption of 15% of regular property taxes. The veteran owns property valued at $240,000. If the property is assessed at 75% of value and the current rate is 1.3%, how much tax is due each six months?

Assessed value: $240,000 × 0.75 = $180,000

Regular taxes: $180,000 × 0.013 = $2,340

Taxes due after exemption: $2,340 × 0.85 (100% − 15%) = $1,989

Taxes due each six months: $1,989 ÷ 2 = $994.50

CONCEPT CHECK 10.4

a. The city of Denton voted to build a new library at a cost of $420 per residence, to be spread over a period of 15 years. If the Douglas family presently has a yearly tax bill of $730, paid semiannually, what will be the amount of their next tax payment?

$420 per residence ÷ 15 years = $28 per year

$730 present yearly tax amount + $28 = $758 new yearly tax amount

$758 ÷ 2 = $379 new semiannual tax amount

b. If a home with an annual tax bill of $620 is sold at the end of the third month of the tax year, after taxes have already been paid, how much will the buyer reimburse the seller when taxes are prorated?

12 − 3 = 9 months prepaid by seller

$620 \times \dfrac{9}{12} = $465 reimbursed by buyer

c. A 70-year-old man lives in a state that grants senior citizens a 10% exemption from property taxes. If his home has a market value of $150,000 and the tax rate is 1.5%, how much will be his yearly taxes? The county in which he resides assesses property at 80% of market value.

$150,000 market value × 0.8 = $120,000 assessed valuation

$120,000 assessed valuation × 0.015 = $1,800 regular taxes

$1,800 regular taxes × 0.10 = $180 reduction

$1,800 regular taxes − $180 reduction = $1,620 revised taxes

COMPLETE ASSIGNMENTS 10.1 AND 10.2.

CyberConnection

You can examine San Francisco City taxes—sales and property—at this Web site. By clicking on the highlighted items, you can see how the city taxes are distributed for school districts, transit, public safety, and other purposes.

cityspan

San Francisco's Five Year Tax Rate History

| TAX TYPE | FY97-98 | FY98-99 | FY99-00 | FY00-01 | FY01-02 |
|---|---|---|---|---|---|
| Property (%) | 1.190 | 1.165 | 1.129 | 1.136 | 1.124 |
| | | | | | |
| **Business Taxes** | | | | | |
| Payroll (%) | 1.50 | 1.50 | 1.50 | 1.50 | 1.50 |
| Gross Receipts (%) | 0.30 | 0.30 | 0.30 | (1) | (1) |
| Exemption Level ($Tax) | 2,500 | 2,500 | 2,500 | 2,500 | 2,500 |
| Registration Fee ($) | (2) | (2) | (2) | (2) | (2) |
| Tax Credit for Registration Fee | No | No | No | No | No |
| | | | | | |
| Sales (%) | 8.50 | 8.50 | 8.50 | 8.25 | 8.25 |
| | | | | | |
| **Utility Users** | | | | | |
| Gas & Electricity (%) | 7.5 | 7.5 | 7.5 | 7.5 | 7.5 |
| Telephone (%) | 7.5 | 7.5 | 7.5 | 7.5 | 7.5 |
| Water (%) | 7.5 | 7.5 | 7.5 | 7.5 | 7.5 |
| Residential Exemption | Yes | Yes | Yes | Yes | Yes |

Source: http://www.ci.sf.ca.us/controller/ratehist.htm. Office of the Controller, City & County of San Francisco.

Chapter Terms for Review

assessed valuation

excise tax

levy

market value

mill

property tax

sales tax

tax rate

Visit the Web

http://www.cbr.gov.pk/—Read about the sales tax in Pakistan

http://www.salestax.org/—The National Retail Sales Tax Alliance supports a national sales tax and the abolishment of income tax.

The following Web sites present more information about sales and property taxes in various locations.

www.salestaxinstitute.com

www.nysalestax.com

www.SuperCitizen.com

www.nationtax.com

www.epropertytax.com

www.calproptax.com

www.timvw.com/maps/atrmenu

www.ttara.org/foundation

www.taxadmin.org

CYBERCONNECTION

Sales Tax

Sales Tax

- The National sales tax, like the flat tax, would scrap the existing tax code in favor of a 15 percent surcharge on most goods and services.

 That would make things a whole lot more simple come April because taxpayers wouldn't need to file a return. The idea is to encourage savings and simplify tax filing.

 But a national sales tax could have all sorts of

Source: http://www.cnn.com/US/9704/tax.special/tax.reform/sales.html. Copyright © 1997 by Cable News Network, Inc.; used with permission.

Read this article and the Cato Institute's proposal for a national sales tax to replace the income tax at http://www.cato.org/pubs/pas/pa-272es.html.

The national sales tax proposal calls for a 15% sales tax on most retail goods and services purchased. It would eliminate the income tax and would apply to the amount of purchases based on income above the poverty level. Think about this proposal and give an example of how it might be less or more costly than the current income tax system.

The Bottom Line

Summary of chapter learning objectives:

| Learning Objective | Summary | Example |
|---|---|---|
| **10.1** | Compute sales taxes, using rate tables and percents | 1. The Farley family lives in a state in which the sales tax rate is 8%. When they purchased a dining room table and chairs regularly priced at $1,400, they were given a discount of 15%. Shipping charges were $50. What was the total cost to the Farleys?

 2. Lucy Carter lives in a state in which the state tax on gasoline is $0.18 a gallon. Federal tax is $0.19 a gallon. If she purchased an average of 8 gallons per week during the 52-week year, how much did she pay in state and federal taxes combined? |
| **10.2** | Compute assessed valuations and property taxes based on assessed valuation | 3. The Nguyen family lives in a town in which the assessed valuation on property is 65% of market value. The Parker family lives in a town in which the assessed valuation on property is 80% of market value. Each home has a market value of $162,000. How much is the assessed valuation of each home? |
| **10.3** | Compute tax rates in percents and mills | 4. The town of Belton has a total assessed valuation of $850,000,000. For the coming year the city must raise $11,730,000 for operating expenses.
 a. What will be the tax rate?
 b. What will the semiannual taxes be on a home with an assessed valuation of $135,000?

 5. a. Convert 400 mills to its dollar equivalent.
 b. Convert $0.90 to mills. |
| **10.4** | Compute tax payments involving special assessments, prorations, and exemptions. | 6. A home with annual tax payments of $420 was sold at the end of the tenth month of the taxable year. What was the amount of tax prorated to the buyer?

 7. A veteran living in Richmond receives a partial exemption of 10% of regular property taxes. The veteran owns property valued at $210,000. If the property is assessed at 80% of value and the current rate is 1.5%, how much tax is due each six months? |

Answers: **1.** $1,335.20 **2.** $153.92 **3.** $105,300 (Nguyen), $129,600 (Parker) **4. a.** 1.38% **b.** $931.50 **5. a.** $0.40 **b.** 900 mills **6.** $70 **7.** $1,134

Self-Check | Review Problems for Chapter 10

1 The Johnson Company is considering the purchase of some equipment from two different suppliers. For a sales tax rate of 6%, which of the following offers should Johnson Company accept?

Company A: Equipment price of $65,000 plus installation and shipping costs of $1,200.

Company B: Equipment price of $73,500 less 10% discount, no additional charge for installation or shipping.

2 Cherry City needs to raise $7,800,000 in property taxes on property with a total market value of $650,000,000.

a. What will the tax rate be if property is assessed at 80% of market value?

b. Determine the amount of semiannual property tax to be paid by each of the following property owners who live in Cherry City.

Juan Garcia's home in Cherry City has a market value of $350,000.

Yolanda Green is a senior citizen who receives a 10% exemption from property tax. Her home in Cherry City has a market value of $215,000.

3 The residents of Hunterville voted to add street lights and sidewalks to their city at a cost per residence of $324 to be spread over 12 years.

a. If Mary Nowitski, a resident of Hunterville, had an annual tax bill of $860 before the special assessment, how much must she now pay semiannually for her property taxes?

b. If Mary Nowitski sells her home at the end of the eighth month of the tax year and has already paid the property taxes for the full year, including the special assessment, how much of the prepaid property tax should be allocated to the purchaser?

Assignment 10.1: Sales Tax

LEARNING OBJECTIVES

Name _____

Date _____ Score _____

A **(50 points) Solve the following problems. (1 point for each correct answer)**

1. Drake's fast-food restaurant is in a state with a tax rate of 7%. Compute the sales tax, the total sale, and the change given for each transaction .

| Amount of Sale | Sales Tax | Total Sale | Cash Paid | Amount of Change |
|---|---|---|---|---|
| $ 5.89 | .41 | _____ | $10.00 | _____ |
| 3.07 | .21 | _____ | 5.00 | _____ |
| 14.12 | .99 | _____ | 20.00 | _____ |
| 7.97 | .56 | _____ | 10.53 | _____ |
| 0.85 | .06 | _____ | 1.00 | _____ |
| 23.15 | 1.62 | _____ | 30.00 | _____ |
| 9.60 | .67 | _____ | 20.00 | _____ |
| 28.35 | 1.99 | _____ | 50.00 | _____ |
| 13.21 | .93 | _____ | 15.00 | _____ |
| 17.49 | 1.23 | _____ | 19.00 | _____ |

2. Giramonte's clothing store is in a city where the state tax is 3% and the city tax is 1%. Determine the sales tax, the total sale, and the change given for each transaction. Then compute the total sales taxes and total sales.

| Amount of Sale | Sales Tax | Total Sale | Cash Paid | Amount of Change |
|---|---|---|---|---|
| $317.20 | 12.69 | 329.89 | $350.00 | 20.11 |
| 29.30 | 1.17 | 30.47 | 31.00 | .53 |
| 72.85 | 2.91 | 75.76 | 80.00 | 4.24 |
| 9.95 | _____ | _____ | 20.35 | _____ |
| 109.40 | _____ | _____ | 120.00 | _____ |
| 299.95 | _____ | _____ | 500.00 | _____ |
| Total | _____ | _____ | | |

Score for A (50)

B **(30 points) Solve the following problems. Use Figure 10.1 for problems 3 and 4. (points for correct answers as marked)**

3. A candy store, operating in a state with a sales tax of 4%, made 868 sales at 10¢; 946 sales at 45¢; 457 sales at 50¢; 921 sales at 75¢; and 815 sales at 90¢. How much did the store receive in taxes? (8 points) _____

4. If the candy store in problem 3 computed the amount of state sales tax submitted to the state based on 4% of gross sales, what would be the difference between the amount of tax the store collected and the amount it submitted to the state? (8 points) _____

5. King Carpets and Oriental Rugs, Inc., each bought a new car. King is located in a state that has a 4% sales tax and paid the regular price of $12,487 plus tax. Oriental Rugs is located in a state that has a 5% sales tax and received a special discount of $150 off the regular $12,487 price.

 a. Including sales tax, which company paid more for its car? (8 points) __King Carpets__

 b. How much more? (6 points) __25.13__

12,487
4% 499.48

12,487
5% 624.35 tax
13,111.35
150⁰⁰

12961.35 Oriental
12,986.48 King

Score for B (30)

C **(20 points) Solve the following problems. (points for correct answers as marked)**

6. The Yellow Rose has stores in four states. Sales tax rates for the four states are as follows: state A, 8%; state B, $6\frac{1}{2}$%; state C, $5\frac{1}{2}$%; and state D, 3%. Annual sales for the four states last year were as follows: state A, $865,000; state B, $837,000; state C, $491,000; and state D, $628,000.

 a. How much did the Yellow Rose collect in sales taxes during the year? (10 points) _____

 b. If all four states had the same lower sales tax rate of 3%, how much would the Yellow Rose have collected in sales taxes during the year? (5 points) _____

 c. If all four states had the same higher tax rate of 8%, how much would the Yellow Rose have collected in sales taxes during the year? (5 points) _____

Score for C (20)

Assignment 10.2: **Property Taxes**

Name

Date Score

A **(40 points) Solve the following problems. (4 points for each correct answer)**

1. Find the assessed valuation for each of the following towns.

| Town | Property Value | Basis for Tax Billing | Assessed Valuation |
|------|----------------|-----------------------|--------------------|
| A | $940,000,000 | 100% | |
| B | 515,625,000 | 80% | 412,500,000 Million |
| C | 424,000,000 | 75% | 318,000,000 million |

2. Find the tax rate for each of the following towns. Show your answer as a percent.

| Town | Assessed Valuation | Amount to Be Raised | Tax Rate |
|------|--------------------|--------------------|----------|
| F | $940,000,000 | $16,920,000 | _____ |
| G | 580,000,000 | 7,250,000 | _____ |
| H | 876,000,000 | 11,826,000 | _____ |

3. Convert the following percentage tax rates into dollars and cents per $100 of assessed valuation.

| Tax Rate | Dollars and cents |
|----------|-------------------|
| 1.4% | _____ |
| 0.95% | _____ |

4. Convert the following percent tax rates into mills per $100 of assessed valuation.

| Tax Rate | Mills |
|----------|-------|
| 1.4% | _____ |
| 0.95% | _____ |

Score for A (40)

Assignment 10.2 *Continued*

B **(24 points) Solve the following problems. (6 points for each correct answer)**

5. The Griffin Company is located in a state in which assessed valuation is 100% of market value. The tax rate this year is $1.35 on each $100 of market value. The market value of the company building is $190,000. How much property tax will Griffin pay this year? _____

6. The Raulston Corp. is located in an area in which assessed valuation is 70% of market value. The tax rate this year is 1.6%. The market value of Raulston's property is $310,000. How much property tax will Raulston pay this year? _____

7. Next year, the assessed valuation in Griffin's area (problem 6) will decrease to 65% of market value and the tax rate will remain the same as this year. How much less tax will Griffin pay next year than it paid this year? _____

8. Perez, Inc., is headquartered in an area in which assessed valuation is 80% of market value. The tax rate this year is $1.40 on each $100 of assessed valuation. Its property has a market value of $320,000. How much property tax will Perez pay this year? _____

<div align="right">Score for B (24)</div>

C **(24 points) Solve the following problems. Round to the nearest dollar. (3 points for each correct answer)**

9a. There are four towns in Jackson county: Hillsdale, Littleton, Greenville, and Byers. Using the total assessed valuations given and the amount of money the town must raise for operating expenses, compute the necessary tax rate for each town.

| Town | Total Assessed Valuation | Money That Must Be Raised | Tax Rate as a Percent |
|---|---|---|---|
| Hillsdale | $200,000,000 | $3,400,000 | _____ |
| Littleton | $340,000,000 | $5,780,000 | _____ |
| Greenville | $280,000,000 | $3,780,000 | _____ |
| Byers | $450,000,000 | $6,750,000 | _____ |

b. Convert each of the percentage rates in part (a) to mills per dollar of assessed valuation.

Hillsdale _____

Littleton _____

Greenville _____

Byers _____

<div align="right">Score for C (24)</div>

Assignment 10.2 *Continued*

D **(12 points) Solve the following problems. Round to the nearest dollar. (6 points for each correct answer)**

10. A home with annual tax payments of $780 was sold at the end of the ninth month of the taxable year. The seller had already paid the entire tax for the year. How much tax was the seller reimbursed on proration of taxes at the time of the sale? _____

11. A senior citizen lives in a state that grants a 20% exemption on property taxes. Her property is valued at $290,000 and is assessed at 75% of value. The current tax rate is 1.6%. How much tax is due each six months? _____

Score for D (12)

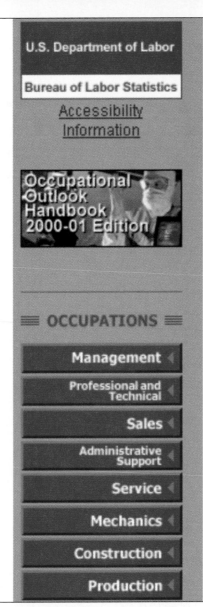

Visit the U.S. Department of Labor's Occupational Outlook Handbook Web site and click on Sales. Many of the sales jobs listed include commissions as part of earnings. For example, earnings of travel agents owning their agencies depend mainly on commissions from airlines and other carriers, cruise lines, tour operators, and lodging places. Commissions for domestic travel arrangements, cruises, hotels, sightseeing tours, and car rentals are about 7–10 percent of the total sale and for international travel are about 10 percent. Travel agents may also charge clients a service fee for the time and expense involved in planning a trip.

Check this site not only for information on commissions, but also to see occupational opportunities, job descriptions and outlooks, median salaries, educational requirements, and much more. Visit their site at http://stats.bls.gov/oco/.

ook

Order

GO

Search Tips

models

nagers

services

Commissions

11

LEARNING OBJECTIVES

By studying this chapter and completing all assignments you will be able to:

 1 LEARNING OBJECTIVES

Calculate sales commissions and gross pay.

 2 LEARNING OBJECTIVES

Calculate graduated sales commissions.

 3 LEARNING OBJECTIVES

Calculate sales and purchases for principals.

Some of the highest earning professionals do it without a salary, working on commission as salespeople. On every sale, they are paid a percentage of the sales price. They can sell real estate, cars, insurance, and stocks and bonds, as well as high-ticket goods that can result in six-figure commissions.

1. What are the advantages to an employer of commission-based pay?
2. What are the advantages to an employee of commission-based pay?
3. What happens when a salesperson is paid a commission on a sale and then the customer returns the merchandise?
4. What is a graduated commission rate?
5. Explain the difference in the calculations of the net proceeds by a commission merchant who is selling merchandise for a principal and the gross cost by a merchant who is buying merchandise for a principal.

A **commission** is a payment to an employee or to an agent for performing a business transaction or service. The most familiar type of commission is that received by a salesperson. Many companies have employees who are paid either totally or partially on a commission basis. People who sell insurance, real estate, and automobiles typically are in this category.

For a business owner, one advantage of using the commission method to pay employees is that the commission is an incentive. Employees are paid on the basis of the volume of business they produce for the company. They can earn more by being more productive.

Besides typical salespeople, other businesspeople provide selling and buying services. These include commission merchants, agents, and brokers, all of whom are paid a commission for their services. The person for whom the services are provided is called the **principal.** A commission merchant will normally take actual possession of the merchandise and make the sales transaction in his or her name. A **broker,** however, will usually make the transaction in the principal's name and does not take possession of the merchandise.

Calculating Sales Commissions and Gross Pay

Calculate sales commissions and gross pay.

A sales commission paid to a salesperson is usually a stated percent of the dollar value of the goods or services sold. Whether the commission is based on the wholesale or retail value of the goods will depend on the type of business and merchandise sold. The rate used to calculate the commission also will vary among different businesses. In some companies, the salesperson receives both a salary and a commission.

STEPS **to Calculate Commission and Total Pay**

1. Multiply the commission rate by the amount sold to get the commission amount.
2. If there is a salary, add it to the commission amount to get the total gross pay.

© PHOTODISC, INC.

EXAMPLE A

Dottie Houston sells yachts and marine equipment for Bay Area Marina. She receives a base salary of $2,500 per month and earns a commission that is 2% of the value of all boating equipment that she sells during the month. Find her commission and total pay during September, a month in which she sold $106,000 worth of equipment.

STEP 1 2% × $106,000 = 0.02 × $106,000 = $2,120 commission

STEP 2 $2,120 commission + $2,500 base salary = $4,620 total pay

Commissions normally are paid only on actual sales. Thus goods that are returned or orders that are canceled are not subject to commission. The reason for this policy is to protect the business owner. Suppose that Bay Area Marina in example A pays the 2% commission whether or not the goods are returned. When Dottie Houston got an order for $10,000, her commission would be 2% × $10,000 = $200. If the goods were all returned but the commission were still paid, the owner would have to pay her $200. Because no goods were sold, the owner actually would lose $200 on this transaction.

STEPS to Calculate Commission When a Sale Involves Returned Goods

1. Subtract the value of the returned goods from the total ordered to determine the amount sold.
2. Multiply the commission rate by the amount sold to get the commission amount.

www.howtosell.net/
html/archives/
commissions.html

www.isp-planet.com/
business/isp_sales_
commissions.html

Read these articles on sales commissions, base salaries, and quotas.

EXAMPLE B

Hobart Hamilton is a salesperson for Turlock Office Supply. He works on commission-only basis—he receives a commission of 2.5% on his monthly sales, but no base salary. What are his commission and total pay during a month when he sells $167,500 worth of office products, but one of his customers cancels an order for $15,000 and returns the merchandise that had already been delivered?

STEP 1 $167,500 − $15,000 = $152,500

STEP 2 2.5% × $152,500 = 0.025 × $152,500 = $3,812.50 commission

Total Pay = $3,812.50, as he is paid on a commission-only basis

 CONCEPT CHECK 11.1

Calculate the commission and gross pay for a salesperson who is paid a $1,600 salary and earns a 4% commission. Total sales were $96,000, but there were returns of $8,000.

$96,000 − $8,000 = $88,000 net sales

$4\% \times \$88,000 = 0.04 \times \$88,000 = \$3,520$ commission
$+ 1,600$ salary
$\overline{\$5,120}$ gross pay

Calculating Graduated Sales Commissions

Commission plans provide incentives for employees because they can earn more money by selling more products. A company can provide additional incentive for even greater productivity by using **graduated commission rates.** As the level of sales increases, so does the commission rate.

LEARNING OBJECTIVES

Calculate graduated sales commissions.

STEPS to Calculate Commission Under a Graduated Rates Plan

1. Calculate the dollar amount at each rate level by using subtraction.
2. Multiply each level's commission rate by the level's sales dollars.
3. Add the products computed in Step 2 to determine the total commission.

CyberConnection

A sales manual designed to provide guidelines to sales brokers interested in selling VA-owned properties is available at this and other VA Web sites. Samples of forms, related documents, and information on commissions are presented.

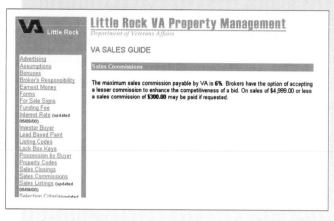

Little Rock VA Property Management
Department of Veterans Affairs

VA SALES GUIDE

Advertising
Assumptions
Bonuses
Broker's Responsibility
Earnest Money
Forms
For Sale Signs
Funding Fee
Interest Rate (updated 05/08/00)
Investor Buyer
Lead Based Paint
Listing Codes
Lock Box Keys
Possession by Buyer
Property Codes
Sales Closings
Sales Commissions
Sales Listings (updated 05/08/00)
Selection Criteria(updated

Sales Commissions

The maximum sales commission payable by VA is **6%**. Brokers have the option of accepting a lesser commission to enhance the competitiveness of a bid. On sales of $4,999.00 or less a sales commission of **$300.00** may be paid if requested.

Source: http://www.vahomes.org/lr/sacomm.htm. Courtesy of Little Rock VA Property Management.

EXAMPLE C

Melody Chiu has a monthly commission plan under which she receives 2% on the first $40,000 of sales during the month and 3% on sales above $40,000 for the month. If Melody has sales of $73,000 during a month, compute her commission for that month.

STEP 1

| | |
|---|---|
| $73,000 | total sales |
| −40,000 | at 2% |
| $33,000 | at 3% |

STEP 2

$40,000 × 0.02 = $ 800
33,000 × 0.03 = 990

STEP 3

Total commission = $1,790

EXAMPLE D

Assume that Melody has a monthly commission plan under which she receives 2% on the first $40,000 of sales during the month, 3% on sales between $40,000 and $80,000, and 4% on all sales over $80,000. If Melody has sales of $106,500 during a month, compute her commission for that month.

STEP 1

| | |
|---|---|
| $106,500 | total sales |
| −40,000 | at 2% |
| $66,500 | |
| −40,000 | at 3% |
| $26,500 | at 4% |

STEP 2

$40,000 × 0.02 = $ 800
40,000 × 0.03 = 1,200
26,500 × 0.04 = 1,060

STEP 3

Total commission = $3,060

The same graduated incentive plan can be defined in terms of bonus rates. The calculations are similar.

EXAMPLE E

Dale Crist has a monthly commission plan under which he receives 2% on all sales during the month. If Dale has sales over $40,000, he receives a bonus of 1% of everything over $40,000. If he sells more than $80,000, he receives a "super bonus" of an additional 1% of everything over $80,000. What is Dale's commission for a month during which he sold $96,500?

| | 0 | $40,000 | $80,000 | $96,500 | |
|---|---|---|---|---|---|
| Base | $40,000 | | | | $0.02 \times$ $96,500 = \$1,930$ |
| Bonus | | $96,500 - \$40,000 = \$56,500$ | | | $0.01 \times$ $56,500 = \quad 565$ |
| Super Bonus | | | $96,500 - \$80,000$ $= \$16,500$ | | $0.01 \times$ $16,500 = \quad 165$ |

Total commission (add the three commission amounts) = $2,660

CONCEPT CHECK 11.2

Calculate the total commission on sales of $182,400. The commission is graduated: 1% on sales to $60,000, 2% on sales from $60,000 to $120,000, and 3% on sales above $120,000.

| | | | |
|---|---|---|---|
| 1% × $60,000 = 0.01 × $60,000 | = | $ 600 | |
| 2% × $60,000 = 0.02 × $60,000 | = | 1,200 | |
| 3% × $62,400 = 0.03 × $62,400 | = | 1,872 | |
| Total commission | = | $3,672 | |

Calculating Sales and Purchases for Principals

A producer may send goods to an agent, called a **commission merchant,** for sale at the best possible price. Such a shipment is a **consignment.** The party who sends the shipment is the **consignor;** the party to whom it is sent—that is, the commission merchant—is the **consignee.**

Whatever price the commission merchant gets for the consignment is the **gross proceeds.** The commission is generally a certain percent of the gross proceeds. Sometimes it is a certain amount per unit of weight or measure of the goods sold. The commission and any other sales expenses (e.g., transportation, advertising, storage, and insurance) are the **charges.** The charges are deducted from the gross proceeds. The resulting amount, which is sent to the consignor, is the **net proceeds.**

LEARNING OBJECTIVES

Calculate sales and purchases for principals.

EXAMPLE F

Ernest Dawkins Ranch has been trying to sell a used truck for hauling livestock and a tractor. Unsuccessful after 3 months, Dawkins consigns the pieces to Carson Equipment Brokers. They agree on commission rates of 6% on the gross proceeds from the truck and 9% on the gross proceeds from the tractor. Carson sells the truck for $34,600 and the tractor for $41,800. Carson also pays $380 to deliver the truck and $625 to deliver the tractor. What are the net proceeds due the Ernest Dawkins Ranch from the sale of the equipment?

| | | | | | |
|---|---|---|---|---|---|
| Truck: | Commission: $0.06 \times \$34,600 = \$2,076$ | | Gross Proceeds: | | $34,600 |
| | Freight: | + 380 | less charges | | − 2,456 |
| | Total Charges | $2,456 | Net Proceeds: | | $32,144 |
| Tractor: | Commission: $0.09 \times \$41,800 = \$3,762$ | | Gross Proceeds: | | $41,800 |
| | Freight: | + 625 | less charges | | − 4,387 |
| | Total Charges | $4,387 | Net Proceeds: | | $37,413 |

$\$32,144 + \$37,413 = \$69,557$ Total Net Proceeds

Along with the net proceeds, the commission merchant sends the consignor a form known as an **account sales.** It is a detailed statement of the amount of the sales and the various deductions. Figure 11-1 shows a typical account sales.

Figure 11.1: Account Sales

CARSON EQUIPMENT BROKERS

August 16, 20-- NO. 67324

309 Sule Road, Wilbraham, MA 01095-2073

BELOW ARE ACCOUNT SALES OF Consignment No. 76
RECEIVED August 1, 20--
and sold for account of Same

NAME Ernest Dawkins Ranch
ADDRESS 127 N. Kaye
Albany, GA 31704-5606

| DATE | CHARGES | AMOUNT | DATE | SALES | AMOUNT |
|---|---|---|---|---|---|
| Aug. 1 | Freight (truck) | $380 | Aug. 10 | Truck | $34,600 |
| 16 | 6% Commission (truck) | 2,076 | | | |
| | Net proceeds (truck) | 32,144 | 13 | Tractor | 41,800 |
| | | | | Gross proceeds | $76,400 |
| | Freight (tractor) | 625 | | | |
| | 9% Commission (tractor) | 3,762 | | | |
| | Net proceeds (tractor) | 37,413 | | | |
| | Total | $76,400 | | | |

When commission merchants purchase goods for their principals, the price they pay for the merchandise is the **prime cost.** The prime cost and all charges are the **gross cost,** or the cost the principal pays.

EXAMPLE G

Asia-Pacific Tours commissioned Specialty Marketing Group to purchase 10,000 vinyl travel bags that will be labeled with Asia-Pacific's logo and used as promotional items. For this size order, Specialty Marketing purchased the bags for $4.29 each. Charges include the commission, which is 6% of the prime cost; storage, $42.50; and freight, $24.40. What is the gross cost that Asia-Pacific should pay to Specialty Marketing?

```
$    4.29                    $42,900    prime cost
× 10,000  units             ×  0.06
$ 42,900  prime cost        $ 2,574    commission
```

$2,574 commission + $42.50 storage + $24.40 freight = $2,640.90 charges

$42,900 prime cost + $2,640.90 charges = $45,540.90 gross cost

An **account purchase** is a detailed statement from the commission merchant to the principal. It shows the cost of goods purchased, including charges. Figure 11-2 shows a typical account purchase, for the transaction in example G.

Figure 11.2: Account Purchase

SPECIALTY MARKETING GROUP

4445 Mission Street
San Francisco, CA 94112

ACCOUNT PURCHASE
NO. 1311

October 26 20 --

Bought on Consignment for

Asia-Pacific Tours
7300 Harbor Place
San Francisco, CA 94104

| DATE | DESCRIPTION | CHARGES | AMOUNT |
|------|-------------|---------|--------|
| Oct. 23 | 10,000 units stock #T805 @ $4.29 | | 42,900.00 |
| 23 | 6% commission | 2,574.00 | |
| | Storage | 42.50 | |
| | Freight | 24.40 | 2,640.90 |
| | Gross Cost | | 45,540.90 |

CONCEPT CHECK 11.3

a. Calculate the commission and the net proceeds on a consignment sale of $6,500. The commission rate is 4%, local delivery charges are $328.16, and storage charges are $125.

 4% × $6,500 = 0.04 × $6,500 = $260 commission

 $6,500 − $260 − $328.16 − $125 = $5,786.84 net proceeds

b. Calculate the commission and gross cost on a $12,500 purchase for a principal. The commission rate is 6%, air freight is $138.70, and local delivery charges are $64.60.

 6% × $12,500 = 0.06 × $12,500 = $750 commission

 $12,500 + $750 + $138.70 + $64.60 = $13,453.30

COMPLETE ASSIGNMENTS 11.1 AND 11.2.

Chapter Terms for Review

account purchase

account sales

broker

charges

commission

commission merchant

consignee

consignment

consignor

graduated commission rates

gross cost

gross proceeds

net proceeds

prime cost

principal

Visit the Web

http://www.howtosell.net/html/archives/commissions.html and **http://www.isp-planet.com/business/isp_sales_commissions.html**—Read articles on sales commission, base salaries, and quotas at these sites.

http://www.one-ifbyland.com/salescommission.htm—Visit this site for information on commissions and other aspects of real estate sales.

CYBERCONNECTION

Home Sellers Hit by Higher Sales Commissions

Business couldn't be better in the real estate industry, but that's not stopping some real estate agencies from raising the standard sales commission from 6% to 7%. That takes a bigger chunk out of a seller's profit—an additional $1,500 on a $150,000 home, for example. It's not good news for home buyers, either, because sellers are free to hike their prices to cover the commission.

Along with higher commissions, some agencies also are charging sellers "administrative fees" of $100 to $500. What's the bottom line for home sellers? Shop around!

The increases follow a wave of industry mergers that shrunk the number of real-estate brokerage firms in the U.S. Consumer advocates are worried that things may get worse with fewer real estate firms competing against each other, making it easier for the industry to raise fees, and harder for home sellers to shop around. Big investment firms and marketing companies are gob-

bling up agencies with household names, such as Coldwell Banker and Century 21. According to industry estimates, the number of residential real estate firms in the U.S. has dropped from 150,000 in 1990 to only 80,000 today.

6% has long been the industry standard, and the rate hikes have been controversial. Fortunately, the commission hikes are not yet an industry-wide trend. Some consumers are deliberately taking their business elsewhere. Perhaps that will be enough incentive for competing firms to stick with 6%. Certain brokerages such as Long & Foster in the Washington, DC area, a firm with $12 billion in sales in 1998, are leading the 7% movement.

Coldwell Banker Parnegg Metro in New Mexico not only raised commissions from 6 to 7% but also added on a "global marketing fee" of $245. The new fee hikes seem to be more widespread. Perhaps it's easier to raise

prices incrementally by a few hundred dollars than by a whole percent of the sales price. Additional charges as high as $500 are appearing on settlement sheets around the country. What's the justification? Some brokers say that paperwork processing and technology costs are getting higher. They say they must invest in computer equipment and software to stay efficient and keep up with the competition. Others just cite the cost of doing business and of providing high-impact marketing to their customers.

The best hope for consumers is that more firms will keep rates at 6% to undercut the competition. Obviously, every time a firm charges 7% or adds on higher processing fees, the more it favors competing firms that charge less. The worst fear for consumers is that more mergers and acquisitions will consolidate the real estate industry, and reduce competition to monopolistic proportions. For Sale by Owner, anyone?

Sources used to create this article include writer Felicia Paik and The Wall Street Journal.
Source: http://www.ahahome.com/articles/99/070799.html, May 15, 2001. Courtesy of AHA, the American Homeowners Association.

A homebuyer is trying to decide between two real estate agencies. Agency A charges 6% plus a $500 administrative fee, and Agency B charges 6.25%. How would the total charges differ on a home that sold for $200,000? Which agency is less expensive if the home sells for less than $200,000?

The Bottom Line

Summary of chapter learning objectives:

| Learning Objective | Summary | Example |
|---|---|---|
| **11.1** | Calculate sales commissions and gross pay | 1. A salesperson gets a $2,240 salary and a 2% commission. Find the commission and the gross pay when sales are $58,200 and returns are $6,500. |
| **11.2** | Calculate graduated sales commissions | 2. A salesperson has a graduated commission rate: 1% on sales up to $100,000; 1.5% on sales from $100,001 to $200,000; and 2% on sales above $200,000. Find the commission when sales are $245,000. |
| **11.3** | Calculate sales and purchases for principals | 3. A broker sells a principal's merchandise at a gross sales price of $15,600 and a commission rate of 3.5%. There are sales costs of $300 for storage and $119 for delivery. Find the commission and net proceeds. |
| | | 4. A commission merchant purchases merchandise for a principal at a prime cost of $6,250. The commission rate is 8%, air freight is $139, and local delivery is $65. Find the commission and gross cost. |

Answers: 1. Commission: $1,034; Gross pay: $3,274 **2.** $3,400 **3.** Commission: $546; Net proceeds: $14,635 **4.** Commission: $500; Gross cost: $6,954

Self-Check | Review Problems for Chapter 11

In problems 1–4, calculate both the commission and the total pay based on the information given.

1 Salary, $3,500; commission rate, 6%; total sales, $48,000; returns, $0

 a. Commission _____ b. Total pay _____

2 Salary, $2,800; commission rate, 5%; total sales, $88,500; returns, $12,450

 a. Commission _____ b. Total pay _____

3 Salary, $4,500; commission rate, 4%; total sales, $74,000; returns, $8,975

 a. Commission _____ b. Total pay _____

4 Salary, $0; commission rate, 8%; total sales, $98,000; returns, $11,425

 a. Commission _____ b. Total pay _____

5 Calculate the total commission on sales of $150,000 if the commission rates are 3% on the first $100,000 and 5% on everything above $100,000. _____

6 Calculate the total commission on sales of $75,000 if the commission rates are 3% on the first $100,000 and 5% on everything above $100,000. _____

7 Calculate the total commission on sales of $250,000 if the commission rates are 2% on the first $75,000; then 3% on the next $75,000; and 4% on everything above $150,000. _____

8 Calculate the total commission on sales of $125,000 if the commission rates are 2% on the first $75,000; then 3% on the next $75,000; and 4% on everything above $150,000. _____

9 Calculate the total commission on sales of $65,000 if the commission rates are 2% on the first $75,000; then 3% on the next $75,000; and 4% on everything above $150,000. _____

10 Calculate the total commission on sales of $125,000 if the commission rates are 4% on the first $35,000; then 6% on the next $45,000; and 8% on everything above $80,000. _____

11 Mike Colangelo is paid 3% on all sales. He is also paid a bonus of an additional 1% on any sales above $75,000. Calculate Mike's total commission on sales of $115,000. _____

12 JoAnn Brandt is paid 4% on all sales. She is also paid a bonus of an additional 2% on any sales above $40,000. Calculate JoAnn's total commission on sales of $95,000. _____

13 Charles Ho sells used logging equipment on consignment. He charges 22% plus expenses. Calculate Charles's commission on a log truck he sold for $38,750. _____

14 For the sale in problem 13, Charles also paid an additional $365 to deliver the truck to the new owner. Calculate the net proceeds that Charles's principal should receive. _____

15 Virginia Hendricks is a designer who purchases furniture for clients. She charges 16% of the price, plus expenses. Calculate her commission on furniture priced at $18,500. _____

16 For the sale in problem 15, calculate the gross cost to the client if Virginia also had expenses of $646.12.

Notes

Assignment 11.1: Commission

Name _____

Date _____ Score _____

A **(24 points) Find the commission and the total gross pay. (2 points for each correct answer)**

| Employee | Monthly Salary | Commission Rate | Monthly Sales | Commission | Gross Pay |
|---|---|---|---|---|---|
| **1.** Sosa, Manuel | $ 0 | 9% | $42,000 | _____ | _____ |
| **2.** Nakano, Beverly | 2,000 | 3% | 36,000 | _____ | _____ |
| **3.** Haliski, Chester | 1,600 | 4.5% | 38,000 | _____ | _____ |
| **4.** Aguilar, Cirano | 2,400 | 2% | 40,000 | _____ | _____ |
| **5.** Fong, Zhilong | 1,200 | 5% | 52,000 | _____ | _____ |
| **6.** Rossi, Gary | 2,200 | 4% | 36,000 | _____ | _____ |

Score for A (24)

B **(36 points) Calculate the total commission for the following commission payment plans. (6 points for each correct answer)**

| Graduated Commission Rates | Sales | Commission |
|---|---|---|
| **7.** 2% on sales to $50,000
4% on sales above $50,000 | $ 92,000 | _____ |
| **8.** 1% on sales to $175,000
2% on sales above $175,000 | $195,000 | _____ |
| **9.** 3% on sales to $40,000
5% on sales above $40,000 | $ 82,200 | _____ |
| **10.** 1% on sales to $80,000
2% on sales from $80,001 to $160,000
3% on sales above $160,000 | $225,000 | _____ |
| **11.** 3% on sales to $40,000
4% on sales from $40,001 to $100,000
5% on sales above $100,000 | $116,000 | _____ |

Assignment 11.1 *Continued*

12. 2% on sales to $65,000 $134,800 _____
 3% on sales from $65,001 to $150,000
 4% on sales above $150,000

C **(20 points) Kathy Maguire is a commission merchant. She charges different commission rates to sell different types of merchandise. During May, she completed the following consignment sales for consignors. Find Kathy's commission on each sale and the net proceeds sent to each consignor. (2 points for each correct answer)**

| | Gross Sales | Comm. Rate | Commission | Local Delivery | Storage | Air Freight | Net Proceeds |
|---|---|---|---|---|---|---|---|
| 13. | $28,400 | 3% | _____ | $68.75 | $ 0 | $173.50 | _____ |
| 14. | 1,600 | 3.5% | _____ | 88.50 | 65.00 | 0 | _____ |
| 15. | 8,400 | 6% | _____ | 284.65 | 0 | 0 | _____ |
| 16. | 12,880 | 5% | _____ | 0 | 0 | 248.00 | _____ |
| 17. | 5,600 | 4% | _____ | 0 | 85.00 | 115.00 | _____ |

D **(20 points) David Klein, a commission merchant in Chicago, buys merchandise exclusively for principals. Listed below are five recent transactions. Calculate David's commission on each purchase and the gross cost. (2 points for each correct answer)**

| | Prime Cost | Comm. Rate | Commission | Local Delivery | Storage | Air Freight | Gross Cost |
|---|---|---|---|---|---|---|---|
| 18. | $14,200 | 5% | _____ | $89.50 | $ 88.00 | $0 | _____ |
| 19. | 4,975 | 12% | _____ | 0 | 0 | 195.00 | _____ |
| 20. | 8,400 | 6% | _____ | 30.00 | 28.00 | 96.00 | _____ |
| 21. | 2,850 | 8% | _____ | 0 | 110.00 | 108.80 | _____ |
| 22. | 19,700 | 9% | _____ | 150.00 | 0 | 0 | _____ |

Assignment 11.2: Applications with Commission

Name

Date Score

A **(56 points) Solve each of the following business application problems involving salespeople who are paid partly or entirely on a commission basis. Solve the problems in order, because some of the questions are sequential. (8 points for each correct answer)**

1. Mike Dimmick sells memberships to an athletic club. He receives a monthly salary of $1,500 plus a commission of 12% on new membership fees. What was Mike's monthly pay for May, when he sold new memberships valued at $32,500? _____

2. Roberta Rivas sells commercial restaurant supplies and equipment. She is paid on a commission-only basis. She receives 2% for her sales up to $60,000. For the next $90,000 of sales, she is paid 3%, and for any sales above $150,000 she is paid 4%. How much commission would she earn in a month when her sales were $165,000? _____

3. Roberta Rivas (problem 2) is not paid commission on any restaurant supplies or equipment that are later returned. If an item is returned, its price is deducted from Roberta's total sales to get her net sales. The commission-only rate is applied to her net sales. Suppose that Roberta sold merchandise worth $165,000 but that $20,000 of that was later returned. What would be Roberta's commission on net sales? _____

4. Peggy Covey works for Southwest Appliance Depot. She receives a monthly salary of $2,500 for which she must sell $20,000 worth of appliances. She also receives a commission of 4% on net sales above $20,000. What will be her pay for October when her net appliance sales were $42,000? _____

5. Southwest Appliance Depot (problem 4) offers service contracts with all appliance sales. To encourage salespeople such as Peggy to sell more service contracts, the company pays a commission of 20% on all service contracts. What will be her total pay for a month if she sells $42,000 worth of appliances and $1,200 worth of service contracts? _____

6. Stockbrokers for companies such as PaineWebber are normally paid a commission on the stocks that they buy and sell for their clients. Suppose that the commission rate is 0.5% of the value of the stock. What will the commission be on 5,000 shares of General Motors stock that is selling for $67.31 per share? _____

7. Judy Taylor works in telemarketing. Her job is to make telephone calls from a computerized list of names and try to convince people to make an appointment with a life insurance salesperson. Judy receives 30¢ for each completed telephone call, $6.00 for each appointment made and kept, and 0.75% of any initial revenue that results from the appointment. How much would Judy earn if she completed 862 calls, 137 persons made and kept appointments, and $23,500 in revenue resulted from the appointments? _____

<div align="right">

Score for A (56)

</div>

B **(24 points) Solve each of the following business applications about consignment sales and commission merchants. (8 points for each correct answer)**

8. Teresa Gomez is a commission merchant specializing in consignment sales of antique furniture. Hank Windell, a furniture owner, sets the price of his antique furniture at $5,000. He agrees to pay Teresa a 15% commission plus $145 shipping from his home to her showroom. He also agrees to reimburse Teresa $188 for shipping the antiques from her showroom to the buyer's home. What will be Hank's net proceeds from the sale? _____

9. What would net earnings for Teresa Gomez (problem 8) be if she were required to pay the shipping expenses from her showroom to the buyer's home? _____

10. Sandy McCulloch makes artistic weavings that are used as wall hangings. She sells her weavings primarily at open-air art shows and street fairs through her agent, Ruth Danielson. Ruth charges 20% on all sales, plus the fees to operate a sales booth and transportation expenses. What will be Sandy's net proceeds if Ruth sold weavings worth $22,400 at four different art shows? Each art show charged a booth fee of $450, and Ruth's total transportation expenses were $375. _____

<div align="right">

Score for B (24)

</div>

C **(20 points) The following problems involve the purchase of a home. (10 points for each correct answer)**

11. JoAnn Ednie has a house that she would like to sell and she asks real estate agent Gene Jenkins to sell it. Gene works for Coldwell Banker Real Estate, which advises JoAnn that she should be able to sell her house for $150,000. The commission rate for selling a house is 6%. If the house sells for the expected price, what will be the total commission amount that JoAnn pays? _____

12. See problem 11. To sell her home, JoAnn Ednie must pay some additional fees for termite inspection and title insurance, as well as fees to the county to record the transaction. These fees total $2,500 and are added to the 6% commission. What will JoAnn's net proceeds be from the sale of her $150,000 home? _____

<div align="right">

Score for C (20)

</div>

Notes

Discounts are used throughout business. Trade discounts are frequently given to buyers "in the trade" who buy large quantities of merchandise. Airlines give discounts to frequent travelers or to those who agree to travel at certain times to certain places.

SAM'S Club is the nation's leading members-only warehouse, with 42 million members who receive exceptional value on name-brand merchandise at "members only" prices for both business and personal use. In addition, Sam's holds weekly auctions. Visit their site at http://www.samsclub.com.

Discounts 12

LEARNING OBJECTIVES

By studying this chapter and completing all assignments you will be able to:

 1 Calculate single trade discounts.

 2 Calculate a series of trade discounts.

 3 Calculate the equivalent single discount rate for a series of trade discounts.

 4 Calculate cash discounts and remittance amounts for fully paid invoices.

 5 Calculate cash discounts and remittance amounts for partially paid invoices.

Smart travelers and theater-goers are able to see top shows for half price by taking advantage of a discount known as the "Twofer," which entitles you to purchase two seats for the price of one. But travelers must be wary: "Discount" and "Going Out of Business" signs can actually mean higher prices.

1. What is a primary purpose of a trade discount?
2. What is one advantage that the complement method has over the discount method?
3. What is a primary purpose of a cash discount?
4. In a practical sense, why are returned goods and/or freight charges subtracted from the invoice before cash discount is calculated?
5. Why might a merchant not want to give any cash discount on an invoice that is only partially paid within the discount period?

When one business sells merchandise to another business, the seller often offers two types of discounts: trade discounts and cash discounts. Trade discounts affect the agreed-upon selling price *before* the sale happens. Cash discounts affect the amount actually paid *after* the transaction.

Calculating Single Trade Discounts

Calculate single trade discounts.

Businesses that sell products want to attract and keep customers who make repeated, large-volume purchases. Manufacturers, distributors, and wholesalers frequently offer **trade discounts** to buyers "in the trade," generally based on the quantity purchased. For example, Federated Food Supply gives a 40% discount to Cajun Hot, a local chain of 34 sidewalk sandwich carts that sell hot dogs and sausages. Another Federated customer is Suzi Wilson, founder and owner of Suzi's Muffins. Suzi's business is still small. She bakes her muffins between 11 P.M. and 2 A.M. in oven space that she leases from a bakery. Federated gives Suzi only a 25% discount because she doesn't do as much business as Cajun Hot does with Federated. Federated also sells to people who aren't "in the trade." These retail customers pay the regular **list price**, or full price without discount.

Large restaurant chains such as McDonald's can go directly to the manufacturer for most items or even do their own manufacturing. They can have items manufactured to their exact specifications for a contracted price. They reduce their costs by eliminating the distributors (the "middle men").

The two traditional methods for calculating trade discounts are the discount method and the complement method. You can use both to find the **net price** that a distributor will charge a customer after the discount. The **discount method** is useful when you want to know both the net price and the actual amount of the trade discount. The **complement method** is used to find only the net price. It get its name because you use the **complement rate,** which is 100% minus the discount rate. Each method has only two steps.

> **STEPS** **to Calculate Net Price with the Discount Method**
>
> **1.** Multiply the discount rate (expressed as a decimal) by the list price to get the discount amount:
>
> Discount = Trade discount rate × List price
>
> **2.** Subtract the discount from the list price to get the net price:
>
> Net price = List price − Discount

© PHOTODISC, INC.

EXAMPLE A

Federated Food Supply sells a set of stainless steel trays to Suzi's Muffins. The list price is $240, and Suzi qualifies for a 25% trade discount. Calculate the net price using the discount method.

STEP 1 Discount = 0.25 × $240 = $60

STEP 2 Net price = $240 − $60 = $180

> **STEPS** **to Calculate Net Price with the Complement Method**
>
> 1. Subtract the discount rate (expressed as a percentage) from 100% to get the complement rate:
>
> Complement rate = 100% − Trade discount rate
> 2. Multiply the complement rate (expressed as a decimal) by the list price to get the net price:
>
> Net price = Complement rate × List price

EXAMPLE B

Using the data in example A, compute the net price, using the complement method.

STEP 1 Complement Rate = 100% − 25% = 75%

STEP 2 Net price = 0.75 × $240 = $180

CONCEPT CHECK 12.1

a. Calculate the trade discount amount and the net price, using the discount method.

 List price = $230 Trade discount = 30%
 Discount amount = 0.30 × $230 = $69
 Net price = $230 − $69 = $161

b. Calculate the complement rate and the net price, using the complement method.

 List price = $230 Trade Discount = 30%
 Complement rate = 100% − 30% = 70%
 Net price = 0.70 × $230 = $161

Calculating a Series of Trade Discounts

LEARNING OBJECTIVES

Calculate a series of trade discounts.

A distributor or manufacturer may give additional discounts to customers who actually buy the largest volumes. Suppose that Federated gives all food preparation businesses a 25% discount for being in the trade. However, if one business buys twice as much from Federated, it may be rewarded with an **additional discount.** For example, Suzi's Muffins may receive its first discount of 25% automatically. Then, Suzi's gets an additional 20% discount if its accumulated purchases were between $10,000 and $25,000 during the previous year and another 10% if accumulated purchases were more than $25,000 during the previous year. Therefore Suzi's Muffins could have discounts of 25%, 20%, and 10%, called a **series of discounts.**

Both the discount method and the complement method can be used to compute the net price for a series of discounts. *The two methods are the same as shown previously, except that the steps are repeated for each discount in the series.* For example, if there are three discounts, repeat the steps three times. Apply the first **discount rate** to the list price. For the second and third discounts, compute intermediate prices and then apply the discount rates to them.

**www.
goodfooddistributors.
co.uk**

Trade discounts are common around the world. In England, Good Food Distributors offers two plans of trade discounts, one based on order totals with full payment on delivery and the other based on monthly purchases. Go to this Web site to learn more about these plans.

EXAMPLE C

Federated Food Supply sells a set of mixing bowls for a list price of $480. Suzi's Muffins qualifies for the series of discounts: 25%, 20%, 10%. Calculate the net price using the discount method.

| | **1st discount** | **2nd discount** | **3rd discount** |
|---|---|---|---|
| STEP 1 | $0.25 \times \$480 = \120 | $0.20 \times \$360 = \72 | $0.10 \times \$288 = \28.80 |
| STEP 2 | $\$480 - \$120 = \$360$ | $\$360 - \$72 = \$288$ | $\$288 - \$28.80 = \$259.20$ |

EXAMPLE D

Using the data in example C, calculate the net price using the complement method.

| | **1st discount** | **2nd discount** | **3rd discount** |
|---|---|---|---|
| STEP 1 | $100\% - 25\% = 75\%$ | $100\% - 20\% = 80\%$ | $100\% - 10\% = 90\%$ |
| STEP 2 | $0.75 \times \$480 = \360 | $0.80 \times \$360 = \288 | $0.90 \times \$288 = \259.20 |

Complement Method Shortcut

When you use complement rates, an efficient shortcut eliminates all the intermediate prices:

Multiply the list price by the product of all the complement rates.

EXAMPLE E

Repeat example D, using the shortcut. The list price is $480, and the discounts are 25%, 20%, and 10%. The complement rates are 75%, 80%, and 90%.
Net price = $\$480 \times 0.75 \times 0.80 \times 0.90 = \259.20

Note: Remember that there should be *no rounding* until you reach the final net price. Then round it to the nearest cent.

CONCEPT CHECK 12.2

a. A wholesaler offers a series of trade discounts: 30%, 25%, and 10%. Find each of the discount amounts and the final net price on a $1,200 purchase.

| | |
|---|---|
| First discount amount: | $\$1,200 \times 0.30 = \360 |
| Second discount amount: | $\$1,200 - \$360 = \$840; \$840 \times 0.25 = \$210$ |
| Third discount amount: | $\$840 - \$210 = \$630; \$630 \times 0.10 = \$63$ |
| Net price: | $\$630 - \$63 = \$567$ |

b. A series of trade discounts is 30%, 25%, and 10%. Find each of the complement rates, and use the shortcut to calculate the final net price on a purchase of $1,200.

| | |
|---|---|
| First complement rate: | $100\% - 30\% = 70\%$ |
| Second complement rate: | $100\% - 25\% = 75\%$ |
| Third complement rate: | $100\% - 10\% = 90\%$ |
| Net price: | $\$1,200 \times 0.70 \times 0.75 \times 0.90 = \567 |

Calculating the Equivalent Single Discount Rate

Suppose that North Coast Hotel Supply Company offers a single discount of 45% to Suzi's Muffins. How does that rate compare to the series of discounts from Federated: 25%, 20%, and 10%? Suzi's accountant could check by calculating the **equivalent single discount rate**, which is the single discount rate that can be used in place of two or more trade discount rates to determine the same discount amount.

 The most efficient way to find the single discount rate that is equivalent to a series of discounts is essentially the shortcut used in example E.

STEPS **to Compute the Equivalent Single Discount Rate**

1. Calculate the complement of each rate.
2. Multiply all the complement rates and write the product as a percent.
3. Subtract the product (Step 2) from 100% to get the equivalent single discount rate.

EXAMPLE F

Find the equivalent single discount rate for Federated's series of discounts: 25%, 20%, and 10%.

STEP 1
| | |
|---|---|
| 1st complement rate | $= 100\% - 25\% = 75\%$ |
| 2nd complement rate | $= 100\% - 20\% = 80\%$ |
| 3rd complement rate | $= 100\% - 10\% = 90\%$ |

STEP 2 Product of complements $= 0.75 \times 0.80 \times 0.90 = 54\%$

STEP 3 Equivalent single discount $= 100\% - 54\% = 46\%$

 CONCEPT CHECK 12.3

A series of trade discounts is 40%, 20%, and 5%. Find the three complement rates and then find the equivalent single trade discount rate.

Complement rates: $100\% - 40\% = 60\%, 100\% - 20\% = 80\%, 100\% - 5\% = 95\%$
Product of the complement rates: $0.60 \times 0.80 \times 0.95 = 0.456$, or 45.6%
Equivalent single discount rate: $100\% - 45.6\% = 54.4\%$

COMPLETE ASSIGNMENT 12.1.

Calculating Cash Discounts for Fully Paid Invoices

LEARNING
OBJECTIVES

Calculate cash discounts and remittance amounts for fully paid invoices.

When a seller sends merchandise to a buyer, the seller usually wants to get its payment quickly and some buyers often try to delay payment as long as possible. Sellers can encourage early payment by offering a **cash discount;** they can discourage late payment by assessing an extra interest payment; or they can do both. These stipulations are called the **terms of payment,** or simply the *terms.* The terms describe details about cash discounts and/or penalty periods.

After shipping merchandise to a buyer, the seller usually sends a document called an invoice, requesting payment. An **invoice** lists each product, its cost (including packaging and freight) and the total cost. The invoice states the terms of payment. The amount the buyer pays is called the **remittance.** The net purchase amount is the price of the merchandise actually purchased, including allowances for returns and excluding handling and other costs.

> **STEPS** **to Calculate the Remittance**
>
> 1. Multiply the discount rate (expressed as a decimal) times the net purchase amount to get the cash discount:
> Cash discount = Discount rate × Net purchase amount
> 2. Subtract the cash discount from the net purchase amount to get the remittance:
> Remittance = Net purchase amount − Cash discount

VIDEO

Cash Discounts

Figure 12-1 shows an invoice from National Automotive Supply, which sold car wax to Speedway Auto Detailers for $480. The wax will be shipped via UPS, and National will pay for the shipping. The invoice lists terms of 2/10, n/30. The **invoice date**, or the beginning of the discount period, is May 23.

Figure 12-1: Sales Invoice

NATIONAL AUTOMOTIVE SUPPLY

INVOICE NO. 782535

SOLD TO Speedway Auto Detailers
730 W. Columbia Dr.
Peoria, IL 62170-1184

DATE May 23, 200–
TERMS 2/10, n/30
SHIP VIA UPS

| QUANTITY | DESCRIPTION | UNIT PRICE | GROSS AMOUNT | NET AMOUNT |
|---|---|---|---|---|
| 24 gals. | Car wax | $20.00 | $480.00 | $480.00 |

The expression 2/10, n/30 means that Speedway can get a 2% discount if it pays the invoice within 10 days of the invoice date. Ten days after May 23 is June 2, which is called the **discount date.** The 10-day period between May 23 and June 2 is called the **discount period.** The n/30 is short for net 30, which means that if Speedway does not pay within 30 days, National will charge an interest penalty. Thirty days after May 23 is June 22, which is called the **due date.** (See Figure 12-2)

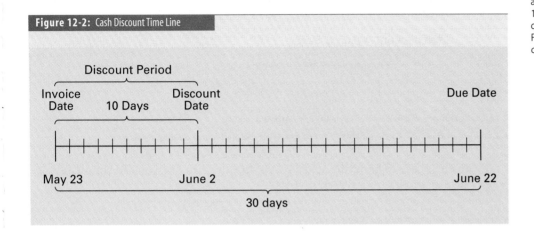

Figure 12-2: Cash Discount Time Line

EXAMPLE G

Calculate the remittance due if Speedway pays National within the 10-day discount period.

STEP 1 Cash discount = 2% of $480 = 0.02 × $480 = $9.60

STEP 2 Remittance = $480 − $9.60 = $470.40

All companies do not use exactly the same notation for writing their terms; 2/10, n/30 is also written as 2/10, net 30 or as 2-10, n-30. Likewise, there can be more than one discount rate and discount period. For example 2/5, 1/15, n/30 means that the seller gets a 2% discount by paying within 5 days, gets a 1% discount by paying between 6 and 15 days, and must pay a penalty after 30 days.

Returned Merchandise and Freight Charges

The seller gives a discount only on merchandise that is actually purchased—the net purchases. For example, there is no discount on returned items. Likewise, there is no discount on charges from a third party, such as freight.

STEPS **to Calculate the Remittance When There Are Merchandise Returns and/or Freight Charges**

1. Net purchase = Invoice amount − Merchandise returns − Freight
2. Cash discount = Discount rate × Net purchase
3. Cost of merchandise = Net purchase − Cash discount
4. Remittance = Cost of merchandise + Freight, if any

EXAMPLE H

National Automotive Supply sells merchandise to Speedway Auto Detailers. The invoice amount is $510, which includes $30 in freight charges. The invoice date is August 13, and the terms are 2/10, n/30. Speedway returns $200 worth of merchandise and pays the rest of the invoice before the discount date. Compute the cash discount and the remittance. Also, determine the discount date and due date.

STEP 1 Net purchase = $510 − $200 − $30 = $280

STEP 2 Cash discount = 0.02 × $280 = $5.60

STEP 3 Cost of merchandise = $280 − $5.60 = $274.40

STEP 4 Remittance = $274.40 + $30 = $304.40

Discount date = August 13 + 10 days = August 23
Due date = August 13 + 30 days = September 12

If you don't need to know the actual cost of the merchandise, you can eliminate Step 3 and calculate the remittance directly:

Remittance = $280.00 − $5.60 + $30.00 = $304.40

There is also a complement method for cash discounts. However, it isn't used as often as the discount method because most businesses want to know the amount of the cash discount before deciding whether to pay the invoice early. In the complement method for cash discounts, only Steps 2 and 3 change.

STEPS to Calculate the Remittance with the Complement Method

1. Net purchase = Invoice amount − Merchandise returns − Freight
2. Complement rate = 100% − Cash discount rate
3. Cost of merchandise = Net purchase × Complement rate
4. Remittance = Cost of merchandise + Freight, if any

EXAMPLE I

Solve example H by using the complement method for cash discounts. The invoice amount is $510, merchandise returns are $200, and freight is $30.

STEP 1 Net purchase = $510 − $200 − $30 = $280

STEP 2 Complement rate = 100% − 2% = 98%

STEP 3 Cost of merchandise = $280 × 0.98 = $274.40

STEP 4 Remittance = $274.40 + $30 = $304.40

a. Use the given information to calculate the discount date, due date, cash discount, and remittance.

| Terms: | 1/10, n/60 | Discount date = July 22 + 10 days = August 1 |
|---|---|---|
| Invoice date: | July 22 | Due date = July 22 + 60 days = September 20 |
| Invoice amount: | $842.17 | |
| Returned goods: | $195.67 | Net purchases = $842.17 − $195.67 − $51.50 = $595.00 |
| Freight: | $51.50 | Cash discount = 0.01 × $595 = $5.95 |
| | | Remittance = $595 − $5.95 + $51.50 = $640.55 |

b. Calculate the remittance for the problem in part (a), using the complement method.

Net purchases = $842.17 − $195.67 − $51.50 = $595.00
Complement rate = 100% − 1% = 99%
Cost of merchandise = $595 × 0.99 = $589.05
Remittance = $589.05 + $51.50 = $640.55

Calculating Cash Discounts for Partially Paid Invoices

Sometimes a buyer wants to take advantage of a cash discount but can afford to pay only part of the invoice within the discount period. The invoice will be reduced by the amount paid (remittance) plus the amount of the discount. The total of the amount paid plus the amount of cash discount is called the **amount credited** to the buyer's account. To calculate the amount credited, you need to know the complement rate: 100% − Discount rate.

Calculate cash discounts and remittance amounts for partially paid invoices.

STEPS to Compute the Unpaid Balance

1. Calculate the complement of the discount rate (100% − Discount rate).
2. Calculate the amount credited by dividing the amount paid (remittance) by the complement rate.
3. Calculate the unpaid balance by subtracting the amount credited (Step 2) from the invoice amount.

Kathi Palmer operates a shop called The Space Place, a do-it-yourself center for closets and storage. Kathi buys shelving materials with an invoice price of $392 and terms of 2/10, net 60. Within the 10-day discount period, she sends in a check for $200. How much credit should Kathi receive, and what is her unpaid balance?

STEP 1 Complement rate = 100% − 2% = 98%

STEP 2 Amount credited = $200 ÷ 0.98 = $204.0816 = $204.08

STEP 3 Unpaid balance = $392.00 − $204.08 = $187.92

Note that, in example J, Kathi receives $1.00 credit for every $0.98 paid. In other words, the $200 actually remitted is 98% of the total amount credited. We check this result with multiplication:

Cash discount = 0.02 × $204.08 = $4.0816, or $4.08
Remittance = $204.08 − $4.08 = $200.00

A slightly different situation, which arises less frequently, is when a buyer decides in advance the total amount that he or she wants to have credited to the account. This problem is exactly like the original cash discount problems.

EXAMPLE K

Kathi Palmer buys $392 worth of shelving materials for use in her closet and storage shop. The terms are 2/10, net 60. Kathi wants to pay enough within the 10-day discount period to reduce her unpaid balance by exactly $200. What amount should she remit to the seller? What will be her unpaid balance?

STEP 1 Cash discount = 2% × $200 = $4

STEP 2 Remittance = $200 − $4 = $196

STEP 3 Unpaid balance = $392 − $200 = $192

CONCEPT CHECK 12.5

a. An invoice for $476 has terms of 1/15, net 25. How much is the unpaid balance after a $350 remittance is made within the discount period?
Complement rate = 100% − 1% = 99%
Amount credited = $350 ÷ 0.99 = $353.54
Unpaid balance = $476.00 − $353.54 = $122.46

b. An invoice for $476 has terms of 1/15, net 25. What size remittance should be made in order to have a total of $350 credited to the account?
Cash discount = $350 × 0.01 = $3.50
Remittance = $350.00 − $3.50 = $346.50

COMPLETE ASSIGNMENT 12.2.

Chapter Terms for Review

| | |
|---|---|
| additional discount | equivalent single discount rate |
| amount credited | invoice |
| cash discount | invoice date |
| complement method | list price |
| complement rate | net price |
| discount date | net purchase amount |
| discount method | remittance |
| discount period | series of discounts |
| discount rate | terms of payment |
| due date | trade discounts |

Visit the Web

http://www.directron.com/discounts.html—Consumers can also obtain discounts as repeat customers and as volume buyers. For more on discounts, go to this Web site.

http://www.ppa.org/ppaonline/about_ppa/member_discounts.asp—PPAI negotiates discounts and special services by pooling the buying power of its member companies. Learn more about this firm by accessing its Web site.

http://www.norweld.lib.oh.us/vendorlist.htm—The NORWELD Regional Library System assists Northwest Ohio public, academic, and school libraries and offers vendor discounts for products and services. For more information about this organization's offerings, visit its Web site.

CYBERCONNECTION

Original Series

The Original Series is a comprehensive line of traditional wall clocks made of high impact plastic. With a variety of colors, sizes and dial face options, you coordinate the clock to your specifications!

Order Below

Size Options:
8" Dial (9-1/16" Total Diameter)
12" Dial (13-1/16" Total Diameter)

Dial Face Options:
Arabic-White
Arabic-Special
Arabic-Brass
Roman-White

Samples:
click on these images for an enlarged view

Taupe
w/Arabic-White

Stone Gray
w/Roman-White

Source: http://www.chicagolighthouse.org/showcases/designers/original.html. Copyright © 2001 The Chicago Lighthouse for People who are blind or visually impaired. Used with permission.

Chicago Lighthouse Industries is a state-of-the-art manufacturing facility that employs people who are blind or visually impaired to manufacture quality clocks, as well as other products. Special trade discounts are available to architects, interior designers, dealers, and corporate buyers. You are a corporate buyer, and your company has just opened new offices. You need to buy 30 wall clocks.

Go to this Web site and select the 12-inch dial face in Taupe w/Arabic-White, which costs $35.31. Assume a series trade discount of 20% off orders of more than $500 and an additional 10% off orders of more than $1,000. Terms are 2/10, net 30. Calculate the total amount of the invoice and the remittance.

The Bottom Line

Summary of chapter learning objectives:

| Learning Objective | Summary | Example |
|---|---|---|
| **12.1** | Calculate single trade discounts | 1. Find the net price on a list price of $450 with a 20% trade discount, using the discount and the complement methods. |
| **12.2** | Calculate a series of trade discounts | 2. Find the net price on a list price of $800 with a series of trade discounts of 25% and 10%. Use both the discount method and the complement method. |
| **12.3** | Calculate the equivalent single discount rate for a series of trade discounts | 3. A series of trade discounts is 20%, 10%, 5%. Use complement rates to find the equivalent single discount rate. |
| **12.4** | Calculate cash discounts and remittance amounts for fully paid invoices | An invoice is dated December 26 and has terms of 1.5/10, n/25. The total amount is $964.24, with $141.34 of returned goods and $82.90 freight.

4. Calculate the discount date, due date, cash discount, and remittance.
5. Calculate the remittance using the complement rate. |
| **12.5** | Calculate cash discounts and remittance amounts for partially paid invoices | An invoice for $500 has terms of 3-10, n-30.

6. Calculate the unpaid balance after a $400 payment within the discount period.
7. Calculate the remittance required within the discount period in order to have $400 credited to the account. |

Answers: **1.** Discount method: $450 − $90 = $360; complement method: 0.80 × $450 = $360 **2.** Discount method: $800 − $200 = $600, $600 − $60 = $540; complement method: 0.75 × 0.90 × $800 = $540 **3.** 31.6% **4.** Discount date: Jan. 5; due date; Jan. 20; cash discount: $11.10; remittance: $811.80 **5.** $811.80 **6.** $87.63 **7.** $388.00

In problems 1 and 2, use the discount method to calculate the missing terms.

In problems 3 and 4, use the complement method to calculate the missing terms.

1 List price, $480; trade discount, 25%

 a. Discount amount _____

 b. Net price _____

3 List price, $640; trade discount, 35%

 a. Complement rate _____

 b. Net price _____

2 List price, $1,200; trade discounts, 40% and 20%

 a. First discount amount _____

 b. Second discount amount _____

 c. Net price _____

4 List price, $1,600; trade discounts, 30% and 10%

 a. First complement rate _____

 b. Second complement rate _____

 c. Net price _____

5 Patty Duncan is a broker of hotel rooms in Europe. To tour directors, she offers a standard trade discount of 40% off the list price. She has additional discounts of 20% and 10%, which are based on the number of tours in a season and the total number of tourists. Calculate the equivalent single discount rate for tour organizer Kristi Atchison who qualifies for all three discounts.

 a. First complement rate _____

 b. Second complement rate _____

 c. Third complement rate _____

 d. Equivalent single discount rate _____

Use the invoice information given in problems 6 and 7 to calculate the missing terms.

6
| | |
|---|---|
| Terms: | 2/10, n/30 |
| Invoice Date: | July 25 |
| Invoice Amount: | $874.55 |
| Freight: | $116.00 |
| Returned Goods: | 0 |

 a. Discount date _____

 b. Due date _____

 c. Discount amount _____

 d. Remittance _____

7
| | |
|---|---|
| Terms: | 3/5, net 15 |
| Invoice Date: | October 29 |
| Invoice Amount: | $2,480 |
| Freight: | 0 |
| Returned Goods: | $642 |

 a. Discount date _____

 b. Due date _____

 c. Complement rate _____

 d. Remittance _____

8 Danny Martin purchased some new pieces of office furniture for his software firm. The invoice amount was $16,540 with terms of 2/10, net 60 and the discount would apply to any partial payment made within the discount period. Danny sent in a check for $10,000 by the discount date. Find (a) the amount credited to Danny's account _____ and (b) the unpaid balance _____ .

Assignment 12.1: Trade Discounts

Name _____

Date _____ Score _____

A (24 points) Problems 1–3: Find the dollar amount of the trade discount and the net price, using the discount method. Problems 4–6: Find the complement rate and the net price, using the complement method. (2 points for each correct answer)

| Trade Discount | List Price | Discount Amount | Net Price |
|---|---|---|---|
| **1.** 30% | $1,250 | _____ | _____ |
| **2.** 40% | $6,475 | _____ | _____ |
| **3.** 35% | $3,400 | _____ | _____ |

| Trade Discount | List Price | Complement Rate | Net Price |
|---|---|---|---|
| **4.** 15% | $1,600 | _____ | _____ |
| **5.** 34% | $3,000 | _____ | _____ |
| **6.** 40% | $4,750 | _____ | _____ |

Score for A (24)

B (16 points) Find the amount of each discount in the given series of trade discounts. Then find the net price. Where a discount doesn't exist, place a dash. (2 points for each correct answer)

| List Price | Trade Discounts | Trade Discount Amounts | | | Net Price |
|---|---|---|---|---|---|
| | | First | Second | Third | |
| **7.** $2,200 | 40%, 25% | _____ | _____ | _____ | _____ |
| **8.** $1,650 | 30%, 20% | _____ | _____ | _____ | _____ |

Score for B (16)

Assignment 12.1 *Continued*

C (20 points) Find the complement rate for each discount in the given series of trade discounts. Then find the net price, using the complement method. Where a complement rate doesn't exist, place a dash. (2.5 points for each correct answer)

| List Price | Trade Discounts | Complement Rates | | | Net Price |
|---|---|---|---|---|---|
| | | First | Second | Third | |
| **9.** $1,800 | 30%, 15% | ____ | ____ | ____ | ____ |
| **10.** $2,050 | 30%, 20%, 10% | ____ | ____ | ____ | ____ |

Score for C (20)

D (20 points) Find the complement rate for each discount in the given series of trade discounts. Then find the equivalent single discount rate, to the nearest $\frac{1}{10}$ of a percent. (2.5 points for each correct answer)

| Trade Discounts | Complement Rates | | | Equivalent Single Discount Rates |
|---|---|---|---|---|
| | First | Second | Third | |
| **11.** 30%, 20%, 5% | ____ | ____ | ____ | ____ |
| **12.** 20%, 10%, 5% | .8 (80%) | .9 (90%) | .95 | ____ |

Score for D (20)

E (20 points) Solve each of the following business applications about trade discounts. Use either the discount method or the complement method. (10 points for each correct answer)

13. Fleming Landscaping, Inc., purchased $425 worth of plants and $180 worth of soil and fertilizer from a garden supply wholesaler. The wholesaler gives Fleming a 20% trade discount on the plants and a 30% trade discount on the other items. Compute the net price that Fleming Landscaping will be required to pay. _____

14. Jack Hackett Roofing is purchasing redwood shakes to reroof a house. The shakes have a list price of $14,400. The Northwest Roofing Supply Company gives Jack the normal trade discount of 20%. In addition, Northwest gives Jack two further trade discounts of 10% and 5% because of the large volume of business that Jack has done with Northwest so far this year. What is Jack's net price on the order of redwood shakes? _____

14,400 × .8 × .9 × .95 =

$9,849.60

Score for E (20)

Assignment 12.2: Cash Discounts

Name

Date Score

A **(64 points) For the following problems, find the discount date, the due date, the amount of the cash discount, and the amount of the remittance. (2 points for each correct date and 6 points for each correct amount)**

1. Terms: 3/5, n/25 Discount date: _____

 Invoice date: March 28 Due date: _____

 Invoice amount: $518.30 Discount amount: _____

 Remittance: _____

2. Terms: 2/10, n/30 Discount date: _____

 Invoice date: October 24 Due date: _____

 Invoice amount: $482.86 Discount amount: _____

 Freight: $25.00 Remittance: _____

Discount

Full Amount

3. Terms: 2.5/15, net 45 Discount date: _____

 Invoice date: May 21 Due date: _____

 Invoice amount: $692.00 Discount amount: _____

 Returned goods: $184.00 Remittance: _____

$692 - 184 = 508$ $508 \times .025 = \$12.70$

 $508 - 12.70 = 495.30$

4. Terms: 1.5/20, N/60 Discount date: _____

 Invoice date: March 29 Due date: _____

 Invoice amount: $1,095.55 Discount amount: _____

 Returned goods: $298.88 Remittance: _____

 Freight: $80.00

Score for A (64)

Assignment 12.2 *Continued*

B **(16 points) For the following problems, find the discount date, the complement rate, and the amount of the remittance. (2 points for each date and rate; 4 points for each correct remittance)**

5. Terms: 1/7, n/25

 Invoice date: January 25

 Invoice amount: $482.80

Discount date: _____

Complement rate: _____

Remittance: _____

6. Terms: 1/25, net 55

 Invoice date: July 9

 Invoice amount: $684.92

 Returned goods: $171.12

 Freight: $45.00

Discount date: _____

Complement rate: _____

Remittance: _____

Score for B (16)

C **(20 points) The following problems involve partial payments made within the discount period. Solve for the items indicated. (5 points for each correct answer)**

7. Terms: 1.5/7, n/30 99 70 = .99 +

 Invoice date: June 2

 Invoice amount: $848 482 50 =

 $4|99 99

Amount credited: _____

Remittance: $500

Unpaid balance: _____

8. Terms: 2/15, net 35

 Invoice date: April 9

 Invoice amount: $532.90

 Returned goods: $184.00

Amount credited: _____

Remittance: $300

Unpaid balance: _____

Score for C (20)

Notes

Source: http://www.bluelight.com/home/index.jsp. Copyright © 2001 by bluelight.com; used with permission.

To compensate employees, pay store rents or mortgages, and earn a profit, owners of retail stores such as Kmart must sell merchandise for more than it costs them. The difference is called margin, and the amount of margin depends on many factors.

Kmart's BlueLight brand online rewards customers with nationally recognized products at unprecedented values. Popular brand names including Martha Stewart Everyday, Panasonic, Sesame Street, Black & Decker, and Nickelodeon are featured on BlueLight.com, along with the registration for Kmart's low-priced Internet Access, BlueLight Unlimited. The site also provides important information about company programs, and initiatives in the store or online. Visit Kmart's Web site at http://www.bluelight.com.

Markup

13

LEARNING OBJECTIVES

By studying this chapter and completing all assignments you will be able to:

1 LEARNING OBJECTIVES — Compute the variables in the basic markup formula.

2 LEARNING OBJECTIVES — Compute the markup variables when the markup percent is based on cost.

3 LEARNING OBJECTIVES — Compute markup percent based on cost.

4 LEARNING OBJECTIVES — Compute the markup variables when the markup percent is based on selling price.

5 LEARNING OBJECTIVES — Compute markup percent based on selling price.

When you read an ad for a sale, you may see the words "Marked down from . . . " This means that the price had been marked up from the store cost and now is lowered to sell goods faster. Goods are marked up high enough to allow for markdowns to stimulate sales.

1. What is the relationship between the cost of goods sold, the markup, and the selling price?
2. Does it make more sense to have markup based on cost or on selling price? Explain.
3. Give one reason why a particular merchant might want to have markup based on selling price.
4. If an item has a 100% markup based on cost, will it also have a 100% markup based on selling price? Explain, or give an example.

Computing Markup Variables

Compute the variables in the basic markup formula.

Some businesses manufacture products and sell them. Other businesses buy products from someone else and then resell them. Both types of businesses must sell their products for more than it costs to produce or purchase them. This price increase is called the **markup.**

Athletes' World is a retail store that sells athletic equipment and athletic clothing. The store buys running shoes directly from a manufacturer. Suppose that the manufacturer charges $43.00 per pair for one particular type of running shoes and an additional $0.50 to deliver them to the store. The total cost of a pair of the shoes, with delivery, is $43.50, called the **cost of goods sold,** or just the **cost.**

If Athletes' World sells the shoes for exactly the cost, $43.50, it will lose money on the sale. The store has many other expenses—such as rent, utilities, and salaries—that aren't part of the cost of acquiring the shoes. Athletes' World must mark up the selling price enough above the cost of the shoes to cover all these additional costs—and also leave some profit for the owners.

The total amount that Athletes' World marks up the selling price is called the **dollar markup.** (Note: Markup is expressed both in dollars and in percents. To eliminate confusion, in this book we use two separate terms: *dollar markup* and *markup percent.*)

Suppose that Athlete's World estimates that $18.80 of additional expenses should be allocated to a pair of running shoes. Also, suppose that the store would like a profit of $16.00 on the pair of shoes. Then the total dollar markup that it should give the shoes is $18.80 + $16.00 = $34.80.

To determine the selling price of the shoes, Athlete's World adds the dollar markup to the cost of goods sold (cost), using the basic markup formula:

Selling price = Cost + Dollar markup = $43.50 + $34.80 = $78.30

Because the dollar markup is the difference between the selling price and the cost of goods sold, it is often useful to rewrite the formula as

Dollar markup = Selling price − Cost = $78.30 − $43.50 = $34.80

Likewise, cost is the difference between selling price and dollar markup. Thus

Cost = Selling price − Dollar markup = $78.30 − $34.80 = $43.50

CONCEPT CHECK 13.1

Compute the missing terms in the basic markup formula: Selling price = Cost + Dollar markup

a. Cost = $417.82; Dollar markup = $204.20

b. Cost = $154.40; Selling price = $392.12

c. Dollar markup = $41.26; Selling price = $93.20

Selling price = Cost + Dollar markup
= $417.82 + $204.20 = $622.02
Dollar markup = Selling price − Cost
= $392.12 − $154.40 = $237.72
Cost = Selling price − Dollar markup
= $93.20 − $41.26 = $51.94

www.sba.gov/library/ pubs/fm-13.txt

The Small Business Administration (SBA) provides information on calculating markups at this Web site.

www.aaamath.com/ pct68-markup.html

www.jaxworks.com/ pricing.htm

More information about markups and practice in computing them is available at these two Web sites.

Computing Markup Based on Cost

Athletes' World computed its markup directly by determining its expenses and the desired profit. However, this method isn't practical when a business has hundreds or thousands of items. Allocating expenses and profit to each item would be too tedious. A more practical method is for the owner, an employee, or an accountant to analyze prior sales of the company or a similar company. The analyst can look at the costs of goods, additional expenses, and desired profit to determine a percent to use to mark up various items, called the **markup percent**.

One company may use different markup percents for different types of items. For example, an appliance store typically also performs repair services and sells replacement parts for the appliances it sells. The store may have one markup percent for the actual appliance, a second markup percent for repair services, and a third markup percent for replacement parts.

In Chapter 6 on percents, we introduced three terms: rate, base, and percentage. In this chapter, rate is the *markup percent*, or **markup rate.** Percentage is the *dollar markup*. Determining the base is more challenging because sometimes *cost* is the base and sometimes *selling price* is the base. For some businesses, cost may be the more logical base for calculating dollar markup. However, calculating dollar markup based on selling price is an advantageous method for many retail stores.

The accountant for Athletes' World says that, in order to pay all expenses and have a reasonable profit, and based upon a cost of $43.50, the company should have an 80% markup. When the cost and the markup percent are known, the dollar markup and the selling price can be computed.

LEARNING OBJECTIVES 2

Compute the markup variables when the markup percent is based on cost.

STEPS to Compute the Selling Price Based on Cost

1. Multiply the cost by the markup percent to get the dollar markup.
2. Add the dollar markup to the cost to get the selling price.

VIDEO

Markup Based on Cost/Selling Price

For Athlete's World's running shoes:

STEP 1 Dollar markup = Markup percent × Cost = 0.80 × $43.50 = $34.80

STEP 2 Selling price = Cost + Dollar markup = $43.50 + $34.80 = $78.30

EXAMPLE A

Using markup based on cost, what are the dollar markup and the selling price on merchandise that costs $40 and has a 25% markup?

STEP 1 Dollar markup = Markup percent × Cost = 0.25 × $40 = $10

STEP 2 Selling price = Cost + Dollar markup = $40 + $10 = $50

Computing Selling Price Directly from Cost

You can compute the selling price directly from the cost, without computing the dollar markup.

www.procrafter.com/ markup2.htm

When professional crafters sell their goods at a crafts show, they need to know how much to mark up their products. For more information, visit this Web site.

© PHOTODISC, INC.

STEPS **to Compute the Selling Price Directly from the Cost**

1. Add 100% to the markup percent.
2. Multiply this sum by the cost to get the selling price.

EXAMPLE B

What is the selling price of an item that has a cost of $200 and a markup percent of 30% based on cost?

STEP 1 Markup percent + 100% = 30% + 100% = 130%

STEP 2 Selling price = (Markup percent + 100%) × Cost = 1.30 × $200 = $260

Computing Cost from Selling Price

When you know the selling price and the markup percent, the procedure for computing cost is just the reverse of that for computing selling price.

STEPS **to Compute the Cost from the Markup Percent**

1. Add the markup percent to 100%.
2. Divide the selling price by this sum to get the cost.

EXAMPLE C

The selling price of a pair of shoes is $70. The markup percent based on cost is 40%. Find the cost.

STEP 1 100% + Markup percent = 100% + 40% = 140%

STEP 2 Cost = Selling price ÷ (100% + Markup percent) = $70 ÷ 1.40 = $50

You can always check your work in markup problems.
Cost is $50, and markup percent is 40%.
Dollar markup = Cost × Markup percent = $50 × 0.40 = $20
Selling Price = Cost + Dollar markup = $50 + $20 = $70

It checks!

CONCEPT CHECK 13.2

Compute the required values when the markup percent is based on cost.

a. Cost = $1,340; Markup percent = 30%
 Find dollar markup, and then find selling price.
b. Cost = $430; Markup percent = 60%
 Find 100% + Markup percent, and then find selling price directly.
c. Selling price = $780; Markup percent = 25%
 Find 100% + Markup percent, and then find cost directly.

Dollar markup = $1,340 × 0.30 = $402
Selling price = $1,340 + $402 = $1,742
100% + Markup percent = 100% + 60% = 160%
Selling price = $430 × 1.60 = $688
100% + Markup percent = 100% + 25% = 125%
Cost = $780 ÷ 1.25 = $624

Computing Markup Percent Based on Cost

In the illustration for Athletes' World, the accountant determined that the markup percent needed to be 80% of cost, which meant that the selling price needed to be $78.30. However, the storeowner may prefer to price the shoes at $78.95. Now, the markup percent is no longer 80% of cost. The **markup percent based on cost** can be computed in two steps.

3
LEARNING OBJECTIVES

Compute markup percent based on cost.

STEPS to Compute the Markup Percent Based on Cost

1. Subtract the cost from the selling price to get the dollar markup.
2. Divide the dollar markup by the cost to get the markup percent.

For the running shoes from Athletes' World, priced at $78.95:

STEP 1 Dollar markup = Selling price − Cost = $78.95 − $43.50 = $35.45

STEP 2 Markup percent = Dollar markup ÷ Cost = $35.45 ÷ $43.50 = 0.815, or 81.5% (rounded to one decimal place)

www. moneymarketingatm. com/default.asp

Did you realize that the money you withdraw from an ATM and spend near the same store can make a difference on the markup of retail goods in the store? For more information, access this Web site.

EXAMPLE D

What is the markup percent based on cost when the selling price is $80 and the cost is $50?

STEP 1 Dollar markup = Selling price − Cost = $80 − $50 = $30

STEP 2 Markup percent = Dollar markup ÷ Cost = $30 ÷ $50 = 0.60, or 60%

EXAMPLE E

What is the markup percent based on cost when the dollar markup is already known to be $15 and the cost is $75? (Step 1 is not necessary.)

STEP 2 Markup percent = Dollar markup ÷ Cost = $15 ÷ $75 = 0.20, or 20%

 CONCEPT CHECK 13.3

Cost = $1,800; Selling price = $3,150
Find the markup percent based on cost.

Dollar markup = $3,150 − $1,800 = $1,350
Markup percent = $1,350 ÷ $1,800 = 0.75, or 75%

COMPLETE ASSIGNMENT 13.1.

Computing Markup Based on Selling Price

LEARNING OBJECTIVES

Compute the markup variables when the markup percent is based on selling price.

Although many businesses base their markup on cost, some retailers commonly use a percent of selling price—that is, they use **markup based on selling price.** That doesn't mean that selling price is determined without considering cost or even before considering cost. It merely means that the dollar markup is computed by multiplying the markup percent by the selling price.

Many people start a business when they observe another successful business selling a product. New owners believe that they can acquire the product, pay all expenses, and still sell it for less than the existing business sells its product. Instead of basing the selling price on costs, expenses, and satisfactory profit, the new owners may price their product just under the competition's price. They base their selling price on the competition rather than marking up from their costs.

Basing markup calculations on selling price can be an advantage in a retail store where the salesperson or sales manager has the authority to lower the sales price immediately in order to make a sale.

STEPS to Compute the Dollar Markup and Cost from the Markup Percent

1. Multiply the selling price by the markup percent to get the dollar markup.
2. Subtract the dollar markup from the selling price to get the cost.

EXAMPLE F

Robert Landles enters an appliance store to buy a washing machine. He finds one with a selling price of $400. He knows that he can buy it for $375 at another store, but he prefers this store because of its reputation for good service. He tells the salesperson, "I would buy it for $375." The manager, Kristin Jorgensen, knows that the markup percent is 25% of the selling price. What is the cost of the washing machine?

STEP 1 Dollar markup = Markup percent × Selling price = 0.25 × $400 = $100

STEP 2 Cost = Selling price − Dollar markup = $400 − $100 = $300

Kristin can then decide whether she prefers no sale or one for which she gets only a $75 markup. Although it would be helpful if Kristin knew how much markup she would need to pay for expenses, at least she would know the cost.

EXAMPLE G

Find the dollar markup and the cost of an item that sells for $120 and has a markup percent that is 40% of the selling price.

STEP 1 Dollar markup = Markup percent × Selling price = 0.40 × $120 = $48

STEP 2 Cost = Selling price − Dollar markup = $120 − $48 = $72

Computing Cost Directly

You can compute the cost directly from the selling price, without computing the dollar markup.

STEPS to Compute the Cost from the Markup Percent and Selling Price

1. Subtract the markup percent from 100%.
2. Multiply this difference by the selling price to get the cost.

EXAMPLE H

What is the cost of an item that has a selling price of $200 and a markup percent of 30% based on selling price?

STEP 1 100% − Markup percent = 100% − 30% = 70%

STEP 2 Cost = (100% − Markup percent) × Selling price = 0.70 × $200 = $140

Computing Selling Price from Cost

When you know the cost and the markup percent, the procedure for computing cost is just the reverse of that for computing selling price.

STEPS to Compute the Selling Price from the Cost

1. Subtract the markup percent from 100%.
2. Divide the cost by this difference to get the selling price.

EXAMPLE I

The cost of a bicycle is $90. The markup percent based on selling price is 25%. Find the selling price.

STEP 1 100% − Markup percent = 100% − 25% = 75%

STEP 2 Selling price = Cost ÷ (100% − Markup percent) = $90 ÷ 0.75 = $120

You can always check your work in markup problems:
Selling price is $120, and markup percent is 25%.
Dollar markup = Selling price × Markup percent = $120 × 25% = $30
Cost = Selling price − Dollar markup = $120 − $30 = $90

It checks!

 CONCEPT CHECK 13.4

Compute the required values when the markup percent is based on selling price.

a. Selling price = $640; Markup percent = 40%
 Find dollar markup, and then find cost.
b. Selling price = $40; Markup percent = 30%
 Find 100% − Markup percent, and then find cost directly.
c. Cost = $160; Markup percent = 25%
 Find 100% − Markup percent, and then find selling price directly.

Dollar markup = $640 × 0.40 = $256
Selling price = $640 − $256 = $384
100% − Markup percent = 100% − 30% = 70%
Cost = $40 × 0.70 = $28
100% − Markup percent = 100% − 25% = 75%
Cost = $160 ÷ 0.75 = $213.33

Computing Markup Percent Based on Selling Price

Compute markup percent based on selling price

In the illustration for Athletes' World, the pair of running shoes had a cost of $43.50. The store owner decided that the selling price of the running shoes would be $78.95. The markup percent based on selling price can be calculated in two steps.

STEPS **to Compute the Markup Percent from the Selling Price**

1. Subtract the cost from the selling price to get the dollar markup.
2. Divide the dollar markup by the selling price to get the markup percent.

For Athletes' World's running shoes:

STEP 1 Dollar markup = Selling price − Cost = $78.95 − $43.50 = $35.45

STEP 2 Markup percent = Dollar markup ÷ Selling price = $35.45 ÷ $78.95 = 0.449, or 44.9% (rounded to one decimal place)

EXAMPLE J

What is the markup percent based on selling price when the selling price is $80 and the cost is $50?

STEP 1 Dollar markup = Selling price − Cost = $80 − $50 = $30

STEP 2 Markup percent = Dollar markup ÷ Selling price = $30 ÷ $80 = 0.375, or 37.5%

EXAMPLE K

What is the markup percent based on selling price when the dollar markup is already known to be $15 and the selling price is $60? (Step 1 is not necessary.)

STEP 2 Markup percent = Dollar markup ÷ Selling price = $15 ÷ $60 = 0.25, or 25%

CONCEPT CHECK 13.5

Cost = $1,800; Selling price = $3,150
Find the markup percent based on selling price.

Dollar markup = $3,150 − $1,800 = $1,350
Markup percent = $1,350 ÷ $3,150 = 0.429, or 42.9%

COMPLETE ASSIGNMENT 13.2.

Chapter Terms for Review

cost of goods sold
dollar markup
markup
markup based on selling price

markup percent
markup percent based on cost
markup rate

Visit the Web

http://www.biznetcenter.com/tips/markup.htm—More information about markups can be found here.

http://www.anjoch-investors.com/new98/indexn.html?../fm13—For information on pricing, visit this Web site.

http://www.kiplinger.com/magazine/archives/2000/November/spending/SHOPPING.html—Read an article about shopping habits at this Web site.

http://www.coconutgroup.com/retailin.htm—Read more on retail profits here.

http://fisher.osu.edu/citm/expa/costing_example.html—Review the costs that can be included in the final price to the consumer by accessing this Web site.

CYBERCONNECTION

◀ Return to the FAS Home Page

Singaporean Pets Pant for U.S. Pet Food

Pet food producers who sell to Singapore animal lovers will have no problem sniffing out profits in this fast growing market. Almost all the pet food is imported, and Singaporean pets are chowing down. For the past five years, retail sales of pet food have doubled, increasing 10 to 20 percent a year.

U.S. exporters supplied over 2,300 tons of dinners in 1996, approaching a C.I.F. (cost, insurance, freight) value of $3 million (U.N. Food and Agriculture Organization's trade data).

While this market may be small by some standards, Singapore serves as a retail product showcase for Southeast Asia. Due to its affluent, ethnically diverse population, it's viewed as a product test center, offering entry into the regional market.

Singapore trade data indicates that 5,300 metric tons of pet food, worth $16-20 million, was imported in 1996. Primary beneficiaries of this largesse--the pampered pooches of Singapore. Though this pet population is not increasing, doting owners provide amply for them.

Appealing to doggie connoisseurs at this time: chicken and rice flavors, which are in short supply because the market is currently flooded with beef-based products. The coming flavor sensation, lamb and rice, is reputedly good for doggie dermatitis, which crops up often in Singapore's warm, humid climate. Another trendy flavor for Fido--turkey and barley.

Singapore's kitty population has traditionally purred for fishy flavors. Most recent trends, however, include chopped liver. And then there are the ever-popular prawns. One humane note here: Tender-hearted Singaporeans feed so many neighborhood strays, that a new market for less expensive dry cat food has sprung up in recent years.

Stiff Markups Take a Bite Out of Profits

While there are no tariffs on imported pet foods, import markups range from 25 to 50 percent, and retail markups add about 30 percent. Those yummy pet treats suffer even higher markups, sometimes as much as 100 percent.

Source: http://www.fas.usda.gov/info/agexporter/1998/singpet.html

On the FAS Website, markups on imported pet foods are discussed. One sentence states "Those yummy pet treats suffer even higher markups, sometimes as much as 100 percent." Do you think this refers to markup based on cost or markup based on selling price? Explain your logic. (Hint: Make up a simple numeric example.)

The Bottom Line

Summary of chapter learning objectives:

| Learning Objective | Summary | Example |
|---|---|---|
| **13.1** | Compute the variables in the basic markup formula | Find the missing variables in the basic formula: Selling price = Cost + Dollar markup

 1. Cost = $231.50; Dollar markup = 109.12
 2. Cost = $35.20; Selling price = $58.95
 3. Dollar markup = $475; Selling price = $900 |
| **13.2** | Compute the markup variables when the markup percent is based on cost | 4. Cost = $750; Markup percent = 40%

 Find the dollar markup and then find the selling price. Find 100% + Markup percent, and then find selling price.

 5. Selling price = $2,400; Markup percent = 50%

 Find 100% + Markup percent and then find cost. |
| **13.3** | Compute the markup percent based on cost | 6. Cost = $80; Selling price = $108

 Find the markup percent based on cost. |
| **13.4** | Compute the markup variables when the markup percent is based on selling price | 7. Selling price = $820; Markup percent = 25%

 Find the dollar markup and then find the cost. Find 100% − Markup percent and then find the cost.

 8. Cost = $1,350; Markup percent = 40%

 Find 100% − Markup percent, and then find the selling price. |
| **13.5** | Compute the markup percent based on selling price | 9. Cost = $440; Selling price = $800
 Find the markup percent based on the selling price. |

Answers: 1. Selling price = $340.62 2. Dollar markup = $23.75 3. Cost = $425 4. $300, $1,050; 140%; 5. 150%, $1,600 6. 35% 7. $205, $615; 75%, $615 8. 60%, $2,250 9. 45%

Self-Check | Review Problems for Chapter 13

1 Find the missing terms.

| | Cost of Goods Sold | Dollar Markup | Selling Price | | | Cost of Goods Sold | Dollar Markup | Selling Price |
|---|---|---|---|---|---|---|---|---|
| a. | $26.50 | $13.80 | _____ | c. | _____ | $1,405 | $2,675 |
| b. | $188.12 | _____ | $395.50 | d. | $426.25 | _____ | $948.88 |

In problems 2–9, the markup percent is based on cost. Find the missing terms. Round all percents to the nearest one-tenth of a percent.

| | Cost | Markup Percent | Dollar Markup | Selling Price | | | Cost | Markup Percent | 100% + Markup Percent | Selling Price |
|---|---|---|---|---|---|---|---|---|---|---|
| **2** | $500 | 50% | a. _____ | b. _____ | **4** | $88 | 40% | a. _____ | b. _____ |
| **3** | $36 | 65% | a. _____ | b. _____ | **5** | $420 | 60% | a. _____ | b. _____ |

| | Selling Price | Markup Percent | 100% + Markup Percent | Cost | | | Selling Price | Cost | Dollar Markup | Markup Percent |
|---|---|---|---|---|---|---|---|---|---|---|
| **6** | $720 | 100% | a. _____ | b. _____ | **8** | $480 | $240 | a. _____ | b. _____ |
| **7** | $84 | 40% | a. _____ | b. _____ | **9** | $2,000 | $1,600 | a. _____ | b. _____ |

In problems 10–13 the markup percent is based on selling price. Find the missing terms. Round all percents to the nearest one-tenth of a percent.

| | Selling Price | Markup Percent | Dollar Markup | Cost | | | Selling Price | Markup Percent | 100% − Markup Percent | Cost |
|---|---|---|---|---|---|---|---|---|---|---|
| **10** | $52 | 25% | a. _____ | b. _____ | **12** | $1,240 | 40% | a. _____ | b. _____ |
| **11** | $168 | 30% | a. _____ | b. _____ | **13** | $528 | 75% | a. _____ | b. _____ |

| | Cost | Markup Percent | 100% − Markup Percent | Selling Price | | | Selling Price | Cost | Dollar Markup | Markup Percent |
|---|---|---|---|---|---|---|---|---|---|---|
| **14** | $960 | 60% | a. _____ | b. _____ | **16** | $480 | $240 | a. _____ | b. _____ |
| **15** | $36 | 25% | a. _____ | b. _____ | **17** | $2,000 | $1,600 | a. _____ | b. _____ |

18 Evelyn Haynes sells toys, specializing in toys for preschool children. She pays $20 for a toy truck made entirely of wood. Evelyn sells the toy truck for $32. a. Find the dollar markup. _____ b. Find the markup percent based on cost. _____ c. Find the markup percent based on selling price. _____

Assignment 13.1: Markup Based on Cost

Name

Date _____ Score _____

A **(12 points) Calculate the missing terms. (2 points for each correct answer)**

| | Cost | Dollar Markup | Selling Price | | Cost | Dollar Markup | Selling Price |
|---|---|---|---|---|---|---|---|
| **1.** | $482.43 | $175.25 | _657.68_ | **2.** | $48.51 | _21.44_ | $69.95 |
| **3.** | _455.48_ | $374.50 | $829.98 | **4.** | $175.50 | $57.50 | _233.00_ |
| **5.** | $629.12 | _280.47_ | $909.59 | **6.** | _399.86_ | $384.25 | $784.11 |

Score for A (12)

B **(32 points) In the following problems, the markup percent is based on *cost*. Find the missing terms. (2 points for each correct answer)**

| | Cost | Markup Percent | Dollar Markup | Selling Price | | Cost | Markup Percent | 100% + Markup Percent | Selling Price |
|---|---|---|---|---|---|---|---|---|---|
| **7.** | $750 | 30% .3 | _225_ | _975.00_ | **8.** | $48 | 75% | _____ | _____ |
| **9.** | $1,200 | 80% .8 | _960_ + | _2160._ | **10.** | $280 | 150% | _____ | _____ |
| **11.** | $840 | 25% .25 | _210_ + | _1050_ | **12.** | $824 | 60% | _____ | _____ |
| **13.** | $2,500 | 100% | _2500_ + | _5000_ | **14.** | $88 | 120% | _____ | _____ |

2465 1.0
 6090 = 1479 + 2465 = 3944
 .6

Score for B (32)

Assignment 13.1 *Continued*

C **(32 points)** In the following problems, the markup percent is based on cost. Find the missing terms. Round all percents to the nearest tenth of a percent. **(2 points for each correct answer)**

| | Selling Price | Markup Percent | 100% + Markup Percent | Cost | | Selling Price | Cost | Dollar Markup | Markup Percent |
|---|---|---|---|---|---|---|---|---|---|
| **15.** | $1,300 | 30% | _____ | _____ | **16.** | $48 | $30 | _____ | _____ |
| **17.** | $110 | 100% | _____ | _____ | **18.** | $1,922 | $1,240 | _____ | _____ |
| **19.** | $840 | 40% | _____ | _____ | **20.** | $675 | $388 | _____ | _____ |
| **21.** | $1,050 | 150% | _____ | _____ | **22.** | $216 | $96 | _____ | _____ |

Score for C (32)

D **(24 points)** Business Applications. In the following problems, the markup percent is based on cost. Round all percents to the nearest tenth of a percent. **(3 points for each correct answer)**

23. Nancy Lo owns a firm that sells office furniture to other businesses. One matched set of six pieces costs Nancy $2,465. To cover her own business expenses and allow a reasonable profit, Nancy marks up this set by 60% of the cost. Find the dollar markup and the selling price.

Dollar markup 1479

Selling price 3944

24. Stan Wegner manufactures a hand-held heart monitoring device. He sells it for $840, which represents a markup of 275% on his production cost. Stan marks it up this much to cover additional business expenses and profit as well as product development. Find Stan's production cost and the dollar markup.

Cost _____

Dollar markup _____

25. Rotweiller Security Systems sells burglar and fire alarm systems for homes and small businesses. One basic system costs Rotweiller $380. Rotweiller charges an installation price and also marks up the alarm system by $190. Find the selling price, and find the markup percent based on cost.

Selling price _____

Markup percent _____

26. After Dave Smith drove his car with no oil, he needed a new engine. Dave's car dealer charged him $1,740 for an engine that cost $1,200. The labor was additional. Compute the dollar markup and the markup percent based on cost.

Dollar markup _____

Markup percent _____

Score for D (24)

Assignment 13.2: Markup Based on Selling Price

Name _____

Date _____ Score _____

A (12 points) Calculate the missing terms. (2 points for each correct answer)

ADD

| | Cost | Dollar Markup | Selling Price | | Cost | Dollar Markup | Selling Price |
|---|---|---|---|---|---|---|---|
| 1. | $171.38 | $82.15 | _253.53_ | 2. | $293.11 | _159.17_ | $452.28 |
| 3. | _1535_ | $1,140 | $2,675 | 4. | $789.75 | $476.50 | _313.25_ |
| 5. | $62.50 | _37.49_ | $99.99 | 6. | _671.80_ | $307.15 | $978.95 |

on Tcost

Score for A (12)

B (32 points) In the following problems, the markup percent is based on selling price. Find the missing terms. (2 points for each correct answer)

subtract

| | Selling Price | Markup Percent | Dollar Markup | Cost | | Selling Price | 100% − Markup Percent | Markup Percent | Cost |
|---|---|---|---|---|---|---|---|---|---|
| 7. | $64 | 45% X .45 | _28.80_ − | _35.20_ | 8. | $180 | 25% | _____ | _____ |
| 9. | $368 X | 60% .60 | _220.80_ | _147.20_ | 10. | $1,260 | 20% | _____ | _____ |
| 11. | $1,900 X | 70% .70 | _1330_ | _570_ | 12. | $72 | 70% | _____ | _____ |
| 13. | $824 X | 50% .5 | _412_ | _412._ | 14. | $926 | 55% | _____ | _____ |

Score for B (32)

Assignment 13.2 *Continued*

C (32 points) In the following problems, the markup percent is based on selling price. Find the missing terms. (2 points for each correct answer)

| | Cost | Markup Percent | 100% − Markup Percent | Selling Price | | Selling Price | Cost | Dollar Markup | Markup Percent |
|---|---|---|---|---|---|---|---|---|---|
| **15.** | $854 | 30% | _____ | _____ | **16.** | $220 | $143 | _____ | _____ |
| **17.** | $64 | 75% | _____ | _____ | **18.** | $45 | $27 | _____ | _____ |
| **19.** | $2,460 | 40% | _____ | _____ | **20.** | $1,400 | $924 | _____ | _____ |
| **21.** | $612 | 25% | _____ | _____ | **22.** | $840 | $462 | _____ | _____ |

Score for C (32)

D (24 points) Business Applications. In the following problems, the markup percent is based on selling price. Round all percents to the nearest tenth of a percent. (3 points for each correct answer)

23. At the end of winter, Westwood Supply features garden equipment specials. One leaf mulcher has a selling price of $248. The markup to cover expenses and profit is 40% of the selling price. Calculate the dollar markup and the cost.

Dollar markup _____

Cost _____

24. T.R. Burke Cyclery is a retail bicycle store. For last Christmas season, T.R. purchased one model of mountain bike to use as a Christmas promotion. The bicycles cost $126 each. For this promotion, T.R.'s markup was only 20% of selling price. Find the selling price and the dollar markup.

Selling price _____

Dollar markup _____

Assignment 13.2 *Continued*

25. Northern Lights TV also sells cordless telephones. A two-line model with a speaker phone base and an answering machine is priced at $182.40. This price includes a markup of $109.44. If a phone sells at this price, what are the cost and the markup percent based on selling price?

Cost _____

Markup percent _____

26. Patio Depot, a warehouse store, purchased a large volume of redwood lounge chairs for $246 each. Upholstered pads were included in the price. To sell the chairs and pads quickly, the store priced the chairs at $328. Compute the dollar markup and the markup percent based on selling price.

Dollar markup _____

Markup percent _____

Score for D (24)

| capital markets | leasing | treasury management |
| credit services | merchant services | trust and investments |
| EFT | retirement plans | industry specializations |
| international | | |

www.53.comSM

corp

what's new

Fifth Third DirectSM
Single-point Internet access to commercial banking products and services. ▶

Wholesale Lockbox Imaging
Fifth Third offers same-day online check images through Fifth Third Direct Wholesale Lockbox Imaging. ▶

Online Foreign Exchange Trading
One of the first Internet-based platforms for foreign exchange trading in the country through fx.53.com. ▶

Select a Solution . . .

--Choose One-- ▾ Go

sign on ▶

⊖ Our Financial Performance
⊖ About Fifth Third
⊖ Site Map

▶ Fifth Third D
▶ FX Internet T
▶ Portfolio Pro

Fifth Third Bank

Banks lend money to individuals, businesses, nonprofit organizations, and governments. In all cases, banks charge interest on these loans. Most of the money that banks lend comes from depositors. To entice customers to deposit their money, banks pay interest. Although banks provide many additional services, borrowing and lending are still their primary activities, and interest is always involved in these transactions.

Visit the Fifth Third Bank Web site at http://www.53.com to learn more about its small business services as well as its domestic and international corporate services.

Simple Interest

14

LEARNING OBJECTIVES

By studying this chapter and completing all assignments you will be able to:

 1 Compute simple interest with time in years or months.

 2 Compute ordinary simple interest, using a 360-day year.

 3 Compute exact simple interest, using a 365-day year.

 4 Compare ordinary simple interest and exact simple interest.

 5 Estimate simple interest computations.

To understand the concept of simple interest, consider the money a bank makes on loaning—or renting—money. The more it can lend, the more it can make. From the borrower's standpoint, interest is an added cost beyond the money being rented.

1. What does the formula $I = PRT$ represent?
2. For a borrower, is exact simple interest cheaper or more expensive than ordinary simple interest? Explain.
3. For a lender, is the use of exact simple interest or ordinary simple interest better? Explain.
4. Give one reason why being able to estimate the amount of interest is important.

**www.learner.org/
exhibits/dailymath/
savings.html**

For more information about interest and credit, visit this Web site.

Most businesses and individuals buy at least some assets without making full payment at the time of the purchase. The seller gives immediate possession to the buyer, but doesn't require payment until some later date. For example, large retailers such as Macy's Department Store may receive merchandise for the Christmas season but may not be required to pay the seller until January. The seller, who *extends credit* to the buyer, may or may not charge for this privilege. The charge is called **interest,** and it is usually quoted as a percent of the amount of credit extended (the principal). When part of the price is paid at the time of purchase, that part is called a **down payment.**

If the seller charges too much interest or doesn't extend credit, the buyer might borrow money from a third party, such as a bank. A retailer such as Macy's could then buy the merchandise and sell it to repay the bank loan. The amount borrowed is called the **principal,** and the interest charged is a percent of the principal. The bank will charge interest between the loan date and the repayment date. This period of **time** is called the **interest period** or the **term of the loan.**

The promise to repay a loan or pay for merchandise may be oral or written. If it is written, it may be in the form of a letter or it could be one of several special documents known collectively as **commercial paper. Short-term credit** transactions are those for between 1 day and 1 year. **Long-term credit** transactions are those for longer than 1 year. Normally, long-term credit transactions involve major items such as new buildings or equipment rather than supplies or merchandise for sale.

Computing Simple Interest

Compute simple interest with time in years or months.

The easiest type of interest to calculate is called **simple interest.** The calculations are the same for both a loan and a purchase on credit. The interest is a percent of the principal for the period of the loan or credit. The quoted percent usually is an *annual* (yearly) rate. A rate of 10% means that the interest payment for 1 year will be 10% of the principal.

To compute the simple interest on a 1-year loan, simply multiply the Principal by the Rate.

EXAMPLE A

Dorothy Smith borrowed $1,200 for 1 year at a rate of 10% simple interest. Compute the interest.

The principal is $1,200. The interest for 1 year is 10% of $1,200, or $0.10 \times \$1,200 = \120.

Most loans, however, are not for a period of exactly 1 year. Loans for longer periods will require the borrower to pay more interest. Likewise, loans for shorter periods will require less interest. To compute the simple interest on loans of any period, multiply the Principal by the Rate and then multiply by the Time, with Time stated in years or in fractions of years. The fundamental formula for simple interest is

Interest = Principal × Rate × Time
abbreviated as $I = P \times R \times T$ or, even more simply, $I = PRT$.

**www.warwick.ac.uk/
fac/cross_fac/nflc/
cal1.html**

Try this simple interest calculator.

EXAMPLE B

Find the simple interest on loans of $1,000 when the rate is 8% and the loan periods are $\frac{1}{2}$ year and 3 years.

| $\frac{1}{2}$ **year** | **3 years** |
|---|---|
| $I = P \times R \times T$ | $I = P \times R \times T$ |
| $= \$1{,}000 \times 0.08 \times \frac{1}{2}$ | $= \$1{,}000 \times 0.08 \times 3$ |
| $= \$40$ | $= \$240$ |

The time period often will be measured in months instead of years. Before computing the interest, change the time into years by dividing the number of months by 12 (the number of months in 1 year).

EXAMPLE C

Compute the interest on credit purchases of $2,000 at 10% for periods of 9 months and 18 months.

| **9 months** | **18 months** |
|---|---|
| $I = P \times R \times T$ | $I = P \times R \times T$ |
| $= \$2{,}000 \times 0.10 \times \frac{9}{12}$ | $= \$2{,}000 \times 0.10 \times \frac{18}{12}$ |
| $= \$150$ | $= \$300$ |

 CONCEPT CHECK 14.1

The principal is $1,500, the rate is 6%, and interest = Principal × Rate × Time, or $I = P \times R \times T$. Find the interest both for 4 years and for 4 months.

a. If Time is 4 years: $I = P \times R \times T = \$1{,}500 \times 0.06 \times 4 = \360
b. If Time is 4 months: $I = P \times R \times T = \$1{,}500 \times 0.06 \times \frac{4}{12} = \30

Computing Ordinary Interest

If the term of the loan is stated as a certain number of days, computing interest involves dividing the number of days by the number of days in one year—either 360 or 365. Before computers and calculators, interest was easier to compute by assuming that every year had 360 days and that every month had 30 days. The 360-day method, called the **ordinary interest method,** is still used by some businesses and individuals.

Compute ordinary simple interest, using a 360-day year.

www.aaamath.com/ pct68-simple-interest. html

For more practice with simple interest problems, go to this Web site.

EXAMPLE D

Compute the ordinary simple interest on $800 at 9% for 90 days.

$I = P \times R \times T$
$= \$800 \times 0.09 \times \frac{90}{360}$
$= \$18$

CONCEPT CHECK 14.2

The Principal is $3,000, the Rate is 7%, and the Time is 180 days. Compute the ordinary simple interest.

Ordinary interest involves use of a 360-day year: $I = P \times R \times T = \$3,000 \times 0.07 \times \frac{180}{360} = \105.00

Computing Exact Interest

LEARNING OBJECTIVES 3

Compute exact simple interest, using a 365-day year.

Banks, savings and loan institutions, credit unions, and the federal government use a 365-day year (366 days for leap years) to compute interest. This method is called the **exact interest method.** The computations are the same as for ordinary simple interest, except 365 days is used instead of 360 days.

EXAMPLE E

Compute the exact simple interest on $800 at 9% for 90 days.

$I = P \times R \times T$

$= \$800 \times 0.09 \times \dfrac{90}{365}$

$= \$17.7534, \text{ or } \17.75

CONCEPT CHECK 14.3

The Principal is $2,400, the Rate is 8%, and the Time is 30 days. Compute the exact simple interest.

Exact interest involves use of a 365-day year: $I = P \times R \times T = \$2,400 \times 0.08 \times \frac{30}{365} = \15.78

Comparing Ordinary Interest and Exact Interest

LEARNING OBJECTIVES 4

Compare ordinary simple interest and exact simple interest.

The 360-day year was very useful before the advent of calculators and computers, so there is a long tradition of using it. However, the 365-day year is more realistic than the 360-day year. Also, the 365-day year is financially better for the borrower because the interest amounts are always smaller. (Why? Because a denominator of 365 gives a smaller quotient than a denominator of 360).

Reexamine examples D and E. The difference between ordinary interest and exact interest is only $18.00 − $17.75, or $0.25. When businesses borrow money, however, the

principal may be very large and then the difference will be more significant. Example F is similar to examples D and E, except that the principal is in millions of dollars rather than hundreds.

EXAMPLE F

Find the difference between ordinary interest and exact interest on $8,000,000 at 9% for 90 days.

Ordinary Interest

$I = P \times R \times T$

$= \$8,000,000 \times 0.09 \times \dfrac{90}{360}$

$= \$180,000$

Exact Interest

$I = P \times R \times T$

$= \$8,000,000 \times 0.09 \times \dfrac{90}{365}$

$= \$177,534.2466$, or $177,534.25

The difference is $180,000 − $177,534.25, or $2,465.75.

 CONCEPT CHECK 14.4

The principal is $6,000, the rate is 12%, and the time is 120 days. Find the difference between the amounts of simple interest calculated by using the ordinary method (360-day year) and the exact method (365-day year).

Ordinary interest: $I = P \times R \times T = \$6,000 \times 0.12 \times \frac{120}{360} = \240.00

Exact interest: $\quad I = P \times R \times T = \$6,000 \times 0.12 \times \frac{120}{365} = \236.71

Difference = Ordinary interest − Exact interest = $240.00 − $236.71 = $3.29

Use of Calculators

Today, calculators or computers are used in almost every interest application. The numbers are often large and are always important. The steps are performed on the calculator in the same order as they are written in the formula.

EXAMPLE G

Write the calculator steps for computing the exact interest on $8,000,000 at 9% for 90 days.

$I = P \times R \times T = \$8,000,000 \times 0.09 \times \dfrac{90}{365}$

\qquad 8 000 000 $\boxed{\times}$.09 $\boxed{\times}$ 90 $\boxed{\div}$ 365

\qquad $\boxed{=}$ 177 534.2466, or $177,534.25

With the percent key $\boxed{\%}$, the steps would be

8 000 000 $\boxed{\times}$ 9 $\boxed{\%}$ $\boxed{\times}$ 90 $\boxed{\div}$ 365 $\boxed{=}$ 177 534.2466, or $177,534.25

Estimating Simple Interest

Although calculators are used to compute interest, approximation remains as useful as ever. The following calculator solution requires a minimum of 19 key entries.

8 000 000 $\boxed{\times}$.09 $\boxed{\times}$ 90 $\boxed{\div}$ 365 $\boxed{=}$ 177 534.2466

Pressing any one of the 19 keys incorrectly will result in a large error. By making an estimate of the interest in advance, you may spot a significant calculator error.

Combinations of Time and Interest That Yield 1%

To make simple mental approximations, you can round the rate and time to numbers that are easy to compute mentally. For ordinary interest, several combinations of rate and time are easy to use because their product is 1%. For example, $12\% \times \frac{30}{360} = 12\% \times \frac{1}{12} = 1\%$ and $6\% \times \frac{60}{360} = 6\% \times \frac{1}{6} = 1\%$.

EXAMPLE H

Approximate the ordinary interest on $1,500 at 6.25% for 59 days.

| | |
|---|---|
| Round | 6.25% to 6% and 59 days to 60 days. |
| Estimate: | $1,500 \times 0.06 \times \frac{60}{360} = \$1,500 \times 0.01 = \$15$ |
| Actual interest: | $1,500 \times 0.0625 \times \frac{59}{360} = \15.3646, or $15.36 |

50
AND OVER
FREE
INTEREST
 CHECKING

© PHOTODISC, INC.

Other Rates and Times

Table 14.1 shows several combinations of rate and time whose products are useful for estimating interest.

Table 14-1: Rate and Time

| | |
|---|---|
| $4\% \times \dfrac{90}{360} = 4\% \times \dfrac{1}{4} = 1\%$ | $10\% \times \dfrac{36}{360} = 10\% \times \dfrac{1}{10} = 1\%$ |
| $6\% \times \dfrac{60}{360} = 6\% \times \dfrac{1}{6} = 1\%$ | $12\% \times \dfrac{30}{360} = 12\% \times \dfrac{1}{12} = 1\%$ |
| $8\% \times \dfrac{45}{360} = 8\% \times \dfrac{1}{8} = 1\%$ | $18\% \times \dfrac{20}{360} = 18\% \times \dfrac{1}{18} = 1\%$ |
| $9\% \times \dfrac{40}{360} = 9\% \times \dfrac{1}{9} = 1\%$ | $6\% \times \dfrac{120}{360} = 6\% \times \dfrac{1}{3} = 2\%$ |
| $12\% \times \dfrac{60}{360} = 12\% \times \dfrac{1}{6} = 2\%$ | $12\% \times \dfrac{90}{360} = 12\% \times \dfrac{1}{4} = 3\%$ |
| $8\% \times \dfrac{90}{360} = 8\% \times \dfrac{1}{4} = 2\%$ | $9\% \times \dfrac{120}{60} = 9\% \times \dfrac{1}{3} = 3\%$ |

Estimation of Exact Interest

The goal in approximating interest is just to get an estimate. Even though exact interest requires 365 days in a year, you can make a reasonable estimate by assuming that the number of days in a year is 360.

EXAMPLE I

Compute the estimate and exact interest on $1,200 at 11.8% for 62 days.

First, round 11.8% to 12% and 62 days to 60 days.

Second, assume that a year has 360 days.

Estimate: \qquad $\$1,200 \times 0.12 \times \dfrac{60}{360} = \$1,200 \times 0.02 = \$24$

Actual interest: \quad $\$1,200 \times 0.118 \times \dfrac{62}{365} = \$24.0526,$ or $\$24.05$

CONCEPT CHECK 14.5

The Principal is $3,650, the Rate is 11.7%, and the Time is 31 days. Calculate the actual exact simple interest. Then make an estimate by using a 360-day year and simpler values for R and T. Compare the results.

Actual value: \quad $I = P \times R \times T = \$3,650 \times 0.117 \times \frac{31}{365} = \36.27

Estimate: \qquad $I = P \times R \times T = \$3,650 \times 0.12 \times \frac{30}{360} = \$3,650 \times 0.01 = \$36.50$

Difference: \qquad Estimate − Actual = $36.50 − $36.27 = $0.23

COMPLETE ASSIGNMENTS 14.1 AND 14.2.

Chapter Terms for Review

| | | |
|---|---|---|
| commercial paper | interest period | short-term credit |
| down payment | long-term credit | simple interest |
| exact interest method | ordinary interest method | term of the loan |
| interest | principal | time |

Visit the Web

http://www.ihtml.com/learning/examples/sint/sint.htm—Try this simple interest calculator.

http://id.mind.net/~zona/mmts/functionInstitute/exponentialFunctions/simpleInterest.html—Additional explanation and practice on simple interest is available here.

http://www.necu.com.au/loans.htm—Visit the site of New England Credit Union—which actually is in Australia—to see how interest rates there on home, auto, personal, and commercial loans compare with those in the United States.

CyberConnection

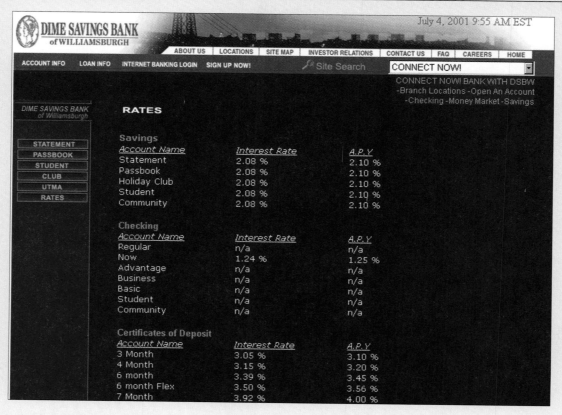

Source: http://www.dimewill.com. Copyright © 2001 Courtesy of Dime Savings Bank.

Banks offer a range of interest rates on savings accounts. Visit the following Web sites for the Dime Savings Bank of Williamsburgh (http://www.dimewill.com), Provident Bank (http://www.providentbank.com), and Bank of America (http://www.bankofamerica.com).

Their regular savings account interest rates in July 2001 were as shown in the following table. Why does the larger bank, Bank of America, pay less interest than the other two?

| Dime Savings Bank | Provident Bank | Bank of America |
|---|---|---|
| 2.08% | 1.49% | 0.75% |

Go to your own bank's Web site and compare its rates with these. Is it better or not as good? Check your bank's site to see whether it offers special savings or checking accounts for students. Many banks, including the Dime Savings Bank of Williamsburgh, New York, have checking and savings account with no fees and no minimum balances required for students.

The Bottom Line

Summary of chapter learning objectives:

| Learning Objective | Summary | Example |
|---|---|---|
| **14.1** | Compute simple interest with time in years or months | Find the simple interest using the basic formula:
Interest = Principal × Rate × Time, or ***I = P × R × T***

1. Principal = $2,500; Rate = 9%; Time = 1.5 years
2. Principal = $975; Rate = 8%; Time = 9 months |
| **14.2** | Compute ordinary simple interest, using a 360-day year | 3. Find the ordinary simple interest for a 360-day year:
Principal = $3,250; Rate = 10%; Time = 90 days |
| **14.3** | Compute exact simple interest, using a 365-day year | 4. Find the exact simple interest for a 365-day year:
Principal = $1,800; Rate = 11%; Time = 45 days |
| **14.4** | Compare ordinary simple interest and exact simple interest | 5. Find the difference between ordinary simple interest and exact simple interest: Principal = $5,000; Rate = 6%; Time = 75 days |
| **14.5** | Estimate simple interest computations | 6. Estimate the exact interest by using a 360-day year and simpler values for rate and time: Principal = $2,100; Rate = 5.8%; Time = 62 days |

Answers: 1. $337.50 **2.** $58.50 **3.** $81.25 **4.** $24.41 **5.** $62.50 − $61.64 = $0.86
6. $2,100 × 0.06 × $\frac{60}{360}$ = $21

In problems 1 and 2, compute the amount of (a) ordinary simple interest and (b) the amount of exact simple interest. Then compute (c) the difference between the two interest amounts

| | Principal | Rate | Time | Ordinary Interest | | Exact Interest | | Difference |
|---|---|---|---|---|---|---|---|---|
| ❶ | $1,260 | 7% | 180 Days | a. _____ | | b. _____ | | c. _____ |
| ❷ | $12,500 | 11% | 30 Days | a. _____ | | b. _____ | | c. _____ |

In problems 3 and 4, change each rate and time to the closest numbers that permit use of the shortcuts shown in Table 14.1. Then compute (a) the estimated amount of ordinary interest and (b) the actual amount of ordinary interest. Finally, compute (c) the difference between the two interest amounts.

| | Principal | Rate | Time | Estimated Ordinary Interest | | Actual Ordinary Interest | | Difference |
|---|---|---|---|---|---|---|---|---|
| ❸ | $4,675 | 5.7% | 62 Days | a. _____ | | b. _____ | | c. _____ |
| ❹ | $6,200 | 8.3% | 87 Days | a. _____ | | b. _____ | | c. _____ |

In problems 5 and 6, change each rate and time to the closest numbers that permit use of the shortcuts shown in Table 14.1. Then compute (a) the estimated amount of *exact* interest and (b) the actual amount of *exact* interest. Finally, compute (c) the difference between the two interest amounts.

| | Principal | Rate | Time | Estimated Exact Interest | | Actual Exact Interest | | Difference |
|---|---|---|---|---|---|---|---|---|
| ❺ | $14,000 | 7.9% | 44 Days | a. _____ | | b. _____ | | c. _____ |
| ❻ | $2,000 | 9.2% | 117 Days | a. _____ | | b. _____ | | c. _____ |

❼ Kaye Mushalik loaned $1,500 to Fay Merritt, a good friend since high school. Because of their friendship, Kaye charged only 4% ordinary simple interest. Two months later, when Fay received her annual bonus, she repaid the entire loan and all the interest. What was the total amount that Fay paid?

❽ Katherine Wu and her sister Madeline have a home decorating and design business. Often, they buy antiques and fine art objects and then resell the items to their clients. They have a line or credit at their bank to provide short-term financing, if necessary, for these purchases. The bank always charges exact simple interest, but the rate varies depending on the economy. Katherine and Madeline need to borrow $25,400 for 75 days to buy a large collection of antique furniture at an estate sale. If the bank charges 7.25%, how much interest would they pay?

Assignment 14.1: **Simple Interest**

Name _____

Date _____ Score _____

A **(20 points) Compute the simple interest. If the time is given in months, let one month be $\frac{1}{12}$ of a year. If the time is in days, let one year be 360 days. (2 points for each correct answer)**

| Principal | Rate | Time | Interest | Principal | Rate | Time | Interest |
|---|---|---|---|---|---|---|---|
| **1.** $400 | 7.0% | 1 year | _____ | **2.** $2,000 | 8% | 4 years | _____ |
| **3.** $1,800 | 8% | 4 months | _____ | **4.** $960 | 4% | 15 months | _____ |
| **5.** $8,250 | 6% | $\frac{1}{2}$ year | _____ | **6.** $3,600 | 12% | 30 months | _____ |
| **7.** $12,800 | 7% | 2.5 years | _____ | **8.** $450 | 5% | $3\frac{1}{2}$ years | _____ |
| **9.** $4,860 | 10% | 21 months | _____ | **10.** $24,000 | 5.5% | 4 months | _____ |

Score for A (20)

B **(40 points) Compute the ordinary interest, the exact interest, and their difference. Round answers to the nearest cent. (2 points for each correct interest; 1 point for each correct difference)**

| | Principal | Rate | Time | Ordinary Interest | Exact Interest | Difference |
|---|---|---|---|---|---|---|
| **11.** | $780 | 9% | 90 days | _____ | _____ | _____ |
| **12.** | $2,350 | 6% | 150 days | _____ | _____ | _____ |
| **13.** | $4,800 | 5% | 75 days | _____ | _____ | _____ |
| **14.** | $12,400 | 7% | 225 days | _____ | _____ | _____ |
| **15.** | $985 | 10% | 180 days | _____ | _____ | _____ |
| **16.** | $1,400 | 15% | 60 days | _____ | _____ | _____ |
| **17.** | $6,500 | 8% | 45 days | _____ | _____ | _____ |
| **18.** | $365 | 4% | 30 days | _____ | _____ | _____ |

Score for B (40)

Assignment 14.1 *Continued*

C **(40 points)** In each problem, make an estimate of the interest. Assume a 360-day year and round each rate and time to the nearest numbers that will permit the shortcuts in Table 14-1. Then find the actual interest (ordinary interest in problems 19–22, exact interest in problems 23–26) and find the difference. If the estimate is larger than the actual interest, write a "+" in front of the difference; if the estimate is smaller than the actual interest, write a "−" in front of the difference. Round answers to the nearest cent. (2 points for each correct estimate and actual interest; 1 point for each correct difference)

| | Principal | Rate | Time | Estimate | Actual Ordinary Interest | Difference |
|---|---|---|---|---|---|---|
| **19.** | $850 | 6% | 60 days | _____ | _____ | _____ |
| **20.** | $7,250 | 5.9% | 61 days | _____ | _____ | _____ |
| **21.** | $700 | 8.25% | 88 days | _____ | _____ | _____ |
| **22.** | $2,500 | 12.1% | 31 days | _____ | _____ | _____ |

| | Principal | Rate | Time | Estimate | Actual Exact Interest | Difference |
|---|---|---|---|---|---|---|
| **23.** | $625 | 8.1% | 44 days | _____ | _____ | _____ |
| **24.** | $5,600 | 3.99% | 92 days | _____ | _____ | _____ |
| **25.** | $2,000 | 8.95% | 123 days | _____ | _____ | _____ |
| **26.** | $10,000 | 6% | 60 days | _____ | _____ | _____ |

Score for C (40)

Assignment 14.2: Simple Interest Applications

Name _____

Date _____ Score _____

A **(50 points) Solve each of the following ordinary simple interest problems by using a 360-day year. Find both the interest dollars and the total amount (i.e., principal plus interest) of the loan. (7 points for each correct interest; 3 points for each correct amount)**

1. Tom Titus plans to lend $750 to his friend Bill White so that Bill can fly with him to Canada for vacation. Tom is charging Bill only 4% ordinary simple interest. Bill repays everything, interest plus principal, to Tom 90 days later. How much does Bill pay?

 Interest _____

 Amount _____

2. John Kwok and Lena Chung are planning to start a business that will export American food to China. They estimate that they will need $50,000 to pay for organizational costs, get product samples, and make three trips to Shanghai. They can borrow the money from their relatives for 5 years. John and Lena are willing to pay their relatives 12% ordinary simple interest. Compute the amount that John and Lena will owe their relatives in 5 years.

 Interest _____

 Amount _____

3. Carolyn Wilfert owns a temporary services employment agency. Businesses call her when they need to hire various types of workers for a short period of time. The businesses pay a fee to Carolyn, who pays the salaries and benefits to the employees. One benefit is that Carolyn will make small, short-term loans to her employees. After a flood, employee Judy Hillstrom needed to borrow $3,600 to have her house cleaned and repainted. Judy repaid the loan in 6 months. If Carolyn charged 5% ordinary simple interest, how much did Judy repay?

 Interest _____

 Amount _____

4. Several years ago, Ben Mahaffy and Lee Oman formed a partnership to rent musical instruments to school districts that do not want to own and maintain the instruments. In the spring, they investigate borrowing $50,000 to buy trumpets and trombones. Because they collect their rental fees in advance, they anticipate being able to repay the loan in 105 days. How much will they need to repay if the ordinary simple interest rate is 9.5%?

 Interest _____

 Amount _____

5. With her husband, Ruby Williams owns and manages a video game arcade. A manufacturer developed a new line of games and offered very low interest financing to encourage arcade operators such as Ruby to install the new games. Ruby was able to finance $75,000 worth of games for 8 months for 3.2% ordinary simple interest. Calculate how much Ruby will repay.

 Interest _____

 Amount _____

 Score for A (50)

B **(50 points) Solve each of the following exact simple interest problems, using a 365-day year. Find both the interest dollars and the total amount (i.e., principal plus interest) of the loan. (7 points for each correct interest; 3 points for each correct amount)**

6. Charles Peterson, managing partner of a local transportation company, thinks that the company should borrow money to upgrade its truck repair facility. After investigating several sources of short-term loans, Charles determines that the company can borrow $250,000 for 180 days at 7.5% exact simple interest. If the company agrees to take out this loan, how much will it need to repay at the end of the 180 days?

Interest _____

Amount _____

7. Keith Smith, a former teacher, now has a business selling educational materials such as books and computer software to parents and schools. In June, he borrowed $25,000 from his bank to buy some new educational computer games that he hopes to sell during August and September. The bank's rate is 8.25% exact simple interest as long as the time does not exceed half a year. If Keith repays everything in 100 days, how much will he pay?

Interest _____

Amount _____

8. After working in construction for five years, Jerry Weekly had saved almost enough money to buy a fishing boat and move to Alaska to become a commercial fisherman. He still needed $9,500, which his wife could borrow from her parents until the end of the first fishing season. The parents charged 5% exact simple interest, and Jerry repaid them after 95 days. How much interest did he pay, and what was the total amount?

Interest _____

Amount _____

9. Bill and Carol Campbell need to purchase two new saws for their retail lumber yard. The company that sells the saws offers them some short-term financing at the relatively high rate of 11% exact simple interest. They decide to accept the financing offer, but only for $5,000 and only for 45 days. How much will Bill and Carol repay at the end of the 45 days?

Interest _____

Amount _____

10. After working for a large accounting firm for ten years, Linda Anderson, C.P.A., decided to open her own office. She borrowed $40,000 at 8.2% exact simple interest. She made enough during the first income tax season to repay the loan in 210 days. How much did Linda repay?

Interest _____

Amount _____

Score for B (50)

Notes

Health Resources and Services Adi

Division of Student Assista

Loan, Scholarship and Faculty Loan Repayment Programs for Students in the

Forms for Campus Based Progran

Current versions of forms are available from this site for the following campus based programs

- **HPSL** - Health Professions Student Loan
- **NSL** - Nursing Student Loan
- **PCL** - Primary Care Loan
- **LDS** - Loans for Disadvantaged Students
- **EFN** - Exceptional Financial Need Scholarship
- **SDS** - Scholarships for Disadvantaged Students
- **FAD** - Financial Assistance for Disadvantaged Health Professions Students Sch

NOTE: You will need Adobe Acrobat Reader to view and print the
If you do not have it, you can **Download Acrobat Reader** without char

| Form (Click on Form Name to View) | Applies t | | |
|---|---|---|---|
| | HPSL | NSL | PCL |
| HPSL Promissory Note | X | | |
| NSL Promissory Note | | X | |
| PCL Promissory Note | | | X |
| LDS Promissory Note | | | |
| Entrance Interview Checklist/Questionnaire | X | X | X |

Source: http://www.bhpr.hrsa.gov/dsa/school_docs/index.htm. Courtesy of the U.S. Dept. of Health and Human Services.

The U.S. Department of Health and Human Services offers financial assistance for students in health care professions. As with other educational loans, students must complete and sign a promissory note. If you have had experience with your school's financial aid department, you may already be familiar with such notes. If not, visit this Web site at http://www.bhpr.hrsa.gov and read one or more of the promissory notes presented there.

Notes and Interest Variables

LEARNING OBJECTIVES

By studying this chapter and completing all assignments you will be able to:

 1 Compute the number of interest days of a note.

 2 Determine the due date of a promissory note.

 3 Compute the maturity value of a promissory note.

 4 Compute the Principal, Rate, and Time from the basic interest formula.

Businesses often take out short-term loans by signing promissory notes. Whether you own a business or work for one, you may have to know how to compute interest on a promissory note in order to make good financing decisions.

1. Explain the difference between a promissory note and a check.
2. A person says, "A promissory note can be more expensive during a leap year than during a non–leap year." Explain why that person might be correct.
3. If you know the maturity value, interest dollars, and the time of a loan, can you solve for the rate? Explain.

www.toerien.com/
advice_column/
corporate_
promissory_notes.htm

For information on corporate promissory notes, visit this Web site.

Businesses use many different methods and documents to charge interest. One such fundamental document is the **promissory note**, which is an unconditional promise made in writing by one party to another. The maker of the promise agrees to pay on demand—or at a particular time—a certain sum of money to the bearer. This form of note is called a **negotiable promissory note** when it may be sold to a third party.

The dollar amount written on the note is called the **face value** of the note. The face value is the same as the **principal (P)**, or amount borrowed. Promissory notes may be either interest bearing or non–interest bearing. If a note is interest bearing, the sum of the face value and the **interest dollars (I)** is the **maturity value (MV)**. It is the total amount that must be paid by the borrower when the note becomes due. The date when a note is due is called the **due date** or **maturity date**. Figure 15-1 illustrates a simple promissory note.

Fig. 15-1: Promissory Note

$ _2,000 00_____ ATLANTA, GEORGIA _March 15_____ 20 --

____— Sixty days —_____ AFTER DATE _____ I, Sylvia Cometta,_____ PROMISE TO PAY TO

THE ORDER OF ____William Dale Crist_____

PAYABLE AT Bank of the South

___Two thousand and 00/100_____ DOLLARS

VALUE RECEIVED WITH EXACT INTEREST AT ___10 %_____ PER ANNUM

NO. _47_____ DUE _May 14, 20--____

_____Sylvia Cometta_____

Computing the Number of Interest Days of a Note

LEARNING OBJECTIVES **1**

Compute the number of interest days of a note.

To define the **interest period**, or **term**, of a promissory note, the lender either specifies the due date of the note or states the number of interest days. When the due date is given, the number of interest days must be computed before the interest charge can be computed.

To do so you need the number of days in each month, as shown in Table 15-1. February has 29 days in leap years. A leap year is any year that is evenly divisible by 4, except for certain years ending in 00 (e.g., 1900 and 2000). In order to be leap years, years ending in 00 must be evenly divisible by 400; thus 2000 was a leap year, but 1900 wasn't.

STEPS to Compute the Number of Interest Days Between Two Dates

1. Determine the number of interest days in the beginning month.
2. Determine the number of interest days in the middle months.
3. Add the numbers from Steps 1 and 2 to the number of interest days in the final month. (For the final month, the number of interest days is equal to the number of the due date.)

A promissory note is made on July 25. The due date is October 8. Use Table 15-1 to help you determine the number of interest days between July 25 and October 8.

Table 15-1: Days in Each Month (non-leap years)

| Month | Number of Days | Month | Number of Days | Month | Number of Days |
|---|---|---|---|---|---|
| January | 31 days | May | 31 days | September | 30 days |
| February | 28 days | June | 30 days | October | 31 days |
| March | 31 days | July | 31 days | November | 30 days |
| April | 30 days | August | 31 days | December | 31 days |

STEP 1

$$
\begin{array}{rl}
31 & \text{days in July} \\
- 25 & \text{days of note} \\
\hline
6 & \text{days of interest in July}
\end{array}
$$

STEP 2

August has 31 days
September has 30 days

STEP 3

$$
\begin{array}{rl}
6 & \text{days in July} \\
31 & \text{days in August} \\
30 & \text{days in September} \\
+ 8 & \text{days in October} \\
\hline
75 & \text{total interest days in the promissory note}
\end{array}
$$

CONCEPT CHECK 15.1

A promissory note is dated October 20. The maturity date (due date) is February 20. Determine the number of interest days.

As October has 31 days and the note is dated October 20, there are $31 - 20 = 11$ days of interest in October. Since the note is due on February 20, there are 20 interest days in February. The total can be expressed as

| October | | November | | December | | January | | February | | Total Interest Days |
|---|---|---|---|---|---|---|---|---|---|---|
| 11 | + | 30 | + | 31 | + | 31 | + | 20 | = | 123 |

Determining the Due Date of a Note

When the promissory note explicitly states the number of interest days, then you must determine the due date. The procedure is somewhat the reverse of finding the number of interest days.

LEARNING OBJECTIVES

Determine the due date of a promissory note.

STEPS to Determine the Due Date

1. Determine the number of interest days in the beginning month.
2. Determine the number of interest days that remain after the first month.
3. Determine the number of interest days remaining at the end of each succeeding month by subtracting. Continue subtracting until less than one month remains. The due date is the number of interest days remaining in the final month.

EXAMPLE B

A promissory note is made on July 25. The note is for 75 days. Determine the due date.

STEP 1

```
  31  days in July
− 25  days of note
   6  days of interest in July
```

STEP 2

```
  75  days of interest in the note
−  6  days of interest in July
  69  days left in term after end of July
```

STEP 3

```
  69  days of interest left after July
− 31  days in August
  38  days of interest left after August
− 30  days in September
   8  days of interest left after end of September,
      or 8 days of interest in October
```

The due date is October 8.

Although the procedure looks somewhat cumbersome on paper, it goes very quickly on a calculator. You can subtract repeatedly to deduct the days of each month, and after each subtraction, the calculator will display the number of interest days remaining. You don't need to write down all the intermediate results.

When the length of the interest period is expressed in months, the date is advanced by the number of months given. The due date is the same date of the month as the date of the note. For example, a 3-month note dated July 3 will be due on September 3. The exact number of interest days must then be computed, as shown previously. If the note is dated the 31st of a month and the month of maturity is April, June, September, or November, the due date is the 30th. If the month of maturity is February, the due date is the 28th (or 29th in a leap year).

EXAMPLE C

Find the due date of a 3-month note dated January 31 (the last day of the month).
Maturity month: April (count "February, March, April")
Last day: 30 (last day of April)
 Therefore the due date is April 30

 CONCEPT CHECK 15.2

a. A 90-day promissory note is dated February 5 in a non-leap year. Determine the due date.
Since February has 28 days, the note has 28 − 5 = 23 days of interest in February.

| Total Interest Days | | February | | March | | April |
|---|---|---|---|---|---|---|
| 90 | − | 23 | − | 31 | − | 30 |
| | = | 67 | = | 36 | = | 6 days remaining after April |

The due date is May 6.

b. A 4-month promissory note is dated April 30. Determine the due date.
Four months after April 30 is August 30. The due date is August 30.

Computing the Maturity Value of a Note

The maturity value (*MV*) of a promissory note is the sum of the face value (principal) of the note and the interest:

Maturity value = Principal + Interest or $MV = P + I$

EXAMPLE D

Compute the maturity value of the interest-bearing promissory note illustrated in Figure 15-1.

The face value (*P*) of the note is $2,000. The interest rate (*R*) is 10% exact interest per year. The loan period of the note is 60 days, so the time in years (*T*) is $\frac{60}{365}$.

$$I = P \times R \times T = \$2,000 + 0.10 \times \frac{60}{365} = \$32.88$$

$$MV = P + I = \$2,000 + \$32.88 = \$2,032.88$$

 CONCEPT CHECK 15.3

A 90-day promissory note has a face value of $2,800 and an exact simple interest rate of 7.5%. Compute the maturity value.

$$I = P \times R \times T = \$2,800 \times 0.075 \times \frac{90}{365} = \$51.78 \qquad MV = P + I = \$2,800 + \$51.78 = \$2,851.78$$

CyberConnection

Visit this site for helpful information on promissory notes, interest rates, and tips to guide you when you must sign such a note.

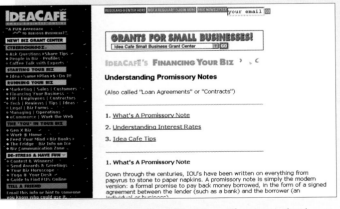

Source: http://www.ideacafe.com/getmoney/fgr_promiss.html.
Copyright © 2001, Courtesy of IdeaCafe.

Computing the Interest Variables

Every simple interest problem has four variables: Interest Amount, Principal, Rate, and Time. In Chapter 14 you solved for the amount of Interest Amount (I) when the Principal (P), Rate (R), and Time (T) were all given. However, so long as any three variables are given, you can always compute the fourth by just changing the formula $I = P \times R \times T$ into one of its possible variations, as shown in Table 15-2.

Table 15-2: PRT formulas

| To find | You must know | Use this formula |
|---|---|---|
| I | $P, R,$ and T | $I = P \times R \times T$ |
| P | $I, R,$ and T | $P = \dfrac{I}{(R \times T)}$ |
| R | $I, P,$ and T | $R = \dfrac{I}{(P \times T)}$ |
| T | $I, P,$ and R | $T = \dfrac{I}{(P \times R)}$ |

You should assume the use of ordinary interest (a 360-day year) unless the use of exact interest (a 365-day year) is indicated. The stated or computed interest rate is the rate for one full year. Also, the length of time used for computing interest dollars must be stated in terms of all or part of a year.

Finding the Interest Amount, Principal, Rate, or Time

When any three variables are known, you can solve for the fourth variable, using a formula from Table 15-2.

EXAMPLE E

Find the Principal if the Interest Amount is $45, the Rate is 9%, and the Time is 60 days.

$$P = \text{?}; \quad I = \$45; \quad R = 9\%; \quad T = \frac{60}{360}\,\text{year}$$

$$P = \frac{I}{(R \times T)} = \frac{\$45}{\left(0.09 \times \dfrac{60}{360}\right)} = \frac{\$45}{0.015} = \$3{,}000$$

EXAMPLE F

Find the Rate if the Interest Amount is $22, the Principal is $2,000, and the Time is 30 days.

$$R = \text{?}; \quad I = \$22; \quad P = \$2{,}000; \quad T = \frac{30}{360}\,\text{year}$$

$$R = \frac{I}{(P \times T)} = \frac{\$22}{\left(\$2{,}000 \times \dfrac{30}{360}\right)} = \frac{\$22}{\$166.67} = 0.132, \text{ or } 13.2\%$$

Find the Time if the Interest amount is $54, the Principal is $1,800, and the Rate is 12%. Express Time in days, based on a 360-day year.

$T = ?;\quad I = \$54;\quad P = \$1,800;\quad R = 12\%$

$$T = \frac{I}{(P \times R)} = \frac{\$54}{(\$1,800 \times 0.12)} = \frac{\$54}{\$216} = 0.25 \text{ year}$$

Based on a 360-day year, 0.25 year = 0.25 × 360 days = 90 days.

CONCEPT CHECK 15.4

Each of the following problems gives three of the four variables. Find the missing variable. All rates are ordinary simple interest (360-day year). Round P and I to the nearest cent; round R to the nearest $\frac{1}{10}\%$; round T to the nearest whole day, assuming that 1 year has 360 days. Use one of the four formulas:

$$I = P \times R \times T, \qquad P = \frac{I}{(R \times T)}, \qquad R = \frac{I}{(P \times T)}, \quad \text{and} \quad T = \frac{I}{(P \times R)}$$

a. Principal = $840; Rate = 8%; Time = 120 days
 Find Interest Amount:

 $$I = P \times R \times T = \$840 \times 0.08 \times \frac{120}{360} = \$22.40$$

c. Principal = $1,280; Interest Amount = $64; Rate = 10%
 Find Time:

 $$T = \frac{I}{(P \times R)} = \frac{\$64}{(\$1,280 \times 0.10)} = 0.5 \text{ year}$$

 In a 360-day year, $T = 0.5$ year = 0.5 × 360 days = 180 days.

b. Principal = $8,000; Interest Amount = $50; Time = 45 days
 Find Rate:

 $$R = \frac{I}{(P \times T)} = \frac{\$50}{\left(\$8,000 \times \frac{45}{360}\right)} = 0.05, \text{ or } 5\%$$

d. Interest Amount = $90; Rate = 12%; Time = 75 days
 Find Principal:

 $$T = \frac{I}{(R \times T)} = \frac{\$90}{\left(0.12 \times \frac{75}{360}\right)} = \$3,600$$

COMPLETE ASSIGNMENTS 15.1 AND 15.2.

CyberConnection

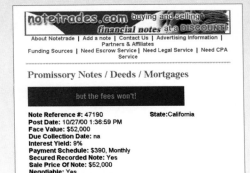

Promissory notes are often bought and sold. Visit this Web site to learn about the kinds of notes traded on the Internet.

Source: http://www.notetrades.com/promissory-notes-deeds-mortgages.asp. Copyright © 2001 Courtesy of Notetrades.com.

Chapter Terms for Review

| | | |
|---|---|---|
| due date | maturity date | principal (P) |
| face value | maturity value (MV) | promissory note |
| interest dollars (I) | negotiable promissory note | term of note |
| interest period | | |

Visit the Web

http://premium.findforms.com/pnotes.htm—Look over some of the promissory note forms for several states presented at this Web site.

http://www.nolo.com/encyclopedia/articles/sb/borrowing.html—Read about some legal aspects of promissory notes here.

http://www.glencoe.com/sec/accounting/teacher/pchtree/promissory.htm—Learn about the accounting aspects of promissory notes used as receivables at this Web site.

http://www.sec.gov/answers/promise.htm—What does the Securities and Exchange Commission (SEC) have to say about promissory notes? Access its Web site to find out.

CYBERCONNECTION

Home | Sign In | Register | MyAccounts | **Investing** | Banking/Bills | Loans | Insurance | Planning | Help

Investing Center | Portfolios | Stocks | Funds | Broker Center | News | Perspectives | Learning

Promissory Notes Promise Far More Than They Deliver

Printer-friendly | E-mail to a friend

By Jordan E. Goodman

Section 1: $170 Million Lost

How would you like to earn a risk-free 13% in just nine months? That's the promise being made by salesmen for a relatively new kind of security called a promissory note.

The only problem with the pitch? It's likely to be a fraud. The North American Securities Administrators Association [(888) 84N-ASAA] has just reported that securities regulators in 29 states have taken scores of actions against 179 individuals and 153 companies selling promissory notes. NASAA estimates that 4600 investors have lost over $170 million in the enforcement actions so far, and regulators are just starting to scratch the surface of this growing scam.

What is a promissory note?

PN's are investments in which the investor loans money to a company for a particular period of time, usually nine months, in exchange for a fixed rate of interest. The interest rate quoted is usually at least 10%, 15%, or sometimes as much as 25%. That seems unusually attractive to people living on a fixed income, particularly the elderly, who are frustrated with today's single-digit CD yields. Unlike a certificate of deposit issued by an FDIC-insured bank, however, there is no federal insurance backing the PN. To give the instrument an air of

Learn More! From the Experts at OnMoney

Table of Contents

$170 Million Lost
▸ Red Flags Abound
▸ Companies to Watch

Feedback
▸ Your Comments

Related Stories
Retirement: How Much You Will Need Jordan Goodman, the author of "Everyone's Money Book," tells how to achieve financial security for your retirement years.

Promissory notes are now being traded as investments—somewhat like when a corporation buys your mortgage or your credit card account—because they offer interest and profits. Unfortunately, some of the "too-good-to-be-true" deals are actually fraudulent promissory notes. After reading this article, read more at the Web sites
http://www.wdfi.org/newsroom/legislative/fyc_fraudulent_notes.htm,
http://www.dbf.state.fl.us/alerts99/alrt0599.html, and
http://www.onmoney.com/editorial/investing/goodman/promissory/promissory_note_step1.html.

The Bottom Line

Summary of chapter learning objectives:

| Learning Objective | Summary | Example |
|---|---|---|
| **15.1** | Compute the number of interest days of a note | 1. Find the number of days between December 20 and February 25. |
| **15.2** | Determine the due date of a promissory note | 2. Find the due date of a 60-day note written on April 20. |
| **15.3** | Compute the maturity value of a promissory note | 3. Find the maturity value of a 75-day promissory note with a face value of $8,100 and an exact interest rate of 11%. |
| **15.4** | Compute the Principal, Rate, and Time from the basic interest formula | Solve for Principal, Rate, and Time by using a 360-day year and the formulas $$P = \frac{I}{(R \times T)}, \quad R = \frac{I}{(P \times T)}, \text{ and } T = \frac{I}{(P \times R)}$$ 4. Interest Amount = $42; Rate = 6%; Time = 105 days
5. Principal = $1,600; Interest Amount = $30; Time = 75 days
6. Principal = $7,200; Interest Amount = $135; Rate = 15% |

Answers: 1. 67 days 2. June 19 3. $8,283.08 4. $2,400 5. 0.09, or 9% 6. 45 days

1 A 120-day promissory note for $2,500 is dated June 17, 2002. Find (a) the due date and (b) the maturity value, if the rate is 6% ordinary simple interest.

2 A 75-day promissory note for $12,000 is dated November 26, 2003. Find (a) the due date and (b) the maturity value, if the rate is 9% ordinary simple interest.

3 A promissory note for $1,400 is dated July 15, 2004 and has a due date of October 15, 2004. Find (a) the number of interest days and (b) the maturity value, if the rate is 8% ordinary simple interest.

4 A promissory note for $8,850 is dated December 10, 2003 and has a due date of April 10, 2004. Find (a) the number of interest days and (b) the maturity value, if the rate is 7% ordinary simple interest.

5 A 90-day promissory note for $25,000 is dated October 24, 2002. Find (a) the due date and (b) the maturity value, if the rate is 5.6% exact simple interest.

6 A 135-day promissory note for $4,675 is dated May 4, 2002. Find (a) the due date and (b) the maturity value, if the rate is 8.1% exact simple interest.

7 A promissory note for $980 is dated April 23, 2003 and has a due date of July 23, 2003. Find (a) the number of interest days and (b) the maturity value, if the rate is 7.5% exact simple interest.

8 A promissory note for $16,290 is dated August 19, 2003 and has a due date of December 19, 2003. Find (a) the number of interest days and (b) the maturity value, if the rate is 6.9% exact simple interest.

9 Vernon Lee borrowed money for 120 days at a rate of 8% ordinary simple interest. How much did Vernon borrow if he paid $112 in interest?

10 Linda Hinajosa loaned $3,000 to one of her employees for 180 days. If the employee's interest amount was $105, what was the ordinary simple interest rate?

11 Danny O'Rourke loaned $10,000 to a machine shop owner who was buying a piece of used equipment. The interest rate was 6% exact simple interest, and the interest amount was $360. Compute the number of days of the loan.

12 To lease a new space for her antique store, Sandra Panowski borrowed money for 150 days at 9% exact simple interest. Sandra paid $540 in interest. What was the maturity value of the loan? (*Hint*: First calculate the principal.)

Notes

Assignment 15.1: Dates, Times, and Interest

Name _____

Date _____ Score _____

A (20 points) Problems 1–5: Find the number of interest days. Problems 6–10: Find the due date. Be sure to check for leap years. (2 points for each correct answer)

| Date of Note | Due Date | Days of Interest |
|---|---|---|
| **1.** August 30, 2002 | December 6, 2002 | _____ |
| **2.** June 14, 2002 | September 13, 2002 | _____ |
| **3.** March 6, 2003 | September 11, 2003 | _____ |
| **4.** December 8, 2003 | April 9, 2004 | _____ |
| **5.** January 9, 2004 | May 10, 2004 | _____ |

| Date of Note | Interest Days | Due Date |
|---|---|---|
| **6.** July 17, 2002 | 180 days | _____ |
| **7.** December 4, 2002 | 90 days | _____ |
| **8.** January 28, 2003 | 45 days | _____ |
| **9.** August 22, 2003 | 75 days | _____ |
| **10.** December 31, 2004 | 2 months | _____ |

Score for A (20) _____

B **(48 points) For each of the following promissory notes, find the missing entry for days of interest or maturity date (due date). Then compute the amount of interest due at maturity and the maturity value. For problems 11–14, use a 360-day year. For problems 15–18, use a 365-day year. (2 points for each correct answer)**

| | Face Value | Date of Note | Days of Interest | Maturity Date | Rate | Interest | Maturity Value |
|---|---|---|---|---|---|---|---|
| **11.** | $8,000 | Oct. 8, 2003 | 90 | _____ | 6% | _____ | _____ |
| **12.** | $960 | Mar. 28, 2003 | _____ | July 8, 2003 | 8% | _____ | _____ |
| **13.** | $24,000 | Jan. 28, 2004 | 75 | _____ | 7% | _____ | _____ |
| **14.** | $1,180 | April 10, 2002 | _____ | Aug. 23, 2002 | 10% | _____ | _____ |
| **15.** | $845 | July 15, 2002 | _____ | Oct. 28, 2002 | 9% | _____ | _____ |
| **16.** | $16,400 | Jan. 28, 2004 | 65 | _____ | 5.1% | _____ | _____ |
| **17.** | $9,250 | Feb. 9, 2004 | _____ | Aug. 27, 2004 | 8.4% | _____ | _____ |
| **18.** | $3,200 | Nov. 5, 2003 | 125 | _____ | 6.5% | _____ | _____ |

Score for B (48)

Assignment 15.1 *Continued*

C **(32 points) Determine the missing variable by using one of the formulas**

$$I = P \times R \times T, \quad P = \frac{I}{(R \times T)}, \quad R = \frac{I}{(P \times T)}, \quad \text{or} \quad T = \frac{I}{(P \times R)}$$

For problems 19–26, use a 360-day year. For problems 27–34, use a 365-day year. Round dollar amounts to the nearest cent. Round interest rates to the nearest $\frac{1}{10}$ of a percent. Find the time in days, rounded to the nearest whole day. (2 points for each correct answer)

| | Principal | Rate | Time | Interest |
|---|---|---|---|---|
| **19.** | $6,000 | 6% | 180 days | _____ |
| **20.** | $840 | _____ | 75 days | $14.00 |
| **21.** | _____ | 11% | 120 days | $352.00 |
| **22.** | $12,000 | 5% | _____ | $50.00 |
| **23.** | $600 | _____ | 60 days | $9.00 |
| **24.** | $2,480 | 6% | 75 days | _____ |
| **25.** | $24,890 | 3% | _____ | $280.00 |

Assignment 15.1 *Continued*

| | Principal | Rate | Time | Interest |
|---|---|---|---|---|
| **26.** | _____ | 8% | 270 days | $510.00 |
| **27.** | $1,280 | 6.7% | 90 days | _____ |
| **28.** | $34,950 | 5.5% | _____ | $395.00 |
| **29.** | $18,600 | _____ | 45 days | $195.00 |
| **30.** | _____ | 4.9% | 135 days | $50.00 |
| **31.** | $2,040 | 4.1% | _____ | $55.00 |
| **32.** | $200,000 | _____ | 50 days | $1,700.00 |
| **33.** | _____ | 6.8% | 105 days | $25.00 |
| **34.** | $8,450 | 7.4% | 120 days | _____ |

Score for C (32)

Assignment 15.2: Using Interest in Business

Name _____

Date _____ Score _____

A **(50 points) The following problems involve ordinary simple interest, using a 360-day year. Find the due date or the number of interest days, as required. Find the missing variable by using one of the formulas:**

$$I = P \times R \times T, \quad P = \frac{I}{(R \times T)}, \quad R = \frac{I}{(P \times T)}, \quad \text{or} \quad T = \frac{I}{(P \times R)}$$

Compute time to the nearest day; compute rates to the nearest $\frac{1}{10}$ of a percent. (5 points for each correct answer)

1. Sandra Miller is opening a new bakery. On June 17, 2002, Sandra signs a promissory note to borrow $9,500 for 60 days to help pay for some new dough-mixing equipment. The rate is 6% ordinary simple interest. Find the amount of interest that Sandra will pay and the due date.

Interest amount _____

Due date _____

2. Kathleen Donaldson and her husband raise game birds—including pheasants and ducks—for sale only to restaurants. On March 12, 2003, they borrow money at 10% ordinary simple interest to pay for some new holding pens. On July 25, 2003, they pay back the loan principal and also $150 in interest. First compute the number of interest days between the two dates. Then compute the principal.

Interest days _____

Principal _____

3. Michelle Fong inherited a jewelry store from her father. On September 23, 2003, she borrowed $36,000 for new items for the Christmas holiday season. Ninety days later, she repaid a total of $36,540, which was the principal plus interest. After determining the amount of interest, compute the ordinary simple interest rate. Then find the due date.

Interest rate _____

Due date _____

4. Terry Alter bought a roofing company. On February 20, 2004, she borrowed $75,000 to help finance a machine to hydraulically lift materials, workers, and equipment onto roofs. On the due date, when Terry repaid the principal, she also paid $1,250 in interest. If the ordinary simple interest rate was 8%, compute the number of interest days. Then determine the due date.

Interest days _____

Due date _____

5. Ramona Cruz is planning to write an economics book for high school students, but to do so she needs a new computer and printer. On April 7, 2004, she borrows money from her father at 5% ordinary simple interest. She repays the principal and $45 interest 180 days later. Compute the principal and the due date.

Principal _____

Due date _____

Score for A (50)

B **(50 points) The following problems involve exact simple interest, using a 365-day year. Find the due date or the number of interest days, as required. Find the missing variable by using one of the formulas**

$$I = P \times R \times T, \quad P = \frac{I}{(R \times T)}, \quad R = \frac{I}{(P \times T)}, \quad \text{or} \quad T = \frac{I}{(P \times R)}$$

Compute time to the nearest day; compute rates to the nearest $\frac{1}{10}$ of a percent. (5 points for each correct answer)

6. Stan Keefer, a contractor, needed a short-term loan of $10,000 to pay one of his suppliers. Stan signed a promissory note at 8.7% exact simple interest on January 18, 2002. He repaid the principal and interest on March 4, 2002. First find the number of interest days. Then compute the dollars of interest that Stan paid.

Interest days _____

Interest amount _____

7. A few months after graduating from medical school, Sharon Sumpter borrowed $16,500. The exact simple interest rate was 6.2%; the date of the promissory note was November 29, 2002. On the due date, when Sharon repaid the principal, she also paid $420.41 interest. Compute the number of interest days. Then determine the due date of the note.

Interest days _____

Due date _____

8. On May 2, 2003, Vikosh Singh signed a promissory note to borrow a small amount of money for 105 days. The rate on the loan was 7.7% exact simple interest. On the due date, Vikosh repaid the entire principal and $32.34 interest. Calculate the amount of money that Vikosh borrowed, and then determine the due date of the note.

Principal _____

Due date _____

9. On October 14, 2002, Melvin Counts loaned $830 to Don Galusha, one of his friends from high school. On the due date, 270 days later, Don repaid the principal and $30.70 interest. Compute the exact simple interest rate and the due date.

Interest rate _____

Due date _____

Assignment 15.2 *Continued*

10. Five days after starting the fall semester, Lola Smith had an automobile accident. She needed to buy a replacement car, so she borrowed $4,000 on September 12, 2005. On November 11, 2005, Lola received a gift from her grandfather and repaid the $4,000, along with $61.15 interest. First, determine the number of interest days; then compute the exact simple interest rate.

Interest days _____

Interest rate _____

Score for B (50)

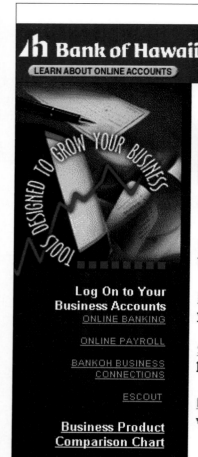

Bank of Hawaii

LEARN ABOUT ONLINE ACCOUNTS

BUSINESS BANKING

Log On to Your Business Accounts
ONLINE BANKING

ONLINE PAYROLL

BANKOH BUSINESS CONNECTIONS

ESCOUT

Business Product Comparison Chart

Business Products & Services
CHECKING & SAVINGS

BUSINESS CONVENIENCE PRODUCTS & SERVICES

CASH MANAGEMENT

Business Borrowing Options

Choose from several different types of loans, all geared to help your business grow.

Business CreditFlex – The flexible credit plan that changes with your business. Apply today!

Small Business Administration (SBA) Guaranty Loans – Business loans guaranteed by the Small Business Administration

Bank of Hawaii Equipment Leasing – Why buy office equipment when you can lease it from Bank of Hawaii Equipment Leasing

What Products & Services are Right For Your Business?

Business Products & Services
CHECKING & SAVINGS
BUSINESS CONVENIENCE PRODUCTS & SERVICES
CASH MANAGEMENT · CREDIT CARD PROCESSING
BUSINESS BORROWING OPTIONS · LEASING · SPECIALIZED FINANCING
RETIREMENT PLANNING AND EMPLOYEE BENEFITS
TRADE & INTERNATIONAL SERVICES · BUSINESS PLANNING

Bank of Hawaii offers financing for commercial enterprise. Its Web site provides information on its various services, whether your business is new or established, if you need a commercial loan to aquire another firm, buy assets, handle inventory and receivables, lease equipment, or other finance functions. It offers cash management, business planning, and trade and international services as well. Visit their site at http://www.boh.com/busbank/business.

Borrowing By Business

LEARNING OBJECTIVES

By studying this chapter and completing all assignments you will be able to:

 Compute the savings from borrowing money to take a cash discount.

 Compute interest with the unpaid-balance method.

 Compute the proceeds and actual interest rate on a bank discount loan.

 Discount a non–interest-bearing note.

 Discount an interest-bearing note.

To stay competitive, businesses must borrow for everything from current inventory to major capital investments in new equipment, buildings, and retooling for new products. Knowing when and how to borrow can make the difference between thriving and closing down.

1. Explain why a business might pay interest to borrow money just to be able to take advantage of a cash discount.
2. In general, what is meant by the phrase *unpaid balance method*?
3. A business accepts an interest-bearing promissory note in exchange for goods. What does it mean to say that "the business sold the promissory note at a discount"?
4. Why is "discounting a non–interest-bearing note" less realistic than "discounting an interest-bearing note"?

Businesses, like individuals, use long-term loans (more than 1 year) to purchase large items such as new equipment or buildings. Unlike individuals, though, businesses often borrow money for very short periods—even for only a few days. When retail businesses such as Stereo SuperStore or DVD Max buy merchandise from manufacturers and wholesalers, often they know in advance that they can't sell the merchandise in time to pay the supplier's invoice. If they don't have sufficient cash on hand, they must borrow money until they have made enough sales to make the payment. If the supplier offers a cash discount (see Chapter 12), money can usually be saved by borrowing enough cash to take advantage of the discount.

In Chapters 14 and 15, all the interest was single payment. The lender gave money to the borrower; after a specified period of time, the borrower made a single payment, repaying the principal and all the interest. For loans that are longer than a few days or weeks, banks and other financial institutions usually require the borrower to make some intermediate partial payments of principal and/or interest. There are several borrowing and repayment methods, some specific to individuals and others more common for businesses. In Chapter 17 we describe some examples of loans that are consumer-oriented; here we discuss borrowing and repayment by businesses.

Borrowing Money to Take a Cash Discount

LEARNING OBJECTIVES

Compute the savings from borrowing money to take a cash discount.

In Chapter 12 we described how wholesalers and manufacturers use cash discounts to encourage their customers to pay quickly. Recall that the terms 2/10, net 30 mean that the buyer will receive a discount of 2% by paying the invoice within 10 days and that the entire invoice is due within 30 days.

EXAMPLE A

DVD Max purchased $100,000 worth of CDs and DVDs. The invoice was dated October 4 with terms of 2/10, net 30. Compute the due date, the discount date, the cash discount, and the total remittance required to get the cash discount. (Review Chapter 12 if necessary.)

Due date = October 4 + 30 days = November 3
Discount date = October 4 + 10 days = October 14
If paid by October 14:
 Cash discount = $100,000 × 0.02 = $2,000
 Total remittance: $100,000 − $2,000 = $98,000

Regardless of whether it takes the discount, DVD Max needs to pay $100,000 by November 3. The company may want to save the $2,000, but perhaps it doesn't have the $98,000 now. Or maybe it has the money but wants to spend it on something else. In either situation, DVD Max might be able to borrow the money from October 14 until November 3. Before borrowing, DVD Max should compare the savings from the cash discount with the interest on a loan.

EXAMPLE B

DVD Max can borrow $98,000 for 20 days (October 14 to November 3) by paying 10% exact simple interest (365-day year). Compute the interest on the loan and the savings for DVD Max if it borrows to take the discount.

$$\text{Interest} = P \times R \times T = \$98,000 \times 0.10 \times \frac{20}{365} = \$536.99$$

Savings = $2,000 discount − $536.99 interest = $1,463.01

CyberConnection

Visit this site to learn about loans and discounts.

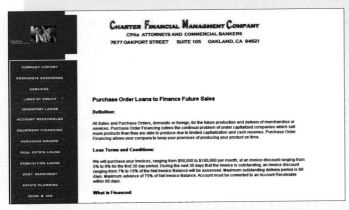

Source: http://www.e-charterfinancial.com/. Copyright © 2001
Courtesy of Charter Financial Management Company.

The reason for borrowing only between the discount date and the due date is to delay making payments as long as possible, whether to get discounts or to avoid penalties. The discount date is the latest possible date to pay and get the discount; the due date is the latest possible date to pay and avoid a penalty.

Although borrowing and taking the discount is almost always cheaper, the actual dollar amount may determine what DVD Max decides. If the original purchase were only $1,000, the savings would be only $14.63. Such an amount may not be worth the effort of getting a loan. However, for borrowing small amounts regularly, businesses often have "revolving lines of credit." These allow them to borrow and repay frequently, without always making a new loan application.

CONCEPT CHECK 16.1

A retailer purchases merchandise under the terms 1.5/20, net 45. The invoice is for $45,000 and is dated July 22. For the cash discount, calculate the due date, the discount date, the amount of the cash discount, and the total remittance required. The retailer borrows enough money to pay the entire remittance. The interest rate is 12% exact simple interest, and the loan is for the length of time between the discount date and the due date. Calculate the amount of the interest and the savings gained by borrowing the remittance to take the discount.

Discount:
| | | |
|---|---|---|
| Due date: | July 22 + 45 days = September 5 |
| Discount date: | July 22 + 20 days = August 11 |
| Cash discount: | $45,000 × 0.015 = $675 |
| Remittance: | $45,000 − $675 = $44,325 |

Loan:
Interest days: August 11 to September 5 = 25 days

$$\text{Interest} = P \times R \times T = \$44,325 \times 0.12 \times \frac{25}{365} = \$364.32$$

Savings: $675 cash discount − $364.32 interest = $310.68

Computing Interest with the
Unpaid-Balance Method

LEARNING OBJECTIVES

Compute interest with the unpaid-balance method.

Recall that lenders charge interest on the *principal*, which is the amount of money that the borrower owes to the lender. Many lenders require the borrower to repay the loan in regular monthly payments. The payments may or may not be equal every month. Usually the borrower pays all the interest for that month and also repays part of the principal. The amount of the principal that has not been repaid is called the **unpaid balance**. As the borrower repays part of the principal, the unpaid balance decreases. The new unpaid balance will be the principal in the interest calculation for the next month. This method of calculating interest is called the **unpaid-balance method**.

In this chapter, we compute interest for each month with the basic unpaid-balance method, using the formula $I = P \times R \times T$. P is the unpaid balance for that month; I is the annual interest rate; and T, the time, is expressed as a fraction of a 365-day year. Recall from Chapter 14 that this interest is *exact interest*.

In Chapter 17 we describe slightly different versions of the unpaid-balance method that are more commonly used in credit purchases and consumer loans.

STEPS **to Compute Interest and the New Unpaid Balance,
Using the Unpaid-Balance Method**

1. Determine $P, R,$ and T.
 a. P is the unpaid balance for the month.
 b. R is the quoted annual interest rate.
 c. T, the time, is a fraction of a 365-day year, where the numerator is the exact number of days in that particular month and the denominator is 365.
2. Compute the interest payment for the month, using $I = P \times R \times T$.
3. Compute the principal payment by subtracting the interest payment from the total monthly payment.
4. Compute the new unpaid balance by subtracting the principal payment from the former unpaid balance (P in Step 1a).
5. Repeat Steps 1 through 4 until the last month. For the last month, compute the total payment by adding the interest due to the unpaid balance.

EXAMPLE C

On October 13, Ladera Service Station borrows $30,000 for some new oil-changing equipment. The interest rate is 11% exact (365-day year). Ladera makes payments of $10,000 on November 13 and December 13. Each payment includes interest on the unpaid balance from the preceding month. The rest of the $10,000 is used to repay the principal. On January 13, Ladera makes a final payment to repay the balance plus the interest since December 13. Compute the interest and principal payment for each month. Also, show the new unpaid balance for each month and the total payment for January 13.

| STEP 1 | P, R & T: | $P = \$30{,}000$ (loan amount); $R = 11\%$; $T = \frac{31}{365}$ (31 days between October 13 and November 13) |
|---|---|---|
| STEP 2 | Interest payment: | $I = P \times R \times T = \$30{,}000 \times 0.11 \times \frac{31}{365} = \280.27 |
| STEP 3 | Principal payment: | $\$10{,}000 - \$280.27 = \$9{,}719.73$ |
| STEP 4 | Unpaid balance: | $\$30{,}000 - \$9{,}719.73 = \$20{,}280.27$ |
| STEP 1 | P, R & T: | $P = \$20{,}280.27$ (unpaid balance from November 13); $R = 11\%$; $T = \frac{30}{365}$ (30 days between November 13 and December 13) |
| STEP 2 | Interest payment: | $I = P \times R \times T = \$20{,}280.27 \times 0.11 \times \frac{30}{365} = \183.36 |
| STEP 3 | Principal payment: | $\$10{,}000 - \$183.36 = \$9{,}816.64$ |
| STEP 4 | Unpaid balance: | $\$20{,}280.27 - \$9{,}816.64 = \$10{,}463.63$ |
| STEP 1 | P, R & T: | $P = \$10{,}463.63$ (unpaid balance from December 13); $R = 11\%$; $T = \frac{31}{365}$ (31 days between December 13 and January 13) |
| STEP 2 | Interest payment: | $I = P \times R \times T = \$10{,}463.63 \times 0.11 \times \frac{31}{365} = \97.76 |
| STEP 3 | Principal payment: | $\$10{,}463.63$ (repay the entire unpaid balance) |
| STEP 4 | Unpaid balance: | $\$10{,}463.63 - \$10{,}463.63 = \$0.00$ |
| STEP 5 | Total payment: | $\$10{,}463.63 + \$97.76 = \$10{,}561.39$ |

CONCEPT CHECK 16.2

On January 24, 2003, a bank loaned $4,000 to a business for two months at 12% exact simple interest (365-day year). On February 24, 2003, the borrower made a total payment of $2,000, which included interest on the unpaid balance. On March 24, 2003, the borrower paid the remaining principal and interest on the balance. Calculate the amount of interest and principal paid each month. (Check for leap year.)

There are 31 days between January 24 and February 24. There are 28 days between February 24 and March 24.

| | February 24, 2003 | | March 24, 2003 | |
|---|---|---|---|---|
| Unpaid balance: | Original principal | $4,000.00 | From February 24 | $2,040.77 |
| Interest payment: | $4{,}000.00 \times 0.12 \times \frac{31}{365} = \40.77 | | $2{,}040.77 \times 0.12 \times \frac{28}{365} = \18.79 | |
| Total payment: | | $2,000.00 | Add: $2,040.77 + $18.79 = | $2,059.56 |
| Principal payment: | $2,000 − $40.77 = | $1,959.23 | | $2,040.77 |
| New balance: | $4,000 − $1,959.23 = | $2,040.77 | $2,040.77 − $2,040.77 = | $0.00 |

COMPLETE ASSIGNMENT 16.1.

Analyzing Bank Discounts

Compute the proceeds and actual interest rate on a bank discount loan.

As you learned in Chapter 14, in the simplest loan situation the borrower goes to a lender, borrows some money, and then repays the money plus some interest. The interest is the lender's fee for making the loan.

EXAMPLE D

Norman Brewer, owner/operator of a McDonald's Restaurant, borrows $80,000 from his bank for 60 days at 9% ordinary interest. Using a 360-day year, compute the interest and amount due at maturity.

$$P = \$80,000; R = 9\%; T = \frac{60}{360}$$

$$\text{Interest} = P \times R \times T = \$80,000 \times 0.09 \times \frac{60}{360} = \$1,200$$

$$\text{Amount due} = P + I = \$80,000 + \$1,200 = \$81,200$$

Norman will keep the entire $80,000 principal for the entire 60 days and then repay $81,200 on the due date.

www.ulsterbank.com/
ni/business/bbusiness/
index.asp

Visit the Web site of Ulster Bank for information on invoice discounting in Ireland.

Recall from Chapter 15 that, in order to get a loan, a borrower is often required to sign a document called a *promissory note*. If there were a promissory note for the loan in example D, the $80,000 would be the face value of the note and the $81,200 would be the maturity value of the note.

Another lending method that banks sometimes use is to *subtract* the loan fee from the face value at the time the loan is made. In example D, instead of giving Norman Brewer $80,000 for 60 days, the bank would subtract $1,200 from $80,000 and give him only $78,800. Then, on the due date, he would repay a total amount of $80,000.

The procedure is called **discounting**. The calculation is the same: $\$80,000 \times 9\% \times \frac{60}{360} = \$1,200$. The 9% is called the *discount rate*, and the 60 days is called the *discount period*, as defined in Chapter 12. The $1,200 is called the **discount amount**, or the **bank discount** in this case. The $78,800 is called the **proceeds** of the loan. The $80,000 is both the *face value* and the *maturity value*, as introduced in Chapter 15. Note that the $80,000 isn't really the "principal," because $80,000 isn't the amount actually borrowed. To distinguish the calculations, we use different symbols:

Discount amount $= FV \times R \times T$, where FV is the face value and R is the discount rate

and

Proceeds $= FV - D$, where D is the discount amount

EXAMPLE E

Compute the discount amount and the proceeds for a 60-day loan that has a face value of $80,000 and a bank discount rate of 9%. (Use a 360-day year.)

$$FV = \$80,000; R = 9\%; T = \frac{60}{360}$$

$$\text{Discount amount} = FV \times R \times T = \$80,000 \times 0.09 \times \frac{60}{360} = \$1,200$$

$$\text{Proceeds} = FV - D = \$80,000 - \$1,200 = \$78,800$$

Some people refer to this type of loan as non–interest-bearing because the borrower repays only the face value. However, the term *non–interest-bearing* is misleading because the borrower isn't really getting an "interest-free loan." There is a charge of $1,200 to borrow $78,800 for 60 days.

Discount Rate Versus Interest Rate

When a bank uses the discount method, the discount rate is not the interest rate. In example E, the discount rate was 9%, and the borrower had to pay $1,200 to borrow $78,800 for 60 days. To compute the actual interest rate, use the formula $R = I/(P \times T)$ from Chapter 15, letting $I = \$1,200$, $P = \$78,800$, and $T = \frac{60}{360}$.

$$R = \frac{I}{(P \times T)} = \frac{\$1,200}{\left(\$78,800 \times \dfrac{60}{360}\right)} = 0.09137, \text{ or } 9.14\%$$

The interest rate is actually 9.14%, whereas the discount rate is 9%. Because the two rates do not represent the same thing, a borrower must understand which rate is used in the loan.

CONCEPT CHECK 16.3

A bank made a 90-day loan on a discount basis. The face value was $64,000, and the discount rate was 11%. Compute the discount amount and the proceeds. Then compute the actual interest rate, using the proceeds as the principal of the loan instead of the face value. Use a 360-day year in all calculations.

Discount amount $= FV \times R \times T = \$64,000 \times 0.11 \times \dfrac{90}{360} = \$1,760$

Proceeds $=$ Face value $-$ Discount amount $= \$64,000 - \$1,760 = \$62,240$

Actual Interest Rate $= \dfrac{I}{(P \times T)} = \dfrac{\$1,760}{\left(\$62,240 \times \dfrac{90}{360}\right)} = 0.1131, \text{ or } 11.31\%$

Discounting Non–Interest-Bearing Notes

In example E, the bank made the loan on a discount basis. Although the bank charged $1,200 for the loan, the promissory note might be said to be a **non–interest-bearing note** because the maturity value equals the face value. The same situation occurs when the loan truly has no interest charge, as might happen when an individual lends money to a relative or close friend.

When a lender holds a non–interest-bearing promissory note, the lender may be able to "sell" the note to someone else. Such a note is said to be a **negotiable note**. A lender might sell a note to get cash immediately instead of having to wait until the maturity date. Of course, whoever buys the note will not pay the entire maturity value to the orig-

LEARNING OBJECTIVES

Discount a non–interest-bearing note.

VIDEO

Discounting Notes

inal lender, because risk is involved. The level of risk depends on both the reputation (credit rating) of the original borrower and the length of time until the maturity date. Because the buyer of the note will pay less than the maturity value, the buyer is said to be discounting the note. The discount amount is calculated just like a bank discount. The original lender, who is now selling the note, receives the proceeds, as computed by the following formulas:

Discount amount $= FV \times R \times T$, where R is the discount rate

Proceeds $= FV - D$, where D is the discount amount

STEPS **to Discount a Non–Interest-Bearing Note**

1. Determine the maturity (due) date of the note.
2. Compute the length of the discount period—the number of days between the discount date and the due date. Divide the number of days by 360 (or 365) to get T.
3. Compute the discount amount, using $D = FV \times R \times T$.
4. Compute the proceeds, using $FV - D$.

EXAMPLE F

Roberto Saenz, owner of a True-Value Hardware Store, is holding a 45-day non–interest-bearing note for $5,000, dated April 2. On April 11, Roberto sells the note to Marshfield Finance Company. The finance company discounts the note at 15%. Find the discount amount and the proceeds. (Use a 360-day year.)

STEP 1 Due date: April 2 + 45 days = May 17

STEP 2 Discount period: April 11 to May 17 = 36 days

STEP 3 Discount amount: Face value × Discount rate × Time

$$= \$5,000.00 \times 0.15 \times \frac{36}{360}$$

$$= \$75$$

STEP 4 Proceeds: Face value − Discount amount

$$= \$5,000 - \$75$$

$$= \$4,925$$

 CONCEPT CHECK 16.4

A 90-day non–interest-bearing promissory note dated March 19 has a face value of $2,500. On April 14, a finance company discounts the note at a rate of 11%. Find the discount amount and the proceeds. Note: The maturity date and the days of discount must first be determined. Use a 365-day year for the discount calculation.

| | |
|---|---|
| Maturity date: | March 19 + 90 days = June 17 |
| Days of discount: | April 14 to June 17 = 64 days |

Discount amount: $\$2,500 \times 0.11 \times \dfrac{64}{365} = \48.22

Proceeds: $\$2,500 - \$48.22 = \$2,451.78$

Discounting Interest-Bearing Notes

A promissory note such as the one in example G is an **interest-bearing note**.

Discount an interest-bearing note.

EXAMPLE G

On September 20, Redbourne Medical Supply borrowed $120,000 from a private investor, John White. In return for the $120,000, Redbourne Medical gave John a 90-day promissory note at an ordinary interest rate of 10%. Compute the due date and maturity value of the note.

Due date: September 20 + 90 days = December 19

Interest: $I = P \times R \times T = \$120,000 \times 0.10 \times \dfrac{90}{360} = \$3,000$

Maturity value: MV = Principal + Interest
 = $120,000 + $3,000 = $123,000

In example G, Redbourne Medical will not pay the $123,000 to John White until December 19. During the 90 days, John has only the promissory note. If the note is negotiable, John can sell it for cash to a third party before December 19.

Suppose that John decides to sell the note on November 19. The note will be worth $123,000 on December 19. Whoever buys the note wants to earn a profit, so the buyer will pay less than $123,000 for the note. We again say that the new buyer discounts the note, or buys the note at a discount. Again, the amount that John White receives is the proceeds, and the decrease in value of the note is the discount amount.

This discount amount is computed the same way as the discount amount on a non–interest-bearing note.

Discount amount = Maturity value × Discount rate × Time
Proceeds = Maturity value − Discount amount

STEPS to Discount an Interest-Bearing Note

1. Compute the interest and maturity value of the original note.
2. Determine the maturity (due) date of the original note.
3. Compute the number of days in the discount period. The time, T, is the number of days in the discount period divided by 360 (or 365).
4. Compute the discount amount, using $D = MV \times R \times T$.
5. Compute the proceeds by subtracting the discount amount (Step 4) from the maturity value (Step 1).

CyberConnection

Source: http://www.sba.gov/starting/finding.html. © 2001 Courtesy of The U.S. Small Business Administration.

Businesses often borrow to pay their bills, expand, invest, and buy assets such as office equipment, vehicles, computer systems, and other materials. Many banks provide their own loans—or facilitate government loans for small businesses. The Small Business Administration provides information on such loans. Visit their site at http://www.sba.gov.

EXAMPLE H

John White held the 90-day promissory note signed by Redbourne Medical Supply on September 20. On November 19, John sold the note to Central Finance Company. Central discounted the note at 15%. Compute the due date of the note, the maturity value, the discount amount, and the proceeds. (Use a 360-day year throughout.)

Maturity value: MV = Principal + Interest (from example G)
= \$120,000 + \$3,000 = \$123,000

Due date: December 19 (from example G)

Discount period: November 19 to December 19 = 30 days

Discount amount: Maturity value × Discount rate × Time

$$= \$123,000 \times 0.15 \times \frac{30}{360} = \$1,537.50$$

Proceeds: Maturity value − Discount amount
= \$123,000 − \$1,537.50 = \$121,462.50

 ## CONCEPT CHECK 16.5

A 75-day promissory note, bearing interest at 10%, is dated December 11 and has a face value of \$5,000. On January 24, the note is discounted at 14%. Find the discount amount and the proceeds. Note: The interest amount, the maturity value, the maturity date, and the days of discount must first be determined. Use a 365-day year for all interest and discount calculations.

amount: $\quad \$5,000 \times 0.10 \times \dfrac{75}{365} = \102.74

urity value: $\quad \$5,000 + \$102.74 = \$5,102.74$

aturity date: \quad Dec. 11 + 75 days = Feb. 24

Days of discount: \quad Jan. 24 to Feb. 24 = 31 days

Discount amount: $\quad \$5,102.74 \times 0.14 \times \dfrac{31}{365} = \60.67

Proceeds: $\quad \$5,102.74 - \$60.67 = \$5,042.07$

COMPLETE ASSIGNMENTS 16.2 AND 16.3.

Chapter Terms for Review

bank discount
discounting a note
interest-bearing note
negotiable note

non–interest-bearing note
proceeds
unpaid balance
unpaid-balance method

Visit the Web

http://www.small-business-grants.com—Learn about grants offered to small businesses at this Web site.

http://www.small-business-plans.com/sbstuffw.htm—Grants and loans are available from many sources for women and minorities. Visit this Web site for information.

http://www.sbls.com/default.asp—Many lending firms specialize in providing loans to all types of eligible small businesses, such as day care centers, dry cleaners, manufacturing companies, bowling alleys, and many more. Visit the Small Business Loan Source Web site for information and to see which businesses are eligible for loans and which are not.

CYBERCONNECTION

QUESTIONS YOUR BANKER WILL ASK

The key questions the banker will be seeking to answer are as follows:

1. Can the business repay the loan? (is cash flow greater than debt service?)
2. Can you repay the loan if the business fails? (is collateral sufficient to repay the loan?)
3. Does the business collect its bills?
4. Does the business control its inventory?
5. Does the business pay its bills?
6. Are the officers committed to the business?
7. Does the business have a profitable operating history?
8. Does the business match its sources and uses of funds?
9. Are sales growing?
10. Does the business control expenses?
11. Are profits increasing as a percentage of sales?
12. Is there any discretionary cash flow?
13. What is the future of the industry?
14. Who is your competition and what are their strenghts and weaknesses?.

(Charlotte Taylor, Venture Concepts, in association with New Jersey Association of Women Business Owners, Inc., 5/97)

Back to Top

COMMERCIAL LOAN EVALUATION FACTORS

The following factors indicate the "ideal" situation for going to a bank for a small business loan. If you cannot respond "yes" to all of these factors, it does not mean that you cannot obtain financing. Lenders look at these factors in the aggregate. In other words, if you are weak with respect to one factor but strong in another, your overall situation may allow you to obtain a loan.

Applicant Factors

Source: http://www.onlinewbc.org/docs/finance/loanqual.html.

Visit the Online Woman's Business Center at the Small Business Administration's Web site. Here you will find a wealth of information for any small business owner or entrepreneur. Answers to many relevant questions, such as "How does my business qualify for a loan?" and "What questions will the banker ask?" are offered.

The factors that indicate the "ideal" situation for going to a bank for a small business loan are also given. These include applicant factors and business/financial factors. Some applicant factors include creditworthiness, net worth, applicant's experience in the business, available cash, and whether the applicant has ever been arrested.

Business factors cover both existing businesses and new, start-up business. If you want to buy a local retail store, a bank will want to know about the store's track record of profitability, its gross sales, the asking price, the market position, and the financial ratios, to name a few. For a new business, the bank will want a market analysis, confidence in your experience, and a detailed business plan.

The Bottom Line

Summary of chapter learning objectives:

| Learning Objective | Summary | Example |
|---|---|---|
| **16.1** | Compute the savings from borrowing money to take a cash discount | 1. A $20,000 invoice dated March 15 has terms of 2/5, net 25. Find the due date, discount date, cash discount, and required remittance. Then calculate the savings realized by borrowing the remittance at 9% exact interest for the time between the discount date and the due date. |
| **16.2** | Compute interest with the unpaid-balance method | 2. A business borrows $20,000 on July 17 at 11% exact simple interest. On August 17, September 17, and October 17, it will make equal payments of $5,000. Each payment will include interest on the unpaid balance. On November 17, it will pay the remaining balance and interest on the unpaid balance. Compute the payments and balances for the first month. |
| **16.3** | Compute the proceeds and actual interest rate on a bank discount loan | 3. A 60-day bank loan with a face value of $3,900 is made on a discount basis at a discount rate of 12%. Use the 360-day year to compute the discount amount, the proceeds, and the actual interest rate, based on the proceeds rather than on the face value. |
| **16.4** | Discount a non–interest-bearing note | 4. A 45-day non–interest-bearing note has a face value of $1,600 and is dated May 13. On June 3, the note is discounted at 13%. Use a 365-day year to find the discount amount and the proceeds. |
| **16.5** | Discount an interest-bearing note | 5. A 30-day note, bearing an interest rate of 9%, is dated November 6 and has a face value of $8,000. On November 15, the note is discounted at 12%. Use a 365-day year to find the discount amount and the proceeds. |

Answers: **1.** Due date, April 9; discount date, March 20; cash discount, $400; remittance, $19,600; savings, $303.34 **2.** Unpaid balance, $20,000; interest payment, $186.85; principal payment, $4,813.15; new balance, $15,186.85 **3.** Discount, $78; proceeds, $3,822; interest rate, 12.24% **4.** Discount, $13.68; proceeds, $1,586.32 **5.** Discount, $55.64; proceeds, $8,003.54

1 A company just received an invoice for $1,600 that has cash discount terms of 2/10, net 30. The company borrows enough money at 10% exact simple interest (365-day year) to take advantage of the cash discount. It borrows the money only for the time period between the due date and the last day to take advantage of the discount. Calculate (a) the amount of the cash discount; (b) the number of interest days; (c) the amount of interest on the loan; and (d) the amount of the savings.

2 A bank made a 120-day loan for $4,500 on a discount basis, using a discount rate of 9%. Using a 360-day year, calculate (a) the discount amount; (b) the proceeds; and (c) the actual interest rate (to two decimal places).

3 A company made a $26,000 loan on October 15 and financed it for two months at 8% exact simple interest (365-day year) with interest calculated on the unpaid balance each month. The company agreed to make one payment of $13,000 (interest and principal) on November 15 and the balance on December 15. Complete the following table.

| Payment Date | Interest Days | Unpaid Balance | Interest Payment | Total Payment | Principal Payment | Ending Balance |
|---|---|---|---|---|---|---|
| Nov 15 | a. _____ | b. _____ | c. _____ | $13,000.00 | d. _____ | e. _____ |
| Dec 15 | f. _____ | g. _____ | h. _____ | i. _____ | j. _____ | k. _____ |

4 A 90 day non–interest-bearing note for $3,100 is dated November 10. On December 10, it is sold at a discount rate of 12%. Using a 365-day year, calculate (a) the maturity date; (b) the days of discount; (c) the discount amount; and (d) the proceeds.

5 A 120-day interest-bearing note for $2,960 is dated May 15 and has a rate of 8% exact simple interest. On July 15, it is sold at a discount rate of 15%. Using a 365-day year, calculate (a) the interest amount; (b) the maturity value; (c) the maturity date; (d) the days of discount; (e) the discount amount; and (f) the proceeds.

Assignment 16.1: Single and Multiple Payment Interest

Name _____

Date _____ Score _____

A (64 points) Joslin & McGraw Home Builders, Inc., made several purchases from vendors, who offered various terms of payment. How much can Joslin & McGraw save on each invoice if it borrows the money to pay the invoice early and receive the cash discount? The loan interest rates are all exact simple interest (365-day year). Assume that the number of interest days is the time between the due date and the last day to take advantage of the cash discount. (2 points for each correct answer)

| Invoice | Terms | Cash Discount | Interest Rate on Loan | Interest Days | Amount of Interest | Savings |
|---------|-------|---------------|----------------------|---------------|--------------------|---------|
| 1. $3,000 | 2/10, n/30 | _____ | 12% | _____ | _____ | _____ |
| 2. $7,500 | 1/15, n/45 | _____ | 6.75% | _____ | _____ | _____ |
| 3. $17,500 | 3/5, n/25 | _____ | 8% | _____ | _____ | _____ |
| 4. $880 | 1.5/15, n/30 | _____ | 11% | _____ | _____ | _____ |
| 5. $9,200 | 1/30, n/60 | _____ | 9% | _____ | _____ | _____ |
| 6. $1,380 | 2/10, n/30 | _____ | 10% | _____ | _____ | _____ |
| 7. $26,000 | 2.5/5, n/25 | _____ | 8.5% | _____ | _____ | _____ |
| 8. $65,400 | 3/10, n/25 | _____ | 7.5% | _____ | _____ | _____ |

Score for A (64)

Assignment 16.1 *Continued*

B (36 points) Complete the tables below. Compute the exact simple interest (365-day year) on the unpaid balance each month. Reduce the principal accordingly, and compute the new unpaid balance. (1 point for each correct answer)

9. On March 20, Gentry Garden Supply financed $12,000 worth of new garden equipment and gardening supplies at an exact simple interest rate of 9.8%. On April 20, it will pay $7,000, which will include the interest for the first month. On May 20, Gentry will repay the entire unpaid balance plus the interest for the last month.

| Payment Date | Interest Days | Unpaid Balance | Interest Payment | Total Payment | Principal Payment | New Balance |
|---|---|---|---|---|---|---|
| April 20 | _____ | _____ | _____ | _____ | _____ | _____ |
| May 20 | _____ | _____ | _____ | _____ | _____ | _____ |

10. Chard Electrical Supply purchased $16,000 worth of electrical supplies from one manufacturer on June 5. The manufacturer offered Chard an exact simple interest rate of 7.5%, and Chard agreed to pay $4,000 on July 5, August 5, and September 5. The $4,000 payments will include the interest on the unpaid balance and reduce the principal accordingly. On October 5, Chard will repay the remaining unpaid balance and the interest for the final month.

| Payment Date | Interest Days | Unpaid Balance | Interest Payment | Total Payment | Principal Payment | New Balance |
|---|---|---|---|---|---|---|
| July 5 | _____ | _____ | _____ | _____ | _____ | _____ |
| August 5 | _____ | _____ | _____ | _____ | _____ | _____ |
| September 5 | _____ | _____ | _____ | _____ | _____ | _____ |
| October 5 | _____ | _____ | _____ | _____ | _____ | _____ |

Score for B (36)

Assignment 16.2: Bank Discount and Discounting Notes

Name

Date _____ Score _____

A (36 points) Suppose that Barbara Finell Lending, a small finance company, made four new loans on a discount basis. Compute the discount amount and the proceeds. Then compute the actual interest rate, based on the proceeds rather than the face value. Use a 360-day year for all calculations. Round the actual interest rate to the nearest $\frac{1}{100}$ of a percent. (3 points for each correct answer)

| Face Value | Discount Rate | Time | Discount Amount | Proceeds | Actual Interest Rate |
|---|---|---|---|---|---|
| **1.** $6,600 | 7% | 60 days | _____ | _____ | _____ |
| **2.** $24,000 | 6% | 90 days | _____ | _____ | _____ |
| **3.** $2,850 | 12% | 30 days | _____ | _____ | _____ |
| **4.** $12,840 | 8% | 120 days | _____ | _____ | _____ |

Score for A (36)

B (16 points) Assume that you are loan officer Carol Swift of Swift Financial Services. Discount the following non–interest-bearing notes using a 360-day year. (2 points for each correct answer)

5. Face value: $5,000
Date of note: July 15
Length of note: 120 days
Discount date: September 10
Discount rate: 11%

Maturity date: _____
Days of discount: _____
Discount amount: _____
Proceeds: _____

6. Face value: $3,750
Date of note: March 1
Length of note: 90 days
Discount date: April 15
Discount rate: 12%

Maturity date: _____
Days of discount: _____
Discount amount: _____
Proceeds: _____

Score for B (16)

Assignment 16.2 *Continued*

C **(48 points) New England Loan Company discounts interest-bearing promissory notes. Compute the requested information for the following notes. Use a 360-day year for all interest and discount computations. (2 points for each correct answer)**

7. Face value: $4,200

 Date of note: June 3

 Interest rate: 8%

 Length of note: 75 days

 Discount date: July 1

 Discount rate: 10%

 Interest amount: _____

 Maturity value: _____

 Maturity date: _____

 Days of discount: _____

 Discount amount: _____

 Proceeds: _____

8. Face value: $960

 Date of note: April 4

 Interest rate: 9%

 Length of note: 60 days

 Discount date: April 20

 Discount rate: 12%

 Interest amount: _____

 Maturity value: _____

 Maturity date: _____

 Days of discount: _____

 Discount amount: _____

 Proceeds: _____

9. Face value: $6,480

 Date of note: October 9

 Interest rate: 7%

 Length of note: 105 days

 Discount date: November 1

 Discount rate: 13%

 Interest amount: _____

 Maturity value: _____

 Maturity date: _____

 Days of discount: _____

 Discount amount: _____

 Proceeds: _____

10. Face value: $1,170

 Date of note: July 17

 Interest rate: 6%

 Length of note: 90 days

 Discount date: August 20

 Discount rate: 11%

 Interest amount: _____

 Maturity value: _____

 Maturity date: _____

 Days of discount: _____

 Discount amount: _____

 Proceeds: _____

Score for C (48)

Assignment 16.3: Discounting Notes, Using a 365-Day Year

LEARNING
OBJECTIVES

Name

Date Score

A **(36 points) Compute the missing information for the following applications, which involve discounting non–interest-bearing notes. Use a 365-day year for all discount calculations. (3 points for each correct discount amount; 2 points for all other correct answers)**

1. Dorothy Sanchez held a 60-day, $2,100 non–interest-bearing note dated July 20. On August 10, she took the note to a finance company, which discounted it at 14%.

 Maturity date: _____

 Days of discount: _____

 Discount amount: _____

 Proceeds: _____

2. Bonnie Bomar was holding a $750, 90-day non–interest-bearing note from her brother. The note is dated October 13. On November 24, Bonnie sold the note to a loan company, which charged a 13% discount rate.

 Maturity date: _____

 Days of discount: _____

 Discount amount: _____

 Proceeds: _____

3. Jim Walter was holding a 75-day, $3,500 non–interest-bearing note dated December 3. He needed money for Christmas, so on December 20 Jim took the note to a local finance company, which discounted it at 12%.

 Maturity date: _____

 Days of discount: _____

 Discount amount: _____

 Proceeds: _____

4. Terry Chung had a 120-day non–interest-bearing note. The face value is $925; the date of the note is March 29. On May 3, Terry sold the note to a loan company at a discount rate of 15%.

 Maturity date: _____

 Days of discount: _____

 Discount amount: _____

 Proceeds: _____

Score for A (36)

Assignment 16.3 *Continued*

B (64 points) Compute the missing information for the following applications, which involve discounting interest-bearing notes. Use a 365-day year for all interest and discount calculations. (4 points for each correct interest amount and discount amount; 2 points for all other correct answers)

5. Bill Starnes was holding a $5,600, 120-day note with an interest rate of 9%. The note is dated May 10. On June 15, a finance bought the note from Bill, discounting it at a rate of 13%.

 Interest amount: _____

 Maturity value: _____

 Maturity Date: _____

 Days of discount: _____

 Discount amount: _____

 Proceeds: _____

6. On July 28, Wayne Runn needed cash and sold a 60-day promissory note at a 14% discount rate. The note had a face value of $820, was dated June 25, and carried an interest rate of 10%.

 Interest amount: _____

 Maturity value: _____

 Maturity Date: _____

 Days of discount: _____

 Discount amount: _____

 Proceeds: _____

7. Sharon Wilder had been holding a 75-day, $1,800 note that had an 8% interest rate and had been written on April 14. Sharon decided to sell the note on May 20 to a loan company, which discounted it at 12%.

 Interest amount: _____

 Maturity value: _____

 Maturity Date: _____

 Days of discount: _____

 Discount amount: _____

 Proceeds: _____

8. On September 9, Evelyn Semanski decided to sell a 150-day note to an agency that used a 15% discount rate. Evelyn's note had a face value of $6,000, was dated July 17, and carried a 9% interest rate.

 Interest amount: _____

 Maturity value: _____

 Maturity Date: _____

 Days of discount: _____

 Discount amount: _____

 Proceeds: _____

Notes

November 19, 2001

SITE SEARCH:

[] GO

Topics

- Cars
- Children
- Computers
- Education
- Employment
- Federal Programs
- Food
- Health
- Housing
- Money
- Small Business
- Travel
- And more...

Resources

- Consumer Help
- Contact Center
- Kids.Gov
- Recalls
- Scams/Frauds

☆☆ Federal Consumer Information Center

| Search | In the News | Press Room | FAQ | Order | Links | About us |

Looking for books advertised in the media?

America Responds to Terrorism

FCIC has put together a list of publications to help you and your family during this difficult time. Read more on **Coping with Crisis**.

Also check out **Suspicious Mail: What You Can Do**.

Publication of the Week:

Buying a Computer
Computers have a language all their own and it's easy to get lost if you're not familiar with the lingo. This guide can help - get information on everything from how much memory you'll need and common computer terms to how to protect your computer and data. You'll also find a handy checklist to take along when shopping for a computer.

Consumer Action Website:

Up in the air about where to get help

Federal Benefi Services
We live in a grea Our government provid variety of benefits and ser Americans. FCIC is proud present our **Federal Ben Services** page. This pag links to sites for Social Se Veteran's Benefits, Medic Children's Health Insuran Heating and Cooling Assis and much, much, more.

Consumer Focus:
Food for the Holidays
Everyone enjoys this time of the year - as long as you're not a turkey. We all look forward to a hearty m our friends and family. Bu it be great if traditional holi could be both good for you and pleasing to your palat more for some tips and id help you plan and prepare

Source: http://www.pueblo.gsa.gov. Courtesy of Federal Trade Commission.

Y ou can find consumer credit information at the Federal Consumer Information Center Web site. Do a search for "credit" and you'll find resources on using credit cards overseas, credit protection laws, maintaining a good credit rating, starting a small business, disputing credit report errors, and much more. Another excellent site is the Federal Trade Commission Web site. Look under Consumer Protection at www.ftc.gov.

Charges for Credit

17

LEARNING OBJECTIVES

By studying this chapter and completing all assignments you will be able to:

 1 Convert between annual and monthly interest rates.

 2 Compute simple interest on a monthly basis.

 3 Compute finance charges for credit account purchases.

 4 Compute costs of installment purchases.

 5 Compute effective rates.

 6 Amortize a loan.

Credit is part of an increasingly "cashless" economic system. As a medium of exchange, it has speeded up the production, exchange, and consumption of goods. That, in turn, has brought about a higher standard of living.

1. What is the relationship between monthly interest rates and annual interest rates?
2. What is the relationship between a *quoted* interest rate and an *effective* interest rate?
3. How can a lender increase the effective interest rate?
4. What does *amortization* of a loan mean? Explain in terms of the size of the interest and principal payments.

Most individuals today can purchase goods or services on credit if they want to. The buyer gets immediate possession or immediate service but delays payment. Either the seller extends the credit or the buyer uses a **credit card,** or loan, from a third party.

Credit may be offered for an interest charge, often called a **finance charge.** This charge is usually computed each month, and a summary (the statement) is sent to the buyer (borrower). Comparing the finance charges of two competing creditors can be difficult. The statement itself may be confusing because more than one interest rate may be mentioned. Also, there may be additional charges such as administrative fees, annual membership fees, and insurance fees. Finally, the borrower may not understand the method the creditor uses to compute the finance charges. Often, the finance charge is based on the *average daily balance,* which is easy for the creditor's computer to compute but difficult for the borrower to reconcile.

Converting Interest Rates

LEARNING
OBJECTIVES

Convert between annual
and monthly interest rates.

The general concept behind charging for credit purchases is to compute finance charges on the unpaid balance each month. The formula is still $I = P \times R \times T$, where P is basically the unpaid balance. However, T is not years or a fraction of a year (as in Chapters 14–16)—T is in months, and R, the rate, is a monthly rate. For example, the rate might be 1.5% *per month.*

Understanding the relationship between monthly and annual rates is important.

> **Rule: To convert an annual rate to a monthly rate, divide the annual rate by 12; to convert a monthly rate to an annual rate, multiply the monthly rate by 12.**

EXAMPLE A

a. Convert 18% per year to the equivalent monthly rate.
 18% annually ÷ 12 = 1.5% monthly
b. Convert 0.5% per month to the equivalent annual rate.
 0.5% monthly × 12 = 6% annually

 CONCEPT CHECK 17.1

a. Convert a 9% annual rate to the equivalent monthly rate.
 Divide the annual rate by 12 to get the monthly rate: 9% ÷ 12 = 0.75% per month
b. Convert a 1.25% monthly rate to the equivalent annual rate.
 Multiply the monthly rate by 12 to get the annual rate: 1.25% × 12 = 15% per year

Computing Simple Interest on a Monthly Basis

In terms of single-payment simple interest, 1.5% *per month* is identical to 18% *per year*.

Rule: If the rate is annual, the time must be in years; if the rate is monthly, the time must be in months.

EXAMPLE B

Compute the simple interest on $1,000 for 2 months at 18% per year, on an annual basis and on a monthly basis.

Annual:　$I = P \times R \times T = \$1{,}000 \times 0.18$ per year $\times \frac{2}{12}$ year $= \$30$

Monthly:　18% per year $= 18\% \div 12 = 1.5\%$ per month

　　　　　$I = P \times R \times T = \$1{,}000 \times 0.015$ per month $\times 2$ months

　　　　　　　　$= \$30$

Reminder: Both computations differ from those in Chapter 14, where you counted the exact number of days and divided by either 360 or 365.

CONCEPT CHECK 17.2

Compute the simple interest on $800 for 3 months at 0.5% per month.

$I = P \times R \times T = \$800 \times 0.5\%$ per month $\times 3$ months $= \$800 \times 0.005 \times 3 = \12

Computing Finance Charges

To enable consumers to compute the total cost of credit, Congress has passed several laws, beginning with the Consumer Credit Protection Act of 1968. Also known as the **Truth in Lending Act**, it requires creditors to tell consumers three things:

1. The total of all finance charges, including interest, carrying charges, insurance, and special fees
2. The annual percentage rate (APR) of the total finance charge
3. The method by which they compute the finance charge

The **annual percentage rate (APR)** is the monthly rate multiplied by 12. However, the quoted APR considers only the interest computation; it doesn't include other charges such as membership fees, loan fees, and insurance. The total of all finance charges is easy to find on the statement, because usually it is well labeled. The method for computing the finance charge is easy to find, but it may not be easy to understand. One difficulty is determining the balance used in the computation. A wide variety of methods may be applied:

1. The finance charge may be based on the amount owed at the beginning of the current month, ignoring payments and purchases.

2. The finance charge may be based on the amount owed at the beginning of the month, after subtracting any payments during the month and ignoring purchases.
3. The finance charge may be based on the **average daily balance**. (Add the unpaid balance each day; divide the total by the number of days in the month.) Payments are usually included; new purchases may or may not be included.
4. A variation of the average daily balance method is to compute the interest charge each day, on a daily basis, and then add all the daily interest charges for the month.

Even with the same interest rate and exactly the same daily payments and purchases, these four methods lead to different results, depending on the timing of purchases and payments. Some methods, although easy for a computer, are tedious for a consumer to check, even with a calculator. The Truth in Lending Act doesn't require creditors to use a particular method, nor does it limit the rate or the amount that the creditor may charge. However, the creditor must tell consumers the annual percentage rate, the total finance charge, and the method by which the finance charge is computed. A borrower who doesn't understand the description on the statement can write to the creditor for a more detailed explanation.

Figure 17-1 is the lower portion of a typical statement of a retail store. Examples C and D illustrate two methods used to compute finance charges.

Fig. 17-1: Retail Statement of Account

| PREVIOUS BALANCE | FINANCE CHARGE | PAYMENTS | CREDITS | PURCHASES | NEW BALANCE | MINIMUM PAYMENT | CLOSING DATE |
|---|---|---|---|---|---|---|---|
| 624.00 | 9.36 | 500.00 | 62.95 | 364.57 | 434.98 | 45.00 | 10-16-99 |

IF WE RECEIVE PAYMENT OF THE FULL AMOUNT OF THE NEW BALANCE BEFORE THE NEXT CYCLE CLOSING DATE, SHOWN ABOVE, YOU WILL AVOID A FINANCE CHARGE NEXT MONTH. THE FINANCE CHARGE, IF ANY, IS CALCULATED ON THE PREVIOUS BALANCE BEFORE DEDUCTING ANY PAYMENTS OR CREDITS SHOWN ABOVE. THE PERIODIC RATES USED ARE 1.5% OF THE BALANCE ON AMOUNTS UNDER $1,000 AND 1% OF AMOUNTS IN EXCESS OF $1,000, WHICH ARE ANNUAL PERCENTAGE RATES OF 18% AND 12% RESPECTIVELY.

CyberConnection

Businesspeople who travel to foreign countries and make purchases in foreign currency can check their expense reports at this Web site.

| Provide Transaction Details | |
|---|---|
| Amount and currency shown on the **receipt** for the purchase: (From receipt or invoice) | [] [(Currency you paid) ▾] |
| Date of purchase: (From receipt or invoice) | [January ▾] [01 ▾] [2001 ▾] |
| Amount and currency shown on your **card** statement for the purchase: (From card statement) | [] [(Currency you were charged) ▾] |
| Type of card used for purchase: | [(Select one) ▾] [(other card type)] |
| | Please DO NOT enter your card number anywhere. |
| Name of the bank who issued the card: (Printed on the card) | [(Select one) ▾] [(other bank)] |
| Your guess of the hidden charges for this transaction (optional): | [(Select) ▾] |
| To see an estimate of the bank charges you paid for this transaction, proceed to the calculation: | [Click here to calculate.] Please be patient — processing may take a few moments |

TIP: If the contents of some or all of the cells of the table above "disappear" or go blank, simply minimize (**NOT** close) your browser window, and then restore it. This problem is due to a bug in some browsers, most notably Netscape Navigator 3 for Windows.

From your card statement, please input the transaction details requested above. An estimate of the charge made by the card issuer for converting the foreign exchange and processing the transaction will be displayed. Please check that you are including the charges when making an expense claim.

Source: http://www.xe.net/ccc/. Copyright © 2001 Courtesy of XE.NET.

EXAMPLE C

Compute the finance charge and the new balance for the statement shown in Figure 17-1 based on the previous balance, $624, ignoring all payments, credits, and purchases.
Finance charge = $624 × 1.5% × 1 month = $9.36
New balance = $624.00 + $9.36 − $500.00 − $62.95 + $364.57 = $434.98

EXAMPLE D

Assume that the finance charge in Figure 17-1 is based on the previous balance, less any payments or credits, but ignore subsequent purchases. Compute the finance charge and the new balance.
The finance charge is based on $624.00 − $500.00 − $62.95 = $61.05.
Finance charge = $61.05 × 1.5% × 1 month = $0.91575, or $0.92
New balance = $624.00 + $0.92 − $500.00 − $62.95 + $364.57 = $426.54

 CONCEPT CHECK 17.3

The finance terms given in the charge account statement of Figure 17-1 indicate that the finance charge, if any, is charged on the previous balance, before deducting payments or credits or adding purchases. Calculate the finance charge and the unpaid balance if the previous balance was $2,425.90, the payment was $1,200, there were no credits, and there were $572.50 in new purchases.

An interest rate of 1.5% applies to the first $1,000 and 1% applies to the excess:
$2,425.90 − $1,000 = $1,425.90
0.015 × $1,000 = $15.00
0.01 × $1,425.90 = $14.26
Finance charge = $15.00 + $14.26 = $29.26
New balance = $2,425.90 − $1,200 + $29.26 + $572.50 = $1,827.66

COMPLETE ASSIGNMENT 17.1.

Computing Costs of Installment Purchases

In a credit sale, the buyer may pay the purchase price plus credit charges in monthly payments, called installments. The method of computing the interest is just as important as the interest rate. The interest may be computed just once on the principal or it may be computed monthly on the unpaid balance. In Chapter 16, we used the unpaid balance method exact interest (365-day year). For consumer credit, the unpaid balance method is normally applied monthly, using a monthly interest rate that is the same for all months.

4 LEARNING OBJECTIVES

Compute costs of installment purchases.

Compute the total interest on purchases of $2,400 for 4 months at 15%, when interest is paid equally over 4 months. The principal is also repaid equally over 4 months, in payments of $600 each.

$$\$2,400 \times 15\% \times \frac{4}{12} = \$120 \text{ total interest}$$

$120 ÷ 4 = $30 per month for interest

| Month | Unpaid Balance | Monthly Interest | Principal Payment | Total Payment | New Balance |
|-------|----------------|------------------|-------------------|---------------|-------------|
| 1 | $2,400 | $30.00 | $ 600.00 | $ 630.00 | $1,800 |
| 2 | 1,800 | 30.00 | 600.00 | 630.00 | 1,200 |
| 3 | 1,200 | 30.00 | 600.00 | 630.00 | 600 |
| 4 | 600 | 30.00 | 600.00 | 630.00 | 0 |
| Total | | $120.00 | $2,400.00 | $2,520.00 | |

Compute the total interest on purchases of $2,400 for 4 months at 15%, when interest is computed on the unpaid balance each month. The principal is repaid at $600 per month. For the monthly interest of 15% per year, 15% ÷ 12 = 1.25% per month.

Month 1: $2,400 × 1.25% = $30 Month 3: $1,200 × 1.25% = $15
Month 2: $1,800 × 1.25% = $22.50 Month 4: $600 × 1.25% = $7.50
Total interest = $30 + $22.50 + $15 + $7.50 = $75

| Month | Unpaid Balance | Monthly Interest | Principal Payment | Total Payment | New Balance |
|-------|----------------|------------------|-------------------|---------------|-------------|
| 1 | $2,400 | $30.00 | $ 600.00 | $ 630.00 | $1,800 |
| 2 | 1,800 | 22.50 | 600.00 | 622.50 | 1,200 |
| 3 | 1,200 | 15.00 | 600.00 | 615.00 | 600 |
| 4 | 600 | 7.50 | 600.00 | 607.50 | 0 |
| Total | | $75.00 | $2,400.00 | $2,475.00 | |

CONCEPT CHECK 17.4

A kitchen stove is priced at $600 and is purchased with a $100 down payment. The $500 remaining balance is paid in two successive monthly payments of $250 each. Compute interest using the following methods.

a. Simple interest is calculated on the entire $500 for 2 months at 1.5% per month (18% annual rate).
 $500 × 0.015 per month × 2 months = $15.00
b. Interest of 1.5% is calculated on the unpaid balance each month (18% annual rate).
 Month 1: $500 × 0.015 = $7.50
 Month 2: New balance is $250. $250 × 0.015 = $3.75
 Total interest = $7.50 + $3.75 = $11.25

CyberConnection

Computing Effective Rates

LEARNING OBJECTIVES 5

Compute effective rates.

The accounts in examples E and F are similar but not identical. Both show purchases of $2,400; both show repayment of the principal in four equal payments of $600; both involve use of 15% to compute the interest dollars. But the total interest in example E is $120, in example F it is $75. The problem is that, although both accounts show a 15% interest rate, in example E, 15% is not the true interest rate. The true interest rate is called the **effective interest rate**. To find the effective interest rate, we can use $R = I/(P \times T)$, where I is the interest dollars, T is the time of the loan, and P is the **average principal**, or the *average unpaid balance*, over the period of the loan.

In Chapter 15, we wrote the formula $I = P \times R \times T$ as $R = I/(P \times T)$ to solve for R, the interest rate. For examples E and F, T is $\frac{4}{12}$ of a year, and I is the interest dollars: $120 in E and $75 in F. In Chapter 15, the borrower had the entire principal for the entire time, T. In the loans of examples E and F, the borrower doesn't have the entire $2,400 for the entire 4 months. The principal decreases monthly, from $2,400 to $600.

EXAMPLE G

Compute the effective interest rate on the purchases of examples E and F.

a. The average P is ($2,400 + $1,800 + $1,200 + $600) ÷ 4 = $1,500. Therefore, in example E, $P = \$1,500$, $I = \$120$, and $T = \frac{4}{12}$ year.

$$R = \frac{I}{(P \times T)} = \frac{\$120}{\left(\$1,500 \times \dfrac{4}{12}\right)} = \frac{\$120}{\$500} = 0.24, \text{ or } 24\%$$

b. The average P is ($2,400 + $1,800 + $1,200 + $600) ÷ 4 = $1,500. Therefore, in example F, $P = \$1,500$, $I = \$75$, and $T = \frac{4}{12}$ year.

$$R = \frac{I}{(P \times T)} = \frac{\$75}{\left(\$1,500 \times \dfrac{4}{12}\right)} = \frac{\$75}{\$500} = 0.15, \text{ or } 15\%$$

When interest is computed on the unpaid balance each month, the quoted rate and the effective rate will be the same. When interest is computed only on the original principal but the unpaid balance decreases each month, the effective (true) interest rate is always higher than the quoted rate.

Increasing the Effective Rate

Many credit purchasers do not even consider the cost of credit when they make credit purchases. Other purchasers consider only the stated interest rate. Examples E, F, and G illustrate how creditors can increase the true interest rate by manipulating the method of computing interest and requiring payments.

Another method of increasing the cost of credit is to charge additional fees for the credit purchase or loan. There may be a credit application fee, a transaction fee, or an insurance fee on the merchandise purchased. Recall that the Truth in Lending Act doesn't require these fees to be included in the computation of the stated APR, but they are certainly part of the increased cost of buying an item on credit instead of paying cash. Essentially, such fees increase the true interest rate and make comparison of credit costs difficult for credit purchasers. The only way to compare them accurately is to include all costs and compute a true rate.

Even when interest is computed on the unpaid balance, if additional fees are charged, the effective rate will be higher than the quoted rate. The result is practically the same whether the extra charges are paid in advance, paid equally over the repayment period, or all paid at the end. Consider the following variation to the credit purchase in example F, which is also part (b) of example G, where interest was computed on the unpaid balance.

EXAMPLE H

A loan of $2,400 is for 4 months at 15% with interest computed on the unpaid balance each month. The principal is repaid in four monthly payments of $600 each. A loan origination fee of 1% of the purchase price and a $1 per month insurance premium on the merchandise are added. Compute the effective rate, using $R = I/(P \times T)$, where P is the average principal over the 4 months and I is the total of all finance-related charges.

The loan origination fee is $1\% \times \$2,400 = \24.00.

The insurance premiums are $1 per month, or $4.00 total.

From example F, the total interest is $75.00.

From part (b) of example G, the average principal, P, is $6,000/4 = $1,500.

The time, T, is $\frac{4}{12}$ year (4 months).

I = Total finance charge = Interest + Loan origination fee + Insurance
$= \$75 + \$24 + \$4 = \103

$$R = \frac{I}{(P \times T)} = \frac{\$103}{\left(\$1,500 \times \frac{4}{12}\right)} = \frac{\$103}{\$500} = 0.206, \text{ or } 20.6\%$$

In part (b) of example G, the effective rate was 15%, the same as the quoted rate. By including additional charges, the creditor has increased the effective rate from 15% to 20.6%.

An Alternative Formula for Estimating the Effective Rate

One difficulty with using the formula $R = I/(P \times T)$ to find the effective rate is that P is the average unpaid balance. Sometimes the payment plan is stated without reference to interest rates, making the calculation of the monthly unpaid balance impossible. The

following formula gives exactly the same results as $R = I/(P \times T)$—if the principal were repaid *equally* during each month of the loan. The P in this new formula is only the original amount financed or borrowed.

$$R = \frac{2 \times M \times I}{[P \times (n + 1)]}$$

where

R = Effective annual interest rate
M = 12 months in 1 year (use 12 for all problems in this book; see note about M)
I = Total finance charge (interest plus additional fees, if any)
P = Amount of the credit purchase or amount of the loan
n = Number of payments in the credit plan

Note about M: M is the number of payment periods in exactly 1 year—regardless of the length of the credit plan. If payments are weekly, $M = 52$; if payments are quarterly, $M = 4$; and so on.

EXAMPLE I

Use the alternative formula to find the effective rate in example H.

$M = 12$
I = Total finance charge = $75 + $24 + $4 = $103
P = Amount of the purchase = $2,400 (instead of the average unpaid balance)
$n = 4$ payments

$$R = \frac{2 \times M \times I}{[P \times (n + 1)]} = \frac{2 \times 12 \times \$103}{[\$2,400 \times (4 + 1)]} = \frac{\$2,472}{\$12,000} = 0.206, \text{ or } 20.6\%$$

This formula gives the same result, without the need to calculate the average unpaid balance. In example H, however, coming up with the total interest charge, I, still required a computation for every month. Therefore not much time was saved.

A more useful application of the alternative formula is when it isn't possible to compute the average unpaid balance but it is possible to find the total finance charge, as in example J. Although we don't know whether the principal is being paid equally each month, the alternative formula will *approximate* the effective rate.

EXAMPLE J

An advertisement reads: "Buy this big-screen television today for $2,200. Or, make a $200 down payment and 11 equal monthly payments of $200 each." The buyer doesn't know how the principal is being repaid.

The total cost of the television on credit equals the down payment plus 11 × $200: $200 + $2,200 = $2,400.

The finance charge is the total cost less the cash price: $I = \$2,400 - \$2,200 = \$200$. The amount that is financed, P, is the cash price less the down payment: $P = \$2,200 - \$200 = \$2,000$. The number of monthly payments is 11: $n = 11$ monthly payments.

$$R = \frac{2 \times M \times I}{P \times (n + 1)} = \frac{2 \times 12 \times \$200}{[\$2,000 \times (11 + 1)]} = \frac{\$4,800}{\$24,000} = 0.2, \text{ or } 20\%$$

R is the approximate effective rate. We don't know whether the $2,000 is being repaid in 11 equal parts, but we assume so in order to estimate R.

CONCEPT CHECK 17.5

a. From Concept Check 17.4, a kitchen stove priced at $600 is purchased with a $100 down payment. The remaining balance of $500 may be financed over 2 months with either of the following installment payment plans.
Plan 1: Two monthly principal payments of $250 each and a total interest amount of $15.00
Plan 2: Two monthly principal payments of $250 each and a total interest amount of $11.25
Calculate the effective annual rate of each plan, using $R = \frac{I}{(P \times T)}$, where P is the average unpaid monthly balance and T is $\frac{2}{12}$ year. In each plan, the monthly unpaid balances are $500 in month 1 and $250 in month 2.
The average unpaid balance is $\frac{(\$500\ +\ \$250)}{2} = \frac{\$750}{2} = \375, so $P = \$375$.

Plan 1: $R = \frac{I}{(P \times T)} = \frac{\$15.00}{(\$375 \times \frac{2}{12})} = \frac{\$15.00}{\$62.50} = 0.24$, or 24% effective annual rate

Plan 2: $R = \frac{I}{(P \times T)} = \frac{\$11.25}{(\$375 \times \frac{2}{12})} = \frac{\$11.25}{\$62.50} = 0.18$, or 18% effective annual rate

b. A kitchen stove priced at $600 can be purchased on an installment plan for a $100 down payment and five equal monthly payments of $105 each. Calculate the approximate effective rate using the formula
$R = \frac{2 \times M \times I}{[P \times (n + 1)]}$, where $M = 12$, the number of months in 1 year; $I =$ the total finance charge; $P = \$500$, the amount financed; and $n = 5$, the actual number of payments.
The total charge to finance the stove is the difference between the total amount paid and the cash price.
Total paid = $100 down payment + 5 × $105 = $100 + $525 = $625; $I = \$625 - \$600 = \$25$

$\frac{2 \times M \times I}{P \times (n + 1)} = \frac{2 \times 12 \times \$25}{\$500 \times (5 + 1)} = \frac{\$600}{\$3,000}$

The effective annual rate is 20%.

COMPLETE ASSIGNMENT 17.2.

Amortizing a Loan

Amortize a loan.

In example F, we computed interest on the unpaid balance, but the total monthly payments were unequal. In example E, the total payments were equal, but we didn't compute the interest on the unpaid balance. It is easier for both borrower and lender if the monthly payments are equal, but the quoted rate doesn't equal the true effective rate unless the interest is computed on the unpaid balance.

By using a computer, a financial calculator, or special tables, you can determine monthly payments that will always be equal and still have the interest computed on the unpaid balance. This method is called **amortization**; it is usually used for loans to purchase homes, cars, and other more expensive items over longer terms. In a credit purchase, the buyer may make a down payment and amortize the balance.

The combined principal and interest are paid in equal monthly payments. Each monthly payment includes the interest on the unpaid balance of the loan and a payment on the principal. With each successive payment, the unpaid balance decreases. Thus the subsequent interest payments decrease, and the corresponding payments on the principal increase. Because of rounding, the final monthly payment may be a few cents different from the other payments.

Computing the Monthly Payment

Companies that amortize loans compute the monthly payment by using computers, financial calculators, or special tables, such as Table 17-1. It has only five interest rates (columns) and twelve loan time periods (rows), but it illustrates the concept. Each number in the table is an **amortization payment factor**. Multiply the amount of the loan or the finance purchase by the amortization payment factor to find the monthly payment.

Table 17-1: Amortization Payment Factors (for monthly payments)

| Term of Loan | Annual Interest Rate | | | | | |
| --- | --- | --- | --- | --- | --- | --- |
| | **6%** | **9%** | **12%** | **15%** | **18%** | **21%** |
| **1 month** | 1.00500 | 1.00750 | 1.01000 | 1.01250 | 1.01500 | 1.01750 |
| **2 months** | 0.50375 | 0.50563 | 0.50751 | 0.50939 | 0.51128 | 0.51316 |
| **3 months** | 0.33667 | 0.33835 | 0.34002 | 0.34170 | 0.34338 | 0.34507 |
| **4 months** | 0.25313 | 0.25471 | 0.25628 | 0.25786 | 0.25944 | 0.26103 |
| **5 months** | 0.20301 | 0.20452 | 0.20604 | 0.20756 | 0.20909 | 0.21062 |
| **6 months** | 0.16960 | 0.17107 | 0.17255 | 0.17403 | 0.17553 | 0.17702 |
| **1 year** | 0.08607 | 0.08745 | 0.08885 | 0.09026 | 0.09168 | 0.09311 |
| **2 years** | 0.04432 | 0.04568 | 0.04707 | 0.04849 | 0.04992 | 0.05139 |
| **3 years** | 0.03042 | 0.03180 | 0.03321 | 0.03467 | 0.03615 | 0.03768 |
| **4 years** | 0.02349 | 0.02489 | 0.02633 | 0.02783 | 0.02937 | 0.03097 |
| **5 years** | 0.01933 | 0.02076 | 0.02224 | 0.02379 | 0.02539 | 0.02705 |
| **10 years** | 0.01110 | 0.01267 | 0.01435 | 0.01613 | 0.01802 | 0.01999 |

EXAMPLE K

Find the monthly payment required to amortize a $4,000 loan over 4 months at 12% (1% per month).

STEP 1 Find the intersection of the 12% column and the 4-month row. The amortization payment factor is 0.25628.

STEP 2 Multiply the amount amortized by the payment factor:
$4,000 × 0.25628 = $1,025.12 monthly payment

EXAMPLE L

Judith Kranz agrees to purchase an automobile for $18,300. Judith will make a $2,000 down payment and amortize the balance over 4 years at 15% (1.25% per month). Determine Judith's monthly payment.

STEP 1 $18,300 − $2,000 = $16,300 amortized

STEP 2 Find the intersection of the 15% column and the 4-year row. The amortization payment factor is 0.02783.

STEP 3 Multiply the amount amortized by the payment factor:
$16,300 × 0.02783 = $453.63 monthly payment

Loan Payment Schedule

After determining the amount of the monthly payments, a lender can prepare a schedule of loan payments called an **amortization schedule**. The payment for the last month is

determined in the schedule, and it may be slightly different from the payment in the other months.

> ### STEPS to Create an Amortization Schedule
>
> For each row except the last:
> 1. Interest payment = Unpaid balance × Monthly interest rate
> 2. Principal payment = Monthly payment − Interest payment
> 3. New unpaid balance = Old unpaid balance − Principal payment
>
> For the last row (i.e., for the final payment):
> 1. Interest payment = Unpaid balance × Monthly interest rate
> 2. Monthly payment = Unpaid balance + Interest payment
> 3. Principal payment = Unpaid balance

EXAMPLE M

Create an amortization schedule for the loan in example K, a $4,000 loan amortized at 12% over 4 months. The interest rate is 1% per month.

| Month | Unpaid Balance | Interest Payment | Principal Payment | Total Payment | New Balance |
|---|---|---|---|---|---|
| 1 | $4,000.00 | $ 40.00 | $ 985.12 | $1,025.12 | $3,014.88 |
| 2 | 3,014.88 | 30.15 | 994.97 | 1,025.12 | 2,019.91 |
| 3 | 2,019.91 | 20.20 | 1,004.92 | 1,025.12 | 1,014.99 |
| 4 | 1,014.99 | 10.15 | 1,014.99 | 1,025.14 | 0 |
| Totals | | $100.50 | $4,000.00 | $4,100.50 | |

Note: In example M, the last monthly payment is 2 cents larger than the others. Because the interest payments need to be rounded, the final payment usually will be slightly different from the previous payments.

The alternative formula for effective rate gives

$$R = \frac{2 \times M \times I}{[P \times (n + 1)]} = \frac{2 \times 12 \times \$100.50}{[\$4{,}000 \times (4 + 1)]} = \frac{\$2{,}412}{\$20{,}000} = 0.1206, \quad \text{or} \quad 12.06\%$$

The rate is 12.06% instead of 12% because the alternative formula requires that the amount paid on the principal be equal every period. The amortization schedule in example M shows that the amounts paid on the principal are slightly different each month. Therefore the rate from this formula is slightly different from the true effective rate of 12%.

CONCEPT CHECK 17.6

A $500 purchase is amortized over 2 months at an annual rate of 18%. First, use Table 17-1 to calculate the monthly payment. Then show the calculations to construct a 2-month amortization schedule.

Amortization payment factor from Table 17-1: 0.51128 × $500 principal = $255.64 monthly payment

| Month | 1 | | 2 | |
|---|---|---|---|---|
| Unpaid balance | Original principal: | $500.00 | From end of month 1: | $251.86 |
| Monthly rate | 0.18 ÷ 12 = 0.015 | | | |
| Interest payment | $500.00 × 0.015 = | $7.50 | $251.86 × 0.015 = | $3.78 |
| Total payment | From table 17.1: | $255.64 | $251.86 + $3.78 = | $255.64 |
| Principal payment | $255.64 − $7.50 = | $248.14 | | $251.86 |
| New balance | $500 − $248.14 = | $251.86 | $251.86 − $251.86 = | $0.00 |

COMPLETE ASSIGNMENT 17.3.

Chapter Terms for Review

amortization

amortization payment factor

amortization schedule

annual percentage rate (APR)

average daily balance

average principal

credit card

effective interest rate

finance charge

Truth in Lending Act

Visit the Web

http://www.mastercard.com—Visit the MasterCard site to learn about the company's products and services.

http://www.zoomsys.com/entexp.html—Software for expense reporting can be viewed and samples downloaded at this Web site.

http://www.cardweb.com—Visit CardWeb, an extension of RAM Research Group, an online publisher of information pertaining to all types of payment cards, including credit cards, debit cards, smart cards, prepaid cards, ATM cards, loyalty cards and phone cards, and more. The company serves all constituencies connected to the payment card business—consumers, institutions, merchants, acquirers, processors, manufacturers, consultants, news reporters, and many others—and is renowned for its independence, credibility, and fairness.

CYBERCONNECTION

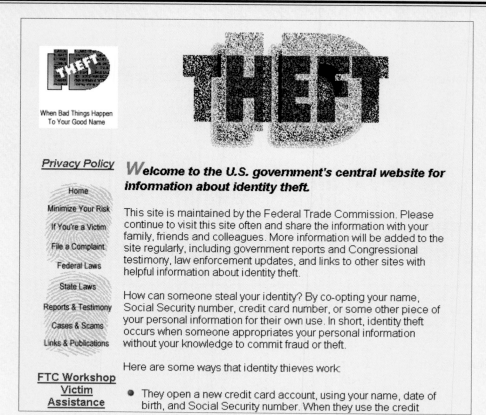

ID THEFT
When Bad Things Happen
To Your Good Name

Privacy Policy

Home
Minimize Your Risk
If You're a Victim
File a Complaint
Federal Laws
State Laws
Reports & Testimony
Cases & Scams
Links & Publications

**FTC Workshop
Victim
Assistance**

Welcome to the U.S. government's central website for information about identity theft.

This site is maintained by the Federal Trade Commission. Please continue to visit this site often and share the information with your family, friends and colleagues. More information will be added to the site regularly, including government reports and Congressional testimony, law enforcement updates, and links to other sites with helpful information about identity theft.

How can someone steal your identity? By co-opting your name, Social Security number, credit card number, or some other piece of your personal information for their own use. In short, identity theft occurs when someone appropriates your personal information without your knowledge to commit fraud or theft.

Here are some ways that identity thieves work:

● They open a new credit card account, using your name, date of birth, and Social Security number. When they use the credit

Source: http://www.consumer.gov/idtheft

How can someone steal your identity? By using your name, Social Security number, credit card number, or other personal information for their own use to commit fraud or theft. Identity thieves can open a new credit card account by using your name, date of birth, and Social Security number. When they use the credit card and don't pay the bills, the delinquent account is reported on your credit report. Or, they call your credit card issuer, pretend to be you, change your mailing address, and then run up charges on your account. Because your bills are being sent to a new address, you may not realize right away that there's a problem.

The Bottom Line

Summary of chapter learning objectives:

| Learning Objective | Summary | Example |
|---|---|---|
| **17.1** | Convert between annual and monthly interest rates | 1. Convert 0.5% per month to an annual rate.
 2. Convert 21% per year to a monthly rate |
| **17.2** | Compute simple interest on a monthly basis | 3. Compute the simple interest on $1,200 for 5 months at 0.75% per month (9% per year). |
| **17.3** | Compute finance charges for credit account purchases | 4. Charge account terms apply a 1.25% finance charge to the previous balance, less any payments and credits, ignoring purchases. Find the finance charge and new balance when the previous balance is $1,683.43, payments plus credits total $942.77, and purchases are $411.48. |
| **17.4** | Compute costs of installment purchases | 5. Furniture worth $1,800 is paid for with a $300 down payment and three payments of $500, plus monthly interest of 1% on the unpaid balance. Find the total interest paid. The monthly balances are $1,500, $1,000, and $500. |
| **17.5** | Compute effective rates | 6. A $2,400 loan is repaid in 12 payments of $210 each. Find the approximate annual effective rate.
 $M = 12$ months in one year; Interest =
 (210×12) − $2,400 = $120; $P = $2,400 financed;
 $n = 12$ payments. |
| **17.6** | Amortize a loan | 7. A $2,400 loan will be amortized over 12 months at an annual rate of 9%. Find the payment, using Table 17-1, and calculate the unpaid balance after the first month. |

Answers: 1. 6% per year. 2. 1.75% per month 3. $45 4. Finance charge, $9.26; new balance, $1,161.40 5. $30 6. 9.2% 7. Payment, $209.88; unpaid balance, $2,208.12

Self-Check Review Problems for Chapter 17

1 Change the monthly rates to annual rates.

a. 0.5% = _____ b. 0.75% = _____ c. 1.25% = _____ d. $\frac{2}{3}$% = _____

2 Change the annual rates to monthly rates.

a. 12% = _____ b. 4% = _____ c. 21% = _____ d. 18% = _____

3 A store offers the following credit terms: "There will be no finance charge if the full amount of the new balance is received on or before the due date. Unpaid balances after the due date will be charged interest based upon the previous balance, less any payments and credits before the due date. The rates are 1.75% on the first $1,000 of the unpaid balance, and 1.25% on the part of the unpaid balance that exceeds $1,000."

Calculate (a) the finance charge and (b) the new balance on account that had a previous balance of $2,752.88; a payment of $800; credits of $215; and purchases of $622.75.

4 Neta Prefontaine buys $3,000 worth of merchandise. She agrees to pay $1,000 per month on the principal. In addition, she will pay interest of 1.25% per month (15% annually) on the unpaid balance. Complete the following table.

| Month | Unpaid Balance | Interest Payment | Principal Payment | Total Payment | New Balance |
|-------|---------------|------------------|-------------------|---------------|-------------|
| 1 | $3,000.00 | a. _____ | $1,000.00 | b. _____ | c. _____ |
| 2 | d. _____ | e. _____ | $1,000.00 | f. _____ | g. _____ |
| 3 | h. _____ | i. _____ | $1,000.00 | j. _____ | $0.00 |

5 Use the results of problem 4 and compute the effective annual interest rate using the formula

$$R = \frac{2 \times M \times I}{P \times (n + 1)},$$ where P is the amount of the credit purchase, I is the total interest paid, M is the number of months in 1 year, and n is the actual number of payments.

6 Use Table 17-1 to find the monthly payment of a $3,000 loan that is amortized over 3 months at 15%.

7 A $3,000 loan is amortized over 3 months at 15%. The first two monthly payments are $1,025.10; the final payment may differ. Complete the following table.

| Month | Unpaid Balance | Interest Payment | Total Payment | Principal Payment | New Balance |
|-------|---------------|------------------|---------------|-------------------|-------------|
| 1 | $3,000.00 | a. _____ | $1,025.10 | b. _____ | c. _____ |
| 2 | d. _____ | e. _____ | $1,025.10 | f. _____ | g. _____ |
| 3 | h. _____ | i. _____ | j. _____ | k. _____ | $0.00 |

Assignment 17.1: Monthly Finance Charges

Name _____

Date _____ Score _____

A **(22 points) Problem 1: Change the rates from annual to monthly. Problem 2: Change the rates from monthly to annual. (1 point for each correct answer)**

1a. 12% = _____

b. 15% _____

c. 16% _____

d. 8% _____

e. 6% _____

f. 20% _____

g. 24% _____

h. 18% _____

i. 9% _____

j. 21% _____

k. 10% _____

2a. 1.5% = _____

b. 1% = _____

c. 1.75% = _____

d. 2% = _____

e. $1\frac{2}{3}$% = _____

f. 1.25% = _____

g. $1\frac{1}{3}$% = _____

h. 0.75% = _____

i. $\frac{7}{12}$% = _____

j. $\frac{2}{3}$% = _____

k. 0.5% = _____

Score for A (22)

Assignment 17.1 *Continued*

B **(30 points)** Lakeside Furniture Store offers the credit terms shown to its retail customers. In problems 3–5 compute the finance charge, if any, and the new balance. Assume that all payments are made within the current billing cycle. In problems 6 and 7 Lelia McDaniel has an account at Lakeside Furniture Store. Compute the missing values in Lelia's accounts summary for the months of July and August. The Previous Balance in August is the same as the New Balance in July. (3 points for each correct answer)

TERMS: There will be no finance charge if the full amount of the new balance is received within 25 days after the cycle-closing date. The finance charge, if any, is based upon the entire previous balance *before* any payments or credits are deducted. The rates are 1.5% per month on amounts up to $1,000 and 1.25% on amounts in excess of $1,000. These are annual percentage rates of 18% and 15%, respectively.

| | Cycle Closing | Previous Balance | Payment Amount | Credits | Finance Charge | Purchases | New Balance |
|---|---|---|---|---|---|---|---|
| **3.** | 3/20/200– | $1,832.50 | $750.00 | $195.50 | _____ | $520.16 | _____ |
| **4.** | 6/20/200– | $743.72 | $0.00 | $15.00 | _____ | $609.88 | _____ |
| **5.** | 9/20/200– | $3,208.46 | $2,500.00 | $158.12 | _____ | $851.72 | _____ |

Assignment 17.1 *Continued*

| Cycle Closing | Previous Balance | Payment Amount | Credits | Finance Charge | Purchases | New Balance |
|---|---|---|---|---|---|---|
| **6.** 7/20/200– | $1,636.55 | $900.00 | $36.00 | _____ | $844.25 | _____ |
| **7.** 8/20/200– | $1,567.76 | $1,250.00 | $88.50 | _____ | $609.88 | _____ |

<div align="right">Score for B (30)</div>

C (48 points) Devlin's Feed & Fuel offers the credit terms shown to its retail customers. In problems 8–11 compute the missing values in the charge accounts shown. Assume that all payments are made within 30 days of the billing date. In problems 10 and 11 compute the missing values in Jimmy Petrasek's charge account summary at Devlin's for the months of January and February. The Previous Balance in February is the same as the New Balance in January. (4 points for each correct answer)

TERMS: Finance Charge is based on the Net Balance, if payment is received within 30 days of the billing date. If payment is made after 30 days, then the Finance Charge is based on the Previous Balance. Net Balance equals Previous Balance less Payments and Credits. In either case, the monthly rate is 1.25% on the first $500 and 1% on any amount over $500. These are annual percentage rates of 15% and 12%, respectively.

| Billing Date | Previous Balance | Payment Amount | Credit | Net Balance | Finance Charge | New Purchases | Balance |
|---|---|---|---|---|---|---|---|
| **8.** 4/25/— | $2,621.05 | $1,700.00 | $0.00 | _____ | _____ | $751.16 | _____ |

| | Billing Date | Previous Balance | Payment Amount | Credit | Net Balance | Finance Charge | New Purchases | Balance |
|---|---|---|---|---|---|---|---|---|
| **9.** | 8/25/— | $1,727.95 | $1,000.00 | $23.75 | _____ | _____ | $499.06 | _____ |
| **10.** | 1/25/— | $1,352.12 | $600.00 | $52.00 | _____ | _____ | $875.21 | _____ |
| **11.** | 2/25/— | $1,583.58 | $750.00 | $0.00 | _____ | _____ | $508.14 | _____ |

Score for C (48)

Assignment 17.2: Installment Sales and Effective Rates

LEARNING OBJECTIVES

Name _____

Date _____ Score _____

A **(24 points) Karen Ann Senter makes a credit purchase of $2,000 for dining room furniture. She agrees to pay simple interest for 4 months at 9% of the purchase price. (Use $T = \frac{4}{12}$ year.) Each month Karen will pay $\frac{1}{4}$ of the total interest and $\frac{1}{4}$ of the $2,000 principal. Compute the answer to each of the following questions. (6 points for each correct answer)**

 1a. What is the simple interest on $2,000 at 9% for 4 months? _____

 b. Over the 4-month period, what is the average unpaid balance? _____

 c. What is the effective annual rate of interest (to the nearest $\frac{1}{10}$%)? Use the formula $R = I/(P \times T)$, where P is the average unpaid balance, I is the total interest paid, and T is $\frac{4}{12}$ year. _____

 2. What is the effective annual rate of interest (to the nearest $\frac{1}{10}$%)? Use the formula $R = 2 \times M \times I/[P \times (n + 1)]$, where P is the amount of the credit purchase, I is the amount of interest, M is the number of months in one year, and n is the number of actual payments. _____

Score for A (24)

B **(40 points) Rework the problem from Part A with a different method of calculating interest. Again, Karen Ann Senter makes a $2,000 credit purchase and agrees to repay the $2,000 principal in four equal monthly payments of $500. This time, however, calculate her interest each month at 0.75% per month (9% annual rate) on the unpaid balance and complete the following table below. (2 points for each correct answer)**

| Month | Unpaid Balance | Interest Payment | Principal Payment | Total Payment | New Balance |
|---|---|---|---|---|---|
| **3.** 1 | _____ | _____ | _____ | _____ | _____ |
| **4.** 2 | _____ | _____ | _____ | _____ | _____ |
| **5.** 3 | _____ | _____ | _____ | _____ | _____ |
| **6.** 4 | _____ | _____ | _____ | _____ | _____ |

Score for B (40)

Assignment 17.2 *Continued*

C **(24 points) Use the information from the table you completed in Part B to solve problems 7 and 8. (6 points for each correct answer)**

7. Compute the effective annual interest rate (to the nearest $\frac{1}{10}$%) for Karen Ann Senter's credit purchase in Part B. Use $R = I/(P \times T)$. Use the table to find P, the average unpaid principal over the 4-month period, and I, the total interest paid; $T = \frac{4}{12}$ year.

 a. Total interest: _____ **b.** Average unpaid balance: _____ **c.** Effective rate: _____

8. For the same credit purchase compute the effective annual rate (to the nearest $\frac{1}{10}$%), using the formula $R = 2 \times M \times I/[P \times (n + 1)]$, where P is the amount of the credit purchase, I is the total amount of interest, M is the number of months in a year, and n is the actual number of payments. _____

<div style="text-align: right">

Score for C (24)

</div>

D **(12 points) Approximate the effective annual rate of interest, using the formula $R = 2 \times M \times I/[P \times (n + 1)]$, where P is the amount of the purchase that is charged, I is the total finance charge, M is the number of months in a year, and n is the actual number of payments. Give the rate to the nearest $\frac{1}{10}$%. (6 points for each correct answer)**

9. Ray Wilson has decided to buy a heavy-duty rototiller from a catalog. He can purchase the rototiller for $1,150 cash, or he can choose a 1-year credit purchase with a $50 down payment and 12 monthly payments of $100 each. What is the approximate effective rate of the credit purchase plan? _____

10. Eastside TV & Stereo is offering a special credit purchase of new big screen television sets. The customer can pay nothing down and make 24 payments of $125 each. Eastside also charges a $10 fee to set up the credit plan and $12 for insurance on the TV set (50 cents per month). The cash price of the television set is $2,495. Find the approximate effective rate of interest. _____

<div style="text-align: right">

Score for D (12)

</div>

Assignment 17.3: **Amortization**

Name

Date Score

A **(20 points)** Southeastern National Bank amortizes most of its personal loans on a monthly basis. The total monthly payments include both interest and principal, and the payments are the same each month. Use Table 17-1 to find the amortization payment factor for the following loans. Then compute the monthly payment for each loan. (2 points for each correct answer)

| Loan and Terms of Amortization | Amortization Payment Factor | Monthly Payment |
|---|---|---|
| **1.** $3,500 over 4 months at 18% | _____ | _____ |
| **2.** $2,750 over 6 months at 15% | _____ | _____ |
| **3.** $12,400 over 2 years at 12% | _____ | _____ |
| **4.** $150,000 over 10 years at 6% | _____ | _____ |
| **5.** $7,650 over 5 years at 9% | _____ | _____ |

Score for A (20)

B **(32 points)** On April 13, Braunda Johannesen borrowed $6,000 from her bank to help pay her federal income taxes for the previous year. The bank amortized the loan over 4 months at a rate of 9%. Use the amortization payment factor from Table 17-1. The interest payment is 0.75% (9%/12) times the unpaid balance each month. Subtract the interest from the total payment to get the principal payment. Subtract the principal payment from the unpaid balance to get the new balance. In the last (fourth) month, add the interest payment to the unpaid balance to get the total payment. (2 points for each correct answer)

6. Amortization factor from Table 17-1: _____
 Multiply the amortization factor by $6,000 to get the total payment shown for months 1, 2, and 3.

| | Month | Unpaid Balance | Interest Payment | Total Payment | Principal Payment | New Balance |
|---|---|---|---|---|---|---|
| **7.** | 1 | $6,000.00 | _____ | $1,528.26 | _____ | _____ |
| **8.** | 2 | _____ | _____ | 1,528.26 | _____ | _____ |
| **9.** | 3 | _____ | _____ | 1,528.26 | _____ | _____ |
| **10.** | 4 | _____ | _____ | _____ | _____ | 0.00 |

Score for B (32)

C (30 points) Refer to Part B, in which Braunda Johannesen borrowed $6,000 to help pay her federal income taxes. Now suppose that Braunda agreed to make payments of $1,200 in months 1, 2, and 3. The bank will compute the interest on the unpaid balance at a rate of 0.75% (9%/12) each month and deduct the interest from the $1,200. In the last (fourth) month, Braunda will pay all of the remaining unpaid balance plus the interest for the last month. Complete the table, using the same procedure as in Part B. (2 points for each correct answer)

| | Month | Unpaid Balance | Interest Payment | Total Payment | Principal Payment | New Balance |
|---|---|---|---|---|---|---|
| 11. | 1 | $6,000.00 | _____ | $1,200.00 | _____ | _____ |
| 12. | 2 | _____ | _____ | 1,200.00 | _____ | _____ |
| 13. | 3 | _____ | _____ | 1,200.00 | _____ | _____ |
| 14. | 4 | _____ | _____ | _____ | _____ | 0.00 |

Score for C (30)

D (18 points) The following questions are about amortized loans. To answer the questions, you will need to compute one or two rows of the amortization schedule. Do only the work necessary to answer the questions. (3 points for each correct answer)

15. Juli Johnson purchased a duplex apartment (two units) so that she could live in one unit and rent the other. To complete the purchase, she decided to borrow $75,000. Her bank amortized the loan over 10 years at 9% interest.

 a. How much did Juli pay in interest during the first month? _____

 b. How much did Juli pay in principal during the first month? _____

 c. What was the unpaid balance at the end of the first month? _____

16. Paul Yeiter bought a new Ford pickup truck. He traded in his old truck, used some of his savings, and got a loan from the Ford agency for $11,000. The Ford agency amortized the loan over 3 years at 12%.

 a. What was Paul's monthly payment? _____

 b. How much total interest did Paul pay during the first 2 months? _____

 c. What was the unpaid balance at the end of the second month? _____

Score for D (18)

Notes

INDIVIDUALS

HOME

- [] auto
- [] home
- [x] business
- [] life
- [] health
- [] disability
- [] disasters
- [] other
- [] en español
- [] ask I.I.I
- [] about I.I.I.
- [] for the media

[]

search

BUSINESS

The success of a business, whether it's a tiny enterprise run out of a ba[sement] corporation, is largely dependent on hard work and ingenuity. However, [no matter how] you are, one disaster can wipe out all your profits and even destroy your [business]. making sure that all the effort and money you have invested in a busines[s before] disaster strikes is to protect it with the appropriate insurance.

BUYING A POLICY & SAVING MONEY

- How do I find the right agent?
- How can I save money on my business insurance?
- How do I insure my home business?

BUSINESS INSURANCE BASICS

- What does a businessowners policy cover?
- Do I need business interruption insurance?
- Are there any disasters my property insurance won't cover?
- How can I disaster-proof my business?
- How do I file a business insurance claim?
- What's the difference between cancellation and non-renewal?
- Do I need workers compensation insurance?

ADDITIONAL COVERAGES

- Can I insure the life of a key employee?
- Do I need professional liability insurance?
- Do I need a commercial auto insurance policy?
- What is employment practices liability insurance (EPLI)?

Source: http://www.iii.org/individuals/business/. Copyright © 2001 Courtesy of Insurance Information Institute.

The mission of the Insurance Information Institute (I.I.I.) is to improve public understanding of insurance—what it does and how it works. For more than 40 years, the I.I.I. has provided definitive insurance information and is recognized today by the media, governments, regulatory organizations, universities and the public as a primary source of information, analysis and referral concerning insurance. Each year, the I.I.I. works on more than 3,700 news stories, handles more than 6,000 requests for information and answers nearly 50,000 questions from consumers. Visit their Web site at www.iii.org.

Business and Personal Insurance | 18

LEARNING OBJECTIVES

By studying this chapter and completing all assignments you will be able to:

 1 Compute costs and savings for auto insurance.

 2 Compute auto insurance premium rates for high- and low-risk drivers.

 3 Compute short-rate refunds.

 4 Compute coinsurance on property losses.

 5 Compute life insurance premiums.

 6 Compute cash surrender and loan values.

 7 Compute medical insurance contributions and reimbursements.

In early colonial times, sea captains and other ocean travelers purchased life insurance. No medical examination was required, and the cost was high. Today, about two out of three Americans have life insurance. Many have property and auto insurance as well.

1. How would a bad driving record affect your insurance premium?

2. What happens if you insure a home for less than its full value and a fire totally destroys your home?

3. Can a business purchase a basic health care plan to cover employees with the provision that the employees will pay part of the premium?

Computing Auto Insurance Costs

Compute costs and savings for auto insurance.

Auto insurance falls into three categories: liability and property damage, comprehensive, and collision. A policy that fully protects the insured will contain all three types.

Auto liability and property damage insurance protects the insured against claims resulting from personal injuries and property damage. Some states require all drivers to carry auto liability and property damage insurance. The amount of protection generally ranges from $50,000 to $1,000,000 per accident.

Auto comprehensive insurance protects the vehicle of the insured against water, theft, vandalism, falling objects, and other damage not caused by collision.

Auto collision insurance protects the vehicle of the insured against collision damage. Such damage may result from a collision with another vehicle or a one-car accident, such as hitting a tree.

The payment for an insurance policy is called a **premium**. Premium rates for auto insurance depend primarily on the coverage included in the policy and the driving record of the insured.

Auto collision insurance policies usually contain a **deductible clause**, which stipulates that the insured will pay the first portion of collision damage, usually $50 to $500, and that the insurance company will pay the remainder up to the value of the insured vehicle. A deductible clause not only reduces the amount of damages for which the insurance company must pay but also keeps the insurance company from having to get involved in and do paperwork for small repairs costing less than the deductible. Therefore a deductible clause lowers the premium for collision insurance.

www.statefarm.com/ insuranc/auto/ auto.htm

Learn more about different types of auto insurance coverage, restrictions, rates, and discounts.

EXAMPLE A

A car was insured for collision damage with a $250 deductible. The premium was $1,250 per year. The insured hit a tree, causing $2,530 damage to his car. How much more did the insured receive than he paid in premiums for that year?

$2,530 damage − $250 deductible = $2,280 paid by insurance

$2,280 received by insured − $1,250 premium paid = $1,030

EXAMPLE B

The driver of car A carried auto liability and property damage insurance only. She struck car B, causing $1,400 damage to car B and $700 in injuries to the driver. Car A suffered $940 damage.

a. How much did the insurance company pay for this accident?

$1,400 for damage to car B + $700 for injuries to driver = $2,100

b. How much did this accident cost the driver of car A?

$940 in uncovered damage to her own car

No-fault insurance requires that the driver of each vehicle involved in an injury accident submit a claim to *his or her own insurance company* to cover medical costs for injuries to the driver and passengers in that person's own vehicle. No-fault insurance is mandatory in some states. No-fault insurance doesn't cover damage to either vehicle involved in an accident.

Drivers A and B live in a state in which no-fault insurance is mandatory. Their two cars collided. Driver A and his passengers incurred medical expenses of $3,500. Driver B and her passengers incurred $1,700 in medical expenses. Car A required $1,400 in repairs. Car B required $948 in repairs. How much did the insurance companies pay under the no-fault insurance coverage?

Driver A's insurance company paid $3,500 in medical expenses.
Driver B's insurance company paid $1,700 in medical expenses.
Car repairs are not covered under no-fault insurance.

CONCEPT CHECK 18.1

Driver A lives in a state in which no-fault insurance is mandatory. He carries all three classifications of insurance to be fully protected. His total insurance premium is $2,400, with a collision deductible of $500. Driver A is involved in a major accident when he loses control of his car and hits two parked cars (cars B and C) before colliding with an oncoming car (car D) containing a driver and three passengers. Driver A is alone.

Damage to Driver A's car is $3,200.
Damages to cars B, C, and D total $8,600.
Medical expenses for driver A are $2,800.
Medical expenses for the driver and passengers of car D are $7,300.

a. How much does driver A's insurance company pay?
 Damage to car A: $3,200 − $500 deductible = $2,700 covered by collision
 Damage to cars B, C, and D: $8,600 covered by liability
 Medical expenses for driver A under no-fault: $2,800
 $2,700 + $8,600 + $2,800 = $14,100 paid by driver A's insurance
b. How much does driver D's insurance company pay?
 Medical expenses paid for driver D and passengers (no-fault): $7,300
c. How much more did driver A's insurance company pay to him and on his behalf for this accident than he paid in insurance expenses for the year? (This is the amount driver A saved this year by being fully insured.)
 $2,400 premium + $500 deductible = $2,900 paid by Driver A
 $14,100 from insurance − $2,900 = $11,200
 Driver A saved $11,200 this year by being fully insured.

Computing Low-Risk and High-Risk Rates

Auto insurance premium rates reflect the risk involved. Insurance companies study the statistics on automobile accidents relative to driving records. Premium rates are adjusted according to the driving record of the insured. A driver with a clear record of long standing is considered to be a **low-risk driver** and may be rewarded with a discount in the premium rate. Conversely, a driver with a record of numerous citations or accidents is considered to be a **high-risk driver** and may pay double or triple the normal premium rate.

2
LEARNING OBJECTIVES

Compute auto insurance premium rates for high- and low-risk drivers.

Drivers A and B have identical automobiles and amounts of insurance coverage. The normal premium rate for each is $1,200 per year. Driver A is a low-risk driver and receives a 15% discount on the premium rate. Driver B is a high-risk driver and must pay double the normal rate. How much more does driver B pay for insurance than driver A?

Driver A pays $1,200 × 85% = $1,020 (100% − 15% discount)

Driver B pays $1,200 × 2 = $2,400

Driver B pays $2,400 − $1,020 = $1,380 more

CONCEPT CHECK 18.2

Driver A, a very careful driver, has had the same insurance company for 5 years and has not had a ticket during that 5-year period. Each year, driver A has received a 10% reduction in her premium. Driver B has a record of speeding tickets. He has had one or more every year for 5 years. His premium for year 1 was normal, for years 2 and 3 it was 150%, and for years 4 and 5 it was 200%. The normal annual premium rate for each driver would be $980.

a. How much did driver A pay in premiums over the 5-year period?

 $980 × 90% = $882

 $882 × 5 = $4,410

b. How much did driver B pay in premiums over the 5-year period?

 Year 1: $980

 Years 2 and 3: $980 × 1.5 × 2 = $2,940

 Years 4 and 5: $980 × 2 × 2 = $3,920

 $980 + $2,940 + $3,920 = $7,840

c. How much more did driver B pay during the 5-year period than driver A?

 $7,840 − $4,410 = $3,430

Computing Short Rates

LEARNING OBJECTIVES

Compute short-rate refunds.

Short rates are rates charged for less than a full term of insurance. If an insurance policy is canceled by the **insured** (the person who receives the benefit of the insurance) before the policy's full term is complete, the insured will receive a short-rate return of premium. If a policy is canceled by the insurance company rather than by the insured, the company must refund the entire unused premium.

A driver paid an annual premium of $980 for auto insurance. After 3 months, the vehicle was sold and the insurance canceled. The insurance company refunded the remaining portion of the premium at the short rate, based on a 10% penalty. What was the refund?

Unused premium: $980 × $\dfrac{3}{4}$ = $735 (9 months canceled = $\frac{9}{12} = \frac{3}{4}$ year)

Penalty: $980 × 10% = $98

Short-rate refund: $735 − $98 = $637

 CONCEPT CHECK 18.3

A company purchased two cars. Each car was insured at an annual premium of $890. At the end of 6 months, the company sold one car and canceled the insurance on that car. At the end of 9 months, the insurance company decided to cancel the insurance on the second car. The insurance company imposes a 10% penalty for short-rate premiums. Compute the refunds the insurance company paid for car 1 and car 2.

Car 1: $890 \times \frac{1}{2}$ year $= 445 unused premium

$890 \times 10\%$ penalty $= 89
$445 - $89 = 356 refunded

Car 2: $890 \times \frac{1}{4}$ year $= 222.50 unused and refunded premium

COMPLETE ASSIGNMENT 18.1.

Computing Coinsurance on Property Losses

Property insurance is insurance against loss of or damage to property. A policy can be written to protect the insured against one or more of the following: fire, casualty, liability, and theft.

Premium rates, which are quoted in terms of dollars per $1,000 of insurance, depend on the nature of the risk, the location of the property, and the length of time covered by the policy. Short rates and short-rate penalties for less than a full term of insurance apply to property insurance as they do to auto insurance.

LEARNING OBJECTIVES

Compute coinsurance on property losses.

EXAMPLE F

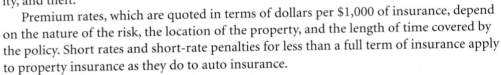

A building worth $350,000 is insured for $210,000. The annual premium for the policy is $5,000. A fire causes $200,000 in damage.

a. How much does the insurance company pay?
 $200,000 in damage is less than the $210,000 policy. The insurance company pays the entire $200,000.

b. How much does the property owner pay?
 The property owner pays no damages.

c. How much does the property owner pay that year in damages and insurance?
 $5,000 for the insurance premium only.

In an ordinary fire insurance policy, the insured will be paid for the loss up to the amount of the insurance. Policies may be obtained at lower rates if they contain a *co-insurance clause*. This clause specifies that if a property is not insured up to a certain percentage of its value, the owner is responsible for part of the insurance and will not be covered for the full amount of damages.

It is common for a policy to have an 80% coinsurance clause. Under this clause, the full amount of the loss will not be paid by the insurance company unless the policy amount equals 80% of the property value. It is important to note that the amount of the loss is not necessarily the total value of the property. For example, suppose that only a part of a building valued at $300,000 were damaged in a fire. The loss would not be the entire value of the building; rather, it would be the cost of repairing the partial damage.

STEPS **to Determine the Owner's Share of Property Loss Under Coinsurance**

1. Compute the amount of insurance required by multiplying the entire value of the property by the coinsurance percentage.

2. Compute the **recovery amount**, the maximum amount the insurance company will pay, by using the formula

$$\frac{\text{Amount of insurance carried}}{\text{Amount of insurance required}} \times \text{Loss} = \text{Recovery amount}$$

3. Compare the recovery amount with the amount of the insurance policy.
 a. If the recovery amount is greater than the amount of the policy, the insurance company will pay the amount of the policy.
 b. If the recovery amount is less than the amount of the policy, the insurance company will pay the recovery amount.
 Note: The insurance company will not pay more than the amount of the loss.

4. Determine the owner's share of the property loss by subtracting the amount the insurance company will pay from the loss amount.

EXAMPLE G

A building valued at $350,000 is insured for $210,000 under a policy with an 80% co-insurance clause. The annual premium is $2,800. A fire causes $200,000 damage to the building.

© PHOTODISC, INC.

a. How much will the insurance company pay the insured?

STEP 1 $350,000 × 80% = $280,000 insurance required

STEPS 2 & 3 $\dfrac{\$210,000}{\$280,000}$ × $200,000 = $150,000 insurance pays

b. How much must the owner pay if the building is repaired for $210,000?

STEP 4 $210,000 − $150,000 = $60,000 paid by owner

c. How much does the property owner pay that year for damages and insurance?
$60,000 damages + $2,800 premium = $62,800

d. How much would the insurance company pay if the fire caused $300,000 damage to the building?

$\dfrac{\$210,000}{\$280,000}$ × $300,000 = $225,000 recovery value

The insurance company would pay $210,000 (the full value of the policy because the recovery value exceeds the policy's coverage).

EXAMPLE H

If the amount of insurance carried in example G had been $280,000, how much would the insured have paid for damages and insurance that year?
$2,800 premium only (the 80% coinsurance requirement would have been met)

CONCEPT CHECK 18.4

A building worth $100,000 is insured for $60,000 with an 80% coinsurance clause. A fire causes $70,000 in damage. How much will the insurance company pay, and how much will the insured pay?

$100,000 × 80% = $80,000 insurance required

$\dfrac{\$60,000}{\$80,000}$ × $70,000 = $52,500 insurance pays

$70,000 − $52,500 = $17,500 insured pays

COMPLETE ASSIGNMENT 18.2.

Computing Life Insurance Premiums

The policies most commonly issued by life insurance companies are straight life (sometimes called ordinary life), limited-payment life, term, endowment, and annuity.

 For **straight (ordinary) life insurance coverage,** a certain premium, or fee, is paid every year *until the death of the insured.* The policy then becomes payable to the **beneficiary.** A policy beneficiary can be a person, a company, or an organization.

LEARNING OBJECTIVES

Compute life insurance premiums.

Limited-payment life insurance (such as 20-payment life) requires the payment of a specified premium each year for a certain number of years or until the death of the insured, whichever comes first. Should the insured live longer than the specified number of years, the policy is then paid up for the remainder of the insured's life and is payable to the beneficiary on the death of the insured.

Term insurance is protection issued for a limited time. A certain premium is paid every year *during the specified time period,* or term. The policy is payable only in case of death of the insured during the term. Otherwise, neither the insured nor the specified beneficiaries receive any payment, and the protection stops at the end of the term.

Endowment insurance provides insurance payable on the insured's death if it occurs within a specified period. If the insured is alive at the end of the specified period, an endowment of the same amount as the policy is payable.

Annuity insurance pays a certain sum of money to the insured every year after the insured reaches a specified age, until the insured's death.

An **additional death benefit** (**ADB**), sometimes referred to as an *accidental death benefit*, accompanies some policies. ADB allows the insured to purchase, at a low rate per thousand dollars of coverage, additional insurance up to the full face value of the policy. In case of death of the insured by accident, both the full value of the policy and the ADB are paid to the beneficiaries. If death occurs other than by accident, the full value of the policy is paid, but no ADB is paid.

Figure 18-1 shows typical annual, semiannual, and quarterly premiums (ages 25–28) for straight life, 20-payment life, and 20-year endowment policies. Premiums vary from company to company because of such differences as the exact type of coverage and participation in dividends.

Figure 18-1: Insurance Premiums per $1,000

| Age | Straight Life | | | 20-Payment Life | | | 20-Year Endowment | | |
|---|---|---|---|---|---|---|---|---|---|
| | Annual | Semi-annual | Quarterly | Annual | Semi-annual | Quarterly | Annual | Semi-annual | Quarterly |
| 25 | $17.20 | $ 8.94 | $4.73 | $31.20 | $16.26 | $8.26 | $52.00 | $27.04 | $14.30 |
| 26 | 17.85 | 9.28 | 4.91 | 31.81 | 16.52 | 8.45 | 52.60 | 27.35 | 14.47 |
| 27 | 18.60 | 9.67 | 5.11 | 32.41 | 16.83 | 8.64 | 53.20 | 27.66 | 14.63 |
| 28 | 19.30 | 10.04 | 5.31 | 33.06 | 17.31 | 8.85 | 53.86 | 28.01 | 14.81 |

EXAMPLE I

Using the premiums shown in Figure 18-1, determine the yearly premiums for each of the following $50,000 life insurance policies purchased at age 27.

| Type of Insurance | Method of Payment | Premium Computation |
|---|---|---|
| Straight Life | Annual | $18.60 × 50 = $930 |
| 20-Year Endowment | Quarterly | $14.63 × 4 × 50 = $2,926 |
| 20-Payment Life | Semiannual | $16.83 × 2 × 50 = $1,683 |
| 20-Year Endowment | Semiannual | $27.66 × 2 × 50 = $2,766 |
| Straight Life | Quarterly | $5.11 × 4 × 50 = $1,022 |

CONCEPT CHECK 18.5

a. If a person at age 28 purchases a straight life insurance policy having a face value of $150,000 with quarterly premiums, what is the yearly premium?

$5.31 × 4 × 150 = $3,186

b. If a person at age 25 purchases a 20-payment life insurance policy having a face value of $100,000 with semiannual premiums, what is the yearly premium?

$16.26 × 2 × 100 = $3,252

c. If a person at age 25 purchases a 20-year endowment insurance policy having a face value of $75,000 with annual premiums, what is the yearly premium?

$52 × 75 = $3,900

Computing Cash Surrender and Loan Values

Except for term insurance, insurance usually has a *cash surrender value,* which is the amount of cash that the company will pay the insured on the surrender, or "cashing in," of the policy. The *loan value* of a policy is the amount that the insured may borrow on the policy from the insurance company. Interest is charged on such loans. The values, often quoted after the third year of the policy, are stated in the policy and increase every year. Figure 18-2 shows typical cash surrender and loan values for policies issued at age 25 per $1,000 of life insurance.

LEARNING OBJECTIVES

Compute cash surrender and loan values.

Figure 18-2: Insurance Values per $1,000

| End of Policy Year | Cash Surrender and Loan Values | | |
| | Straight Life | 20-Payment Life | 20-Year Endowment |
| --- | --- | --- | --- |
| 3 | $ 10 | $ 43 | $ 88 |
| 4 | 22 | 68 | 130 |
| 5 | 35 | 93 | 173 |
| 10 | 104 | 228 | 411 |
| 15 | 181 | 380 | 684 |
| 20 | 264 | 552 | 1,000 |

EXAMPLE J

Use the figures shown in Figure 18-2 to determine the cash surrender or loan value for each of the following policies.

| Policy Year | Type of Policy | Amount of Policy | Cash Surrender or Loan Value |
| --- | --- | --- | --- |
| 10 | Straight Life | $ 75,000 | $104 × 75 = $7,800 |
| 5 | 20-Year Endowment | $ 15,000 | $173 × 15 = $2,595 |
| 10 | 20-Payment Life | $ 50,000 | $228 × 50 = $11,400 |
| 20 | Straight Life | $100,000 | $264 × 100 = $26,400 |
| 15 | 20-Year Endowment | $ 50,000 | $684 × 50 = $34,200 |

 CONCEPT CHECK 18.6

Use the figures shown in Figure 18-2 to determine the cash surrender or loan value for each of the following policies.

a. Third policy year of a $50,000 20-year endowment policy
 $88 × 50 = $4,400

b. Twentieth policy year of a $100,000 straight life policy
 $264 × 100 = $26,400

c. Tenth policy year of a $25,000 20-payment life policy
 $228 × 25 = $5,700

Computing Medical Insurance Contributions and Reimbursements

Compute medical insurance contributions and reimbursements.

Most large employers and many small employers subscribe to a group plan on behalf of their employees. *Group insurance* plans provide medical insurance coverage to large numbers of people at lower premium rates than individuals could obtain separately. Employers generally pay all the premium for employees and a portion of the premium for family members of employees. Most employers now use a group plan known as a **health maintenance organization (HMO)** or a **preferred provider organization (PPO)**.

EXAMPLE K

Employer A selected a basic health care plan to cover employees who want to participate. Monthly premiums are as follows: employee only, $350; employee with one dependent, $450; and employee with multiple dependents, $530. Employees pay a portion of the premium as follows: employee only, $0; employee with one dependent, $80; and employee with multiple dependents, $120. How much does the employer pay during the year for each category of employee?
Employee only: $350 × 12 = $4,200
Employee with one dependent: ($450 − $80) × 12 = $4,440
Employee with multiple dependents: ($530 − $120) × 12 = $4,920

EXAMPLE L

Employer B selected a total care health plan to cover employees who want to participate. Monthly premiums are as follows: employee only, $300; employee with one dependent, $400; and employee with multiple dependents, $480. The employer pays most of the premium, but employees pay a portion as follows: employee only, $30; employee with one dependent, $80; and employee with multiple dependents, $120. What percent of the premium will be paid by a single employee, an employee with 1 dependent, and an employee with 6 dependents?
A single employee: $30 ÷ 300 = 0.10, or 10%
An employee with 1 dependent: $80 ÷ 400 = 0.20, or 20%
An employee with 6 dependents: $120 ÷ 480 = 0.25, or 25%

Many group plans include a provision for an annual deductible, which is the cost that must be paid by the employee before any cost is paid by the insurance company. Group medical plans also frequently provide for the payment by the insurance company of a percent of costs over the deductible, usually 70% to 90%, with the remaining 30% to 10% paid by the insured.

EXAMPLE M

Employer C provides group health coverage that includes a $500 annual deductible per family and payment of 70% of the medical charges exceeding the deductible. How much would an employee with three dependents pay if her year's medical bills were $1,500?
$1,500 − $500 deductible = $1,000
$1,000 × 30% paid by employee = $300
$500 deductible + $300 = $800 paid by the employee

 CONCEPT CHECK 18.7

An employer provides group health coverage that includes a $300 annual deductible per family and payment of 80% of costs over the deductible.

a. How much would an employee with two dependents pay if his year's medical bills were $460?
$460 − $300 deductible = $160
$160 × 20% = $32
$300 deductible + $32 = $332 paid by the employee

b. How much would that employee have paid if total medical bills for the year had been $4,300?
$4,300 medical costs − $300 deductible = $4,000
$4,000 × 20% = $800
$300 deductible + $800 = $1,100

c. How much of the $4,300 in medical bills would that employee have paid if his employer did not provide medical insurance?
$4,300

d. How much did the employer pay if the monthly premium for an employee with multiple dependents was $480?
$480 × 12 = $5,760

COMPLETE ASSIGNMENT 18.3.

Chapter Terms for Review

additional death benefit (ADB)

annuity insurance

auto collision insurance

auto comprehensive insurance

auto liability and property damage insurance

beneficiary

cash surrender value

coinsurance clause

deductible clause

endowment insurance

group insurance

health maintenance organization (HMO)

high-risk driver

insured

limited-payment life insurance

loan value

low-risk driver

no-fault insurance

preferred provider organization (PPO)

premium

property insurance

recovery amount

short rates

straight (ordinary) life insurance

term insurance

Visit the Web

http://www.housingcenter.com/insur/coverage/property.html—Many factors must be considered when a company is planning to meet commercial property insurance needs. Learn what some of them are by accessing this Web site.

Read about President George W. Bush's proposal for seniors' insurance breaks at **http://www.freep.com/news/politics/bush11_20000511.htm**.

http://www.tdi.state.tx.us/consumer/consum89.html—Information provided by the State of Texas Department of Insurance is relevant not only to Texas, but also answers general questions common to citizens across the United States. Visit its Web site to learn more.

http://www.fema.gov—The Federal Emergency Management Administration (FEMA) provides information on risks and insurance issues for individuals and businesses. Check out its Web site.

http://www.cutcomp.com—Another aspect of insurance in business is workers' compensation insurance. Some companies use firms that audit compensation claims. For more information on workers' compensation insurance, visit this Web site.

CYBERCONNECTION

THE NINTH ANNUAL
OCTOBER 22-24, 2001
Business Insurance
Coronado Island Marriott and Resort
San Diego, California
Best Practices and Case Studies
presented in conjunction with
International Business Forum
IBF Conferences
THE BUSINESS INSURANCE WORKERS COMPENSATION and DISABILITY MANAGEMENT CONFERENCE

Business Insurance. ▸ **Cyberbeat**
Reporting on Corporate Risk, Employee Benefit and Managed Health Care News

Site Map / Search businessinsurance.com: [] Go!

EDITORIAL
◂ Updates
◂ Opinions
◂ Commentary
◂ BI Stock Ticker
◂ Directories
◂ Cyberbeat
◂ Editorial Staff
◂ About BI

BI SERVICES
◂ BI Editorial Calendar
◂ Datebook
◂ Services
◂ BI Events
◂ Online Forum
◂ Subscription Services
◂ Links
◂ Contact BI
◂ BI Updates by E-mail

ARCHIVES
◂ Search Archives
◂ Subscribe to Archives
◂ Renew Subscription

October 30, 2000

Technology has shaped claims administration
by By: Joanne Wojcik

MARINA DEL REY, Calif.--Workers compensation claims administration has come a long way since the advent of the computer.

``When I first began in claims, there wasn't a computer in sight," recalled Rick Duggan, manager of workers compensation at Walt Disney Co.'s West Coast operations in Anaheim, Calif.

``If you wanted to research a claim, you went to a long file cabinet of cards, like in the library. You'd physically have to go to find the files, etc. There was no e-mail; letters and correspondence were received the old-fashioned way, via U.S. Mail. You could hear typewriters, and we stocked whiteout by the gallon," Mr. Duggan said.

``Then came the fax machine," he said, and today, ``technology has a foothold in claims management. We

Read this article—and others at Business Insurance—to learn about the varieties of uses, terms, requirements, and kinds of insurance available to businesses. After reading this article, think about how industries other than insurance have changed in response to technological advances.

Notes

The Bottom Line

Summary of chapter learning objectives:

| Learning Objective | Summary | Example |
|---|---|---|
| 18.1 | Compute costs and savings for auto insurance | Drivers A and B live in a state in which no-fault insurance is mandatory. Both drivers carry all three classifications of insurance. Driver A has a deductible of $500; driver B has a deductible of $200. Driver A crashes into driver B. Neither auto has any passengers. Car A has $1,800 in damages; car B has $2,000 in damages. Driver A is not hurt; driver B has $900 in medical bills.
1. How much does driver A's insurance company pay?
2. How much does driver B's insurance company pay? |
| 18.2 | Compute auto insurance premium rates for high- and low-risk drivers | 3. Juan has an excellent driving record and receives a 10% discount on his annual premium. Dave has a record of numerous tickets and must pay one and a half times the normal annual premium rate. If the normal premium for each driver is $1,500, how much more does Dave pay for his insurance than Juan pays? |
| 18.3 | Compute short-rate refunds | 4. XYZ company purchased a delivery truck and paid an annual insurance premium of $3,600. XYZ company sold the truck at the end of 8 months and canceled the policy. The insurance company charges a 10% penalty for short-rate refunds. What was the amount of the short-rate refund to XYZ company? |
| 18.4 | Compute coinsurance on property losses | 5. A building worth $200,000 is insured for $150,000 with an 80% coinsurance clause. Fire causes $140,000 in damage. How much does the insurance company pay? |
| 18.5 | Compute life insurance premiums | 6. Premiums per $1,000 of straight life insurance at the age of 25 are as follows: annual, $17.20; semiannual, $8.94; and quarterly, $4.73. What will be the total premiums for the following three policies: $50,000, annual; $25,000, semiannual; and $20,000, quarterly? |
| 18.6 | Compute cash surrender and loan values | 7. If cash surrender values for year 15 of a policy are $181 per thousand dollars of coverage for straight life and $380 per thousand dollars of coverage for 20-payment life, what is the total cash surrender value of these two policies: $50,000 straight life and $30,000 20-payment life? |
| 18.7 | Compute medical insurance contributions and reimbursements | 8a. An employer provides group health coverage that includes a $200 annual deductible per family and payment of 80% of costs over deductible. How much would an employee with four dependents pay if his year's medical bills were as follows: self, $240; dependent 1, $170; dependent 2, $30; dependent 3, $460; and dependent 4, $2,200?
b. How much would the employee pay if the annual deductible were $50 per person? |

Answers: 1. $3,300 **2.** $900 **3.** $900 **4.** $840 **5.** $131,250 **6.** $1,685.40 **7.** $20,450 **8a.** $780 **8b.** $804

1 Drivers Jim Olson and Joshua Stein live in a state having no-fault auto insurance. Joshua causes an accident by hitting Jim's car. Joshua isn't hurt. Jim spends 3 days in the hospital at a cost of $5,300. Compute the amount each that driver's insurance company pays toward medical expenses. _____

2 IXP insured an office building for $290,000 for 1 year at a premium rate of $7.20 per thousand. At the end of 9 months IXP sold the building and canceled the policy. If the insurance company has a short-rate refund policy that includes a 10% penalty, how much refund did IXP receive? _____

3 Driver Devon Cooper has a poor driving record and pays double the usual premium as a high-risk driver. The regular premium would be $490 for a year. If Devon must pay the high-risk premium every year for 5 years, how much more will he pay for insurance premiums than a low-risk driver receiving a 10% discount over the same 5-year period? _____

4 Insurance company A has a standard 90% coinsurance clause for all fire insurance coverage. Insurance company B has a standard 75% coinsurance clause for all fire insurance coverage. A building is valued at $195,000. How much more insurance coverage would insurance company A require than insurance company B for full coinsurance coverage? _____

5 The Morgan Company warehouse was valued at $425,000. The building was insured for $170,000. The policy contained an 80% coinsurance clause. A fire caused $180,000 in damages. Compute the amount of the fire damage the Morgan company had to pay. _____

6 Mike Jankowski, age 27, purchased a $35,000, 20-payment life policy with premiums payable annually. John Jamison, also age 27, purchased a $35,000 straight life policy with premiums payable semi-annually. Both Mike and John lived 40 more years. How much more in premiums did John pay the insurance company during his lifetime than Mike paid during his? _____

7 Sally Munson, age 25, purchased a $35,000, 20-payment life policy. Five years later she needed cash. Compute the maximum amount she could borrow on the policy. _____

8 An employer provided group health coverage that includes a $300 annual deductible per family and payment of 90% of costs exceeding the deductible amount. An employee with no dependents incurs $1,600 in medical expenses during the year. How much of the medical costs must the employee pay? _____

Assignment 18.1: Auto Insurance

LEARNING
OBJECTIVES

Name _____

Date _____ Score _____

A **(50 points) Solve the following problems. (5 points for each correct answer)**

1. Mary Johnson had full insurance coverage. Her liability and property damage coverage was $100,000 per accident. Her collision insurance had a $500 deductible clause. She struck two cars. Damages to the cars were $640 and $320. Damage to her own car was $470. Her annual insurance premium was $1,180.

 a. What are the total costs to the insurance company for Mary's accident? _____

 b. If this was the only accident that Mary had this year, how much money did the insurance company make on her? _____

 c. What are Mary's total costs this year for insurance and the accident? _____

 d. What would Mary's total costs for the accident have been without insurance? _____

2. Renaldo Garcia paid an annual premium of $1,380 for auto collision insurance with a $500 deductible clause. His steering went out, and he hit a tree causing $2,135 damage to his car. How much did he save this year by having insurance? _____

3. Sean O'Day received his driver's license one year ago. He has had three citations for speeding, but no accidents. His insurance premium last year was $1,250. This year his premium will be 50% higher because of his driving record.

 a. What will be the amount of his premium this year? _____

 b. Four months into the next year, Sean has continued his unsafe driving habits. The insurance company is canceling his policy. What will be the amount of the refund? _____

 c. Sean O'Day has found an insurance company that will insure him as a high-risk driver at triple the standard annual rate of $1,425. What will be his average monthly insurance premium for the first 28 months of his driving career? (Round your answer to the nearest dollar.) _____

 d. If Sean had been a careful driver and kept the amount of his premium unchanged, how much would he have saved in these first 28 months? (Round your computations to the nearest dollar.) _____

4. Drivers A and B have identical insurance coverage. Driver A has an excellent driving record and receives a 15% discount on the standard premium. Driver B has numerous citations and pays 50% above the standard rate. The standard rate in both cases is $1,430. How much more does driver B pay for insurance than driver A? _____

Score for A (50)

B **(50 points) Solve the following problems. (5 points for each correct answer)**

5. Tom Barton carries liability and property damage insurance coverage up to $50,000 per accident, comprehensive insurance, and collision insurance with a $100 deductible clause. He lost control of his car and drove through the display window of a furniture store. Damage to the building was $17,200 and to the inventory was $34,300. Damage to a bike rack on the sidewalk and three bicycles in the rack was $1,840. Damage to his own car was $6,100.

 a. What was the total property damage, excluding damage to Tom's car? _____

 b. How much did the insurance company pay for property damage, excluding damage to Tom's car? _____

 c. How much did the insurance company pay for damage to Tom Barton's car? _____

 d. How much did the accident cost Tom Barton? _____

 e. If Tom Barton had been in a previous accident this year in which there had been property damage to a parked car of $12,700, how much would the insurance company have paid for damages to everything in the current accident, including Tom Barton's car? _____

6. Amy Tan and John Rogers live in a state in which no-fault insurance is mandatory. They have identical full coverage of $50,000 liability and property damage per accident, comprehensive insurance, and collision insurance with a $350 deductible. John lost control of his car on an icy street and struck Amy's car, a parked motorcycle, and a fence. Amy had medical expenses of $780. John had medical expenses of $560. Amy's car had damages of $1,350. John's car had damages of $1,750. Damage to the parked motorcycle was $650 and to the fence was $320.

 a. What did Amy's insurance company pay under the no-fault provision? _____

 b. What did John's insurance company pay under the no-fault provision? _____

 c. How much did John's insurance company pay under his liability and property damage coverage? _____

 d. How much did John's insurance company pay under his comprehensive coverage? _____

 e. How much would John's insurance company have paid under his liability and property damage if he had hit Amy's car and five parked cars, with total damage to the six cars of $56,700? _____

Score for B (50)

Assignment 18.2: Property Insurance

Name _____

Date _____ Score _____

A **(42 points) Solve the following problems. (6 points for each correct answer)**

1. A building valued at $380,000 is insured for its full value. The annual premium is $9.80 per thousand dollars of coverage.

 a. How much does the insured pay to insure his building? _____

 b. If the insurance company cancels the policy at the end of 3 months, how much refund does the insured

 receive? _____

 c. If the insurance company has a 10% penalty clause for short-rate refunds and the insured cancels the policy

 after 9 months, how much refund does the insured receive? _____

2. If a company pays an annual premium of $2,400 and the insurance company charges $12 per thousand dollars of insurance, how much insurance does the company carry? _____

3. A company carries property insurance of $200,000. A fire causes $170,000 in damage. How much does the insurance company pay the insured? _____

4. A company carries property insurance of $100,000. A fire causes $120,000 in damage.

 a. How much does the insurance company pay the insured? _____

 b. What is the amount of the uninsured loss? _____

Score for A (42)

B (58 points) Solve the following problems. (points for correct answers as marked)

5. A building worth $300,000 is insured for $180,000, and the policy carries an 80% coinsurance clause. A fire causes $220,000 in damage.

 a. How much will the insurance company pay? (10 points) _____

 b. How much will the insured pay if the building is repaired for $220,000? (6 points) _____

 c. How much would the insurance company pay if damage to the building totaled $300,000? (10 points) _____

 d. If the damage totaled $300,000, how much would the insured pay if the building were rebuilt for $300,000? (6 points) _____

6. A building worth $350,000 is insured for $280,000, and the policy carries an 80% coinsurance clause. A fire causes $300,000 in damage.

 a. How much does the insurance company pay if the building is repaired for $300,000? (10 points) _____

 b. How much does the insured pay? (6 points) _____

7. If an insurance company issues insurance on property valued at $400,000 with a 90% coinsurance clause, what is the amount required to be carried by the insured? (5 points) _____

8. If an insurance company issues insurance on property valued at $200,000 with a 70% coinsurance clause, what is the amount required to be carried by the insured? (5 points) _____

Score for B (58)

Assignment 18.3: Life and Medical Insurance

Name _____

Date _____ Score _____

A (50 points) Refer to Figures 18-1 and 18-2 in solving the following problems. Assume that every year is a full 12 months long. (points for correct answers as marked)

1. Find the rates per thousand dollars and the premiums for the following policies. (1 point for each correct answer)

| Age | Type | Payments Made | Face Value of Policy | Rate per $1,000 | Premium Paid Each Year |
|-----|------|---------------|----------------------|-----------------|------------------------|
| 28 | Straight Life | Annually | $ 15,000 | _____ | _____ |
| 25 | 20-Payment Life | Quarterly | 60,000 | _____ | _____ |
| 25 | 20-Year Endowment | Semiannually | 20,000 | _____ | _____ |
| 26 | Straight Life | Quarterly | 100,000 | _____ | _____ |
| 27 | 20-Payment Life | Semiannually | 50,000 | _____ | _____ |
| 28 | 20-Year Endowment | Annually | 75,000 | _____ | _____ |

2. Find the cash surrender or loan value for each of the following policies issued at age 25. (1 point for each correct answer)

| Policy Year | Type of Policy | Amount of Policy | Cash Surrender or Loan Value |
|-------------|----------------|------------------|------------------------------|
| 10 | Straight Life | $50,000 | _____ |
| 15 | 20-Payment Life | $25,000 | _____ |
| 10 | 20-Year Endowment | $50,000 | _____ |
| 3 | Straight Life | $20,000 | _____ |
| 5 | 20-Payment Life | $75,000 | _____ |
| 4 | 20-Year Endowment | $60,000 | _____ |

3. When Sue Adams was 27 years old, she took out a $60,000, 20-year endowment policy. She paid the premiums annually and survived the endowment period. How much more did she pay in annual premiums than she received from the insurance company at the end of 20 years? (4 points) _____

4. Roger Johnson purchased a $50,000 ordinary life policy and an ADB for 50% of the value of the policy. In addition, he purchased a 5-year, $50,000 term policy. He died in an accident 3 years later.

 a. How much money did Roger's beneficiaries receive? (4 points) _____

 b. How much money would Roger's beneficiaries have received if he had died in an accident 7 years after purchasing the policies? (4 points) _____

 c. How much money would Roger's beneficiaries have received if he had died of natural causes 10 years after purchasing the policies? (4 points) _____

5. At the age of 25, Carlos Baker purchased a $50,000 straight life policy, with premiums payable annually. He also purchased a $25,000 20-payment life policy, with premiums payable semiannually. At the end of 15 years, he decided to cash in both policies.

 a. How much did Carlos receive for the straight life policy? (4 points) _____

 b. How much did Carlos receive for the 20-payment life policy? (4 points) _____

 c. How much more did Carlos pay in premiums than the total amount received for both policies? (8 points)

 Score for A (50)

B **(50 points) Solve the following problems. (10 points for a correct answer to problem 6; 8 points for each other correct answer)**

6. An employer provides group health coverage that includes a $250 annual deductible per family and payment of 80% of costs exceeding the deductible. How much would an employee with two dependents pay if her year's medical bills were $550 for herself; $920 for dependent 1; and $230 for dependent 2? _____

7. An employer provides group health coverage that includes a $400 annual deductible per family and 70% of costs over the deductible.

 a. How much would an employee with no dependents pay if his medical bills were $980 this year? _____

 b. How much would that employee have paid this year if his medical bills were $7,480? _____

8. An employer provides group health coverage with the following monthly premiums: employee only, $350; employee with one dependent, $450; and employee with multiple dependents, $550.

 a. How much does the employer pay over a 5-year period for an employee with multiple dependents? _____

 b. If that employee had a dependent with a catastrophic illness that cost $97,000 for hospitalization and treatments during that 5-year period, how much did the insurance company lose on that employee, assuming that she had no other medical claims? _____

 c. If an employee with no dependents had no illnesses during that same 5-year period, how much did the insurance company make on that employee? _____

 Score for B (50)

Notes

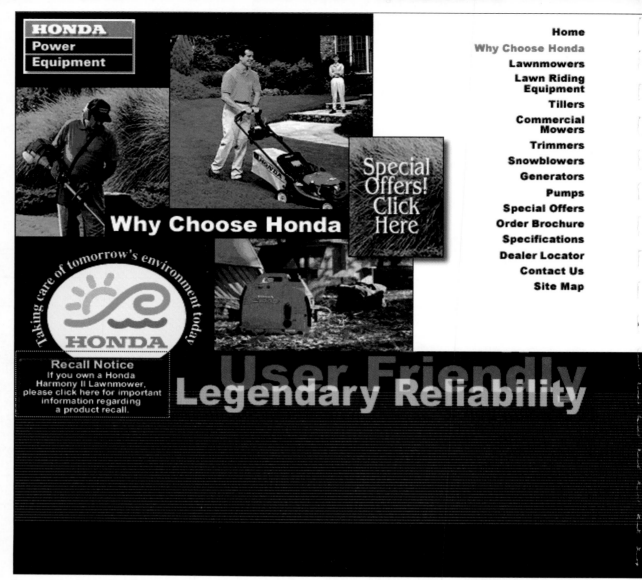

Honda makes other things besides cars! Honda began in the United States as a small motorcycle company in 1959. Today it is a top international manufacturer of motorcycles, power equipment, ATVs, generators, marine engines, and, of course, automobiles. Visit the company's site at http://www.hondapowerequipment.com to learn about its commitment to the environment, the community, and safety. Inventory management is a vital part of Honda's business—the ability to track, produce, and sell is essential to any successful producer and retailer of goods.

Business Inventory and Turnover

19

LEARNING OBJECTIVES

By studying this chapter and completing all assignments you will be able to:

 1 Account for inventory by inventory sheets and reports from a perpetual inventory system.

 2 Compute inventory value by the average cost, LIFO, and FIFO methods.

 3 Compute inventory by using the lower of cost or market value.

 4 Estimate inventory by computing an estimated cost of goods sold.

 5 Compute inventory turnover by using cost or retail values.

Large retail outlets have dominant positions in retailing because they carry huge inventories. Computing inventory and the rate of turnover is vital to these enterprises. Many methods are used to keep track of inventories.

1. If a store maintains an inventory system, does it have a choice of more than one way to keep track of inventory on hand?
2. What is inventory turnover and how is it computed?
3. Are cost of inventory and market value two ways of saying the same thing?
4. In the computation of inventory, what do FIFO and LIFO mean?

A company's inventory is the amount of goods it has on hand at any particular time. Retailers and wholesalers have only one kind of inventory—*merchandise,* which are the goods they sell that have been manufactured by some other firm.

Accounting for Inventory

LEARNING OBJECTIVES 1

Account for inventory by inventory sheets and reports from a perpetual inventory system.

Inventory Sheets

At least once each year, businesses undertake a **physical inventory**—an actual counting of the merchandise on hand. Some stores that require close control take a physical inventory every six months, quarterly, or even monthly. Sometimes retail stores use outside firms that specialize in taking inventory.

When inventory is counted, a description of each item, the quantity, the unit cost or retail price, and the **extension** (quantity × price) are recorded on an **inventory sheet**, as shown in Figure 19-1. The inventory value is then compared with accounting records, and any needed adjustments are made.

Figure 19-1: Inventory Sheet

WARREN'S AUTO PARTS
Inventory Sheet
April 30, 20—

| Description | Quantity | Unit Price (Average Cost) | Extension |
|---|---|---|---|
| Ignition terminals—#746083 | 318 | $36.14 | $11,492.52 |
| Odometer cables—#007614 | 73 | 9.97 | 727.81 |
| Wiper blades, compact—#417654 | 38 | 4.71 | 178.98 |
| Spark plugs, 0.14—#772034 | 354 | 2.34 | 828.36 |
| Hood/truck latches—#476508 | 58 | 13.42 | 778.36 |
| Total | | | $14,006.03 |

Perpetual Inventory Systems

Some firms keep a **perpetual inventory**—a running count of all inventory items, based on tracking each item as it comes into and goes out of inventory. In businesses that handle high-cost items, such as cars or large appliances, the perpetual system keeps track of each item by serial number and price.

Businesses that handle small items, such as candy bars or shoes, have difficulty identifying the cost or serial number of each specific item. Their perpetual inventory systems keep a count of the number of units on hand, not individual prices and serial numbers. The use of electronic scanners has greatly simplified application of this type of perpetual inventory system.

Data for a perpetual inventory system are usually kept on a computer. Figure 19-2 illustrates a computer printout of an inventory record sheet. The last item in the Balance on Hand column shows how many units are on hand on the 4/30 recording date—354 Quickstart spark plugs: 0.14, part number 772034.

Figure 19-2: Inventory Sheet

WARREN'S AUTO PARTS
Inventory Record Sheet

ITEM: QUICKSTART SPARK PLUG: 0.14

ORDER FROM:
Northwest Distributors
2337 Colfax Avenue
Milbrae, CA 93233
Phone—(415) 345-7654

PART NUMBER: #772034

LOCATION: Aisle 72, Bin 4, Box C

MINIMUM STOCK: 200 MAXIMUM STOCK: 800

ORDER: 100–800

| Purchase Orders (PO) | | | Inventory Control | | | | | |
|---|---|---|---|---|---|---|---|---|
| Date | PO No. | Quantity | Date | Source Code | Units In | Unit Cost | Units Out | Balance on hand |
| 2/03 | F0129 | 400 | 1/01 | — | | $2.10 | | 350 |
| 3/15 | M1678 | 300 | 1/31 | SJ01 | | | 120 | 230 |
| 3/22 | M2076 | 200 | 2/28 | SJ02 | | | 58 | 172 |
| 4/26 | A3210 | 400 | 3/02 | F0129 | 400 | 2.36 | | 572 |
| | | | 3/31 | SJ03 | | | 315 | 257 |
| | | | 4/03 | M1678 | 300 | 2.40 | | 557 |
| | | | 4/20 | M2076 | 200 | 2.64 | | 757 |
| | | | 4/30 | SJ04 | | | 403 | 354 |

Note: The 400 units ordered 4/26 have not yet been received.

CONCEPT CHECK 19.1

The CompuParts wholesale computer store maintains a perpetual inventory of computer parts received and removed or shipped out. The following inventory record sheet shows the data for May. Compute the balance on hand after each transaction.

COMPUPARTS WHOLESALE
Inventory Record Sheet

Order From:
Myers Distributors
1422 Oak Drive
Stockton, CA 97777
Fax: 209-775-7823

ITEM: Monitor Cord #A718
Location: Bin #C7
Minimum Stock: 250 Maximum Stock: 1,000

| | Units In | Unit Cost | Units Out | Balance on Hand |
|---|---|---|---|---|
| 5/01 | | $15.40 | | 390 |
| 5/11 | 470 | $15.80 | | 860 |
| 5/15 | | | 260 | 600 |
| 5/28 | 320 | $15.90 | | 920 |
| 5/31 | | | 410 | 510 |

Computing Inventory, Using the Average Cost, FIFO, and LIFO Methods

LEARNING OBJECTIVES

Compute inventory value by the average cost, LIFO, and FIFO methods.

In all inventory systems, the cost of the inventory on hand at the end of the period is called **ending inventory (EI)**. The ending inventory must be computed before financial statements can be prepared.

To compute ending inventory, a business usually adopts one of three cost methods: average cost, first-in, first-out (FIFO) or last-in, first-out (LIFO). Once selected, the method must be followed consistently. We use the cost data from Figure 19-2 to illustrate computations for the three cost methods.

The Average Cost Method

The **average cost method** is based on the assumption that the cost for each item on hand is the average cost for items from the opening inventory and items purchased during the period.

EXAMPLE A

The average cost of the units on the inventory record sheet for stock part #772034 (Quickstart spark plugs: 0.14) is computed as follows:

| Date | Units Purchased | Cost | Extension |
|------|-----------------|------|-----------|
| 1/01 | 350 | $2.10 | $ 735.00 |
| 3/02 | 400 | 2.36 | 944.00 |
| 4/03 | 300 | 2.40 | 720.00 |
| 4/20 | 200 | 2.64 | 528.00 |
| | 1,250 | | $2,927.00 |

Average cost per unit: $2,927 ÷ 1,250 = $2.34
From Figure 19-2, ending inventory at average cost: 354 units × $2.34 = $828.36

The FIFO Method

The **first-in, first-out (FIFO) costing method** is based on the assumption that the cost for units sold is determined in the order in which the units were purchased. Thus the cost of the inventory remaining is assumed to be based on the price of the units received most recently.

EXAMPLE B

Under the FIFO method, the inventory of 354 units shown in Figure 19-2 would consist of the 200 units last purchased plus 154 units from the preceding purchase.

Date

| 4/20 | 200 units × $2.64 = $528.00 |
|------|------------------------------|
| 4/03 | 154 units × $2.40 = $369.60 |
| | 354 $897.60 Ending inventory at FIFO cost |

The LIFO Method

The **last-in, first-out (LIFO) costing method** is based on the assumption that the cost of the inventory remaining is determined by the cost of the units purchased the earliest.

EXAMPLE C

Under the LIFO method, the 354 units shown on the inventory record in Figure 19-2 would consist of the 350 units on hand on 1/01 plus 4 units from the first purchase on 3/02.

Date

| | | |
|---|---|---|
| 1/01 | 350 units × $2.10 = | $735.00 |
| 3/02 | 4 units × $2.36 = | $ 9.44 |
| | 354 | $744.44 Ending inventory at LIFO cost |

CONCEPT CHECK 19.2

The inventory record sheets for Hairbrushes at Debbie's Beauty Supply show 5,000 units purchased (or on hand) at a total cost of $10,240. The inventory at year's end was 1,500 units. Compute the value of the ending inventory by each of the three methods: average cost, FIFO, and LIFO.

| Date | Units Purchased | Cost | Extension |
|---|---|---|---|
| 1/01 | 2,000 | $2.00 | $ 4,000 |
| 1/30 | 200 | 2.10 | 420 |
| 2/20 | 700 | 2.10 | 1,470 |
| 3/17 | 1,100 | 2.00 | 2,200 |
| 10/30 | 500 | 2.20 | 1,100 |
| 11/17 | 500 | 2.10 | 1,050 |
| | 5,000 | | $10,240 |

Average Cost: $10,240 ÷ 5,000 = $2.048
1,500 × $2.05 (rounded) = $3,075

FIFO: (500 × $2.10) + (500 × $2.20) + (500 × $2.00) = $3,150

LIFO: (1,500 × $2.00) = $3,000

Computing Inventory at the Lower of Cost or Market Value

Financial statements usually present the ending inventory at its cost value, computed by using the average, FIFO, or LIFO costing method. However, in some cases the **market value** (current replacement cost) of goods is lower than their original or average cost. Most companies prefer to show the **lower of cost or market value** in their inventories. When market value exceeds the cost, the cost is used; when the cost exceeds market value, market value is used.

LEARNING OBJECTIVES

Compute inventory by using the lower of cost or market value.

STEPS to Determine the Lower of Cost or Market (LCM) Inventory Value

1. Compute the unit or total cost for each type of inventory item, using the average, FIFO, or LIFO costing method.
2. Determine the market value for each inventory item.
3. Compare the cost value from Step 1 with the market value from Step 2 and choose the lower of the two.
4. Compute the extension amount for each item based on the lower amount.
5. Sum the amounts in Step 4 to determine the total inventory value under LCM.

EXAMPLE D

Under LCM, using the average cost method illustrated in example A, the total inventory shown in Figure 19-1 would be valued at $13,802.13.

| | | STEP 1 | STEP 2 | STEP 3 | STEP 4 | |
| | **(A)** | **(B)** Unit Price (Average Cost) | **(C)** Unit Price at Market | **(D)** Lower of (B) or (C) | | |
| **Description** | **Quantity** | | | | **Extension (A × D)** | |
| Ignition terminals—#746083 | 318 | $36.14 | $35.50 | $35.50 | $11,289.00 | Market |
| Odometer cables—#007614 | 73 | 9.97 | 11.00 | 9.97 | 727.81 | Cost |
| Wiper blades, compact—#417654 | 38 | 4.71 | 4.70 | 4.70 | 178.60 | Market |
| Spark plugs, 0.14—#772034 | 354 | 2.34 | 2.64 | 2.34 | 828.36 | Cost |
| Hood/trunk latches—#476508 | 58 | 13.42 | 14.10 | 13.42 | 778.36 | Cost |
| Total | | | | | $13,802.13 | STEP 5 |

EXAMPLE E

Under LCM, using the FIFO cost method illustrated in example B, the FIFO cost for the inventory for Quickstart spark plugs would be $897.60. Combining LCM with FIFO for the Quickstart spark plugs illustrated in example B, the ending inventory for this one item would be valued at $897.60 because the market value ($934.56) is higher than the FIFO cost.

| | | | **Market Value** | | |
| | | | **(C)** | **(D)** | |
| | **(A)** | **(B)** | Unit Price | Total | Lower of |
| **Description** | **Quantity** | **FIFO Cost** | **at Market** | **(A × C)** | **(B) or (D)** |
| Spark plugs | 354 | $897.60 | $2.64 | $934.56 | $897.60 Cost |

CONCEPT CHECK 19.3

L & L Records' inventory shows the following. Compute the inventory value at the lower of cost or market.

| **Description** | **Quantity** | **Cost** | **Market** | **Extension** | |
| Classical #3 | 300 | $ 7.07 | $10.10 | $2,121.00 | Cost |
| Western #8 | 180 | 9.10 | 8.07 | 1,452.60 | Market |
| Modern—light #11 | 410 | 11.17 | 12.08 | 4,579.70 | Cost |
| Rock—new #4 | 89 | 12.10 | 12.10 | $1,076.90 | Cost/market |
| Total | | | | $9,230.20 | |

Estimating Inventory Value

For monthly financial statements, inventory frequently is estimated without a physical count or a perpetual inventory system. The method usually used to estimate month-end inventory is called the **gross profit method**. This method involves estimating the cost of goods sold and subtracting this amount from the sum of the opening inventory and purchases made during the month. Note that **beginning inventory (BI)** is the ending inventory from the month before and **purchases (P)** are those goods for sale that have been purchased during the current month. The gross profit method is based on the formula

LEARNING OBJECTIVES

Estimate inventory by computing an estimated cost of goods sold.

> Beginning inventory (BI)
> + Purchases (P)
> ————————————————
> Cost of goods available for sale
> − Cost of goods sold (CGS) (estimated)
> ————————————————
> Ending inventory (EI) (estimate)

Without a physical inventory, a precise cost of goods sold can't be determined. In this case, it is estimated by applying a markup percentage rate to **net sales** (total sales less sales returned and adjustments for the period). The net sales (100%) less this markup rate (percent) equals the cost of goods sold (percent). For instance, if the markup rate were 30%, the cost of goods sold would be 100% − 30% = 70%. If the rate of markup were 40%, the cost of goods sold would be 100% − 40% = 60%.

EXAMPLE F

Assume that Warren's Auto Parts had a beginning inventory of $80,000. During the month, the company purchased and received $50,000 in goods and had net sales of $90,000. Throughout the month, Warren's maintained a 40% markup on all sales. Its cost of goods sold would be computed as follows.

| | | |
|---|---|---|
| Net sales for the month | $90,000 | |
| Cost of goods sold (estimated) | $54,000 | [$90,000 × (100% − 40%) = $90,000 × 0.60] |

Warren's Auto Parts would then determine its ending inventory (estimated) as follows.

| | |
|---|---|
| Inventory, beginning of month | $ 80,000 |
| Purchases for month | + 50,000 |
| Goods available for sale | $130,000 |
| Cost of goods sold (estimated) | − 54,000 |
| Ending inventory (estimated) | $ 76,000 |

Sometimes a company's markup rate is based on cost rather than selling price. In this case, if the markup on cost were 30%, the cost of goods sold would be net sales divided by 130%. If the markup on cost were 40%, the cost of goods sold would be net sales divided by 140%.

Assume that Warren's Auto Parts had a beginning inventory of $80,000. During the month, it had purchases of $50,000 and net sales of $90,000. Throughout the month, Warren's maintained a markup of 50% based on cost. What were Warren's cost of goods sold and ending inventory?

| | |
|---|---|
| Beginning inventory | $80,000 |
| Purchases | + 50,000 |
| Cost of goods available for sale | $130,000 |
| Cost of goods sold (estimated) | − 60,000 ($90,000 ÷ 150%) |
| Ending inventory (estimated) | $70,000 |

 CONCEPT CHECK 19.4

C & S Electronics records show the following. Compute the estimated ending inventory at cost.

| | | | |
|---|---|---|---|
| Beginning inventory | $24,000 | Net sales for period | $60,000 |
| Purchases for period | $33,000 | Markup based on retail | 40% |

$24,000 + $33,000 = $57,000 cost of goods available
$60,000 × 60% = $36,000 cost of goods sold
$57,000 − $36,000 = $21,000 ending inventory

Computing Inventory Turnover

LEARNING OBJECTIVES 5

Compute inventory turnover by using cost or retail values.

Inventory turnover is the number of times the average inventory is converted into sales during the year. Inventory turnover is very high for a grocery store or ice cream parlor; it is very low for a specialty jewelry store or an antique shop.

Before turnover can be determined, average inventory must be calculated. **Average inventory** is the average of the inventories taken over a specific period of time—annually, semiannually, quarterly, or monthly.

| Inventory is taken | Average inventory (at retail or cost) |
|---|---|
| Annually (once a year) | (BI + EI) ÷ 2 |
| Semiannually (every six months) | (BI + end of 6 months + EI) ÷ 3 |
| Quarterly (every three months) | (BI + 3 quarterlys + EI) ÷ 5 |
| Monthly (every month) | (BI + 11 monthlys + EI) ÷ 13 |

Computation of inventory turnover can be based on either retail (selling) price or cost. Inventory turnover at retail is net sales divided by average inventory.

STEPS to Compute Inventory Turnover at Retail

1. Determine net sales.
2. Compute average inventory using retail price.
3. Compute inventory turnover at retail: Net sales ÷ Average inventory at retail.

EXAMPLE H

Assume that inventories for the year, based on selling price, are as follows: beginning, $90,000; end of month 3, $80,000; end of month 6, $100,000; end of month 9, $70,000; and end of month 12 (ending), $60,000. Net sales for the year are $520,000. Compute the inventory turnover at retail.

STEP 2

$$\text{Average inventory} = (\$90,000 + \$80,000 + \$100,000 + \$70,000 + \$60,000) \div 5$$
$$= \$400,000 \div 5 = \$80,000$$

STEP 3

$$\text{Inventory turnover at retail} = \$520,000 \text{ net sales} \div \$80,000 \text{ average inventory} = 6.5 \text{ times}$$

Note: The value of net sales and average inventory must both be figured at retail.

Some retailers prefer to express their rate of inventory turnover in terms of cost. **Inventory turnover at cost** is obtained by dividing the **cost of goods sold (CGS)** during a period by the average inventory for the same period computed at cost prices. (CGS is simply net sales at cost.)

© PHOTODISC, INC.

STEPS to Compute Inventory Turnover at Cost

1. Compute the cost of goods sold using the formula BI + P − EI = CGS.
2. Compute the average inventory at cost.
3. Compute inventory turnover at cost: Cost of goods sold ÷ Average inventory at cost.

www.muextension. missouri.edu/xplor/ miscpubs/mp0676.htm

Read more about inventory turnover at this Web site.

EXAMPLE I

Assume that beginning inventory cost $60,000, purchases cost $300,000, and ending inventory cost $80,000. Compute the inventory turnover at cost.

STEP 1

| Cost of goods sold: | Inventory at beginning of year | $ 60,000 |
|---|---|---|
| | Purchases during year | + 300,000 |
| | Goods available for sale | $ 360,000 |
| | Inventory at end of year | − 80,000 |
| | Cost of goods sold | $ 280,000 |

STEP 2

$$\text{Average inventory} = (\$60,000 \text{ BI} + \$80,00 \text{ EI}) \div 2$$
$$= \$140,000 \div 2 = \$70,000$$

STEP 3

$$\text{Inventory turnover at cost} = \$280,000 \text{ cost of goods sold} \div \$70,000 \text{ average inventory} = 4.0 \text{ times}$$

Note: The value of goods sold and average inventory must both be figured at cost.

 CONCEPT CHECK 19.5

Brinkman Scooter Shop has two branches (A and B), each using a markup of 50% of retail. Compute ending inventory, average inventory, and inventory turnover based on retail for each branch from the following data.

| | Branch A | Branch B |
|---|---|---|
| Net sales | $1,400,000 | $1,200,000 |
| Beginning inventory | 220,000 | 300,000 |
| Inventory (March 31) | 190,000 | 400,000 |
| Inventory (June 30) | 280,000 | 350,000 |
| Inventory (September 30) | 280,000 | 360,000 |
| Inventory (December 31) | 200,000 | 300,000 |

Average inventory: A—($220,000 + 190,000 + 280,000 + 280,000 + 200,000) ÷ 5 = $234,000
B—($300,000 + 400,000 + 350,000 + 360,000 + 300,000) ÷ 5 = $342,000

Retail turnover: A—$1,400,000 ÷ $234,000 = 5.98
B—$1,200,000 ÷ $342,000 = 3.51

COMPLETE ASSIGNMENTS 19.1 and 19.2.

Chapter Terms for Review

average cost method
average inventory
beginning inventory (BI)
cost of goods sold (CGS)
ending inventory (EI)
extension
first-in, first-out (FIFO) costing method
gross profit method
inventory sheet
inventory turnover

inventory turnover at cost
inventory turnover at retail
last-in, first-out (LIFO) costing method
lower of cost or market value (LCM)
market value
net sales
perpetual inventory
physical inventory
purchases (P)

Visit the Web

http://www.anjoch-investors.com/new98/indexn.html?../mp22 and http://www.bizmove.com/general/m6m.htm—These Web sites contain interesting and useful articles on inventory turnover.

http://www.osc.state.ct.us/manuals/ActivityWelfare/manual.htm#Inventory—The State of Connecticut's Web site specifies how to account for new inventory and how to report losses of state property.

CYBERCONNECTION

Inventory Turnover
Keep your business on track by keeping track of inventory turnover.

When you have replaced 100 percent of your original inventory, you have "turned over" your inventory. If you have, on the average, a 12-week supply of inventory and turn it over four times a year, the count cycle plus the order cycle plus the delivery cycle add up to your needs period. Expressed as an equation, it would read:

Count Cycle + Order Cycle + Delivery Cycle = Needs Period

For instance, suppose you decided to count inventory once every four weeks (the count cycle). Processing paperwork and placing orders with your vendors take two weeks (the order cycle). The order takes six weeks to get to you (delivery cycle). Therefore, you need 12 weeks' worth of inventory from the first day of the count cycle to stay in operation until your merchandise arrives.

You can improve your inventory turnover, however, if you count inventory more often—every two weeks instead of every four—and work with your suppliers to improve delivery efficiency. Alternate ways of distributing goods to the store could cut the delivery cycle down to three weeks, which would cut inventory needs to six weeks. As a result, inventory turnover could increase from four times a year to eight times.

Another way to look at turnover is by measuring sales per square foot. Taking the average retail value of inventory and dividing it by the number of square feet devoted to a particular product will give you your average sales per square foot. You should know how many sales per square foot per year you need to survive. Calculate your sales per square foot once a month to make sure they are in line with your expectations.

Source: http://www.entrepreneurmag.com/Your_Business/YB_SegArticle/0,1314,265119,00.html?. Excerpted from *Start Your Own Business: The Only Start-Up Book You'll Ever Need,* by Rieva Lesonsky and the Staff of Entrepreneur Magazine. Copyright © 1998 by Entrepreneur Press; used with permission.

Read the article. If a company with a 1-month order/delivery cycle switched from a quarterly to a monthly inventory cycle, how much would it improve its annual inventory turnover?

CYBERCONNECTION

MJ Clothiers, Inc.
A retail inventory management simulation
Montana State University-Bozeman College of Business

As the retail manager at MJ Clothiers your ultimate goal is to maximize profits which you accomplish through ordering the appropriate stock in anticipation of demand, and setting prices at a level sufficient to earn an acceptable profit. In addition to general administrative overhead, other costs are incurred when you purchase stock (includes order costs), hold inventory, stock out and pay interest on flooring loans.

Use the inventory button below to review current inventory levels, review historical demand and make ordering decisions. Use the prices button to check your current costs and make decisions about what prices to charge for your products. Use the results button to e-mail the simulation results to your instructor. Weekly orders will be delivered, demand will be met and your profitability will be calculated when you use the next button to move forward in time.

Your customers are expecting you to provide the products they want at acceptable prices and will shop elsewhere if they aren't satisfied. You are able to order stock and set retail prices as you see appropriate, but your inventory and pricing policies will affect current and future demand for your products. Other instructions are provided as necessary throughout the simulation.

 Go to the COB home page Go to the MSU Home Page

Send responses, comments and/or suggestions to Bruce C. Raymond *via e-mail*

Week # 1 | Introduction | Inventory | Prices | Results | Next Wk | Scores |

Source: http://www.montana.edu/~wwwdb/BUS331/MJFrame.htm. Courtesy of Bruce C. Raymond and Montana State University College of Business.

Visit this site for a retail inventory simulation created by Bruce C. Raymond and the Montana State University College of Business.

The Bottom Line

Summary of chapter learning objectives:

| Learning Objective | Summary | Example |
|---|---|---|
| **19.1** | Account for inventory by inventory sheets and re-ports from a perpetual inventory system | 1. Compute the Balance on Hand after each transaction: |
| **19.2** | Compute inventory value by the average cost, LIFO, and FIFO methods | 2. From the data shown, compute the ending inventory by the average cost, FIFO, and LIFO methods for Redwood Stove Company's stove part #717. The ending inventory, by physical count, was 300. |
| **19.3** | Compute inventory by using the lower of cost or market value | 3. Compute Redwood Stove Company's inventory value at the lower of cost or market value. |
| **19.4** | Estimate inventory by computing an estimated cost of goods sold | 4. Redwood Stove Company has a markup of 50% of retail. Last year it had total sales of $400,000. It had a beginning inventory of $150,000 based on cost. It purchased merchandise for $180,000 during the year. Compute the ending inventory at cost. |
| **19.5** | Compute inventory turnover by using cost or retail values | 5. Two years ago Redwood Stove Company used a markup of 65% of cost. That year's data are shown. Compute ending inventory, average inventory, and inventory turnover at retail. |

Example 1.

| Date | Units In | Units Out | Balance on Hand |
|---|---|---|---|
| 12/01 | | | 34,768 |
| 12/17 | 7,789 | | _____ |
| 12/19 | | 17,072 | _____ |
| 12/20 | 11,789 | | _____ |
| 12/31 | | 14,490 | _____ |

Example 2.

Stove Part #717

| Date | Units In | Cost | Extension | Ending Inventory Value: |
|---|---|---|---|---|
| 1/12 | 200 | $3.00 | $600 | |
| 1/14 | 300 | 3.20 | 960 | Average cost: _____ |
| 1/15 | 500 | 3.00 | 1,500 | FIFO: _____ |
| 1/17 | 200 | 3.10 | 620 | LIFO: _____ |
| 1/18 | 400 | 3.00 | 1,200 | |
| Total | 1,600 | | $4,880 | |

Example 3.

| Description | Quantity | Cost | Market | Extension |
|---|---|---|---|---|
| Stoves | 24 | $277.50 | $350.50 | _____ |
| Piping | 90 | 34.50 | 27.00 | _____ |
| Hearths | 75 | 78.00 | 78.00 | _____ |
| Screens | 50 | 105.00 | 125.00 | _____ |
| Tool Sets | 28 | 65.50 | 55.00 | _____ |
| Total | | | | _____ |

Example 5.

| | | | |
|---|---|---|---|
| Net sales | $900,000 | Purchases (cost) | 600,000 |
| Beginning inventory—retail | 300,000 | Inventory—retail (June 30) | 450,000 |

Answers: 1. 42,557; 25,485; 37,274; 22,784 **2.** Average cost, $915; FIFO, $900; LIFO, $920 **3.** $6,660; $2,430; $5,850; $5,250; $1,540; $21,730 **4.** Inventory $130,000 **5.** Ending inventory at retail, $390,000; average inventory $380,000; turnover 2.37

Self-Check | Review Problems for Chapter 19

1 The D&D Company has 45 units on hand January 1. During the month, units in total 320 and units out total 285. What is the balance on hand January 31? _____

2 According to physical count, Dawson Lumber had 3,250 units in inventory March 31. Dawson Lumber's beginning inventory and purchases for the first quarter were as follows.

| Jan. 1 | Beginning Inventory | 2,500 units @ $25.00 |
|--------|---------------------|----------------------|
| Jan. 15 | Purchased | 5,000 units @ $27.50 |
| Feb. 5 | Purchased | 6,000 units @ $26.25 |
| Mar. 10 | Purchased | 3,000 units @ $27.00 |

Calculate the value of the inventory March 31 and cost of goods sold for the quarter based on the average, FIFO, and LIFO costing methods.

| | Inventory Value | Cost of Goods Sold |
|---|---|---|
| a. Average cost: | _____ | _____ |
| b. FIFO cost: | _____ | _____ |
| c. LIFO cost: | _____ | _____ |

3 Lansky Company's inventory January 1 was valued at $41,000. During the first quarter $365,000 of goods were purchased, and sales totaled $550,000. Estimate the inventory March 31 if Baxter's markup is 40% based on selling price. _____

4 Compute the average inventory, cost of goods sold, and turnover based on cost, using the following data. Kelly Pet Supplies takes inventory every 6 months and had inventory of $35,000 on January 1, $42,600 on June 30, and $38,200 on December 31. Kelly's purchased goods totaling $275,000 during the year and had sales of $390,000.

a. Average inventory: _____

b. Cost of goods sold: _____

c. Turnover: _____

Assignment 19.1: Inventory Cost

A **(40 points) Compute the extensions and totals. (1 point for each correct answer)**

1. The inventory of Michelle's Clock Shop shows the following items, at both costs and market prices. Determine the total value of the inventory at the lower of cost or market price for each item.

| Description | Quantity | Unit Cost Price | Unit Market Price | Extension at Lower of Cost or Market |
|---|---|---|---|---|
| Quartz clock and pen set | 22 | $36.00 | $34.80 | _____ |
| Travel alarm clock | 42 | 15.60 | 19.20 | _____ |
| Ultrasonic travel clock | 16 | 23.00 | 23.70 | _____ |
| Digital alarm clock | 40 | 19.80 | 18.60 | _____ |
| AM/FM clock radio | 85 | 21.00 | 21.00 | _____ |
| Digital clock radio | 9 | 54.00 | 57.50 | _____ |
| Total | | | | _____ |

2. A retail furniture dealer counted the following goods in inventory on December 31. An accountant recommended that the inventory items be valued at the lower of cost or market price. Compute the total value of the inventory based on the lower of cost or market price.

| Article | Quantity | Unit Cost Price | Extension at Cost | Unit Market Price | Extension at Market | Inventory Value at Lower of Cost or Market |
|---|---|---|---|---|---|---|
| Armchairs, wood | 24 | $ 40.00 | _____ | $ 68.50 | _____ | _____ |
| Armchairs, tapestry | 6 | 75.00 | _____ | 105.00 | _____ | _____ |
| Armchairs, Windsor | 12 | 115.00 | _____ | 85.00 | _____ | _____ |
| Beds, bunk | 8 | 85.50 | _____ | 75.00 | _____ | _____ |
| Bedroom suites | 3 | 297.50 | _____ | 410.00 | _____ | _____ |
| Tables, coffee | 30 | 63.00 | _____ | 62.00 | _____ | _____ |
| Chairs, kitchen | 24 | 23.00 | _____ | 32.00 | _____ | _____ |
| Dining tables | 8 | 117.40 | _____ | 95.70 | _____ | _____ |
| Dining suites | 5 | 288.80 | _____ | 395.00 | _____ | _____ |
| Sofa sets | 9 | 479.60 | _____ | 325.00 | _____ | _____ |
| Total | | | _____ | | _____ | _____ |

Score for A (40)

Assignment 19.1 *Continued*

B **(60 points) Compute the value of ending inventory. (Problems 3 and 4—8 points for each correct answer. Problem 5(a)—1 point each; problem 5(b)—4 points)**

3. Garcia Manufacturing Company made purchases of a material as shown in the following listing. The inventory at the end of the year was 3,500 units. Compute the value of the inventory by each of the three methods: (a) average cost; (b) first-in, first-out; and (c) last-in, first-out.

| Date | Units | Unit Cost | Total Cost |
|------|-------|-----------|------------|
| Jan 5 | 3,600 | $6.20 | $ 22,320 |
| Mar. 11 | 3,000 | 5.80 | 17,400 |
| May 14 | 5,300 | 6.00 | 31,800 |
| July 8 | 1,600 | 6.30 | 10,080 |
| Sept. 7 | 4,000 | 6.20 | 24,800 |
| Nov. 10 | 2,500 | 6.40 | 16,000 |
| Total | 20,000 | | $122,400 |

a. Average cost: _____

b. First-in, first-out: _____

c. Last-in, first-out: _____

4. The Willand Company had 320 units on hand at the beginning of the year, with a unit cost of $4.20. The number and unit cost of units purchased and the number of units sold during the year are shown. What would be the value of the ending inventory of 380 units based on the (a) average cost; (b) first-in, first-out; and (c) last-in, first-out costing methods?

| Date | Units Purchased | Unit Cost | Units Sold | Units on Hand |
|------|-----------------|-----------|------------|---------------|
| Jan. 1 | | $4.20 | | 320 |
| Feb. 2 | | | 190 | 130 |
| Apr. 16 | 200 | $4.32 | | 330 |
| June 10 | 300 | $4.40 | | 630 |
| Aug. 5 | | | 280 | 350 |
| Oct. 12 | 250 | $4.48 | | 600 |
| Nov. 27 | | | 220 | 380 |

a. Average cost: _____

b. First-in, first-out: _____

c. Last-in, first-out: _____

Assignment 19.1 *Continued*

5a. Complete the following inventory control sheet for Item Number 237P by determining the balance on hand after each transaction.

Inventory Control

| Date | Units In | Unit Cost | Units Out | Balance on Hand |
|------|----------|-----------|-----------|-----------------|
| Jan. 01 | | $25.00 | | 32 |
| Jan. 08 | | | 12 | _____ |
| Jan. 12 | | | 8 | _____ |
| Jan. 15 | 50 | $27.50 | | _____ |
| Jan. 18 | | | 17 | _____ |
| Jan. 21 | | | 22 | _____ |
| Jan. 27 | | | 15 | _____ |
| Jan. 30 | 65 | $26.00 | | _____ |
| Jan. 31 | | | 25 | _____ |

b. If the inventory sheet shows 45 units of item 237P on the shelf by physical count on January 31, will the number of units on the control sheet need to be adjusted? _____

If so, by how many units? _____

Score for B (60)

Notes

Assignment 19.2: Inventory Estimating and Turnover

Name _____

Date _____ Score _____

A **(50 points) Solve the following problems. (2 points for each correct answer)**

1. Fill in the blanks in each of the following calculations with the correct amount. Use the formulas

 Beginning inventory + Purchases = Goods available for sale

 Goods available for sale − Cost of goods sold = Ending inventory

 | | Store A | Store B | Store C | Store D | Store E |
 |---|---|---|---|---|---|
 | Beginning inventory | $ 80,000 | _____ | $ 37,000 | _____ | $42,000 |
 | Purchases | _____ | 90,000 | _____ | 21,000 | _____ |
 | Goods available for sale | 200,000 | 210,000 | 109,000 | 117,000 | 89,000 |
 | Less cost of goods sold | 125,000 | 128,000 | _____ | 27,000 | 74,000 |
 | Ending inventory | _____ | _____ | $ 23,000 | _____ | _____ |

2. Each of the five stores in problem 1 had the net sales shown. What was the average percent of markup, based on cost, for each of the five stores? What was the average percent of markup, based on selling price, for each of the five stores?

 | | Store A | Store B | Store C | Store D | Store E |
 |---|---|---|---|---|---|
 | Net sales | $200,000 | $150,000 | $172,000 | $40,000 | $100,000 |
 | Markup—cost | _____ | _____ | _____ | _____ | _____ |
 | Markup—selling price | _____ | _____ | _____ | _____ | _____ |

Assignment 19.2 *Continued*

3. The Country Kitchen takes inventory at retail sales price every 3 months. Its inventory at the beginning of last year was $40,500; at three months, $48,925; at six months, $50,680; at nine months, $52,830; and at the end of the year, $40,645. Net sales for the year were $285,927.

 a. What was the average inventory? _____

 b. What was the turnover? _____

4. Steve's Auto Shop began the year with an inventory of $78,250. Purchases during the year totaled $278,182. The inventory at the end of the year was $65,130 (all at retail).

 a. What was the cost of goods sold? _____

 b. What was the average inventory? _____

 c. What was the turnover? _____

Score for A (50)

B **(50 points) Solve the following problems. (Points for each correct answer as marked)**

5. Jackson Wholesalers' records showed these figures.

| | Cost | Retail Price | | |
|---|---|---|---|---|
| Beginning inventory | $19,793 | $32,990 | Net sales for the year | $61,450 |
| Purchases for the year | $47,200 | $78,665 | Markup based on sales | 40% |

 Compute the ending inventory:

 a. At cost: _____

 b. At retail price: _____

Assignment 19.2 *Continued*

6. The JM Clothing store kept all merchandise records in terms of selling price. On July 1, the JM books showed the following information.

| | |
|---|---|
| Beginning inventory, January 1: | $23,500 |
| 6-month purchases: | 99,000 |
| 6-month net sales: | 87,800 |

What was the estimated ending inventory on July 1? (5 points) _____

7. The Kid's Land Clothing Store kept all purchase and inventory records on a cost basis. The owner marked up all goods at 37.5% of the cost price. On July 1, the Kid's Land books showed the following information.

| | |
|---|---|
| Beginning inventory, January 1: | $104,500 |
| 6-month purchases: | 215,500 |
| 6-month net sales: | 220,000 |

What was the estimated inventory, at cost, on July 1? (5 points) _____

8. Amy's Art Shop kept all inventory and sales records on the basis of retail prices. It recorded purchases at cost and marked up its merchandise at 120% of cost. On January 1, its inventory of art was $260,000. During the year its purchases were $300,000, and net sales were $730,000. What was its ending inventory? (5 points) _____

9. From the information given, calculate the estimated cost of goods sold and ending inventory. Round to the nearest dollar. (1 point for each correct answer)

| | Cost of Goods Available for Sale | Net Sales | Markup Based on Cost | Markup Based on Sales | Estimated Cost of Goods Sold | Estimated Ending Inventory |
|---|---|---|---|---|---|---|
| **a.** | $204,000 | $260,000 | 30% | | | |
| **b.** | 268,000 | 260,000 | | 30% | | |
| **c.** | 444,000 | 350,000 | | 27% | | |
| **d.** | 444,000 | 350,000 | 27% | | | |
| **e.** | 37,500 | 36,000 | 50% | | | |
| **f.** | 368,000 | 400,000 | | 60% | | |
| **g.** | 420,000 | 600,000 | | 40% | | |
| **h.** | 440,000 | 360,000 | 15% | | | |
| **i.** | 125,000 | 180,000 | 60% | | | |
| **j.** | 130,000 | 200,000 | 100% | | | |

10. Maurice Company sells hair products. From the following inventory record sheets for Baby Soft Shampoo, determine the total units in, total amount, and the value of the ending inventory of 300 bottles based on average cost, FIFO, and LIFO. (2 points for each correct answer)

| Date | Units In | Cost | Amount | |
|---|---|---|---|---|
| 1/11 | 400 | $3.40 | $1,360 | Average cost: _____ |
| 1/23 | 50 | 3.00 | 150 | |
| 2/10 | 100 | 3.20 | 320 | FIFO: _____ |
| 2/20 | 200 | 3.30 | 660 | |
| 2/25 | 80 | 3.50 | 280 | LIFO: _____ |

Score for B (50)

Notes

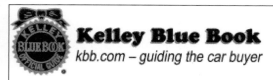

Kelley Blue Book

kbb.com – guiding the car buyer

The Perfect <u>Stocking Stuffer!</u>
(under $10!)

**New Car Pricing
Build a Car
Incentives**

**My Car's Value
Used Car Retail**

**Buy a New Car
Buy a Used Car
Sell Your Car
Motorcycles**

**Financing
Insurance
Lemon Check
Warranties
Accessories**

**Car Reviews
Car Previews
Decision Guides
Advice**

**About kbb
Home**

WHAT'S MY BLUE BOOK VALUE?

Before buying a new car, be sure to get the Blue Book value of the car you own now. Choose <u>Trade-In Value</u> if you intend to sell your car to a dealer, <u>Private Party Value</u> if you intend to sell your car to another consumer.

USED CAR VALUES FOR SHOPPERS

Shopping for a used car? Check out Blue Book values before you buy. Choose <u>Used Car Retail Value</u> if you intend to buy from a dealer. Choose <u>Private Party Value</u> if you intend to buy from another consumer.

NEW CAR PREVIEWS

<u>Previews</u> of brand new models <u>coming soon</u> to a dealership near you. Also, <u>new and improved</u> models. Check out the latest <u>concept cars</u>.

MOTORCYCLES

Free <u>trade-in and retail values</u> on used street and dirt bikes, ATVs, scooters, mopeds and sidecars. Also features used personal watercraft & snowmobiles.

NEW CAR PRICING

Free <u>new car pricing</u>, photos, invoice, MSF rebates. <u>Select option</u> exact car you want.

BUYING & SELLIN

Once you've got you car online or <u>buy a us car</u>. Also, get quotes <u>insurance</u>, <u>inspection</u> Visit the <u>auto parts st</u>

TOOLS·TIPS·ADV

Automotive <u>articles</u> buying new and used information about <u>rec safety ratings</u>. Get a report from Carfax. C new & used? Try our get <u>recommendations</u> based on your prefer <u>side-by-side comparis</u> <u>frequently asked que</u>

WHAT'S NEW

- Car lover's <u>gift idea</u>
- <u>Luxury car</u> reviews
- our 75th <u>anniversary</u>
- Editor's note - <u>cur</u>

In 1918, Les Kelley began his own car dealership, and in the early 1920s, to help acquire new inventory, he distributed to other dealers and to banks a list of automobiles he wanted to buy and the prices he was willing to pay for them. Members of the automotive community began to trust his judgment so much as an accurate reflection of current values that they started to request the list for their own use.

In 1926, Kelley decided to expand his list of auto values and published the first *Blue Book of Motor Car Values*. He showed factory list price and cash value on thousands of vehicles, from Cadillacs to Duesenbergs, from Pierce-Arrows to Hupmobiles. A 1926 Packard sedan limousine, with balloon tires, might be worth as much as $3,825, but a 1921 Nash touring car, even with a clock, was only worth $50, showing how much cars can depreciate. Visit this site at http://www.Kbb.com.

Business Depreciation

20

LEARNING OBJECTIVES

By studying this chapter and completing all assignments you will be able to:

 1 Compute depreciation, using the straight-line method.

 2 Compute depreciation, using the declining-balance method.

 3 Compute depreciation, using the sum-of-the-years-digits method.

 4 Compute depreciation for income tax purposes, using the Modified Accelerated Cost Recovery System (MACRS).

 5 Compute partial-year depreciation, using the four different depreciation methods covered.

Over time and with use, a "usable" asset loses some of its original value, a process called *depreciation*. Similar to depreciation, *depletion* is the term applied to the use of natural resources such as oil, gas, and mineral rights.

1. How many years does the Internal Revenue Service (IRS) give a business to depreciate its computer equipment?
2. Does a business have the right to choose the method it wants to use to depreciate its computer equipment?
3. Is there a method of depreciation in which the amount of depreciation is greater the first year than in succeeding years?
4. What does the term *scrap value* mean?

Depreciation is the decrease in the value of assets owned by a business, such as automobiles, buildings, and computers. Depreciation is caused by wear and tear or by **obsolescence** (becoming out-of-date). In the toy manufacturing industry, some dies and tools last only 1 or 2 years because of changing fads. An automobile will wear out after a number of years or miles of use. Buildings lose value as wood, electrical wiring, and fixtures deteriorate and as design characteristics and owners' needs change. A business computer frequently becomes obsolete in 3 to 5 years.

In business, depreciation is figured on almost all physical assets owned and in use. Depreciation is deducted from gross profits as an expense. In this chapter we present four common methods of calculating depreciation: the straight-line, declining-balance, sum-of-the-years-digits, and Modified Accelerated Cost Recovery System method.

Computing Depreciation with the Straight-Line Method

Compute depreciation, using the straight-line method.

The **straight-line (SL) method** of determining depreciation is the easiest and most frequently used method. It distributes depreciation evenly over the useful life of an asset, assigning equal amounts to designated units (miles, number of items made, etc.) or periods (usually months or years). It is based on the assumption that wear and obsolescence occur evenly over the life of the property. The straight-line method requires that three factors be used to compute depreciation:

1. The **original cost**, which includes not only the price paid for an item but also any freight charges and expenses for installation. Cost includes anything necessary to get the asset to where it is to be used and in a condition to be used.
2. The **estimated physical life** or **estimated service life**, which is the length of time the buyer expects to be able to use an asset or the amount of usefulness the buyer expects to get from the asset. The estimated service life may be stated in terms of years or number of units (miles) that normally may be expected during the life of the asset.
3. The estimated **scrap value (SV)**, which is the amount the owner of an asset expects to receive upon disposing of it at the end of its estimated service life.

The basic formula for computing the amount of depreciation under the straight-line method is

> Depreciation amount for 1 unit or period = (Original cost − Scrap value)
> ÷ Estimated (physical or service) life in units or periods of time

EXAMPLE A

An office computer costing $12,500 has an estimated life of 5 years and an estimated scrap value of $900. What is the annual depreciation amount?
$12,500 cost − $900 SV = $11,600 estimated total depreciation
$11,600 ÷ 5 estimated total years = $2,320 annual depreciation

Note: An annual percentage rate of depreciation can be computed by dividing the annual depreciation amount by the cost. For this example, the rate would be 18.56% ($2,320 ÷ $12,500).

Example B shows depreciation based on the number of hours used. First you must find the hourly depreciation and then multiply it by the number of hours operated during a particular month or year.

EXAMPLE B

A machine costing $10,000 has an estimated life of 60,000 hours of operation and an estimated scrap value of $400. If it was operated for 2,800 hours during the first year, how much depreciation expense will be shown for the first year?

$10,000 cost − $400 SV = $9,600 estimated total depreciation
$9,600 ÷ 60,000 estimated total hours = $0.16 hourly depreciation
2,800 hours operated × $0.16 = $448 first year's depreciation.

If the life of the machine is stated in terms of the number of units that it will produce during its lifetime, divide that number into the estimated total depreciation amount to get the depreciation per unit.

EXAMPLE C

A press that costs $145,000 will produce an estimated 3,500,000 units in its life and has an estimated scrap value of $5,000. If it produced 235,000 units this year, how much depreciation will be shown for the year?

$145,000 cost − $5,000 SV = $140,000 estimated total depreciation
$140,000 ÷ 3,500,000 estimated total units = $0.04 depreciation per unit
235,000 units produced × $0.04 = $9,400 first year's depreciation

Book Value

The **book value** of an asset is the original cost minus the **accumulated depreciation**, or the total of all the depreciation to that time.

EXAMPLE D

At the end of the first year, the book value of the press in example C would be

$145,000 cost − $9,400 accumulated depreciation = $135,600

The book value can be determined at any time in the life of an asset.

EXAMPLE E

At the end of the third year, the book value of the computer in example A would be computed as follows:

$2,320 annual depreciation × 3 years = $6,960 accumulated depreciation
$12,500 cost − $6,960 = $5,540.

**www.calcbuilder.com/
cgi-bin/calcs/AUT2.cgi/
FinanCenter**

Visit the depreciation calculator at
the FinanCenter Web site.

CONCEPT CHECK 20.1

On January 1, Oakdale Appliances bought a new delivery truck for $48,000. Oakdale's accountant estimated a truck life of 200,000 miles and a scrap (trade-in) value of $4,000. In the first year, the truck was driven 38,000 miles; in the second year, it was driven 46,000 miles. Compute the depreciation and book value for the first 2 years.

$48,000 cost − $4,000 SV = $44,000 estimated total depreciation
$44,000 ÷ 200,000 miles = $0.22 depreciation per mile
Year 1: 38,000 miles × $0.22 per mile = $8,360 depreciation
 $48,000 cost − $8,360 = $39,640 book value
Year 2: 46,000 miles × $0.22 per mile = $10,120 depreciation
 $39,640 year 1 book value − $10,120 = $29,520 new book value
 or
 $48,000 cost − ($8,360 + $10,120) accumulated depreciation = $29,520 book value

Computing Depreciation with the Declining-Balance Method

LEARNING OBJECTIVES 2

Compute depreciation, using the declining-balance method.

The **declining-balance (DB) method** is based on the theory that depreciation is greatest in the first year and less in each succeeding year.

STEPS **to Compute Depreciation, Using the DB Method**

1. Divide 100% by the estimated years of useful life to determine the **basic depreciation rate**.
2. Multiply the basic depreciation rate by 2 (**double-declining-balance**) or by 1.5 (**150%-declining-balance**) to determine the **declining-balance depreciation rate**.
3. Multiply the declining-balance depreciation rate by the book value of the asset at the beginning of the year to determine the depreciation amount for that year. (For the first year, the book value at the beginning of the year equals the asset cost. Do not subtract the scrap value.)

 Note that Step 3 must be repeated each year, using the new (declined) book value (last year's beginning book value minus last year's depreciation amount). The same rate is used each year. The declining-balance rate continues to apply until the scrap value is reached. The item may not be depreciated below its scrap value.

EXAMPLE F

Use the declining-balance method with an annual double-declining balance to depreciate the office computer in example A.

STEP 1 100% ÷ 5 years = 20%

STEP 2 20% × 2 = 40% annual double-declining-balance rate.

| Year | | Beginning Book Value | Rate | STEP 3 Depreciation |
|---|---|---|---|---|
| 1 | | $12,500 × | 40% = | $5,000 |
| 2 | $12,500 − $5,000 = | 7,500 × | 40% = | 3,000 |
| 3 | 7,500 − 3,000 = | 4,500 × | 40% = | 1,800 |
| 4 | 4,500 − 1,800 = | 2,700 × | 40% = | 1,080 |
| 5 | 2,700 − 1,080 = | 1,620 × | 40% = | 648 |
| 6 | 1,620 − 648 = | 972 × | 40% = | 388.80 $72* |

*As book value ($972) is larger than estimated scrap value ($900), there is some depreciation in the sixth year. However, the calculated depreciation ($388.80) is greater than book value minus scrap value ($972 − $900 = $72). Thus depreciation is limited to the smaller amount, $72.

CONCEPT CHECK 20.2

On January 1, Oakdale Appliances bought a new delivery truck for $48,000. Oakdale's accountant estimated a truck life of 4 years and a scrap value of $4,000. Compute the depreciation for the first 2 years using the 150%-declining-balance method.

$48,000 cost
100% ÷ 4 years = 25%
25% × 1.5 = 37.5% annual 150%-declining-balance rate
Year 1: $48,000 × 37.5% = $18,000 depreciation
 $48,000 − $18,000 = $30,000 book value
Year 2: $30,000 × 37.5% = $11,250 depreciation

Computing Depreciation with the Sum-of-the-Years-Digits Method

The **sum-of-the-years-digits (SYD) method** also is used to compute a greater depreciation amount in the earlier years of an asset's life. The book value decreases more slowly than under the declining-balance method. This method's name comes from the calculation done in Step 1.

Compute depreciation, using the sum-of-the-years-digits method.

EXAMPLE G

Under the sum-of-the-years-digits method, the office computer in example A would be depreciated as follows.

STEP 1 $\quad \dfrac{(5 + 1) \times 5}{2} = 15$ (or $1 + 2 + 3 + 4 + 5 = 15$)

| Year | STEP 2 Fraction | | Depreciation Total Amount | | STEP 3 Depreciation |
|---|---|---|---|---|---|
| 1 | $\frac{5}{15}$ | × | $11,600 | = | $ 3,866.67 |
| 2 | $\frac{4}{15}$ | × | 11,600 | = | 3,093.33 |
| 3 | $\frac{3}{15}$ | × | 11,600 | = | 2,320.00 |
| 4 | $\frac{2}{15}$ | × | 11,600 | = | 1,546.67 |
| 5 | $\frac{1}{15}$ | × | 11,600 | = | $ 773.33 |

Total depreciation $11,600.00

CONCEPT CHECK 20.3

On January 1, Oakdale Appliances bought a new delivery truck for $48,000. Oakdale's accountant estimated a truck life of 4 years and a scrap value of $4,000. Compute the depreciation for the first 2 years using the sum-of-the-years-digits method.

$48,000 cost − $4,000 SV = $44,000 to be depreciated
$\dfrac{(4 + 1) \times 4}{2} = 10$ (or $1 + 2 + 3 + 4 = 10$)
Year 1: $\frac{4}{10} \times \$44,000 = \$17,600$ depreciation
Year 2: $\frac{3}{10} \times \$44,000 = \$13,200$ depreciation

Computing Depreciation with the Modified Accelerated Cost Recovery System

Businesses use the depreciation methods previously described for financial reporting. However, federal tax laws regulate how depreciation must be taken for income tax purposes. The IRS requires that the **Modified Accelerated Cost Recovery System (MACRS)** be used for depreciation of property purchased and put into service after 1986. MACRS "recovers" the entire cost of depreciable property over the allowable period. No scrap value is permitted.

For common business assets, MACRS provides depreciation periods of 3, 5, 7, 10, 15, and 20 years. Examples of assets from each of these categories are as follows.

3 years: Property with a life of 4 years or less—some types of equipment used for research and development, some machine tools, some tractors, and racehorses more than 2 years old when placed in service.

5 years: Property with a life of 4 to 10 years—computers, cars and light trucks, office machines, semiconductor manufacturing equipment, certain telephone equipment, and general-purpose heavy-duty trucks.

7 years: Property with a life of 10 to 15 years—office furniture and fixtures, some agricultural and horticultural structures, and railroad track.

10 years: Property with a life of 16 to 19 years—tugboats, vessels, and barges.

15 years: Property with a life of 20 to 24 years—this category usually contains certain municipal, public utility, and telephone distribution plants.

20 years: Property with a life of 25 or more years—farm buildings and certain municipal infrastructure items such as sewers.

Figure 20-1 shows IRS annual percentages used to compute depreciation by MACRS.

LEARNING OBJECTIVES 4

Compute depreciation for income tax purposes, using the Modified Accelerated Cost Recovery System (MACRS).

www.mdv-seia.org/ federal_incentives.htm

If business invests in or uses solar energy, the federal government offers a 5-year accelerated depreciation for all solar energy equipment.

Figure 20-1: MACRS Depreciation Schedule

| Year | Appropriate Percentage | | | | | |
| | 3-Year Class | 5-Year Class | 7-Year Class | 10-Year Class | 15-Year Class | 20-Year Class |
|---|---|---|---|---|---|---|
| 1 | 33.33 | 20.00 | 14.29 | 10.00 | 5.00 | 3.750 |
| 2 | 44.45 | 32.00 | 24.49 | 18.00 | 9.50 | 7.219 |
| 3 | 14.81 | 19.20 | 17.49 | 14.40 | 8.55 | 6.677 |
| 4 | 7.41 | 11.52 | 12.49 | 11.52 | 7.70 | 6.177 |
| 5 | | 11.52 | 8.93 | 9.22 | 6.93 | 5.713 |
| 6 | | 5.76 | 8.92 | 7.37 | 6.23 | 5.285 |
| 7 | | | 8.93 | 6.55 | 5.90 | 4.888 |
| 8 | | | 4.46 | 6.55 | 5.90 | 4.522 |
| 9 | | | | 6.56 | 5.91 | 4.462 |
| 10 | | | | 6.55 | 5.90 | 4.461 |
| 11 | | | | 3.28 | 5.91 | 4.462 |
| 12 | | | | | 5.90 | 4.461 |
| 13 | | | | | 5.91 | 4.462 |
| 14 | | | | | 5.90 | 4.461 |
| 15 | | | | | 5.91 | 4.462 |
| 16 | | | | | 2.95 | 4.461 |
| 17 | | | | | | 4.462 |
| 18 | | | | | | 4.461 |
| 19 | | | | | | 4.462 |
| 20 | | | | | | 4.461 |
| 21 | | | | | | 2.231 |

Note: The MACRS percentage for the first year is applicable to a partial or full year.

Use the MACRS Depreciation Schedule shown in Figure 20-1 to depreciate the office computer in example A for tax purposes.

| Year | Rate (%) | Cost | Depreciation (Rounded) | Beginning Book Value | Current Depreciation | Ending Book Value |
|------|----------|------|------------------------|----------------------|----------------------|-------------------|
| 1 | 20.00 × | $12,500 = | $2,500 | $12,500 − | $2,500 = | $10,000 |
| 2 | 32.00 × | 12,500 = | 4,000 | 10,000 − | 4,000 = | 6,000 |
| 3 | 19.20 × | 12,500 = | 2,400 | 6,000 − | 2,400 = | 3,600 |
| 4 | 11.52 × | 12,500 = | 1,440 | 3,600 − | 1,440 = | 2,160 |
| 5 | 11.52 × | 12,500 = | 1,440 | 2,160 − | 1,440 = | 720 |
| 6 | 5.76 × | 12,500 = | 720 | 720 − | 720 = | 0 |

CONCEPT CHECK 20.4

On April 10, Oakdale Appliances bought a new delivery truck for $48,000. Compute the depreciation for the first year and for the second year, using the MACRS table (5-year class).

MACRS depreciation first year: $48,000 × 20.00% = $9,600 (effectively allows for $\frac{1}{2}$ year's depreciation)
MACRS depreciation second year: $48,000 × 32.00% = $15,360 (full year's depreciation)

Computing Partial-Year Depreciation

5
LEARNING OBJECTIVES

Compute partial-year depreciation, using the four different depreciation methods covered.

Frequently, businesses are faced with the need to compute depreciation for only part of the year. Partial-year depreciation is easily computed with any of the methods described in this chapter.

With the straight-line method, compute the depreciation amount for a partial year by dividing the annual depreciation amount by 12 and then multiplying that result by the number of months of use.

With the declining-balance method, find the current year's annual depreciation and then divide by 12; multiply that result by the number of months of use.

With the sum-of-the-years-digits method, first consider the overlapping years. To find the annual depreciation for the first partial year, divide by 12 and multiply the result by the number of months of use. From then on, every year will include the remaining fraction of the prior year's depreciation and the partial-year depreciation for the remainder of the current year.

MACRS tables automatically consider partial-year depreciation for the first and last years regardless of the date the item was placed in service.

EXAMPLE I

Office furniture costing $18,000 and put in use on May 1 is expected to have a useful life of 10 years. Its estimated resale value is $1,500. Using each of the four methods, compute the depreciation expense for May 1 through December 31 of the first tax year and all 12 months of the second year.

| Method | Year | Calculation (rounded to the nearest dollar) |
|--------|------|---|
| SL | 1st | $(\$18{,}000 - \$1{,}500) \div 10 \times \dfrac{8}{12} = \$1{,}100$ |
| | 2nd | $(\$18{,}000 - \$1{,}500) \div 10 = \$1{,}650$ |
| DB (200%) | 1st | $\left(\dfrac{100\%}{10}\right) \times 2 \times \$18{,}000 \times \dfrac{8}{12} = \$2{,}400$ |
| | 2nd | $(\$18{,}000 - \$2{,}400) \times 20\% = \$3{,}120$ |
| SYD | 1st | $\dfrac{(10 + 1) \times 10}{2} = 55$ |
| | | $(\$18{,}000 - \$1{,}500) \times \dfrac{10}{55} \times \dfrac{8}{12} = \$2{,}000$ |
| | 2nd | $(\$18{,}000 - \$1{,}500) \times \dfrac{10}{55} \times \dfrac{4}{12} = \$1{,}000$ ⎫ |
| | | $(\$18{,}000 - \$1{,}500) \times \dfrac{9}{55} \times \dfrac{8}{12} = \$1{,}800$ ⎭ $\$2{,}800$ |
| MACRS | 1st | $\$18{,}000 \times 14.29\% = \$2{,}572.20$ |
| (7-year class) | 2nd | $\$18{,}000 \times 24.49\% = \$4{,}408.20$ |

CONCEPT CHECK 20.5

In October, Oakdale Appliances bought a new mid-size van for $34,000. It had an estimated scrap value of $4,000 and useful life of 5 years. Compute the depreciation expense for the 3 months of the first year and for the full second year, using the 150%-declining-balance and the sum-of-the-years-digits methods.

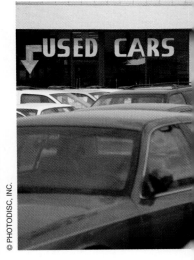

$34,000 cost − $4,000 scrap value = $30,000 to be depreciated

Declining Balance
100% ÷ 5 years × 1.5 = 30%
30% × $30,000 = $9,000

Year 1: $9{,}000 \times \dfrac{3}{12} = \$2{,}250$ (3 months)

Year 2: $(\$30{,}000 - \$2{,}250) \times 30\% = \$8{,}325$ (full year)

Sum-of-the-Years-Digits
$\dfrac{(5 + 1) \times 5}{2} = 15$ (or 1 + 2 + 3 + 4 + 5 = 15)

$\dfrac{5}{15} \times \$30{,}000 = \$10{,}000$

Year 1: $\$10{,}000 \times \dfrac{3}{12} = \$2{,}500$ (3 months)

Year 2: $\$10{,}000 \times \dfrac{9}{12} = \$7{,}500$ (9 months)

$\dfrac{4}{15} \times \$30{,}000 = \$8{,}000$

$\$8{,}000 \times \dfrac{3}{12} = \$2{,}000$ (3 months)

$7,500 + $2,000 = $9,500 in year 2

COMPLETE ASSIGNMENTS 20.1 AND 20.2.

Chapter Terms for Review

accumulated depreciation
basic depreciation rate
book value
declining-balance (DB) method
declining-balance depreciation rate
depreciation
double-declining-balance
estimated physical life
estimated service life

Modified Accelerated Cost Recovery System (MACRS)
obsolescence
150%-declining-balance
original cost
scrap value (SV)
straight-line (SL) method
sum-of-the-years-digits (SYD) method

Visit the Web

www.depr.org—This Web site is for the national Society of Depreciation Professionals, people who work with and are interested and involved in depreciation analysis and application.

http://encarta.msn.com/find/concise.asp?z=18&pg=28&ti=761557233—Depreciation is defined and how depreciation is used in business is described here.

www.fixedassetsoftware.com—Go to this Web site for a description of one of the software programs available to manage the depreciation of assets.

www.calcbuilder.com/cgi-bin/calcs/AUT2.cgi/aol_digitalcities—Access this Web site to learn how to figure depreciation for a personal automobile.

www.multimedia.calpoly.edu/development/busen/alpha/d/depreciation/html—Access this site for a description of the bases for depreciation. It also illustrates federal tax depreciation requirements and other methods commonly used to calculate depreciation.

http://www.lowendmac.com/musings/01/depreciate.html—Read about computer depreciation and the IRS here.

http://www.smbiz.com/sbrl012.html—Learn about Federal depreciation rates and MACRS at this Web site.

http://www.calcbuilder.com/cgi-bin/calcs/AUT2.cgi/FinanCenter—A calculator and information on automobile depreciation is available at this Web site.

CYBERCONNECTION

Five years ago, you started your own landscaping business. Now you're considering buying a pickup truck, so you visit the Kelley Blue Book Web site to see what your options are regarding a new or a used truck. You think you'd like a Dodge truck, and you go to the following Internet pages.

First you complete the following information for a *used vehicle*.

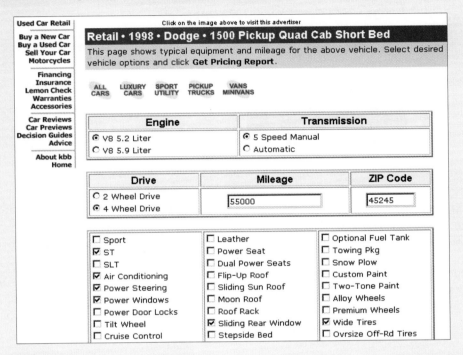

Next, you find that the used truck costs $17,620.

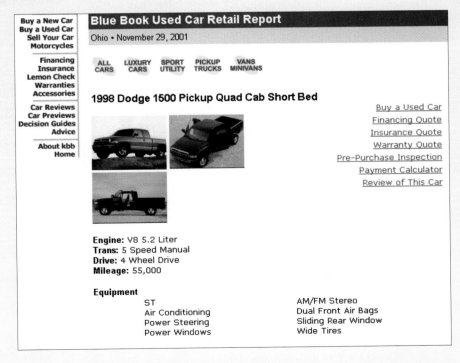

Continued

CYBERCONNECTION *Continued*

Then you look up information on the new truck that you're considering. (We've used a 2001 truck here, but you can go to the site and look up other models.)

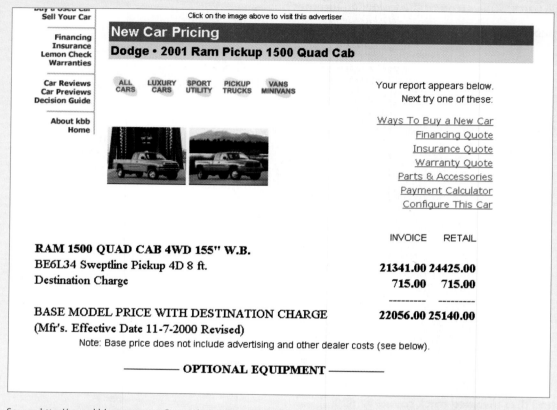

Source: http://www.kbb.com pages. Copyright © 2001 by Kelley Blue Book Company, Inc., used with permission.

Assume that the older 1998 vehicle with 55,000 miles on it was the same retail price when new as the 2001. How much did the 1998 vehicle depreciate, both in dollars and percent? Go to the Kelley Blue Book Web site and try this exercise on a 1998 pickup truck and a brand new truck of the same model. What's the depreciation on the 1998 truck now? Assuming that the company that owned the 1998 Dodge depreciated it at $0.40 per mile, what was the total depreciation? If the company had paid full retail price, what was its current book value?

The Bottom Line

Summary of chapter learning objectives:

| Learning Objective | Summary | Example |
|---|---|---|
| **20.1** | Compute depreciation, using the straight-line method | 1. On January 1, 2000, the local Pepsi-Cola bottling franchise purchased a bottling machine for $320,000. Freight was added for $12,000. The cost of installation was $68,000. It was estimated that the machine could be used for 800,000 hours, after which there would be no resale value. The machine was used 92,000 hours the first year, 86,000 hours the second year, and 104,000 hours the third year. Determine the straight-line depreciation per year based on the hours of use and the book value at the end of each year. |
| **20.2** | Compute depreciation, using the declining-balance method | 2. For $56,000, a Gap clothing store bought display racks with an estimated life of 20 years and a scrap value of $4,000. After 3 years, this store closed and sold the display racks for $32,000. If the racks were depreciated by the declining-balance method (150% annual rate), how much less than the book value did the company receive? Round to the nearest dollar. |
| **20.3** | Compute depreciation, using the sum-of-the-years-digits method | 3. A local Ford dealership purchased, for $60,000, a hydraulic lift unit with an estimated life of 7 years and a scrap value of $4,000. Compute the depreciation for each of the first 2 years using the sum-of-the-digits method. Round to the nearest dollar. |
| **20.4** | Compute depreciation for income tax purposes, using the Modified Accelerated Recovery Systems (MACRS) | 4. Bank One bought new calculators in July for $12,000. Using the MACRS method (5-year class), show the rate, depreciation, and ending book value for the first two years. |
| **20.5** | Compute partial-year depreciation using the four different depreciation methods covered | 5. On October 1, 2001, Corner Grocery bought and installed a new cash register for $1,400. It has an estimated service life of 6 years and an estimated scrap value of $200. The company decided to use the straight-line method of depreciation. What was the depreciation for 2001? What was it for 2002? |

Answers: 1. 1996: $46,000/$354,000; 1997: $43,000/$311,000; 1998: $52,000/$259,000/$259,000 **2.** $12,321 **3.** First year: $14,000; Second year: $12,000 **4.** First year: 20.00% rate, $2,400 depreciation, $9,600 EBV; Second year: 32.00% rate, $3,840 depreciation, $5,760 EBV **5.** 2001: $50; 2002: $200

Self-Check | Review Problems for Chapter 20

1 Determine the annual declining balance depreciation rate to be used for each of the following.

 a. 150% declining balance, 12-year life _____

 b. 200% declining balance, 8-year life _____

 c. 125% declining balance, 5-year life _____

 d. 200% declining balance, 5-year life _____

2 What fraction is to be used each year for sum-of-the-years-digits depreciation for an asset with a useful life of 4 years? _____

3 For which depreciation method(s) is salvage value *not* subtracted to calculate depreciation? _____

4 Lopez Construction Company purchased construction equipment for $116,000 at the beginning of the year. It is estimated that the equipment will have a useful life of 12 years and will have a scrap value of $8,000.

 a. Calculate the annual depreciation if Lopez uses straight-line depreciation. _____

 b. Calculate the book value of the equipment at the end of 5 years, assuming that Lopez uses straight-line depreciation. _____

 c. Compute the depreciation for the first year ending December 31 if Lopez purchased the equipment September 1. _____

 d. Determine the depreciation per hour if Lopez uses the straight-line method based on 120,000 hours of useful life and an $8,000 scrap value. _____

 e. Using the rate determined in (d), what is the depreciation for the year if the equipment is used for 2,360 hours? _____

5 Jurgenson Manufacturing uses the double-declining balance method of depreciation. A piece of equipment costing $37,500 has an estimated useful life of 5 years and an estimated scrap value of $2,700.

 a. Compute the amount of depreciation taken in the second year. _____

 b. What is the book value at the end of the second year? _____

6 Young Manufacturing uses the sum-of-the-years-digits method of depreciation. Equipment costing $37,500 has an estimated life of 5 years and an estimated scrap value of $2,700

 a. Compute the amount of depreciation expense for the second year. _____

 b. What is the book value at the end of the second year? _____

7 Calculate the depreciation expense for tax purposes using MACRS for each asset. Use Figure 20-1 on page 421 to determine the proper life and rate for each asset. (Round to the nearest dollar.)

 a. Computer equipment purchased this year for $5,200. _____

 b. Office furniture purchased 2 years ago for $8,500. (This is the third year.) _____

Assignment 20.1: Business Depreciation

Name _____

Date _____ Score _____

A **(30 points) Solve the following depreciation problems. (points for correct answers as marked)**

1. An ice cream company has fast-freeze yogurt machines on which it estimates depreciation by the straight-line method. The following table shows cost, estimated life, years used, and scrap value for each machine. Find the annual depreciation, total depreciation, and book value after the indicated number of years of use. ($\frac{1}{2}$ point for each correct answer)

| | Original Cost | Estimated Life (years) | Years Used | Scrap Value | Annual Depreciation | Total Depreciation to Date | Book Value |
|---|---|---|---|---|---|---|---|
| **a.** | $ 6,000 | 10 | 4 | $ 350 | _____ | _____ | _____ |
| **b.** | 6,400 | 7 | 4 | 800 | _____ | _____ | _____ |
| **c.** | 8,400 | 12 | 3 | none | _____ | _____ | _____ |
| **d.** | 34,600 | 15 | 13 | 1,000 | _____ | _____ | _____ |

2. Ace Delivery Service bought two new trucks. The following table shows the cost, scrap value, estimated life (in miles), and mileage for the first year. Using the straight-line method based on mileage driven, compute the first year's depreciation and the book value at the end of the first year for each truck. (2 points for each correct depreciation amount and 1 point for each correct book value)

| | Original Cost | Scrap Value | Estimated Life (miles) | Mileage for First Year | Depreciation for First Year | Book Value after 1 Year |
|---|---|---|---|---|---|---|
| **a.** | $49,500 | $1,500 | 150,000 | 21,700 | _____ | _____ |
| **b.** | $23,000 | $ 600 | 80,000 | 9,500 | _____ | _____ |

3. Dole Fruit Company's equipment cost $214,000. Its useful life is estimated to be 15 years, and its scrap value is $4,000. The company uses straight-line depreciation. (2 points for each correct answer)

 a. What is the annual depreciation? _____

 b. What is the book value of the equipment at the end of 14 years? _____

Assignment 20.1 *Continued*

4. Carlucci and sons purchased a machine for $13,645 at the beginning of the year. Additional costs included $250 freight and $175 for installation. It was estimated that the machine could be operated for 30,000 hours, after which its resale value would be $570. Determine the straight-line depreciation based on hours of operation and the book value at the end of each of the first seven years. (1 point for each correct answer)

| Year | Hours of Operation | Depreciation | Book Value |
|---|---|---|---|
| 1 | 2,300 | | |
| 2 | 2,750 | | |
| 3 | 2,500 | | |
| 4 | 2,480 | | |
| 5 | 2,800 | | |
| 6 | 3,100 | | |
| 7 | 2,950 | | |

Score for A (30)

B **(56 points) Solve the following depreciation problems. Round dollar amounts to two decimal places. (points for correct answers as marked)**

5. Anderson Tool and Die Company owns a group of machines, the details of which are shown in the following table. Anderson uses the double-declining-balance method of calculating depreciation. Compute the depreciation for the specific years indicated. (2 points for each correct answer)

| | Original Cost | Estimated Life (years) | Scrap Value | Year | Depreciation | Year | Depreciation |
|---|---|---|---|---|---|---|---|
| a. | $32,000 | 16 | $1,200 | 1 | | 3 | |
| b. | $25,800 | 5 | $3,000 | 3 | | 5 | |
| c. | $ 8,000 | 4 | $ 300 | 2 | | 3 | |
| d. | $15,000 | 10 | — | 3 | | 5 | |
| e. | $12,600 | 8 | $1,200 | 2 | | 4 | |
| f. | $95,000 | 20 | — | 3 | | 5 | |

Part 6 Basic Business Applications

Assignment 20.1 *Continued*

6. Machinery purchased from Telecom, Inc., by Blazedales cost $69,800. Depreciation was determined by the double-declining-balance method for an estimated life of 16 years. Compute the following.

 a. Book value after 4 years (8 points): _____

 b. Total depreciation after 6 years (4 points): _____

7. The Dugan Manufacturing Company bought an engine for $31,500. The engine had an estimated life of 20 years and a scrap value of $5,250. After 6 years, the company went out of business and sold the engine for $15,200. If the machine was depreciated by the double-declining-balance method, how much did the company lose on the sale (the difference between the book value and the selling price)? (20 points) _____

Score for B (56)

Assignment 20.1 *Continued*

C **(14 points) Solve the following depreciation problems. (1 point for each correct answer)**

8. The Western Salvage Service bought three trucks. The following table shows the cost, estimated life, and resale estimate for each truck. Use the sum-of-the-years-digits method to find each truck's depreciation for the first and second years of use. Round answers to the nearest dollar.

| Original Cost | Estimated Life | Resale Estimate | Depreciation for First Year | Depreciation for Second Year |
|---|---|---|---|---|
| **a.** $46,000 | 6 yr | $4,000 | _____ | _____ |
| **b.** 54,000 | 5 yr | 6,000 | _____ | _____ |
| **c.** 88,000 | 7 yr | 8,000 | _____ | _____ |

9. Use the information in problem 8b to compute the amount of depreciation for years 3−5.

Year 3: _____

Year 4: _____

Year 5: _____

10. Use the information in problem 8 to compute the amount of depreciation for each vehicle for the first 2 years, using the straight-line method. Round to the nearest dollar.

a. _____ **b.** _____ **c.** _____

11. Which method of depreciation would give the smaller amount of write-off, and how much less would it be for the three vehicles for the first two years? _____

Score for C (14)

Assignment 20.2: Business Depreciation

Name _____

Date _____ Score _____

A **(43 points) Solve the following depreciation problems. Round dollar amounts to two decimal places. (points for correct answers as marked)**

1. An architect bought drafting equipment for $7,500. Its estimated life was 6 years, and its scrap value was $300. At the end of 4 years, the equipment wears out and is sold for scrap for $225. (4 points for each correct answer)

 a. By the straight-line method, how much difference is there between the book value and the cash value of the equipment on the date of the sale? _____

 b. In April 2000, a computer and software costing $18,000 are purchased. Its estimated life is 5 years. What is the book value of the new computing equipment on December 31, 2001? Use MACRS. _____

2. E, F, and G were partners in a small textile company. They spent $54,000 for equipment that they agreed would last 8 years and have a resale value of 5% of cost. The three partners couldn't agree on the depreciation method to use. E was in favor of using the double-declining-balance system, F insisted on the 150%-declining-balance method, and G was sure that the sum-of-the-years-digits method would be better. Show the depreciation for the first 4 years for each method in the following table. At the end of 4 years, what would be the book value under each of the three methods? (2 points for each correct depreciation amount, 1 point for each correct total, and 1 point for the correct book value)

| Year | Double-DB | 150%-DB | SYD |
|------|-----------|---------|-----|
| 1 | _____ | _____ | _____ |
| 2 | _____ | _____ | _____ |
| 3 | _____ | _____ | _____ |
| 4 | _____ | _____ | _____ |
| Total | _____ | _____ | _____ |
| Book value | _____ | _____ | _____ |

3. Baxter Company owned assets that cost $250,000. Depreciation was figured at a straight-line rate of 3% per year. After 12 years, the company sold the assets for $200,000. How much greater was the selling price than the book value at the time of the sale? (5 points) _____

Score for A (43)

B **(57 points) Solve the following depreciation problems. (points for correct answers as marked)**

4. On March 1, Jarvis Realty spent $16,000 for a new company car with an estimated life of 4 years and an estimated scrap value of $4,000. Jarvis Realty elected to use the straight-line method for depreciation. On the same date, Carter Realty bought an identical car at the same price and also estimated the car's life and scrap value to be 4 years and $4,000, respectively. Carter Realty, however, chose the sum-of-the-years-digits method for depreciation.

 a. At the end of the first year (10 months of use) and second year, how much depreciation did each company calculate? (3 points for each correct answer)

 Jarvis: Year 1 _____ Carter: Year 1 _____
 Year 2 _____ Year 2 _____

 b. At the end of the second year, which company had more recorded accumulated depreciation, and what was the difference in the amounts? (5 points) _____

 c. True or false: At the end of the fourth year, Carter Realty will have recorded more accumulated depreciation than Jarvis Realty. Explain your answer. (4 points) _____

5. In May 2001, Jian & Ming bought a light-duty truck for $20,800. One year later, they bought an additional truck for $21,800. In June 2003, a third truck was purchased for $23,500. Use MACRS (5-year class) to determine the total allowable cost recovery for 2003. (12 points) _____

6. David Marcus purchased new office furniture July 15, 2000, for $28,100. Use MACRS (7-year class) to show the rate, depreciation, and beginning and ending book values for 2000, 2001, and 2002. Round to the nearest dollar. (2 points for each correct answer)

| Year | Rate | | Cost | | Depreciation | Beginning Book Value | Ending Book Value |
|------|------|---|------|---|--------------|---------------------|-------------------|
| 2000 | _____ | × | $28,100 | = | _____ | _____ | _____ |
| 2001 | _____ | × | $28,100 | = | _____ | _____ | _____ |
| 2002 | _____ | × | $28,100 | = | _____ | _____ | _____ |

Score for B (57) _____

Notes

FINANCIAL ACCOUNTIN
STANDARDS BOARD

The mission of the Financial Accounti
Board is to establish and improve sta
financial accounting and reporting for
and education of the public, including
auditors, and users of financial inform

- **Business Reporting Resea**
 Improving Business Repor
 Insights into Enhancing Vo
 Disclosures

- **Business Combinations Pr**
 Summary
 Including goodwill decisior

- **Chairman Jenkins Respon**
 from Senator Abraham

Since 1973, the Financial Accounting Standards Board has been the designated organization in the private sector for establishing standards of financial accounting and reporting. Those standards govern the preparation of financial reports and are officially recognized as authoritative by the Securities and Exchange Commission and the American Institute of Certified Public Accountants.

Accounting standards are essential to the efficient functioning of the economy because decisions about the allocation of resources rely heavily on credible, concise, and understandable financial information. Financial information about the operations and financial position of individual entities also is used by the public in making various other decisions, such as investment decisions. Visit the site of the FASB at http://www.fasb.org or http://www.rutgers.edu/Accounting/raw/fasb/.

Business Financial Statements

LEARNING OBJECTIVES

By studying this chapter and completing all assignments you will be able to:

 1 LEARNING OBJECTIVES Analyze balance sheets, comparing items and periods.

 2 LEARNING OBJECTIVES Analyze income statements, comparing items and periods.

 3 LEARNING OBJECTIVES Compute commonly used business operating ratios.

Physicians and dentists use physical examinations to determine the condition of their patients. Auto mechanics use diagnostic tests to determine the condition of cars. Business people use financial statements to measure the health of a company. Knowing how to read and interpret financial statements is a vital business skill.

1. What monthly statements does a business commonly prepare to observe how well the business is doing?
2. Should you ever prepare a personal balance sheet? Why or why not?
3. What are assets and liabilities?
4. What is equity? How does a business (or a person) know how much equity it has?

Financial statements provide information that allows owners, managers, and others interested in a business to evaluate its current condition and past operating results. Two important financial statements are the balance sheet and income statement. The **balance sheet** shows the current condition of a business at a definite point in time. It lists what a business owns (**assets**), how much it owes (**liabilities**), and the difference between the two (**net worth**), usually referred to as owners' or shareholders' equity. The **income statement** shows the past operating results for a given period of time. It lists the revenues, the expenses, and the net income or loss for the period.

Financial statement data are typically analyzed three ways. The first, called *horizontal analysis*, is a comparison of data from year to year. This analysis shows the dollar amount of change and the percent of change for each item on the statement from one year to the next. The second, called *vertical analysis*, compares all other data on a statement with one figure for that same year. On the balance sheet, for example, each asset, liability, and equity amount is calculated as a percent of total assets (or total liabilities and owners' equity). The third type of analysis compares selected related data for the year such as current assets to current liabilities. These analyses are used by managers, owners, investors, creditors, and others to help them analyze and simplify the complex data and make decisions concerning the business.

Analyzing Balance Sheets

LEARNING OBJECTIVES

Analyze balance sheets, comparing items and periods.

On a balance sheet, total assets must always equal total liabilities plus owners' or shareholders' equity. Balance sheets are analyzed to compare individual items with other items and with the same item on different dates, usually 1 year apart. Many businesses use the form of balance sheet illustrated in Figure 21-1.

Figure 21-1: Balance Sheet

| | **Sun & Surf Ski Shop** | | | | | |
| | **Balance Sheet as of December 31, 2002 and 2001** | | | | | |
| | **2002 Amount** | **2002 Percent** | **2001 Amount** | **2001 Percent** | **Increase/Decrease Amount** | **Percent** |
|---|---|---|---|---|---|---|
| ASSETS | | | | | | |
| Current assets: | | | | | | |
| Cash | $ 90,000 | 12.64% | $ 86,000 | 13.15% | $ 4,000 | 4.65% |
| Accounts receivable | 134,000 | 18.82% | 98,000 | 14.98% | 36,000 | 36.73% |
| Notes receivable | 28,000 | 3.93% | 32,000 | 4.89% | (4,000) | −12.50% |
| Merchandise inventory | 180,000 | 25.28% | 148,000 | 22.63% | 32,000 | 21.62% |
| Total current assets | $432,000 | 60.67% | $364,000 | 55.66% | $68,000 | 18.68% |
| Fixed assets: | | | | | | |
| Equipment | $220,000 | 30.90% | $190,000 | 29.05% | $ 30,000 | 15.79% |
| Less depreciation | (60,000) | −8.43% | (50,000) | −7.65% | (10,000) | 20.00% |
| Equipment net | $160,000 | 22.47% | $140,000 | 21.41% | $ 20,000 | 14.29% |
| Buildings | 300,000 | 42.13% | 300,000 | 45.87% | — | 0.00% |
| Less depreciation | (180,000) | −25.28% | (150,000) | −22.94% | (30,000) | −20.00% |
| Buildings net | $120,000 | 16.85% | $150,000 | 22.94% | $(30,000) | −20.00% |
| Total fixed assets | $280,000 | 39.33% | $290,000 | 44.34% | $(10,000) | −3.45% |
| TOTAL ASSETS | $712,000 | 100.00% | $654,000 | 100.00% | $ 58,000 | 8.87% |

Figure 21-1: Balance Sheet (continued)

| | 2002 Amount | 2002 Percent | 2001 Amount | 2001 Percent | Increase/Decrease Amount | Increase/Decrease Percent |
|---|---|---|---|---|---|---|
| **LIABILITIES AND SHAREHOLDERS' EQUITY** | | | | | | |
| Current liabilities: | | | | | | |
| Accounts payable | $ 18,000 | 2.53% | $ 24,000 | 3.67% | $ (6,000) | −25.00% |
| Accrued payroll | 38,000 | 5.34% | 30,000 | 4.59% | 8,000 | 26.67% |
| Payroll taxes payable | 6,000 | 0.84% | 4,000 | 0.61% | 2,000 | 50.00% |
| Notes payable | 42,000 | 5.90% | 48,000 | 7.34% | (6,000) | −12.50% |
| Total current liabilities | $104,000 | 14.61% | $106,000 | 16.21% | $ (2,000) | −1.89% |
| Long-term liabilities: | | | | | | |
| Mortgage payable | $ 90,000 | 12.64% | $120,000 | 18.35% | $(30,000) | −25.00% |
| Notes payable (over 1 year) | 36,000 | 5.06% | 30,000 | 4.59% | 6,000 | 20.00% |
| Total long-term liabilities | $126,000 | 17.70% | $150,000 | 22.94% | $(24,000) | −16.00% |
| Total liabilities | $230,000 | 32.30% | $256,000 | 39.14% | $(26,000) | −10.16% |
| Shareholders' equity: | | | | | | |
| Common stock | $359,000 | 50.42% | $359,000 | 54.89% | — | 0.00% |
| Retained earnings | 123,000 | 17.28% | 39,000 | 5.96% | 84,000 | 215.38% |
| Total shareholders' equity | $482,000 | 67.70% | $398,000 | 60.86% | $84,000 | 21.11% |
| **TOTAL LIABILITIES AND SHAREHOLDERS' EQUITY** | $712,000 | 100.00% | $654,000 | 100.00% | $58,000 | 8.87% |

In Figure 21-1, the amounts for various items such as cash and accounts payable are compared to total assets and total liabilities and shareholders' equity. Also, the amounts for 2002 are compared to the corresponding amounts for 2001, and the amounts and percents of increase or decrease are shown. When two statements are compared, the earlier period, usually a prior year, is *always* used as the base. The changes in balance sheet items between two periods measure the growth or decline of the business.

The first step in analyzing a balance sheet is to compute the percent each item is of the total assets or of the total liabilities and shareholders' equity (net worth). The second step is to compute the amount and percent of change between two dates being compared. Note three facts:

1. The totals for assets equal the totals for liabilities and shareholders' equity.
2. The percent listed for each item under assets is of the total assets; the percent listed for each item under liabilities and shareholders' equity is of the total liabilities and shareholders' equity.
3. The percent of increase or decrease between the two years is based on 2001, the **earlier** year.

**http://bus.colorado.
edu/faculty/shane/
teaching/3220/notes/
s99/order.html**

Go to this Web site for accounting help.

**http://didit.bussvc.wisc.
edu/ acct/aclman1.html**

Every month at the University of Wisconsin, Accounting Control Ledger Statements are printed for each active grant or contract or departmental budget of state funds. Read about the details prepared for these statements at this Web site.

CONCEPT CHECK 21.1

In its next year, 2003, Sun & Surf Ski Shop had total assets of $750,000, total liabilities of $255,000, cash of $120,000, and mortgage payable of $62,000. Determine the following amounts and percents.

a. What was its total shareholders' equity in 2003?
 $750,000 − $255,000 = $495,000
b. What was its balance sheet percent of cash?
 $120,000 ÷ $750,000 = 16.00%
c. What was its balance sheet percent of mortgage payable?
 $62,000 ÷ $750,000 = 8.27%
d. What was its percent of increase in cash?
 ($120,000 − $90,000) ÷ $90,000 = 33.33%
e. What was its percent of decrease in mortgage payable?
 ($90,000 − $62,000) ÷ $90,000 = 31.11%

Analyzing Income Statements

LEARNING
OBJECTIVES

Analyze income statements, comparing items and periods.

The income statement shows revenue, expenses, and the difference between the two, or net income. Income statements are analyzed by comparing all other statement items with the **net revenue**, which is total revenue less any returns and allowances. Net revenue (frequently called **net sales**) is always 100%. All other items on the income statement are reported as a percent of net revenue/sales. The resulting percents are extremely important for all businesspeople. They are compared to budgeted amounts, to percents for competing businesses, and to percents for past periods.

Figure 21-2 shows a typical income statement for 1 year, in which dollar amounts are converted to percents based on net sales. Percents are rounded to two decimal places, and dollar amounts are rounded to the nearest whole dollar. Cents are seldom used in reporting annual figures.

Figure 21-2: Income Statement

| | | |
|---|---|---|
| **Sun & Surf Ski Shop** | | |
| **Income Statement for Year Ended December 31, 2002** | | |
| | **2002 Amount** | **2002 Percent** |
| Revenue from sales: | | |
| Sales | $988,900 | 101.43% |
| Less returns | 13,900 | 1.43% |
| NET SALES | $975,000 | 100.00% |

Figure 21-2: Income Statement (continued)

| | 2002 Amount | 2002 Percent |
|---|---|---|
| Cost of goods sold: | | |
| Inventory, January 1 | $211,000 | 21.64% |
| Purchases | 423,000 | 43.38% |
| Available for sale | $634,000 | 65.03% |
| Inventory, December 31 | 226,000 | 23.18% |
| Cost of goods sold | $408,000 | 41.85% |
| Gross profit on sales | $567,000 | 58.15% |
| | | |
| Operating expenses: | | |
| Salary and benefits | $290,000 | 29.74% |
| Rent and utilities | 62,000 | 6.36% |
| Advertising | 32,400 | 3.32% |
| Depreciation | 40,000 | 4.10% |
| Equipment and supplies | 15,800 | 1.62% |
| Administrative | 12,500 | 1.28% |
| Total operating expense | $452,700 | 46.43% |
| | | |
| Income before tax | $114,300 | 11.72% |
| Income tax | 30,300 | 3.11% |
| NET INCOME | $ 84,000 | 8.62% |

Most businesses want to compare the operations of the current year with those of the preceding year. The statement shown in Figure 21-3 has information for both the current and the preceding year. It also shows the amount and percent of increase or decrease from the preceding year. Frequently, comparative statements will include these differences, as the percent of net change—especially if it is significantly lower—will affect future business decisions.

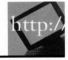

www.businesstown. com/

Another source of information and help is available here.

Figure 21-3: Comparative Income Statement

| | Sun & Surf Ski Shop Income Statement for the Years Ended December 31, 2002 and 2001 | | | | | |
|---|---|---|---|---|---|---|
| | 2002 Amount | 2002 Percent | 2001 Amount | 2001 Percent | Difference Amount | Difference Percent |
| Revenue from sales: | | | | | | |
| Sales | $988,900 | 101.43% | $850,000 | 104.81% | $138,900 | 16.34% |
| Less returns | 13,900 | 1.43% | 39,000 | 4.81% | (25,100) | −64.36% |
| NET SALES | $975,000 | 100.00% | $811,000 | 100.00% | $164,000 | 20.22% |
| Cost of goods sold: | | | | | | |
| Inventory, January 1 | $211,000 | 21.64% | $193,000 | 23.80% | $ 18,000 | 9.33% |
| Purchases | 423,000 | 43.38% | 401,000 | 49.45% | 22,000 | 5.49% |
| Available for sale | $634,000 | 65.03% | $594,000 | 73.24% | $ 40,000 | 6.73% |
| Inventory, December 31 | 226,000 | 23.18% | 211,000 | 26.02% | 15,000 | 7.11% |
| Cost of goods sold | $408,000 | 41.85% | $383,000 | 47.23% | $ 25,000 | 6.53% |
| Gross profit on sales | $567,000 | 58.15% | $428,000 | 52.77% | $139,000 | 32.48% |

Figure 21-3: Comparative Income Statement (continued)

| | 2002 Amount | 2002 Percent | 2001 Amount | 2001 Percent | Difference Amount | Difference Percent |
|---|---|---|---|---|---|---|
| Operating expenses: | | | | | | |
| Salary and benefits | $290,000 | 29.74% | $242,000 | 29.84% | $48,000 | 19.83% |
| Rent and utilities | 62,000 | 6.36% | 61,400 | 7.57% | 600 | 0.98% |
| Advertising | 32,400 | 3.32% | 25,700 | 3.17% | 6,700 | 26.07% |
| Depreciation | 40,000 | 4.10% | 32,000 | 3.95% | 8,000 | 25.00% |
| Equipment and supplies | 15,800 | 1.62% | 10,300 | 1.27% | 5,500 | 53.40% |
| Administrative | 12,500 | 1.28% | 14,200 | 1.75% | (1,700) | −11.97% |
| Total operating expense | $452,700 | 46.43% | $385,600 | 47.55% | $67,100 | 17.40% |
| Income before tax | $114,300 | 11.72% | $42,400 | 5.23% | $71,900 | 169.58% |
| Income tax | 30,300 | 3.11% | 24,400 | 3.01% | 5,900 | 24.18% |
| NET INCOME | $ 84,000 | 8.62% | $18,000 | 2.22% | $66,000 | 366.67% |

Another analysis carried out by many businesses is a comparison between actual results and budgeted figures. Owners and managers note differences between budgeted and actual amounts and make adjustments where necessary. Most businesses and virtually all government entities use monthly and annual budgets to guide and monitor their operations. Figure 21-4 illustrates a monthly and year-to-date budget comparison at the end of June, the sixth month of the year.

To find the percent change, the budgeted amount is subtracted from the actual amount and the difference is divided by the *budgeted* amount.

Figure 21-4: Monthly/Year-to-Date Budget Comparison

Sun & Surf Ski Shop

Income Statement for the Month and the Six-Month Period Ended June 30, 2002

| | June 2002 Budget | Actual | Amount Difference | Percent Difference | Six Months Year-to-Date Budget | Actual | Amount Difference | Percent Difference |
|---|---|---|---|---|---|---|---|---|
| Revenue from sales: | | | | | | | | |
| Sales | $85,000 | $86,500 | $ 1,500 | 1.76% | $510,000 | $480,000 | $(30,000) | −5.88 |
| Sales returns | 5,000 | 3,500 | $(1,500) | −30.00% | 10,000 | 6,000 | $ (4,000) | −40.00 |
| NET SALES | $80,000 | $83,000 | $ 3,000 | 3.75% | $500,000 | $474,000 | $(26,000) | −5.20 |
| Cost of goods sold | 35,000 | 38,000 | $ 3,000 | 8.57% | 225,000 | 230,000 | $ 5,000 | 2.22 |
| Gross profit | $45,000 | $45,000 | $ — | 0.00% | $275,000 | $244,000 | $(31,000) | −11.27 |
| Operating expenses | 31,000 | 39,000 | $ 8,000 | 25.81% | 185,000 | 196,000 | $ 11,000 | 5.95 |
| Income before tax | $14,000 | $ 6,000 | $(8,000) | −57.14% | $ 90,000 | $ 48,000 | $(42,000) | −46.67 |
| Income tax | 6,000 | 1,000 | $(5,000) | −83.33% | 40,000 | 16,000 | $(24,000) | −60.00 |
| NET INCOME | $ 8,000 | $ 5,000 | $(3,000) | −37.50% | $ 50,000 | $ 32,000 | $(18,000) | −36.00 |

CONCEPT CHECK 21.2

In its next year, 2003, Sun & Surf Ski Shop had total sales of $1,320,000, net sales of $1,200,000, gross profit of $650,000, and advertising expense of $45,000. In 2003, the company budgeted gross profit of $800,000. Determine the following amounts and percents.

a. Amount of sales returns in 2003
 $1,320,000 − $1,200,000 = $120,000

b. Amount of cost of goods sold in 2003
 $1,200,000 − $650,000 = $550,000

c. Percent of net sales increase from 2002 to 2003
 ($1,200,000 − $975,000) ÷ $975,000 = 23.08%

d. Percent of advertising expense in 2003
 $45,000 ÷ $1,200,000 = 3.75%

e. Difference between percent gross profit and 2003 budgeted amount
 ($650,000 − $800,000) ÷ $800,000 = −18.75%

Computing Business Operating Ratios

In addition to comparing dollar amounts and percents on financial statements, business managers and owners frequently want to study relationships between various items on their income statements and balance sheets. These relationships generally are expressed by ratios. A **ratio** is the relation of one amount to another. Thus the ratio of one dollar to one quarter, or $1 to $0.25, is a ratio of 4 to 1, or 4:1.

In analyzing financial statements, six important financial analysis ratios are commonly used: the working capital ratio, the acid test ratio, the ratio of accounts receivable to net sales, the inventory turnover rate, the relation of net income to net sales, and the rate of return on investment (equity).

LEARNING OBJECTIVES

Compute commonly used business operating ratios.

Working Capital Ratio

Working capital and the working capital ratio come from the balance sheet. **Working capital** is the amount of current assets less current liabilities. It tells the amount of current assets that would remain if all the company's current liabilities were paid immediately. The **working capital ratio** shows the relationship between current assets and current liabilities. It calculates the amount of current assets per dollar of current liabilities. The working capital ratio helps the reader of the balance sheet understand how well the company is able to pay its current debts.

www.iasc.org.uk/

If you want to read about international accounting standards, go to this Web site.

Working capital ratio = Total current assets ÷ Total current liabilities

EXAMPLE A

The working capital ratio for Sun & Surf Ski Shop for 2002 from Figure 21-1 is
$432,000 ÷ $104,000 = 4.2 = 4.2 : 1

The ratio 4.2 to 1, or 4.2 : 1, means that the business has $4.20 in current assets to pay for each $1 in current liabilities.

Acid Test Ratio

The **acid test ratio** is used to determine the relationship between assets that can be quickly turned into cash and current liabilities. Usually, these assets are cash and accounts receivable. **Accounts receivable** are amounts owed to a business for services performed or goods delivered.

Acid test ratio = (Total of cash + Accounts receivable) ÷ Total current liabilities

EXAMPLE B

The acid test ratio for Sun & Surf Ski Shop for 2002 from Figure 21-1 is computed as follows:

| | |
|---|---|
| Cash | $ 90,000 |
| Accounts receivable | 134,000 |
| Total cash and receivables | $224,000 |

$224,000 ÷ $104,000 = 2.2 = 2.2 : 1

Ratio of Accounts Receivable to Net Sales

When businesses sell on credit, they need to be alert to the amount and quality of their accounts receivable. They need to compare the amount of their current receivables to the amounts for prior years and compare the extent of their receivables to those of similar companies. By computing the **ratio of accounts receivable to net sales** every year, management and investors can keep track of the percent of sales that have not yet been paid for by customers. An increasing ratio over the years can indicate problems with collecting payments and should be investigated.

Ratio of accounts receivable to net sales = Accounts receivable ÷ Net sales

EXAMPLE C

The Sun & Surf Ski Shop ratio for 2002 is

Figure 21-1 **Figure 21-3**
$134,000 ÷ $975,000 = 0.137 = 0.14 : 1

Inventory Turnover

In retail stores, the cost of inventory often is very high. One way to control inventory costs and increase profit is to maintain a high level of inventory turnover. Recall from Chapter 19 that *inventory turnover* lets management and others know the average number of times inventory is sold during the year. The higher the turnover number, the better is the movement of inventory. Recall also that *average inventory,* found by averaging monthly, quarterly, or yearly inventory amounts, must be computed first. Inventory turnover is given as the number of times instead of as a ratio to 1.

Average inventory = (Beginning inventory + Ending inventory) ÷ 2 (annual)
Inventory turnover = Cost of goods sold ÷ Average inventory

EXAMPLE D

Based on the information given in Figures 21-3 and 21-1, the 2002 inventory turnover for Sun & Surf Ski Shop is found as follows:

January 1 December 31
($211,000 + $226,000) ÷ 2 = $437,000 ÷ 2 = $218,500 average inventory
$408,000 cost of merchandise sold ÷ $218,500 = 1.9 times inventory turnover rate

Relationship of Net Income to Net Sales

An increase in total sales volume doesn't necessarily mean that a business is improving because expenses may be increasing at an equal or greater rate than revenues. Thus looking at the **relationship of net income to net sales** is important. The relationship is given as a percentage.

Relationship of net income to net sales = Net income ÷ Net sales

EXAMPLE E

Based on information from Figure 21-3 the Sun & Surf Ski Shop's 2002 relationship of net income to net sales is $84,000 ÷ $975,000 = 8.6%. Comparison with the relationship for 2001 of 2.2% ($18,000 ÷ $811,000) indicates an improvement.

Rate of Return on Investment

Shareholders and owners want a reasonable return on their investment (equity). A ratio that measures the **rate of return on investment** is the ratio of net income to shareholders'/owners' equity. The rate is given as a percentage.

Rate of return on investment = Net income ÷ Shareholders'/owners' equity

EXAMPLE F

Based on Figures 21-3 and 21-1, the rate of return on the shareholders' investment for Sun & Surf Ski Shop for 2002 is
$84,000 ÷ $482,000 = 0.1742 = 17.4% rate of return

 CONCEPT CHECK 21.3

Atwall Electronics' financial statements showed the following.

| Cash | $ 85,000 | Current liabilities | $320,000 | Net sales | $950,000 |
|---|---|---|---|---|---|
| Accounts receivable | 260,000 | Total liabilities | 560,000 | Inventory 1/1/2002 | 240,000 |
| Total current assets | 580,000 | Net income | 80,000 | Inventory 12/31/2002 | 200,000 |
| Total assets | 990,000 | Shareholders' equity | 430,000 | Purchases for 2002 | 630,000 |

Using the above numbers, compute the following ratios:
a. Working capital ratio $580,000 ÷ $560,000 = 1.04 : 1
b. Acid test ratio ($85,000 + $260,000) ÷ $560,000 = 0.62 : 1
c. Average inventory ($240,000 + $200,000) ÷ 2 = $220,000
d. Inventory turnover $240,000 + $630,000 − $200,000 = $670,000
 $670,000 ÷ $220,000 = 3.05 turnovers
e. Net income to net sales ratio $80,000 ÷ $950,000 = 0.084, or 8.4%
f. Rate of return on investment $80,000 ÷ $430,000 = 0.186, or 18.6%

COMPLETE ASSIGNMENTS 21.1, 21.2, AND 21.3.

Chapter Terms for Review

accounts receivable
acid test ratio
assets
balance sheet
financial statements
income statement
liabilities
net revenue

net sales
net worth
rate of return on investment
ratio
ratio of accounts receivable to net sales
relationship of net income to net sales
working capital
working capital ratio

Visit the Web

http://moneycentral.msn.com/home.asp—You can look at the financial statements of various companies by doing a search for "financial statement" via this Web site.

http://www.accountantsworld.com—Learn about the field of accounting here.

http://www.aicpa.org—This is the Web site of the American Institute of Certified Public Accountants.

http://www.gao.gov/—The Web site of the U. S. General Accounting Office (GAO) provides many financial reports on the performance of the government, legal information, other financial information, and many other links.

http://www.iasc.org.uk—This Web site presents international perspectives of accounting issues.

http://www.businesstown.com—For information on accounting and other aspects of starting a small business, visit this Web site.

CYBERCONNECTION

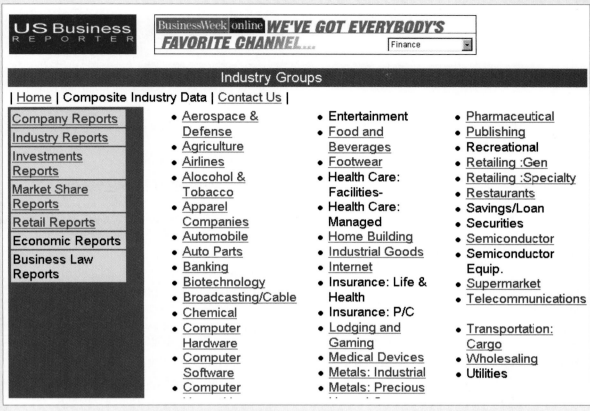

Source: http://www.activemedia-guide.com/indregister.htm. Copyright © 2001 by Centrium, Inc.; used with permission.

Identify the types of financial data available on the Internet, including balance sheets, financial statements, and more by accessing this Web site. Then go to http://www.activemedia-guide.com/5300005mx.htm.

Select two companies in the same industry—for example, under Restaurants, select Applebee's and Lone Star; or under Retailing: Gen, select J.C. Penney and Target; or under Food and Beverages, select General Mills and H.J. Heinz—and look at their financial reports. How do they compare—which company is doing best with regard to sales?

The Bottom Line

Summary of chapter learning objectives:

| Learning Objective | Summary | Example |
|---|---|---|
| **21.1** | Analyze balance sheets, comparing items and periods | 1. A modified balance sheet for Sun & Surf Ski Shop for December 2002 and 2001 is shown. Compute the percents for 2002 and the percents of increase/decrease between 2002 and 2001. |

Sun & Surf Ski Shop

Balance Sheet as of December 31, 2002 and 2001

| | 2002 Amount | 2002 Percent | 2001 Amount | 2001 Percent | Increase/Decrease Amount | Increase/Decrease Percent |
|---|---|---|---|---|---|---|
| ASSETS | | | | | | |
| Current assets: | | | | | | |
| Cash | $ 90,000 | _____ | $ 86,000 | 15.03% | $ 4,000 | _____ |
| Accounts receivable | 134,000 | _____ | 98,000 | 17.13% | 36,000 | _____ |
| Merchandise inventory | 180,000 | _____ | 148,000 | 25.87% | 32,000 | _____ |
| Total current assets | $404,000 | _____ | $332,000 | 58.04% | $ 72,000 | _____ |
| Fixed assets: | | | | | | |
| Equipment | $220,000 | _____ | $190,000 | 33.22% | $ 30,000 | _____ |
| Less depreciation | (60,000) | _____ | (50,000) | −8.74% | (10,000) | _____ |
| Equipment net | $160,000 | _____ | $140,000 | 24.48% | $ 20,000 | _____ |
| Buildings | 100,000 | _____ | 100,000 | 17.48% | — | _____ |
| Total fixed assets | $260,000 | _____ | $240,000 | 41.96% | $ 20,000 | _____ |
| TOTAL ASSETS | $664,000 | _____ | $572,000 | 100.00% | $ 92,000 | _____ |
| LIABILITIES AND SHAREHOLDERS' EQUITY | | | | | | |
| Current liabilities: | | | | | | |
| Accounts payable | $ 18,000 | _____ | $ 24,000 | 4.20% | $ (6,000) | _____ |
| Accrued payroll | 38,000 | _____ | 30,000 | 5.24% | 8,000 | _____ |
| Payroll taxes payable | 6,000 | _____ | 4,000 | 0.70% | 2,000 | _____ |
| Total current liabilities | $ 62,000 | _____ | $ 58,000 | 10.14% | $ 4,000 | _____ |
| Long-term liabilities: | | | | | | |
| Mortgage payable | 90,000 | _____ | 120,000 | 20.98% | (30,000) | _____ |
| Total liabilities | $152,000 | _____ | $178,000 | 31.12% | $ (26,000) | _____ |
| Shareholders' equity: | | | | | | |
| Common stock | $359,000 | _____ | $359,000 | 62.76% | — | _____ |
| Retained earnings | 153,000 | _____ | 35,000 | 6.12% | 18,000 | _____ |
| Total shareholders' equity | $512,000 | _____ | $394,000 | 68.88% | $118,000 | _____ |
| TOTAL LIABILITIES AND SHAREHOLDERS' EQUITY | $664,000 | _____ | $572,000 | 100.00% | $ 92,000 | _____ |

Summary of chapter learning objectives:

| Learning Objective | Summary | Example |
|---|---|---|
| **21.2** | Analyze income statements, comparing items and periods | 2. A modified income statement for Sun & Surf Ski Shop for the years 2002 and 2001 is shown. Compute the percents for 2002 and the percents of difference between 2002 and 2001. |

Sun & Surf Ski Shop

Income Statement for the Years Ended December 31, 2002 and 2001

| | 2002 Amount | 2002 Percent | 2001 Amount | 2001 Percent | Difference Amount | Difference Percent |
|---|---|---|---|---|---|---|
| **Revenue from sales:** | | | | | | |
| Sales | $988,900 | _____ | $850,000 | 104.81% | $138,900 | _____ |
| Less returns | 13,900 | _____ | 39,000 | 4.81% | (25,100) | _____ |
| NET SALES | $975,000 | _____ | $811,000 | 100.00% | $164,000 | _____ |
| **Cost of goods sold:** | | | | | | |
| Inventory, January 1 | $211,000 | _____ | $193,000 | 23.80% | $18,000 | _____ |
| Purchases | 423,000 | _____ | 401,000 | 49.45% | 22,000 | _____ |
| Available for sale | $634,000 | _____ | $594,000 | 73.24% | $40,000 | _____ |
| Inventory, December 31 | 226,000 | _____ | 211,000 | 26.02% | 15,000 | _____ |
| Cost of goods sold | $408,000 | _____ | $383,000 | 47.23% | $25,000 | _____ |
| Gross profit on sales | $567,000 | _____ | $428,000 | 52.77% | $139,000 | _____ |
| **Operating expenses:** | | | | | | |
| Salary and benefits | $221,000 | _____ | $225,000 | 27.74% | $(4,000) | _____ |
| Rent and utilities | 62,000 | _____ | 61,400 | 7.57% | 600 | _____ |
| Advertising | 32,400 | _____ | 25,700 | 3.17% | 6,700 | _____ |
| Depreciation | 40,000 | _____ | 32,000 | 3.95% | 8,000 | _____ |
| Equipment and supplies | 15,800 | _____ | 10,300 | 1.27% | 5,500 | _____ |
| Administrative | 12,500 | _____ | 14,200 | 1.75% | (1,700) | _____ |
| Total operating expense | $383,700 | _____ | $368,600 | 45.45% | $15,100 | _____ |
| Income before tax | $183,300 | _____ | $59,400 | 7.32% | $123,900 | _____ |
| Income tax | 30,300 | _____ | 24,400 | 3.01% | 5,900 | _____ |
| NET INCOME | $153,000 | _____ | $35,000 | 4.32% | $118,000 | _____ |

| **21.3** | Compute commonly used business operating ratios | Using the Balance Sheet and Income Statement for 2002 from The Bottom Line problems 1 and 2, compute the following ratios:
3. Acid test
4. Average inventory
5. Net income to net sales
6. Rate of return on investment |

Answers: 1. 2002 percent: 13.55%; 20.18%; 27.11%; 60.84%; 33.13%; −9.04%; 24.10%; 15.06%; 39.16%; 100.00%; 2.71%; 5.72%; 0.90%; 9.34%; 13.55%; 22.89%; 54.07%; 7.98%; 77.11%; 100.00%. Increase/decrease percent: 4.65%; 36.73%; 21.62%; 21.69%; 15.79%; 20.00%; 14.29%; 0%; 8.33%; 16.08%; −25.00%; 26.67%; 50.00%; 6.90%; −25.00%; −14.61%; 0%; 51.43%; 29.95%; 16.08% **2.** 2002 percent: 101.43%; 1.43%; 100%; 21.64%; 43.38%; 65.03%; 23.18%; 41.85%; 58.15%; 22.67%; 6.36%; 3.32%; 4.10%; 1.62%; 1.28%; 39.35%; 18.80%; 3.11%; 15.69%. Difference percent: 16.34%; −64.36%; 20.22%; 9.33%; 5.49%; 6.73%; 7.11%; 6.53%; 32.48%; −1.78%; 0.98%; 26.07%; 25.00%; 53.40%; −11.97%; 4.10%; 208.59%; 24.18%; 337.14% **3.** 3.61 **4.** $218,500 **5.** 15.7% **6.** 29.9%

1 Quality Construction Company, Inc., had total assets of $620,000 and total liabilities of $335,000 on December 31, 2002. On December 31, 2003, Quality Construction has total assets of $712,000 and total liabilities of $330,000.

 a. What was the amount of the owners' equity as of December 31, 2002?

 b. What is the amount of the owners' equity as of December 31, 2003?

 c. Calculate the percent of increase or decrease in total assets, total liabilities, and owners' equity. (Round to one decimal place.)

2 Quality Construction Company, Inc., had net sales of $460,250 and cost of merchandise sold of $320,600. Compute the gross profit amount and the percent of gross profit based on net sales.

3 The comparative income statement of Benson Electronics, Inc., showed sales of $425,000 in 2001 and $494,450 in 2002. Compute the percent of change in sales. (Round answer to one decimal place.)

4 Calculate the percent of increase or decrease for each of the following balance sheet items. If any percent cannot be calculated, give a brief explanation. (Answers correct to two decimal places.)

| Item | 2003 | 2002 | Percent of Increase/Decrease |
|---|---|---|---|
| a. Cash | $35,000 | $30,000 | _____ |
| b. Supplies | 1,200 | 1,600 | _____ |
| c. Notes Receivable | 2,000 | -0- | _____ |
| d. Merchandise Inventory | 16,500 | 16,500 | _____ |
| e. Accounts Receivable | -0- | 1,500 | _____ |

5 Selected figures from the Balance Sheet and the Income Statement of Multimedia, Inc., are shown below. Use the data to calculate the ratios listed. (Give answers accurate to two decimal places.)

| From the Balance Sheet | | From the Income Statement | |
|---|---|---|---|
| Cash | $210,734 | Net Sales | $244,750 |
| Accounts Receivable | $138,126 | Cost of Merchandise Sold | $190,000 |
| Merchandise Inventory: | | Net Income | $26,406 |
| End of this year | $184,500 | | |
| End of last year | $178,300 | | |
| Total Current Assets | $533,360 | | |
| Total Current Liabilities | $324,152 | | |
| Total Stockholders' Equity | $149,000 | | |

 a. Working capital ratio _____

 b. Acid test ratio _____

 c. Inventory turnover _____

 d. Rate of return on investment _____

 e. Net income as a percent of sales _____

 f. Ratio of accounts receivable to net sales _____

Assignment 21.1: **Balance Sheet Analysis**

Name

Date Score

A **(50 points) Solve the following balance sheet problems. (points for correct answers as marked)**

1. In the following balance sheet, find the percent for each 2002 and 2001 item. Then find the amount and percent of change. Round percents to two decimal places. ($\frac{1}{2}$ point for each correct answer)

Kingston Products Company—Balance Sheet—As of December 31, 2002 and 2001

| | 2002 Amount | 2002 Percent | 2001 Amount | 2001 Percent | Increase/Decrease Amount | Increase/Decrease Percent |
|---|---|---|---|---|---|---|
| ASSETS | | | | | | |
| Current assets: | | | | | | |
| Cash | $ 215,000 | _____ | $ 210,000 | _____ | _____ | _____ |
| Accounts receivable | 250,000 | _____ | 175,000 | _____ | _____ | _____ |
| Inventory | 300,000 | _____ | 275,000 | _____ | _____ | _____ |
| Total current assets | 765,000 | _____ | 660,000 | _____ | _____ | _____ |
| Fixed assets: | | | | | | |
| Machinery | 280,000 | _____ | 280,000 | _____ | _____ | _____ |
| Less depreciation | 120,000 | _____ | 100,000 | _____ | _____ | _____ |
| Machinery net | 160,000 | _____ | 180,000 | _____ | _____ | _____ |
| Building | 350,000 | _____ | 270,000 | _____ | _____ | _____ |
| Land parcel holdings | 235,000 | _____ | 190,000 | _____ | _____ | _____ |
| Total fixed assets | 745,000 | _____ | 640,000 | _____ | _____ | _____ |
| TOTAL ASSETS | $1,510,000 | _____ | $1,300,000 | _____ | _____ | _____ |
| LIABILITIES | | | | | | |
| Current liabilities: | | | | | | |
| Accounts payable | $ 90,000 | _____ | $ 62,000 | _____ | _____ | _____ |
| Accrued payroll | 45,000 | _____ | 35,000 | _____ | _____ | _____ |
| Payroll taxes payable | 15,000 | _____ | 20,000 | _____ | _____ | _____ |
| Total current liabilities | 150,000 | _____ | 117,000 | _____ | _____ | _____ |
| Long-term liabilities: | | | | | | |
| Mortgages payable | 300,000 | _____ | 310,000 | _____ | _____ | _____ |
| Notes payable—long term | 180,000 | _____ | 210,000 | _____ | _____ | _____ |
| Total long-term liabilities | 480,000 | _____ | 520,000 | _____ | _____ | _____ |
| Total liabilities | 630,000 | _____ | 637,000 | _____ | _____ | _____ |
| Shareholder's equity: | | | | | | |
| Common stock | 400,000 | _____ | 400,000 | _____ | _____ | _____ |
| Preferred stock | 330,000 | _____ | 220,000 | _____ | _____ | _____ |
| Retained earnings | 150,000 | _____ | 43,000 | _____ | _____ | _____ |
| Total shareholders' equity | 880,000 | _____ | 663,000 | _____ | _____ | _____ |
| TOTAL LIABILITIES AND SHAREHOLDERS' EQUITY | $1,510,000 | _____ | $1,300,000 | _____ | _____ | _____ |

2. Kingston Products' bookkeeper overlooked the fact that $15,000 cash had been paid to employees but not deducted from the cash account. Assume that the balance sheet in problem 1 was adjusted to reflect the correction. (1 point for each correct answer)

a. What would be the adjusted amount for 2002 cash? _____

b. What would be the adjusted amount for 2002 accrued payroll? _____

Score for A (50)

Assignment 21.1 *Continued*

(50 points) Solve the following balance sheet problems. (points for correct answers as marked)

3. In the following balance sheet, find the percent for each 2002 and 2001 item. Then find the amount and percent of change. Round percents to *one* decimal place. *Note that totals will sometimes be different from individual amounts because of rounding.* ($\frac{1}{2}$ point for each correct answer)

Blackstone Print Shop—Balance Sheet—As of December 31, 2002 and 2001

| | 2002 Amount | 2002 Percent | 2001 Amount | 2001 Percent | Increase/Decrease Amount | Increase/Decrease Percent |
|---|---|---|---|---|---|---|
| **ASSETS** | | | | | | |
| Current assets: | | | | | | |
| Cash | $ 44,000 | | $34,000 | | | |
| Accounts receivable | 28,500 | | 24,800 | | | |
| Inventory | 33,800 | | 32,500 | | | |
| Total current assets | 106,300 | | 91,300 | | | |
| Fixed assets: | | | | | | |
| Equipment | 57,500 | | 52,000 | | | |
| Less depreciation | 14,600 | | 13,300 | | | |
| Equipment net | 42,900 | | 38,700 | | | |
| Building | 200,000 | | 200,000 | | | |
| Land parcel holdings | 65,800 | | 50,000 | | | |
| Total fixed assets | 308,700 | | 288,700 | | | |
| TOTAL ASSETS | $415,000 | | $380,000 | | | |
| **LIABILITIES** | | | | | | |
| Current liabilities: | | | | | | |
| Accounts payable | $ 12,600 | | $11,000 | | | |
| Accrued payroll | 9,200 | | 7,000 | | | |
| Payroll taxes payable | 1,200 | | 1,000 | | | |
| Total current liabilities | 23,000 | | 19,000 | | | |
| Long-term liabilities: | | | | | | |
| Mortgages payable | 70,000 | | 60,000 | | | |
| Notes payable—long term | 26,000 | | 30,000 | | | |
| Total long-term liabilities | 96,000 | | 90,000 | | | |
| Total liabilities | 119,000 | | 109,000 | | | |
| Shareholders' equity: | | | | | | |
| Common stock | 180,000 | | 180,000 | | | |
| Preferred stock | 88,000 | | 84,000 | | | |
| Retained earnings | 28,000 | | 7,000 | | | |
| Total shareholders' equity | 296,000 | | 271,000 | | | |
| TOTAL LIABILITIES AND SHAREHOLDERS' EQUITY | $415,000 | | $380,000 | | | |

4. Show what changes would have been made in the cash and preferred stock amount in 2002 if Blackstone Print Shop had sold an additional $6,000 in preferred stock. ($\frac{1}{2}$ point for each correct answer)

| | Amount | Percent |
|---|---|---|
| Cash | | |
| Preferred stock | | |

Score for B (50)

Assignment 21.2: Income Statement Analysis

Name _____

Date _____ Score _____

A **(50 points) Solve the following income statement problems. (points for correct answers as marked)**

1. In the following income statement, find the percent for each 2002 and 2001 item. Then find the amount and percent of change. Round percents to two decimal places. ($\frac{1}{2}$ point for each correct answer)

Shasta Boat Sales
Income Statement
For the Years Ended December 31, 2002 and 2001

| | 2002 Amount | 2002 Percent | 2001 Amount | 2001 Percent | Difference Amount | Difference Percent |
|---|---|---|---|---|---|---|
| Revenue from sales: | | | | | | |
| Sales | $850,000 | _____ | $725,000 | _____ | _____ | _____ |
| Less returns | 30,000 | _____ | 25,000 | _____ | _____ | _____ |
| NET SALES | $820,000 | _____ | $700,000 | _____ | _____ | _____ |
| Cost of goods sold: | | | | | | |
| Inventory, January 1 | 210,000 | _____ | 190,000 | _____ | _____ | _____ |
| Purchases | 440,000 | _____ | 390,000 | _____ | _____ | _____ |
| Available for sale | 650,000 | _____ | 580,000 | _____ | _____ | _____ |
| Inventory, December 31 | 250,000 | _____ | 210,000 | _____ | _____ | _____ |
| Cost of goods sold | 400,000 | _____ | 370,000 | _____ | _____ | _____ |
| Gross profit | $420,000 | _____ | $330,000 | _____ | _____ | _____ |
| Operating expenses: | | | | | | |
| Wages | 125,000 | _____ | 98,000 | _____ | _____ | _____ |
| Rent | 82,000 | _____ | 78,000 | _____ | _____ | _____ |
| Advertising | 15,000 | _____ | 20,000 | _____ | _____ | _____ |
| Insurance | 4,000 | _____ | 3,500 | _____ | _____ | _____ |
| Depreciation | 3,500 | _____ | 2,800 | _____ | _____ | _____ |
| Equipment rental | 1,400 | _____ | 900 | _____ | _____ | _____ |
| Administrative | 7,000 | _____ | 5,000 | _____ | _____ | _____ |
| Miscellaneous | 2,100 | _____ | 1,800 | _____ | _____ | _____ |
| Total operating expenses | 240,000 | _____ | 210,000 | _____ | _____ | _____ |
| Income before tax | 180,000 | _____ | 120,000 | _____ | _____ | _____ |
| Income tax | 30,000 | _____ | 20,000 | _____ | _____ | _____ |
| NET INCOME | $150,000 | _____ | $100,000 | _____ | _____ | _____ |

2. Assume that the ending inventory was $220,000 in 2002. Compute the following items. (2 points for each correct answer)

2002 Gross profit amount _____ 2001 Gross profit percent _____

2002 NET INCOME amount _____ 2001 NET INCOME percent _____

Score for A (50)

Assignment 21.2 *Continued*

B **(100 points) Solve the following income statement problems. (points for correct answers as marked)**

3. In the following income statement, find the percent for each 2002 and 2001 item, then find the amount and percent of change. Round percents (no decimal place). (84 points, 1 point for each correct answer)

Dixon Field Enterprises

Income Statement

For the Years Ended December 31, 2002 and 2001

| | 2002 Amount | 2002 Percent | 2001 Amount | 2001 Percent | Difference Amount | Difference Percent |
|---|---|---|---|---|---|---|
| Revenue from sales: | | | | | | |
| Sales | $85,000 | _____ | $73,500 | _____ | _____ | _____ |
| Less Returns | 2,000 | _____ | 1,500 | _____ | _____ | _____ |
| NET SALES | $83,000 | _____ | 72,000 | _____ | _____ | _____ |
| Cost of goods sold: | | | | | | |
| Inventory, January 1 | 24,500 | _____ | 17,500 | _____ | _____ | _____ |
| Purchases | 38,500 | _____ | 37,000 | _____ | _____ | _____ |
| Available for sale | 63,000 | _____ | 54,500 | _____ | _____ | _____ |
| Inventory, December 31 | 26,000 | _____ | 24,500 | _____ | _____ | _____ |
| Cost of goods sold | 37,000 | _____ | 30,000 | _____ | _____ | _____ |
| Gross profit | $46,000 | _____ | $42,000 | _____ | _____ | _____ |
| Operating expenses: | | | | | | |
| Salary | 11,400 | _____ | 10,100 | _____ | _____ | _____ |
| Rent | 7,100 | _____ | 6,900 | _____ | _____ | _____ |
| Advertising | 800 | _____ | 1,860 | _____ | _____ | _____ |
| Delivery | 200 | _____ | 340 | _____ | _____ | _____ |
| Depreciation | 750 | _____ | 200 | _____ | _____ | _____ |
| Equipment rental | 350 | _____ | 160 | _____ | _____ | _____ |
| Administrative | 980 | _____ | 530 | _____ | _____ | _____ |
| Miscellaneous | 190 | _____ | 220 | _____ | _____ | _____ |
| Total operating expenses | 21,770 | _____ | 20,310 | _____ | _____ | _____ |
| Income before tax | 24,230 | _____ | 21,690 | _____ | _____ | _____ |
| Income tax | 2,630 | _____ | 1,690 | _____ | _____ | _____ |
| NET INCOME | $21,600 | _____ | $20,000 | _____ | _____ | _____ |

4. Assume that the beginning inventory was $21,500 in 2001 and $30,500 in 2002 and that the rent was $6,400 in 2001 and $6,800 in 2002. Compute the following amounts and percents to reflect the revised beginning inventory and rent numbers. (8 points for each correct row)

| | 2002 Amount | 2002 Percent | 2001 Amount | 2001 Percent | Difference Amount | Difference Percent |
|---|---|---|---|---|---|---|
| Gross profit | _____ | _____ | _____ | _____ | _____ | _____ |
| NET INCOME | _____ | _____ | _____ | _____ | _____ | _____ |

Score for B (100)

Assignment 21.3: Financial Statement Ratios

Name _____

Date _____ Score _____

A **(26 points) Solve the following financial statement ratio problems. ($\frac{1}{2}$ point for each correct answer)**

1. Karen Bailey was considering investing in a business. She used the following statements in analyzing the Bear Country Boutique. Compute the net changes in the balance sheet and income statement. Round to one decimal place.

Bear Country Boutique

Comparative Balance Sheet

As of December 31, 2002 and 2001

| | 2002 | 2001 | Increase/Decrease Amount | Percent |
|---|---|---|---|---|
| ASSETS | | | | |
| Current assets: | | | | |
| Cash | $107,400 | $104,200 | _____ | _____ |
| Accounts receivable | 138,200 | 112,800 | _____ | _____ |
| Merchandise inventory | 172,000 | 133,000 | _____ | _____ |
| Total current assets | $417,600 | $350,000 | _____ | _____ |
| Fixed assets: | | | | |
| Building improvements | $ 40,000 | $ 46,500 | _____ | _____ |
| Equipment | 136,000 | 115,000 | _____ | _____ |
| Total fixed assets | $176,000 | $161,500 | _____ | _____ |
| TOTAL ASSETS | $593,600 | $511,500 | _____ | _____ |
| LIABILITIES | | | | |
| Current liabilities: | | | | |
| Salaries payable | $ 34,600 | $ 28,800 | _____ | _____ |
| Accounts payable | 121,400 | 113,200 | _____ | _____ |
| Total current liabilities | $156,000 | $142,000 | _____ | _____ |
| Long-term liabilities: | | | | |
| Notes payable | $159,000 | $190,000 | _____ | _____ |
| Total liabilities | $315,000 | $332,000 | _____ | _____ |
| Owner's equity: | | | | |
| B. A. Bear, capital | $278,600 | $179,500 | | _____ |
| TOTAL LIABILITIES AND OWNER'S EQUITY | $593,600 | $511,500 | _____ | _____ |

Assignment 21.3 *Continued*

Bear Country Boutique—Comparative Income Statement—For the Years Ended December 31, 2002 and 2001

| | 2002 | 2001 | Difference Amount | Percent |
|---|---|---|---|---|
| Net sales | $750,000 | $835,000 | _____ | _____ |
| Cost of goods sold: | | | | |
| Merchandise inventory, January 1 | $133,000 | $130,000 | _____ | _____ |
| Purchases | 574,000 | 615,000 | _____ | _____ |
| Merchandise available for sale | $707,000 | $745,000 | _____ | _____ |
| Merchandise inventory, December 31 | 172,000 | 133,000 | _____ | _____ |
| Cost of goods sold | $535,000 | $612,000 | _____ | _____ |
| Gross profit on sales | $215,000 | $223,000 | _____ | _____ |
| Expenses: | | | | |
| Selling | $ 81,000 | $ 78,200 | _____ | _____ |
| Other | 27,200 | 28,800 | _____ | _____ |
| Total expenses | $108,200 | $107,000 | _____ | _____ |
| Net income | $106,800 | $116,000 | _____ | _____ |

Score for A (26)

B **(24 points) Solve the following problems. (2 points for each correct answer)**

2. Provide the following information for Karen Bailey's consideration. When the ratio is less than 1, give the ratio to three decimal places; otherwise, round to one decimal place.

| | 2002 | 2001 |
|---|---|---|
| **a.** Working capital ratio | _____ | _____ |
| **b.** Acid-test ratio | _____ | _____ |
| **c.** Ratio of accounts receivable to net sales | _____ | _____ |
| **d.** Inventory turnover | _____ | _____ |
| **e.** Ratio of net income to net sales | _____ | _____ |
| **f.** Rate of return on investment | _____ | _____ |

Score for B (24)

C **(26 points) Solve the following problems. ($\frac{1}{2}$ point for each correct answer)**

3. Karen Bailey was offered a second business. She received the following statements for 2002 and 2001. Complete calculations for a comparative balance sheet and a comparative income statement for Banner Boutique, showing the amount and the percent of change.

Assignment 21.3 *Continued*

Banner Boutique—Comparative Balance Sheet—As of December 31, 2002 and 2001

| | 2002 | | 2001 | | Increase/Decrease | |
|---|---|---|---|---|---|---|
| | Amount | Percent | Amount | Percent | Amount | Percent |
| ASSETS | | | | | | |
| Current assets: | | | | | | |
| Cash | $ 25,000 | 18.2% | $16,000 | 14.7% | _____ | _____ |
| Accounts receivable | 12,000 | 8.8% | 8,000 | 7.3% | _____ | _____ |
| Merchandise inventory | 46,000 | 33.6% | 31,000 | 28.4% | _____ | _____ |
| Total current assets | $ 83,000 | 60.6% | $ 55,000 | 50.5% | _____ | _____ |
| Fixed assets: | | | | | | |
| Store fixtures | $ 39,000 | 28.5% | $ 43,000 | 39.4% | _____ | _____ |
| Office equipment | 15,000 | 10.9% | 11,000 | 10.1% | _____ | _____ |
| Total fixed assets | $ 54,000 | 39.4% | $ 54,000 | 49.5% | _____ | _____ |
| TOTAL ASSETS | $137,000 | 100.0% | $109,000 | 100.0% | _____ | _____ |
| LIABILITIES | | | | | | |
| Current liabilities: | | | | | | |
| Sales tax payable | $ 4,500 | 3.3% | $ 5,500 | 5.0% | _____ | _____ |
| Accounts payable | 9,500 | 6.9% | 6,000 | 5.5% | _____ | _____ |
| Total current liabilities | $ 14,000 | 10.2% | $ 11,500 | 10.6% | _____ | _____ |
| Long-term liabilities: | | | | | | |
| Note payable | $ 30,000 | 21.9% | $ 38,000 | 34.9% | _____ | _____ |
| Total liabilities | $ 44,000 | 32.1% | $ 49,500 | 45.4% | _____ | _____ |
| Owner's equity | | | | | | |
| R. A. Banner, capital | $ 93,000 | 67.9% | $ 59,500 | 54.6% | _____ | _____ |
| TOTAL LIABILITIES AND OWNER'S EQUITY | $137,000 | 100.0% | $109,000 | 100.0% | _____ | _____ |

Banner Boutique—Comparative Income Statement—For the Years Ended December 31, 2002 and 2001

| | 2002 | | 2001 | | Difference | |
|---|---|---|---|---|---|---|
| | Amount | Percent | Amount | Percent | Amount | Percent |
| Net sales | $205,000 | 100.0% | $120,000 | 100.0% | _____ | _____ |
| Cost of goods sold: | | | | | | |
| Merchandise inventory, January 1 | $ 31,000 | 15.1% | $ 27,500 | 22.9% | _____ | _____ |
| Purchases | 154,000 | 75.1% | 84,500 | 70.4% | _____ | _____ |
| Merchandise available for sale | $185,000 | 90.2% | $112,000 | 93.3% | _____ | _____ |
| Merchandise inventory, December 31 | 46,000 | 22.4% | 31,000 | 25.8% | _____ | _____ |
| Cost of goods sold | $139,000 | 67.8% | $ 81,000 | 67.5% | _____ | _____ |
| Gross profit on sales | $ 66,000 | 32.2% | $ 39,000 | 32.5% | _____ | _____ |
| Expenses: | | | | | | |
| Selling | $ 31,000 | 15.1% | $ 21,500 | 17.9% | _____ | _____ |
| Other | 13,000 | 6.3% | 7,250 | 6.0% | _____ | _____ |
| Total expenses | $ 44,000 | 21.5% | $ 28,750 | 24.0% | _____ | _____ |
| Net income | $ 22,000 | 10.7% | $ 10,250 | 8.5% | _____ | _____ |

Any differences of 0.1% from individual items are due to rounding.

Score for C (26)

Assignment 21.3 *Continued*

D **(24 points) Solve the following problems. (points for correct answers as marked)**

4. Use the figures from Part C to compute the following ratios and comparisons for Banner Boutique. Give ratios to two decimal places. (2 points for each correct answer)

| | **2002** | **2001** |
|---|---|---|
| **a.** Working capital ratio | _____ | _____ |
| **b.** Acid-test ratio | _____ | _____ |
| **c.** Ratio of accounts receivable to net sales | _____ | _____ |
| **d.** Relationship of net income to net sales | _____ | _____ |
| **e.** Rate of return on investment | _____ | _____ |

Refer to sections B and D of this assignment to answer the following questions. (2 points for each correct answer)

5a. As of December 31, 2002, which of the two businesses had the better (larger number) return on the owner's equity?

b. As of December 31, 2002, which of the two businesses had the stronger (larger number) ability to provide working capital for new investment? _____

Score for D (24)

Notes

There are approximately 10,000 Employee stock ownership plans (ESOPs) in the United States covering 11.5 million employees (10% of the workforce). These employees draw in excess of 3% of their total compensation from ESOP contributions.

The growth of ESOP formation has been influenced by federal legislation. Although the rapid increase in new ESOPs in the late 1980s subsided after Congress removed certain tax incentives in 1989, the overall number has remained steady with new plans replacing terminated ESOPs. Visit this ESOP site at http://www.the-esop-emplowner.org/ for more information.

Corporate Stocks

22

LEARNING OBJECTIVES

By studying this chapter and completing all assignments you will be able to:

1 Compute the costs and proceeds of stock buy-and-sell transactions.

2 Compute the costs and proceeds of round and odd lots.

3 Compute rates of yield and gains or losses on the purchase and sale of stocks.

4 Compute comparative earning potential of the major classes of corporate stocks.

Buying and selling stocks is a common practice by companies, banks, investors, and the general public. The number of individuals who own stocks is increasing every year. Computing rates of yield and gains and losses on the purchase and sale of stocks is a necessary skill in today's world.

1. What is a share of stock? Does it have monetary value?
2. Who owns the stock of U.S. and foreign corporations?
3. Is there a fee for purchasing and selling stock?
4. Can you make a profit if you buy shares of a stock that goes up and sell it for more than you paid for it? Can you lose money if you buy shares of a stock that then goes down? Can you afford to play the stock market, or should you put your money in the bank?
5. What is the difference between *preferred* and *common* stock?

Many companies operate as corporations. A **corporation** is a body that is granted a charter by a state legally recognizing it as a separate entity, having its own rights, privileges, and liabilities distinct from those of its owners. A corporation acquires assets, enters into contracts, sues or is sued, and pays taxes in its own name. Two primary reasons for forming a corporation are to limit liability and facilitate broadening the ownership base. A corporation raises capital by selling shares of ownership, which increases its assets without increasing its debt.

The general term applied to the shares of a corporation is **capital stock**. Each share of capital stock is a share of the ownership of the company's net assets (assets minus liabilities). The number of shares that a corporation is authorized to *issue*, or offer for sale, is set forth in its **charter,** the basic approval document issued by the state, under which the corporation operates. Ownership of stock is evidenced by a **stock certificate**.

Frequently, the shares of capital stock are assigned a value known as **par**, which is stated on the stock certificate. For example, a company incorporated with capital stock of $1,000,000 and 100,000 shares has a par value of $10 per share. Stock issued without par value is known as **no-par** *stock*. The par value may differ from the market price. In the marketplace, stock may be sold for any amount agreed upon by the buyer and seller.

www.bbandt.com/ investments/products_ corporatestocks.html

For detailed information on corporate stocks—definitions of terms as well as data—visit this Web site.

Computing the Costs and Proceeds of Stock Transactions

LEARNING OBJECTIVES **1**

Compute the costs and proceeds of stock buy-and-sell transactions.

After purchasing stock, a buyer may sell that stock at any price on the open market, regardless of the par value. Stocks are usually bought and sold on *stock exchanges,* the formal marketplaces set up for the purpose of trading stocks. Major exchanges in the United States are the New York Stock Exchange (NYSE), the American Stock Exchange (AMEX), and the National Association of Securities Dealers Automated Quotations (NASDAQ). A **stockbroker** usually handles **stock transactions**—the purchase and sale of stocks for clients. Today, many people also trade via the Internet.

The trading of shares of stock is published daily in newspapers. Figure 22-1 shows a sample stock market report, in which stocks are quoted in the traditional manner— dollars and fractions of a dollar. The NYSE, NASDAQ, and AMEX quote prices in hundredths. Consequently, the smallest increase or decrease in a stock price that will be reported is .01.

Both the buyer and the seller of stock pay commissions to the stockbroker. The total amount paid by a buyer to purchase a stock includes the market price of the stock and the stockbroker's commission (charge). The **total cost** paid by the purchaser is equal to the purchase price plus a broker's commission. The **proceeds** received by the seller are equal to the selling price minus the commission.

Broker commissions may be a flat rate per transaction, a percent of the value of the stock, an amount per share traded, or an amount negotiated between the client and the broker. Generally, commissions for brokers are less than 1% of the value of the stock, ranging from $0.02 to $0.50 per share bought or sold. A number of discount brokerages operating on the Internet now charge $9.95 to $24.95 per transaction, normally for up to 5,000 shares. Figure 22-2 shows a broker's confirmation report of a stock purchase with a commission rate of $50 and a transaction fee of $3.

We use a transaction charge of $0.20 per share or a flat fee of $19.95 per transaction in computing the cost of commissions in this chapter.

Figure 22-1: Daily Stock Report from the NYSE

| 52 weeks | | | | | | Yld | Vol | | | | |
|---|---|---|---|---|---|---|---|---|---|---|---|
| High [1] | Low [2] | Stock [3] | Sym [4] | Div [5] | % [6] | PE [7] | 100s [8] | Hi [9] | Low [10] | Close [11] | Chg. [12] |
| 46.75 | 31.63 | AT&T | T | 1.32 | 3.3 | 11 | 33943 | 39.63 | 39 | 39.63 | + .88 |
| 36.13 | 10.25 | AdvMicro | AMD | | | dd* | 29826 | 34.5 | 31.75 | 34.25 | + 2.25 |
| 71 | 22.38 | AmOnline | AOL | | | dd* | 3253 | 37.25 | 34 | 37.13 | + 3.88 |
| 37.25 | 29.75 | Ford Motor | F | 1.54 | 4.7 | 9 | 25371 | 32.75 | 32.13 | 32.75 | +.5 |
| 28.25 | 89.13 | IBM | IBM | 1.40 | 1.0 | 14 | 46288 | 146.63 | 143 | 145 | −.25 |
| 107 | 77.63 | Exxon | XON | 3.16 | 3.1 | 17 | 15997 | 103.38 | 101.75 | 103 | + 1.38 |
| 87 | 66.75 | DeltaAir | DAL | .20 | .2 | 50 | 4934 | 84.25 | 82.25 | 83.63 | + 2.25 |

*dd = Loss in the most recent four quarters.

[1] The highest price per share in the previous 52 weeks.
[2] The lowest price per share in the previous 52 weeks.
[3] Company names, often abbreviated to fit in stock tables, are listed alphabetically.
[4] The symbol is a stock's designation on databases and quote machines.
[5] The dividend shown usually is the annual rate based on the company's latest payout.
[6] The dividend divided by the closing share price gives the stock's yield.
[7] One measure of a stock's value is its price/earnings ratio (PE). It is based on the per-share earnings as reported by the company for the four most recent quarters. The PE number is found by dividing the current price by those most recent four-quarter earnings.
[8] Volume is the number of shares traded that day, shown in hundreds of shares.
[9] The high for the day's trading range.
[10] The low for the day's trading range.
[11] The closing price on that day.
[12] The net change in price lets you calculate something that isn't in the stock table: the previous day's closing price.

Figure 22-2: Confirmation Report of a Stock Purchase

A.G. Edwards & Sons, Inc. INVESTMENTS SINCE 1887 ONE NORTH JEFFERSON ST. LOUIS, MISSOURI 63103 (314) 955-3000

WE CONFIRM THE FOLLOWING TRANSACTION SUBJECT TO THE AGREEMENT ON THE REVERSE SIDE

| YOU | QUANTITY | PRICE | SECURITY DESCRIPTION | CUSIP NUMBER |
|---|---|---|---|---|
| BOUGHT | 50 | 99.88 | GENERAL ELECTRIC CO | |

| ACCOUNT NUMBER | IB | T | TRF | MKT | OFFICE PHONE NUMBER | SYMBOL |
|---|---|---|---|---|---|---|
| | 47 | 1 | 4 | 3 | 714-756-0353 | GE |

WHEN COMMUNICATING WITH US PLEASE REFER TO YOUR ACCOUNT NUMBER

| TRADE DATE | | | SETTLEMENT DATE | | |
|---|---|---|---|---|---|
| 12 | 31 | 01 | 01 | 06 | 02 |

PLEASE PAY OR DELIVER BY THIS DATE

| PRINCIPAL | STATE TAX | ACCRUED INTEREST | COMMISSION | SEC FEE | TRANSACTION CHARGE | AMOUNT |
|---|---|---|---|---|---|---|
| 4,994.00 | | | 50.00 | | 3.00 | 5,047.00 |

© INDEX STOCK IMAGERY

EXAMPLE A

Lori Stephens bought 200 shares of Sears stock at 32. What was her cost, including commission of $0.20 per share?

| 200 shares × $32 price | = $6,400 | purchase price |
|---|---|---|
| 200 shares × $0.20 commission = | + 40 | commission |
| | $6,440 | total cost |

EXAMPLE B

Joseph Lankowski sold 1,200 shares of Compaq Computer Corporation (COMPAQ) at 20.25, less commission of $0.20 per share. What were the proceeds of the sale?

| 1,200 shares × $20.25 | = $24,300 | selling price |
|---|---|---|
| 1,200 shares × $0.20 commission = | − 240 | commission |
| | $24,060 | proceeds |

EXAMPLE C

Juan Hernandez bought 500 shares of Pepsico stock at 45.375. What was his cost, including a flat fee of $19.95?

| 500 shares × $45.375 price = | $22,687.50 | purchase price |
|---|---|---|
| commission | = + 19.95 | flat fee |
| | $22,707.45 | total cost |

CONCEPT CHECK 22.1

Shane Ovitz purchased 200 shares of Safeway at 18.25. He later sold the stock at 20.5. What was his gain/loss on the purchase and sale, after counting commissions of $0.20 per share on the purchase and the sale?

| Purchase: 200 shares × $18.25 price | = $3,650 | purchase price |
|---|---|---|
| 200 shares × $0.20 commission = | + 40 | commission |
| | $3,690 | total cost |
| Sale: 200 shares × $20.50 price | = $4,100 | selling price |
| 200 shares × $0.20 commission | = − 40 | commission |
| | $4,060 | proceeds |

$4,060 proceeds − $3,690 cost = $370 gain

Computing the Costs and Proceeds of Round and Odd Lots

2
LEARNING OBJECTIVES

Compute the costs and proceeds of round and odd lots.

Stocks are sold in round lots, odd lots, or a combination of the two. A **round lot** usually is 100 shares. An **odd lot** consists of any number of shares less than 100 (1 to 99 shares is an odd lot for a stock with a 100-share round lot). When odd lots are purchased, a small extra charge, or **odd-lot differential,** is commonly added to the round-lot price.

The differential is added to the price for a purchaser and deducted from the price for the seller. In this book, we use a differential of 12.5 cents as the odd-lot rate.

EXAMPLE D

Charles Wang bought 140 shares of Pixar at 28. What was his cost?

Odd-lot purchase price = $28 + $0.125 = $28.125 per odd-lot share

| | |
|---|---|
| 100 shares × $28.00 round-lot price = $2,800.00 | round-lot total cost |
| 40 shares × $28.125 odd-lot price = 1,125.00 | odd-lot total cost |
| 140 shares × $0.20 commission = + 28.00 | commission |
| $3,953.00 | total cost |

EXAMPLE E

Charles sold 140 shares of Pixar at 28. What was the amount of his net proceeds?

Odd-lot selling price = $28 − $0.125 = $27.875

| | |
|---|---|
| 100 shares × $28.00 round-lot price = $2,800.00 | round-lot price |
| 40 shares × $27.875 odd-lot price = 1,115.00 | odd-lot price |
| 140 shares × $0.20 commission = − 28.00 | commission |
| $3,887.00 | net proceeds |

CONCEPT CHECK 22.2

James O'brien bought 160 shares of PG&E at 25.5. What was his total cost?

Odd-lot purchase price = $25.50 + $0.125 = $25.625
100 shares × $25.50 round-lot price = $2,550.00
60 shares × $25.625 odd-lot price = 1,537.50
160 shares × $0.20 commission = + 32.00
Total cost $4,119.50

Sarah Loeb sold 220 shares of Intel at 153.25. What was the amount of her net proceeds?

Odd-lot selling price = $153.25 − $0.125 = $153.125
200 shares × $153.25 round-lot price = $30,650.00
20 shares × $153.125 odd-lot price = + 3,062.50
220 shares × $0.20 commission = − 44.00
Net proceeds $33,668.50

Computing the Rate of Yield and Gains or Losses

The Rate of Yield

The **board of directors** is a group of people elected by shareholders to oversee the operations of the corporation. The board has sole authority to distribute earnings to shareholders. When such action is taken, the directors are said to **declare a dividend.** The rate of dividend is either a certain percent of the par value of the stock or a flat amount of

LEARNING OBJECTIVES

Compute rates of yield and gains or losses on the purchase and sale of stocks.

money per share. Thus a dividend of 8% on a stock with a par value of $100 would be $8.00 per share. Most large corporations pay dividends quarterly.

The **rate of yield** from an investment in stock is the ratio of the dividend to the total cost of the stock.

EXAMPLE F

Daniel Garcia bought 200 shares of MTV stock at 72 and paid a $19.95 commission. A dividend of $3.75 per share was paid this year. What was the rate of yield?

$200 \times \$72 = \$14,400.00$ purchase price
$\underline{+\quad\quad 19.95}$ commission
$\$14,419.95$ total cost

$200 \times \$3.75 \quad\quad = \750 dividend for first year
$\$750 \div \$14,419.95 = 5.2\%$ rate of yield

Gain or Loss on Sale of Stock

For income tax and accounting purposes, the amount of gain or loss on a sale of stock is determined by comparing the sale proceeds to the total cost.

EXAMPLE G

Refer back to example F. If Daniel sold his stock after 3 years at $81.5, less $19.95 commission, what were the amount and the percent of gain or loss?

$200 \times \$81.50 = \$16,300.00$ selling price
$\underline{-\quad\quad 19.95}$ commission
$\$16,280.05$ proceeds

$\$16,280.05$ proceeds $- \$14,419.95$ cost (example F) $= \$1,860.10$ net gain
$\$1,860.10 \div \$14,419.95 = 12.9\%$ gain on sale

EXAMPLE H

Suppose that Daniel held his stock for 3 years and received a $750 dividend each year. Then to determine the total change in value (example G) he would need to add to his proceeds the $2,250 in dividends received.

| Proceeds | Total Dividends | Total Cost | |
|---|---|---|---|
| ($16,280.05 | + $2,250) | − $14,419.95 | = $4,110.10 total gain in value |

$\$4,110.10$ total gain $\div \$14,419.95$ initial cost $= 28.5\%$ gain in value

CONCEPT CHECK 22.3

a. Lila Habib owns 700 shares of stock with a par value of $100. If she receives a dividend of 5%, how much will her total dividend be?

$100 par value \times 5% per share = $5.00 per-share dividend
700 shares \times $5.00 per share = $3,500 total dividend

b. Lila also owns 300 shares of a stock without a stated par value. If she receives a dividend of $2.00 per share, what will her total dividend be?

300 shares \times $2.00 per share = $600 total dividend

c. Madelyn Kline bought 200 shares of Clorox at 32.25. A dividend of $0.45 per share was paid this year. What was the rate of yield?

200 shares × $32.25 = $6,450 purchase price
200 shares × $0.20 = + 40 commission
 $6,490 total cost

200 shares × $0.45 dividend = $90 for first year
$90 dividend ÷ $6,490 total cost = 1.39% rate of yield

d. After four years, Madelyn sold the Clorox stock for 32.50. What were the amount and percent of gain or loss on the sale?

200 shares × $32.50 selling price = $6,500 selling price
200 shares × $0.20 commission = − 40 commission
 $6,460 proceeds

$6,460 proceeds − $6,490 total cost = $(30) loss
$(30) loss ÷ $6,490 total cost = 0.46% loss

e. If Madelyn held the Clorox stock for 4 years, receiving the same $90 dividend each year, what was the total change in the value over the four years?

| Proceeds + | Total Dividends (4 years) | − | Total Cost | = | Gain in Value |
|---|---|---|---|---|---|
| ($6,460 + | $360) | − | $6,490 | = | $330 |

$330 gain in value ÷ $6,490 total cost = 5.08% gain

Computing Comparative Earning Potential

Common stock is the usual type of stock issued by a corporation. Another type frequently issued, **preferred stock,** gives holders a right to share in earnings and liquidation before common shareholders do. For example, a company that has a 7% preferred stock must first pay dividends of 7% of the par value to the holders of preferred stock before anything is paid to the holders of common stock. Preferred stock may be designated as **cumulative—** that is, if the corporation doesn't pay the specified percentage, the unpaid amount, called a dividend in arrears, carries over to the following year or years. If dividends aren't paid on noncumulative preferred stock during one year, the unpaid amount doesn't carry over to the next year.

LEARNING OBJECTIVES

Compute comparative earning potential of the major classes of corporate stocks.

EXAMPLE I

The ABC Company earned $48,000 last year. The capital stock of the company consists of 10,000 shares of 7% preferred stock, with a par value of $40 per share, and 50,000 shares of no-par common stock. If the board of directors declared a dividend of the entire earnings, what amount would be paid in total to the preferred and common shareholders and how much would each common shareholder receive?

Preferred: 10,000 shares × $40 par value = $400,000 total value
$400,000 value × 0.07 = $28,000 paid to preferred

Common: $48,000 total earnings − $28,000 paid to preferred = $20,000
$20,000 ÷ 50,000 shares = $0.40 paid per share to common

EXAMPLE J

Assume in example I that the preferred stock is cumulative and that for the preceding year the company had declared a dividend of only $16,000, or enough to pay a 4% dividend on preferred stock. The earnings of $48,000 for this year would be divided as follows.

Unpaid dividend from preceding year: 7% − 4% = 3%
$400,000 preferred par value × 0.03 = $12,000 cumulative (dividend in arrears)
$400,000 × 0.07 dividend for current year = $28,000
Total paid on preferred stock = $40,000
$48,000 total earnings − $40,000 paid to preferred = $ 8,000
$8,000 ÷ 50,000 common shares = $0.16 dividend per common share

Another feature that sometimes makes preferred stock an attractive investment is the possibility of converting the preferred stock into common stock. **Convertible preferred stock** gives the owner the option of converting those preferred shares into a stated number of common shares. For example, a stated conversion of 1 to 3 means that 1 share of preferred stock could be changed into 3 shares of common stock. The conversion feature combines the safety of preferred stock with the possibility of growth through conversion to common stock.

EXAMPLE K

Lucas Conroy owned 100 shares of ABC convertible preferred stock at $10 par value. He converted each share of preferred into 3 shares of common. How many shares of common stock did Lucas receive when he converted?

100 × 3 = 300 shares of common stock

If common stock was selling at $17 per share on the date of conversion, how much was Lucas's common stock worth?

$17 × 300 shares = $5,100 common stock value

If Lucas paid $38 per share for his preferred stock, how much had his investment increased?

$38 × 100 preferred = $3,800 preferred stock value
$5,100 − $3,800 = $1,300 increase in value

If the convertible stock pays 9% annually and the common stock usually pays $0.50 per share, how much more dividend might Lucas expect to receive annually?

$10 par value × 100 shares = $1,000
$1,000 × 0.09 = $90 preferred stock dividend
300 shares × $0.50 = $150 common stock dividend
$150 − $90 = $60 more dividend annually

CONCEPT CHECK 22.4

a. The XYZ Corporation had a net profit of $116,000 in the fiscal year just ended. The capital stock consists of 6,000 shares of 8% convertible preferred stock with a par value of $50 per share and 20,000 shares of no-par common stock. If the board of directors declared a dividend of the entire earnings, what amount would be paid to preferred and common shareholders?

Preferred: 6,000 shares × $50 per share = $300,000 total par value
$300,000 par value × 8% = $24,000 paid to preferred shareholders
Common: $116,000 total earnings − $24,000 paid to preferred = $92,000 to be paid to common shareholders

b. James Wert owns 1,000 shares of convertible preferred stock in the XYZ Corporation, with a current market price of $52.00 per share. The preferred stock is convertible to common stock at the rate of 2 shares of common for each share of preferred. After the end of the year in part (a), common stock was selling for $32 per share. What would be the current market value of his stock before and after a conversion?

Preferred: 1,000 shares × $52 per share = $52,000 current value
Common: 1,000 shares preferred × 2 = 2,000 shares common
2,000 shares × $32 per share = $64,000 current value

COMPLETE ASSIGNMENTS 22.1 AND 22.2.

Chapter Terms for Review

<table>
<tr><td>board of directors</td><td>par</td></tr>
<tr><td>capital stock</td><td>P/E ratio</td></tr>
<tr><td>charter</td><td>preferred stock</td></tr>
<tr><td>common stock</td><td>proceeds (from sale of stock)</td></tr>
<tr><td>convertible preferred stock</td><td>rate of yield</td></tr>
<tr><td>corporation</td><td>round lot</td></tr>
<tr><td>cumulative preferred stock</td><td>stockbroker</td></tr>
<tr><td>declare a dividend</td><td>stock certificate</td></tr>
<tr><td>dividend in arrears</td><td>stock exchanges</td></tr>
<tr><td>no-par stock</td><td>stock transactions</td></tr>
<tr><td>odd lot</td><td>total cost (for purchaser of stock)</td></tr>
<tr><td>odd-lot differential</td><td></td></tr>
</table>

Visit the Web

http://www.datachimp.com/—Lookup information on stocks, portfolios, and other financial information at this Website.

http://www.sec.gov—Information on government regulations and oversight of securities markets, and more, can be found here.

http://www.stock1.com/—This Web site gives information about buying stocks directly.

http://www.nyse.com, http://www.nasd.com, http://www.csfbdirect.com, http://www.schwab.com, http://www.moodys.com, http://www.etrade.com, and **http://www.barrons.com**—These are other important and useful Web sites dealing with stocks and bonds.

CYBERCONNECTION

RedHerring.com is a leading provider of in-depth online investment information and customized investor relations technologies and services. Its Web site provides individual investors with the ability to analyze and monitor their investments using the most advanced Web technologies and tools available. Services include stock quotes, comprehensive charts, portfolio tracking, research, company profiles, earnings history, custom index, news headlines, and more.

Go to the Red Herring Web site and look at the data tables for Most Active, Big Gainers, and Big Losers. Next, visit http://www.bloomberg.com, another investor site, and click on "Stocks on the Move." Are the same stocks listed at both of these sites? Compare the two sites.

CYBERCONNECTION

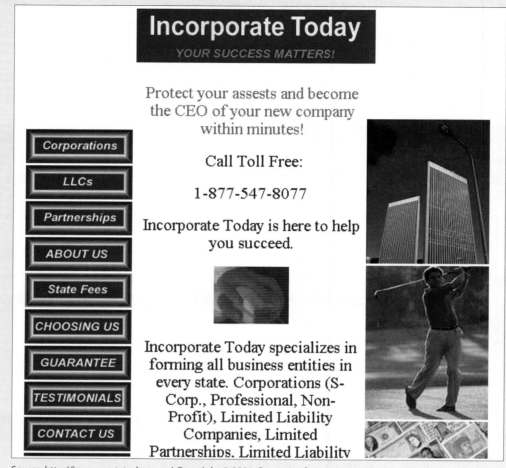

Incorporate Today
YOUR SUCCESS MATTERS!

Corporations

LLCs

Partnerships

ABOUT US

State Fees

CHOOSING US

GUARANTEE

TESTIMONIALS

CONTACT US

Protect your assests and become the CEO of your new company within minutes!

Call Toll Free:

1-877-547-8077

Incorporate Today is here to help you succeed.

Incorporate Today specializes in forming all business entities in every state. Corporations (S-Corp., Professional, Non-Profit), Limited Liability Companies, Limited Partnerships. Limited Liability

Source: http://incorporatetoday.com/. Copyright © 2001. Courtesy of Incorporation Today.

Examine this Web site to find out how to go about incorporating your own business. The site offers information on filing fees for each state and on the different types of corporate formations that exist—partnerships, C-corporations, S-corporations, and more.

The Bottom Line

Summary of chapter learning objectives:

| Learning Objective | Summary | Example |
|---|---|---|
| **22.1** | Compute the costs and proceeds of stock buy-and-sell transactions | For calculations throughout, use $0.20 a share for commissions and $0.125 for the odd-lot differential. Round all percents to two places.

1. Leonard Butler bought 100 shares of Avon at 41.75. What was the total cost of the purchase of common stock?
2. Leonard sold 200 shares of Hasbro at 11.5. What were the proceeds of the sale? |
| **22.2** | Compute the costs and proceeds of round and odd lots | Erin Michaels purchased 1,000 shares of Sysco common stock at 56 and 550 shares of preferred stock at 85.

3. What was the total cost of the purchase of common stock?
4. What was the total cost of the purchase of preferred stock? |
| **22.3** | Compute rates of yield and gains or losses on the purchase and sale of stocks | 5. Jeff Morris purchased 320 shares of IBM at 86 and sold them 1 year later at 92.35. What were his total cost, net proceeds, and amount of gain on these two transactions? |
| **22.4** | Compute comparative earning potential of the major classes of corporate stocks | The Southern Mills Corporation earned $150,000 last year. The capital stock of the company consists of 10,000 shares of 6% preferred stock, with a par value of $50 per share, and 20,000 shares of no-par common stock. The board of directors declared a dividend of $100,000.

6. What amount will be paid to the preferred shareholders?
7. What amount per share will be paid to the common shareholders?
8. Domingo Jura owned 250 shares of Metro Software convertible preferred stock with a $50 par value. He converted each share of preferred into 2 shares of common. How many shares of common stock did he receive?
9. If the common stock was selling at 38.50 on the day of the conversion, how much was his common stock worth? |

Answers: 1. $4,195 **2.** $2,260 **3.** $56,200 **4.** $46,866.25 **5.** Total cost, $27,520.00; net proceeds, $29,552.00; gain $2,032.00 **6.** $30,000 **7.** $3.50 per share **8.** 500 shares **9.** $19,250

Self-Check | Review Problems for Chapter 22

1 The following stock quotes are from the NYSE.

| High | Low | Stock | Div | % | Yld PE | Vol 100s | Hi | Low | Close | Chg. |
|------|-----|-------|-----|---|--------|----------|-----|------|-------|------|
| 72.50 | 63 | Boeing | .56 | .8 | 18 | 2560 | 69.60 | 67 | 68.14 | +2.18 |
| 97 | 80 | Chevron | 2.60 | 3.2 | 16 | 3651 | 83 | 81.32 | 82.50 | −3.40 |

(with a "52 Weeks" label spanning the High and Low columns)

 a. How many shares of Chevron were traded?

 b. What was the closing price per share of Boeing in dollars and cents?

 c. What was the previous day's closing price for each stock?

 d. By how much has the price of one share of Boeing stock changed over the last 52 weeks?

 e. Use the PE ratio to calculate the earnings per share for the last four quarters for Chevron.

2 Determine the total cost or proceeds of each purchase or sale. Include regular commission of $0.20 per share and an odd lot differential of $0.125 per share.

 a. Purchased 500 shares of Hershey at 64.10.

 b. Purchased 650 shares of Nabisco at 29.

 c. Sold 300 shares of PG&E at 27.60.

3 Zachery Taylor purchased 400 shares of XYZ stock at 20.25. One year later he sold the 400 shares at 22. He paid a transaction fee of $19.95 for each transaction.

 a. What was the amount of the gain or loss on the sale?

 b. What was the rate of the gain or loss?

4 Refer back to problem 3. Zachery received dividends of $1.80 per share during the year that he owned the stock.

 a. What was the rate of the dividend yield?

 b. What was the total rate of gain or loss, including the dividend?

5 Marie Dupre owned 300 shares of LX convertible preferred stock with a $20 par value. She converted all 300 shares to common stock at the rate of 4 to 1 (4 shares of common stock for each share of preferred). How many shares of common stock did she receive?

6 The same LX preferred stock paid an annual dividend of 8%. LX paid annual dividends on its common stock of $0.60 per share. How much more will Marie receive each year in dividends by converting her stock from preferred to common?

7 Alpha Company's capital consists of 6,000 shares of $50 par 7% preferred stock and 50,000 shares of no-par common stock. The board of directors declared a dividend of $65,000. What is the dividend per share for preferred and common stock?

8 Refer back to problem 7. Assume that the preferred stock is cumulative and that no dividends were declared the year before. Determine the dividend to be paid for each share of preferred and common if the board declares a total dividend of $75,000 for the current year.

Assignment 22.1: Buying and Selling Stock

Name _____

Date _____ Score _____

A **(41 points) For calculations, use $0.20 a share for commissions unless the problem gives a flat fee and $0.125 for the odd-lot differential. Round all percents to two places. (5 points for a correct answer to problem 3; 4 points for each other correct answer)**

1. Fred Jacobi purchased 2,000 shares of EKC common stock at 26 and 190 shares of preferred stock at 55.

 a. What was the total cost of the purchase of common stock? _____

 b. What was the total cost of the purchase of preferred stock? _____

2. Three months later, Fred sold his 2,000 shares of EKC common stock at $28 and his 190 shares of preferred stock at $58.50.

 a. What were the proceeds on the sale of common stock? _____

 b. What were the proceeds on the sale of preferred stock? _____

 c. How much did Fred gain or lose on the purchase and sale of all of his EKC stock? _____

3. Susan Lu purchased 200 shares of Telmart common stock at $88.50 and paid a $19.95 transaction fee. A dividend of $7.00 per share was paid the first year. What was the rate of yield? _____

Assignment 22.1 *Continued*

4. Jennifer Knutsen purchased stock for a total cost of $14,800, including commission. She sold the stock a month later for $15,555, after commission.

 a. What was her net gain on the sale? _____

 b. What was her percent of gain on the sale? _____

 c. If Jennifer had held her stock another week and sold for $14,250 after commission, what would her percent of loss on the sale have been? _____

5. If Jennifer hadn't sold her stock for $14,250 but had waited another 3 months while the stock fell to a price where she could have realized net proceeds of $11,780, what would have been her percent of loss? _____

<div align="right">

Score for A (41)

</div>

B **(59 points) Solve the following problems. (points for correct answers as marked)**

6. Peter Roncalio, Paul Stevens, and Mary Petrakas each invested $10,000 in different areas. Calculate the value of each $10,000 investment at the end of 2 years. (5 points for each correct answer)

 a. Peter put his $10,000 in a savings account that paid 6.2% interest annually. (Add interest on the savings account the first year to the principal before figuring interest for the second year.) _____

 b. Paul bought 9%, $50 par value preferred stock at $62.50 a share, including commission. He received his full dividend at the end of each year. He sold his stock at the end of the second year. The sales proceeds, after commission, were $62.50 a share. _____

Assignment 22.1 *Continued*

 c. Mary bought common stock at $40 a share, including commission. Her stock paid quarterly dividends of 90 cents per share. In two years, the stock decreased to a value of $38.50 a share. _____

7. Find the amount of the dividend per share and the rate of yield per share for each of the following preferred stocks. The cost per share includes all commissions. (2 points for each correct answer)

 a. Cost per share $58; dividend declared $4.00.

 Amount of dividend _____
 Rate of yield _____

 b. Cost per share $80; par value $100; dividend declared 6%.

 Amount of dividend _____
 Rate of yield _____

 c. Cost per share $44.50; dividend declared $2.00.

 Amount of dividend _____
 Rate of yield _____

 d. Cost per share $98; par value $100; dividend declared 4.5%.

 Amount of dividend _____
 Rate of yield _____

 e. Cost per share $58; par value $50; dividend declared 6.5%.

 Amount of dividend _____
 Rate of yield _____

8. Determine the amount and percent of gain or loss for each of the following transactions. Show an amount of loss in parentheses (). The purchase costs and the sale proceeds include commissions. Round percents to two decimal places. (3 points for each correct answer)

| | Number of Shares | Per-Share Purchase Cost | Per-Share Sale Proceeds | Amount of Gain or Loss | Percent of Gain or Loss |
|---|---|---|---|---|---|
| a. | 100 | $25.50 | $29.00 | _____ | _____ |
| b. | 250 | 12.00 | 14.50 | _____ | _____ |
| c. | 140 | 22.30 | 20.70 | _____ | _____ |
| d. | 980 | 19.80 | 12.75 | _____ | _____ |

Score for B (59)

Assignment 22.2: Capital Stock

Name _____

Date _____ Score _____

A **(34 points) The information in problem 1 also applies to problems 2 and 3. (2 points for each correct answer)**

1. The Pequod Company was incorporated with 8% preferred capital stock of $600,000 and common stock of $1,600,000. The par value of the preferred stock was $100, and the par value of the common stock was $50. How many shares of each kind of stock were there?

 Preferred stock _____

 Common stock _____

2. Last year, dividends were declared by the Pequod Company, which had earnings totaling $534,400.

 a. What was the total amount of the preferred stock dividend? _____

 b. What amount would have been paid on each share of common stock if all the earnings had been distributed?

3. The directors of the Pequod Company actually declared four quarterly dividends of $3.20 a share on the common stock and $\frac{1}{4}$ of the amount due annually on the preferred stock.

 a. What was the total amount paid by Pequod to all common shareholders for each quarterly dividend?

 b. What was the total amount paid to preferred shareholders each quarter? _____

 c. What was the quarterly per-share payment to preferred shareholders? _____

 d. What was the year's total amount of the common stock dividends? _____

 e. What was the total amount of all dividends paid by Pequod during the year? _____

 f. How much more in dividends was paid to each share of common than to each share of preferred? _____

4. The capital stock of the Canwell Company consists of 200,000 shares of preferred stock and 5,500,000 shares of common stock. Last year, a dividend of $3.60 a share was declared on preferred stock and four quarterly dividends of $0.35 a share on common stock. How much was the total dividend for the year on each class of stock?

Preferred stock _____

Common stock _____

5. ComputerMart has 150,000 shares of 6.5% preferred stock at $1 par value and 1,500,000 shares of common stock. ComputerMart declared total dividends of $250,000 for the current year. How much was the total dividend for preferred stock and how much was the dividend per share on the common stock?

Preferred stock _____

Common stock _____

(per share)

6. Sue Chastain bought 300 shares of IBB 8% preferred stock, $10 par value, when it was selling at $12 per share, including commission.

 a. What was Sue's stock worth at the time of purchase? _____

 b. What was the amount of Sue's quarterly dividend? _____

 c. What was Sue's dividend yield? _____

Score for A (34)

B **(66 points) Do not consider commission in the following problems. (points for correct answers as marked)**

7. Inland Sales, Inc., issued 25,000 shares of 8%, $20 par, cumulative preferred stock and 50,000 shares of common stock. The board of directors declare 50% of net income each year as dividends. Inland Sales had net income of $76,000 for 2000, $112,000 for 2001, and $130,000 for 2002. Compute the annual dividends per share for preferred and common stock for each of the 3 years. (2 points per answer)

| Year | Preferred Dividends/share | Common Dividends/share |
|------|---------------------------|------------------------|
| 2000 | _____ | _____ |
| 2001 | _____ | _____ |
| 2002 | _____ | _____ |

Assignment 22.2 *Continued*

8. John Stevens owned 200 shares of Eastern 6.5% convertible preferred stock, $50 par value, for which he paid $56 per share, including commission. Two years later, after receiving preferred dividends each year, he converted to 600 shares of Eastern common stock, valued at $23.50 a share at the time of conversion. (4 points for each correct answer)

 a. What was the cost to John of the preferred stock? _____

 b. How much did John receive in dividends from the preferred stock? _____

 c. What was the value of the common stock that John received? _____

 d. If he sells the 600 common shares immediately, how much gain will John realize, including his dividend?

 e. What would be John's percent of gain? _____

9. Texas Air Corporation issued 5,000,000 shares of 7% preferred stock at $100 par value and 10,000,000 shares of no-par common stock. Bob Thruston owned 100 shares of preferred. Barbara Beck owned 500 shares of common. In 2001, Texas Air paid $25,000,000 in dividends to its common shareholders. How much more than Bob did Barbara receive? (10 points) _____

10. Sonia Revas owned 800 shares of PIE 8% convertible stock, $50 par value, for which she paid $45 a share. She received a dividend for 1 year. She then converted the preferred stock to 500 shares of common stock valued at $98.50 a share. (4 points for each correct answer)

 a. What was the cost to Sonia for her preferred stock? _____

 b. How much did Sonia receive as a dividend for her preferred stock? _____

 c. What was the value of her common stock at the time of conversion? _____

 d. If the common stock paid an annual dividend of $7.00 a share, how much more dividend would she receive annually? _____

 e. What was Sonia's percent of increase in annual return as a result of conversion to common stock? _____

11. Determine the price to earnings ratio (PE) of each of the following stocks. (2 points each)

a. KB common stock has a current market price of $35 and has had earnings per share of $0.72 each quarter for the last four quarters. _____

b. The current market price of CLX common stock is 62.60. CLX has paid dividends of $1.05 per quarter for each of the last four quarters. _____

Score for B (66)

Notes

THE BOND MARKET ASSOCIATION

InvestingBon.co

Search:

Bond Prices
Municipal Bonds
Corporate Bonds
Treasurys

Features
Updated:
Tax-Free vs. Taxable Yield Comparison Calculator
Bond Glossary
Where Wall Street Meets Main Street
Investor's Checklist
Taxation of Municipals
Q&A on Buying and Selling Munis
Putting Compound Interest to Work with Zeros
Reading Bond Prices in the Newspaper

gloss

more...

7 simple steps

Getting Started: Educate Yourself About Investing in Bonds!

1. Read "An Investor's Guide to Bond Basics," An overview of the concepts that investors should know about bonds.

2. Try the "Investor's Checklist" of questions to ask yourself when considering an investment in bonds.

VITAL STATIS

Our free **Daily Muni B Summary** includes full previous-day, frequently muni bonds. We also of price and other data on **investment-grade corp bonds**; and composite r **US Treasurys**, updated throughout the day.

From the muni price rep find links to Thomson F database of Official Stat well as MSRB data on i trades. This information helpful in assessing relat bonds you own, or are t buying or selling. For ot prices, visit our "Gatewa

The U.S. bond market represents the largest and most liquid securities market in the world. The Bond Market Association publishes reports analyzing trends, examines supply and demand factors that affect bond markets, reviews trends in new issuance, and reports on supply and trading activity in the U.S. fixed-income markets. Visit its Web site at http://www.investinginbonds.com/ for additional research, statistical, and other information.

Corporate and Government Bonds

23

LEARNING OBJECTIVES

By studying this chapter and completing all assignments you will be able to:

 1 Compute gains and losses on convertible and callable corporate bond transactions.

 2 Compute annual interest on bonds.

 3 Compute accrued interest on bond transactions made between interest payment dates.

 4 Compute annual yield on bonds selling at a premium or a discount.

 5 Compute a rate of yield to maturity.

Corporate bonds allow companies to raise money for new factories and equipment. State, municipal, and special-district bonds help build new schools and hospitals and make other public improvements. U.S. government bonds have tax advantages, are backed by the U.S. government, and are considered the safest investment in the world.

1. What is the difference between corporate and government bonds?
2. What is a *callable* corporate bond? A *convertible* corporate bond?
3. Why would a company decide to issue corporate bonds?
4. Why would a state or municipality decide to issue government bonds?

When a corporation or government entity needs cash for a long period of time, usually ten years or more, it often will issue long-term notes known as **bonds**. Bonds are bought and sold on the open market, much like stocks.

Two main types of **government bonds** are treasury bonds and municipal bonds. **Treasury bonds** are issued by the United States government. These bonds are fully guaranteed by the full faith and credit of the United States government. Bondholders are protected against default unless the federal government becomes insolvent. **Municipal bonds** are issued by states, cities, school districts, and other public entities. Unlike treasury bonds, municipal bonds pose a risk that the issuer might fail to repay the principal. Interest paid on municipal bonds generally is exempt from federal and state income taxes.

There are many kinds of corporate bonds, two of which are convertible bonds and callable bonds. **Convertible bonds** have a provision that they may be converted to a designated number of shares or a designated value of the corporation's stock. **Callable bonds** have a provision that the issuer can repurchase, or call in the bonds, at specified dates if the board of directors authorizes the retirement (payoff) of the bonds before their maturity date. Such action by the board of directors would be appropriate if interest rates fell significantly below the interest rate of the callable bond.

Computing Gains and Losses on Corporate Bonds

LEARNING OBJECTIVES 1

Compute gains and losses on convertible and callable corporate bond transactions.

EXAMPLE A

Gary Lee bought one ABC Corporation convertible bond for $1,000. The bond was convertible to 100 shares of stock. At the time of the purchase, the stock was selling for $10 per share. At the end of 1 year, the stock was selling for $15 per share. Gary converted his bond. Assuming that the market value of the bond hadn't changed, how much profit did Gary realize by converting?

100 shares of stock × $15 per share = $1,500
$1,500 stock value − $1,000 bond value = $500 profit

EXAMPLE B

XYZ Corporation issued $1,000,000 worth of callable bonds paying 8% interest. The maturity date for the bonds was in 10 years. Two years later, interest rates fell to 6%. The bonds were called, and new bonds were sold at the 6% rate. How much did XYZ Corporation save by calling the bonds?

10 years to maturity at issue − 2 years = 8 years remaining to maturity
8% − 6% = 2% savings per year
$1,000,000 × 2% = $20,000
$20,000 × 8 years = $160,000 saved

www.finpipe.com/ sue21.htm

Answers to financial questions, including questions about bonds, can be found here.

CyberConnection

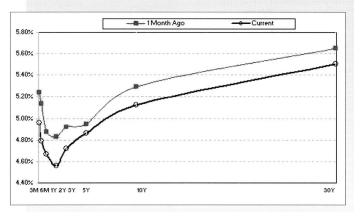

Visit the Bondpage site for graphs of bond yields, such as this one. Its Info Center also provides many helpful items, including ratings, definitions, and a training center.

Source: www.bondpage.com. Copyright © 2001 courtesy of bondpage.com.

 CONCEPT CHECK 23.1

a. What would be the "stock" value of a bond that was convertible to 50 shares of stock if the stock was priced at 48.25?

50 shares × $48.25 = $2,412.50

b. If a company issued a callable bond at $7\frac{1}{2}$% interest, would it be likely to call the bond if the current rate of interest was 8%?

No, because it could invest the cash at an extra $\frac{1}{2}$% interest.

Computing Annual Interest on Corporate and Government Bonds

LEARNING OBJECTIVES

Compute annual interest on bonds.

When first issued, bonds are sold either through brokerage houses or directly to investors at or near the price of $1,000, called face value. **Face value** represents the amount that will be paid to the holder when the bonds are redeemed at maturity. If the market value becomes less than the face value, the bond sells at a **discount**. If the market value becomes more than the face value, the bond sells at a **premium**. (The discount or premium amount is the difference between the market value and the face value.)

Bonds are rated. By checking a bond's rating, buyers can have some indication of how safe their bond investment is. **Bond ratings** are information based on experience and research; they are not a guarantee. One major firm rating bonds is Standard & Poor's.

In Standard & Poor's system, the ratings include AAA (the highest rating), AA, A, BBB, BB, B, CCC, CC, C, and D. A bond with a low rating is a higher risk bond and sometimes is known as a **junk bond**. The lower a bond's rating, the higher are its yield and its risk.

EXAMPLE C

Joan Anderson purchased a $1,000 bond with a rating of B, paying 14% per year. Mary Baker purchased a $1,000 bond with a rating of AAA, paying 5% per year. Jean Carlson purchased a $1,000 junk bond, paying 25% per year. Each bond was to mature in 10 years.

Joan's B-rated bond paid faithfully for 4 years. Then the company filed for bankruptcy and paid 60 cents on the dollar. Mary's AAA-rated bond paid interest during its entire 10-year life and paid face value on maturity. Jean's junk bond paid interest for 3 years. Then the company filed for bankruptcy and paid 30 cents on the dollar.

Compute how much each investor received for her $1,000 investment.

Joan: $1,000 × 14% = $140 annual interest
 $140 × 4 years = $560 interest
 $560 interest + (0.60 × $1,000) redemption = $1,160 total

Mary: $1,000 × 5% = $50 annual interest
 $50 × 10 years = $500 interest
 $500 interest + $1,000 redemption = $1,500 total

Jean: $1,000 × 25% = $250 annual interest
 $250 × 3 years = $750 interest
 $750 interest + (0.30 × $1,000) redemption = $1,050 total

How much would Joan and Jean have received on their investments if the bonds had paid full interest for the 10-year period and face value on maturity?

Joan: $1,000 × 14% × 10 years = $1,400
 $1,400 + $1,000 = $2,400

Jean: $1,000 × 25% × 10 years = $2,500
 $2,500 + $1,000 = $3,500

Newspaper Information on Bonds

Information about the market value and sale of bonds on the major exchanges is reported daily in financial newspapers. Figure 23-1 shows information usually included in a bond report.

Figure 23-1: Bond Market Report

| Bonds | Current Yield | Volume | Close | Net Change |
|-------|---------------|--------|-------|------------|
| ATT 7$\frac{1}{2}$s06 | 7.2 | 10 | 104 | +1 |
| Aetna 6$\frac{3}{8}$s09 | 6.6 | 25 | 96.80 | . . . |
| ClrkOil 9$\frac{1}{2}$s04 | 9.1 | 33 | 104.25 | +.25 |
| Hertz 7s12 | 7.0 | 13 | 99.70 | +.70 |
| IBM 7s25 | 7.4 | 102 | 94.50 | +.80 |
| RJR Nb 8s10 | 7.9 | 15 | 101.50 | . . . |

Prices of bonds are quoted in percents of face value. For example, a $1,000 bond quoted at 104 would sell at a premium price of $1,040 ($1,000 × 104%). If quoted at 87, the bond would sell at a discounted price of $870 ($1,000 × 87%).

Rule: Prices over 100 (100%) include a premium. Those under 100 (100%) include a discount.

The two main factors that influence the market price are the interest rate and the bond rating. For example, if a bond pays 8% interest and the current market rate of interest is greater than 8% for similarly rated bonds, the bonds will sell at a discount sufficient to make up for the difference in interest rates over the term of the bond.

Printed bond reports generally give a letter abbreviation for the company, the interest rate, a small s to designate *semiannual* (every 6 months) interest payments, and the maturity date, followed by the current yield, the number of bonds sold that day, the closing price of the bond, and the net change in price from the prior day.

The first line of the bond market report in Figure 23-1 would be interpreted as ATT (designating American Telephone and Telegraph), a $7\frac{1}{2}$ interest rate based on the face value of the bond, and interest paid semiannually. The bond matures in 2006. The current yield (average annual interest rate based on the current price of the bond) is 7.2%. The day's volume of bonds sold was 10. The closing price was 104, up 1 from the prior day.

EXAMPLE D

Calculate the amount of the semiannual interest check for a $1,000 bond reported in a financial paper as R&S Corp $7\frac{1}{2}$s21.

$1,000 face value × $7\frac{1}{2}$% = $75

$75 ÷ 2 = $37.50 semiannual interest payment

Commissions for Buying and Selling Bonds

The charge for buying and selling bonds varies among brokers, but there is no standard commission. Commissions are very small and thus comprise only a negligible part of the bond transaction. We do not use commission costs for problems in this textbook.

 CONCEPT CHECK 23.2

If John Coran purchased 22 triple-A bonds that pay 7.3% and mature in 7 years, what amount of interest income could he expect annually?

$1,000 × 0.073 × 22 = $1,606

If John holds the bonds until maturity, how much will he receive on redemption of the bonds?

$1,000 × 22 = $22,000 total face value

Computing Accrued Interest on Bond Transactions

LEARNING OBJECTIVES

Compute accrued interest on bond transactions made between interest payment dates.

Most bonds specify that interest is payable quarterly, semiannually, or annually. The interest payment dates—such as January 1 (for interest through December 31) and July 1 (for interest through June 30)—are stated on the bond. When a bond is purchased between these dates, it is customary to add the **accrued interest** (interest earned from the last payment date to the purchase date). This interest is calculated by finding the number of days from the day on which interest was last paid through the day before the purchase and dividing this number by 360.

The buyer pays the seller for the interest accumulated or accrued on the bond since the last interest payment date. On the next regular interest payment date, the new owner receives the interest for the full interest period. This procedure allocates the interest correctly between the buyer and the seller for the split interest period because the corporation that issued the bond will pay the entire amount to whoever owns the bond as of each interest date.

EXAMPLE E

A $1,000 bond, with interest at 9% payable semiannually on January 1 and July 1, was purchased on September 23 at 104 plus accrued interest. What is the number of days for which the accrued interest is paid?

Purchase date: September 23
Days of accrued interest: (July) 31 + (August) 31 + (September) 22 = 84

What is the purchase payment for the bond?

$1,000 × 104% = 1,040 market value
$1,000 × 0.09 interest × $\frac{84}{360}$ accrued days = $21 accrued interest
$1,040 + $21 = $1,061 purchase payment for bond

In example E, although the accrued interest is an additional payment by the buyer, the buyer will get it back in the $45 ($1,000 × 9% × $\frac{1}{2}$) interest payment on January 1.

www.mycititrade.com/ research/bonds/bonds_ corporate.html

For information on bond basics, risks and yields, and much more, access this Web site.

CONCEPT CHECK 23.3

Ann Ahn purchased two Hertz 7s08 bonds at 97.50 on March 7. What amount did she pay her broker?

$2,000 × 0.975 = $1,950.00
Purchase date: March 7
(January) 31 + (February) 28 + (March) 6 = 65 days
$2,000 × 0.07 × $\frac{65}{360}$ = $25.28 accrued interest
$1,950.00 + $25.28 = $1,975.28 paid to her broker

Computing the Rate of Yield for Bonds

Interest on bonds provides income to bondholders. This income is referred to as **yield**. Newspapers and bond brokers refer to the annual yield of a bond as its **current yield**. Many newspaper bond reports include a column showing current yield. To calculate the current yield from an investment in bonds, use the following formula.

Annual interest ÷ Current purchase price = Current yield

When a bond is purchased at a discount, the current yield is greater than the face rate. For example, a $1,000 bond, purchased at 90, pays 7% interest and matures in 10 years. Interest of $70 ($1,000 × 7%) is paid annually, but as the bond was purchased for $900 ($1,000 × 90%), the effective rate, or yield, as a percent of cost is 7.8% ($70 ÷ $900).

When a bond is purchased at a premium, the current yield is less than the face rate. The reason is that the interest paid is calculated on the face value, and the yield is based on the higher market price.

EXAMPLE F

Five $1,000 Levi Straus $9\frac{1}{2}$s19 bonds were purchased at 80. What was the current yield on the bonds?

$1,000 × 5 = $5,000 face value
$5,000 × 80% = $4,000 purchase price
$5,000 × 0.095 = $475 annual interest
$475 ÷ $4,000 = 0.11875 = 11.9 % current yield
or 9.5 ÷ 0.80 = 11.875 = 11.9% current yield

In example F, the bonds sold at a discount of $1,000 ($5,000 − $4,000) because the investor paid that much less for them than the maturity (face) value. Therefore the current yield of 11.8% is more than the stated interest rate of $9\frac{1}{2}$%.

 CONCEPT CHECK 23.4

The RJR Nb bonds listed in Figure 23-1 recently rose to a price of 107.50. Zelda Morantz purchased four at 107.50. What will be her annual current yield?

$4,000 × 107.5% = $4,300 purchase price
$4,000 × 0.08 = $320 annual interest
$320 ÷ $4,300 = 0.0744, or 7.44% annual yield
or 0.08 ÷ 1.075 = 0.0744 = 7.44% annual yield

4 LEARNING OBJECTIVES

Compute annual yield on bonds selling at a premium or a discount.

www.bondresources. com/Corporate/ Education/Guide_to_ Corporate_Bonds

Bond Resources is another good site for bond education, information, analyses, and data.

Computing the Rate of Yield to Maturity

LEARNING OBJECTIVES

Compute a rate of yield to maturity.

Careful investors calculate the **rate of yield to maturity**, or the rate of interest they will earn if they hold the bond to its maturity date. The yield to maturity calculation involves use of the true annual interest by adding a part of the discount or subtracting a part of the premium and basing the rate on the average principal invested (the average of the investor's purchase price and the bond's maturity value).

STEPS to Compute the Rate of Yield to Maturity

1. Compute the annual interest: multiply the face value times the stated (face) rate.
2. Determine the **annual discount** (or **premium**) **amortization**: Divide the discount (or premium) by the number of years from purchase to maturity.
3. Determine the **average principal invested**: Add the maturity value and the cost price and then divide by 2.
4. The following formula computes the rate:

$$\frac{\text{Annual Interest} + \text{Annual discount amortization (or} - \text{Annual premium amortization)}}{\text{Average principal invested}}$$

Again, because brokerage charges are such a small part of the cost, they usually are omitted from the calculations of yield to maturity.

EXAMPLE G

Assume that the Levi Straus bonds in example F matured 20 years after the purchase date.

| STEP 1 | $5,000 × 0.095 = $475 annual interest |
| STEP 2 | $1,000 ÷ 20 years = $50 annual discount amortization |
| STEP 3 | ($5,000 + $4,000) ÷ 2 = $4,500 average principal invested |
| STEP 4 | ($475 + $50) ÷ $4,500 = 0.1167 = 11.67% yield to maturity |

This rate is somewhat less than the 11.8% current yield, but it is more accurate with respect to actual income if the bond is held to maturity.

EXAMPLE H

To calculate the yield to maturity on bonds sold at a premium, assume that five IntTT $8\frac{1}{2}$s16 bonds were bought at a premium price of 124 and that the bonds will mature in 15 years. The market value of the five bonds is $6,200 ($5,000 × 124%).

| STEP 1 | $5,000 × 0.085 = $425 annual interest |
| STEP 2 | ($6,200 − $5,000) ÷ 15 years = $80 annual premium amortization |
| STEP 3 | ($5,000 + $6,200) ÷ 2 = $5,600 average principal invested |
| STEP 4 | ($425 − $80) ÷ $5,600 = 0.0616 = 6.16% yield to maturity |

This rate is less than the stated rate of $8\frac{1}{2}$% on the premium bonds.

CONCEPT CHECK 23.5

If the four RJR Nb 8s10 bonds Zelda Morantz purchased at 107.50 (concept check 23.4) had 12 years to maturity, what would be her rate of yield to maturity?

$4,000 × 0.08 = $320 annual interest
$300 premium ÷ 12 years = $25 annual premium amortization
($4,000 + $4,300) ÷ 2 = $4,150 average principal invested
($320 − $25) ÷ $4,150 = 0.071084 = 7.11% yield to maturity

Complete Assignments 23.1 and 23.2.

Chapter Terms for Review

accrued interest
annual discount amortization
annual premium amortization
average principal invested
bond ratings
bonds
callable bonds

convertible bonds
corporate bonds
current yield
discount
face value
government bonds
junk bond

municipal bonds
premium (bond)
rate of yield to maturity
treasury bonds
yield

Visit the Web

http://www.bloomberg.com/markets/japan.html,
http://www.banqueducanada.ca/en/securities.htm, and
http://www.chinaonline.com—Other countries also offer government bonds, as indicated by these Web sites. Visit some of them.

http://www.idfa.com/bondlocalgov.htm—This Web site covers local government bonds in Illinois.

http://www.ameritrade.com/educationv2/fhtml/stocksfunds/govbonds.fhtml—This excellent Web site provides a variety of information on all types of bonds.

http://www.e-analytics.com/bonds/bond10.htm—Find earnings information about individual companies here.

http://www.cnbc.com/news/bonds/index.html—Another good Web site for the latest on bonds is this one.

http://www.publicdebt.treas.gov/bpd/bpdhome.htm—At the Bureau of Public Debt Web site, you can visit the many links that provide information and data on savings bonds, T-bills, Notes, and Bonds.

CYBERCONNECTION

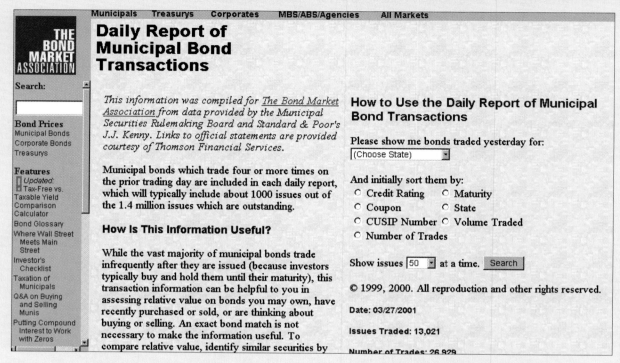

Source: http://www.investinginbonds.com/. © 2000, 2001. Used with permission, The Bond Market Association. All reproduction and other rights reserved.

Look up five bonds for your state. Note what credit rating is given to each of these bonds. Figure the amount of money you would need to invest *today* to buy one of each of the five bonds. Wait one week. Then look up the price of each of the bonds and determine the amount you would receive if you sold them. Compute your "gain" or "loss".

The Bottom Line

Summary of chapter learning objectives:

| Learning Objective | Summary | Example |
|---|---|---|
| **23.1** | Compute gains and losses on convertible and callable corporate bond transactions | 1. Mark Allen bought five DVC bonds at $1,000 per bond. Each bond was convertible after 3 years to 40 shares of stock. At the end of 3 years, shares of DVC stock were selling at $38. The bond price had risen to 120. Should Mark exercise his option to convert? |
| | | 2. Meredith Marketing Corporation issued $1,000,000 worth of callable bonds paying 9% interest. The maturity date for the bonds was in 20 years. Three years later, interest rates fell to $7\frac{1}{2}$ percent. The bonds were called, and new bonds sold at the $7\frac{1}{2}$% rate. How much did Meredith Marketing Corporation save by calling the bonds? |
| **23.2** | Compute annual interest on bonds | 3. Sue Benson purchased three 12-year, $1,000 bonds: one Boeing at 7%, one U.S. Treasury at 4.5%, and one Water World Sports at 12%. If the Water World Sports bond defaulted after 5 years and paid holders 60%, which bond produced the most income in the 5-year period, assuming that the $400 loss on the WWS bond was considered to be a reduction in income? How much did it produce? |
| **23.3** | Compute accrued interest on bond transactions made between interest payment dates | 4. One GMA 8s20 bond was purchased at 102 on January 25. What was the amount of accrued interest if interest is paid January 1 and July 1? |
| **23.4** | Compute annual yield on bonds selling at a premium or a discount | 5. Four Snap On 8.6s19 bonds were purchased at 95. What was the current yield? |
| **23.5** | Compute a rate of yield to maturity | 6. Three NYR 8s10 bonds were purchased at 120. The bonds will mature in 14 years. What is the rate of yield to maturity? |

Answers: 1. The stock has $1,600 greater value; yes, he should convert **2.** $255,000 **3.** Boeing; $350 **4.** $5.33 **5.** 9.05% **6.** 5.97%

1 Maggie Peterson purchased 12 JBC $8\frac{1}{2}$s14 bonds at 103.
 a. What was the cost of the bonds?
 b. How often will interest be paid?
 c. How much interest will Maggie receive each interest period?
 d. Assuming that the bonds pay interest on April 1 and October 1, calculate the accrued interest if the bonds were purchased May 13.
 e. What is the total amount Maggie paid for the bonds including accrued interest?
 f. Were the bonds purchased at a premium or a discount?
 g What was the amount of the premium or discount?
 h When do the bonds mature?
 i. What is the current yield on the bonds?
 j. Assume that the bonds mature in 12 years. Calculate the yield to maturity.

2 Bill Grady purchased six Xerox $1,000 convertible bonds at 92. Each bond was convertible to 25 shares of common stock. After 5 years, when the stock was selling at $52\frac{1}{4}$, Bill converted all six bonds.
 a. How many shares of stock did he receive?
 b. What was the value of the stock on conversion?
 c. What was Bill's gain on conversion of the bonds?
 d. If the bonds are now selling at 100, should Bill convert his bonds to stock if the stock's current market price is $38 per share? Why or why not?

3 Avis, Inc., issued $50,000,000 of $9\frac{1}{2}$%, 20-year, callable bonds. After 6 years the interest rate fell to 8%. How much interest would Avis save by calling the bonds and reissuing bonds at the lower rate?

Assignment 23.1: Corporate and Government Bonds

Name _____

Date _____ Score _____

A **(38 points) Solve the following problems. (points for correct answers as marked)**

1. Sheila Davis purchased seven IBM $1,000 convertible bonds at 96. Each bond was convertible to 20 shares of IBM stock in 7 years. At the end of 7 years, IBM stock was selling at $56\frac{1}{4}$. If Sheila converted, what would be her 7-year capital gain? (4 points) _____

2. Return to problem 1 and assume that the stock price after 7 years was 42. How much more money would Sheila get by cashing in the bonds rather than converting to stock? (4 points) _____

3. The city of Jamestown, Virginia, issued $27,000,000 worth of callable bonds at 9% on January 1, 2000. The bonds were due in 2015. If interest rates were to fall to 6.5% on January 1, 2007, how much could Jamestown save by reissuing the bonds at the 6.5% rate on January 1, 2007? (4 points) _____

4. Assume that an investor had purchased $500,000 worth of the Jamestown bonds referred to in problem 3. How much interest would he lose from having the bonds called if he reinvested in the new bond issue? (4 points) _____

5. Devi Sharma purchased 22 corporate bonds, as shown. What was her total cost, and how much interest income would she realize annually? (1 point for each correct answer)

| Bond | Number Purchased | Price | Total Cost | Annual Interest |
|------|------------------|-------|------------|-----------------|
| **a.** Apex $7\frac{1}{2}$s09 | 4 | 100 | _____ | _____ |
| **b.** DukeP $7\frac{7}{8}$s02 | 3 | 98 | _____ | _____ |
| **c.** PGE $10\frac{1}{8}$s12 | 9 | 86 | _____ | _____ |
| **d.** IBM $9\frac{3}{8}$s08 | 6 | 109 | _____ | _____ |
| Total | 22 | | _____ | _____ |

Assignment 23.1 *Continued*

6. What is the dollar amount of interest per year and the maturity date for each of the following $1,000 bonds? (1 point for each correct answer)

| Bond | Interest | Maturity date | Bond | Interest | Maturity date |
|------|----------|---------------|------|----------|---------------|
| **a.** AEP 7s09 | _____ | _____ | **d.** Fldcst $12\frac{1}{2}$s12 | _____ | _____ |
| **b.** Avnet 8s13 | _____ | _____ | **e.** OwCor 12s10 | _____ | _____ |
| **c.** CPoWV 9s15 | _____ | _____ | **f.** PacSun $9\frac{1}{2}$s07 | _____ | _____ |

Score for A (38)

B **(50 points) Solve the following problems. (points for correct answers as marked)**

7. In each of the following problems, determine the number of days for which accrued interest is paid and the total purchase payment made for the bonds. (5 points for each correct answer)

 a. On August 16, Joe Leu bought, at 103 plus accrued interest, two IBM 8s08 bonds with interest paid on January 1 and July 1.

 Number of days accrued interest: _____ Total payment: _____

 b. On October 9, Ben Blue bought, at 95 plus accrued interest, three IBM $7\frac{1}{2}$s09 bonds with interest paid on January 1 and July 1.

 Number of days accrued interest: _____ Total payment: _____

8. Jack Mueller purchased a $1,000 corporate bond with a rating of AAA, paying 8% per year. Tom Bronkowski purchased a $1,000 junk bond paying 20%. Each bond was to mature in 10 years. Jack's bond paid interest for the 10-year period and face value at maturity. Tom's junk bond paid interest for 3 years before the company filed for bankruptcy and paid 45 cents on the dollar to its bond holders. How much more did Jack receive from his investment than Tom received from his? (10 points)

9. Compute the current yield for the following bonds. (5 points for each correct answer)

| Bond | Price | Current yield |
|------|-------|---------------|
| **a.** PepsiCo 9s08 | 104 | _____ |
| **b.** IBM $7\frac{3}{8}$s08 | 96.50 | _____ |
| **c.** AlaP 11s08 | 102.50 | _____ |
| **d.** ABM 6s08 | 82 | _____ |

Score for B (50)

Assignment 23.2: Bond Rate of Yield

LEARNING
OBJECTIVES

Name _____

Date _____ Score _____

A **(52 points) Solve the following problems. (points for correct answers as marked)**

1. An investor bought an 8.6% bond at 92. The bond would mature in 5 years. Round answers to two decimal places. (4 points for each correct answer)

 a. What was the average annual yield? _____ **b.** What was the rate of yield to maturity? _____

2. In 1998, Jim Ayers bought six LTV 5s13 bonds for which he paid 82. Three years later, he sold the bonds at 84 and bought six Southern Electric $9\frac{1}{2}$s21 bonds at 93. Did he increase or decrease the original rate of yield to maturity and by how much? Round yields to one decimal place. (14 points) _____

3. On July 29, Rose Mehrer purchased four AMAX $9\frac{1}{2}$s03 bonds at 92. Interest was payable March 1 and September 1. Included in Rose's cost was accrued interest for 150 days . (4 points for each correct answer)

 a. What was the total purchase cost? _____ **b.** What was the average annual yield? Do not consider accrued interest when calculating this rate of yield.

4. In 2000, Benito Cooper planned to purchase twenty $1,000 bonds and hold them to maturity. He had two choices: The first was EM&E $8\frac{1}{2}$s13 at 106.50. The second was Standard of California 6s10 at 80. Benito purchased the issue that provided the higher rate of yield to maturity.

 a. Which issue did Benito purchase? (12 points) _____

Assignment 23.2 *Continued*

b. How much income would Benito have earned monthly if Standard of California had been purchased? (3 points)

c. If, in 2003, Benito had purchased EM&E $8\frac{1}{2}$s13 bonds at a price of 89, what would have been the yield to maturity? (6 points) _____

d. Which company's bonds would be the better buy: EM&E at 89 or Standard of California? (1 point)

Score for A (52)

B **(48 points) Complete the following table. Show yield to maturity to one decimal place. (2 points for each correct answer)**

| | Number Purchased | Price Paid | Discount or Premium | Years to Maturity | Interest Rate | Annual Interest | +Discount −Premium Amortization | Average Principal Invested | Yield to Maturity |
|---|---|---|---|---|---|---|---|---|---|
| **a.** | 5 | 96 | $+200.00 | 5 | 8 | | | | |
| **b.** | 2 | 104.50 | −90.00 | 15 | $11\frac{1}{2}$ | | | | |
| **c.** | 20 | 73 | +5,400.00 | 10 | $5\frac{1}{2}$ | | | | |
| **d.** | 7 | 110 | −700.00 | 3 | 11 | | | | |
| **e.** | 3 | 90 | +300.00 | 15 | 7 | | | | |
| **f.** | 1 | 104.50 | −45.00 | $7\frac{1}{2}$ | $6\frac{1}{2}$ | | | | |

Score for B (48)

Notes

Headquartered in Warren, Michigan, **General Motors** (GM) has many facilities across our continent and Europe. GM operations are expanding in the rapid growth economies of China, Taiwan, India, and Southeast Asia, as well as in Latin America, where there are manufacturing and assembly facilities.

GM is the world's largest full-line vehicle manufacturer. According to a recent survey, if General Motors international operations alone were ranked in terms of sales, they would be 26th out of the top 500 global companies. See GM's web page at www.GM.com.

ɔwer

sword: Nuovo Uter

International Business | 24

LEARNING OBJECTIVES

By studying this chapter and completing all assignments you will be able to:

1
LEARNING
OBJECTIVES

Compute currency exchange rates.

2
LEARNING
OBJECTIVES

Compute the effects of exchange rate changes on business transactions with other countries.

3
LEARNING
OBJECTIVES

Compute duties on imports.

4
LEARNING
OBJECTIVES

Convert between U.S. weights and measures and metric weights and measures.

Today everyone living in the United States is closely tied to the international marketplace. Businesses cannot succeed by focusing on their own back yards—the explosion in information technology and the interrelationships of global stocks and currencies have required development of a global perspective.

1. What does the term *currency exchange rate* mean?
2. Do currency exchange rates remain constant, or do they change frequently?
3. Where can you find out how many U.S. dollars you would need to purchase 100 British pounds?
4. What is a *duty*, and why does it exist?
5. What is a *foreign trade zone*?

Businesses in the United States **import** goods made in other countries and **export** domestic goods made in the United States. International business transactions amount to billions of dollars annually and constitute an important part of the economies of most nations in the world.

International trade between U.S. companies and those in other countries is under the jurisdiction of the International Trade Administration (ITA), a branch of the Department of Commerce. All international trade is subject to a set of ITA rules and regulations known as the **Export Administration Regulations.** Any company in the United States planning to sell goods to companies in other countries must have an ITA export license for the transactions.

Computing Currency Exchange Rates

LEARNING OBJECTIVES

Compute currency exchange rates.

In order to conduct international trade, U.S. companies must exchange U.S. dollars for other currencies and vice versa. Figure 24-1 lists the names of the currencies used in the major countries involved in international trade

Figure 24-1: Currencies

| Country | Currency | Country | Currency |
|---|---|---|---|
| Argentina | Peso | Jordan | Dinar |
| Australia | Dollar | Kuwait | Dinar |
| Austria | Schilling | Lebanon | Pound |
| Bahrain | Dinar | Malaysia | Ringgit |
| Belgium | Franc | Mexico | Peso |
| Brazil | Real | Netherlands | Guilder |
| Britain | Pound | New Zealand | Dollar |
| Canada | Dollar | Norway | Krone |
| Chile | Peso | Pakistan | Rupee |
| China | Renminbi | Peru | New Sol |
| Colombia | Peso | Philippines | Peso |
| Czech Republic | Koruna | Poland | Zloty |
| Denmark | Krone | Portugal | Escudo |
| Ecuador | U.S. Dollar | Russia | Ruble |
| European Union | Euro | Saudi Arabia | Riyal |
| Finland | Markka | Singapore | Dollar |
| France | Franc | Slovak Republic | Koruna |
| Germany | Mark | South Africa | Rand |
| Greece | Drachma | South Korea | Won |
| Hong Kong | Dollar | Spain | Peseta |
| Hungary | Forint | Sweden | Krona |
| India | Rupee | Switzerland | Franc |
| Indonesia | Rupiah | Taiwan | New Taiwan Dollar |
| Iran | Rial | Thailand | Baht |
| Ireland | Punt | Turkey | Lira |
| Israel | Shekel | United Arab Emirates | Dirham |
| Italy | Lira | Uruguay | New Peso |
| Japan | Yen | Venezuela | Bolivar |

Most major newspapers print a list of exchange rates daily. These exchange rates change from day to day. Figure 24-2 is a recent listing of exchange rates for the world's major currencies.

Figure 24-2: Foreign Currency/U.S. Dollars Exchange Rate

| Country | Foreign Currency in Dollars | | Dollars in Foreign Currency | |
|---|---|---|---|---|
| | Friday | Thursday | Friday | Thursday |
| Argentina | 1.002 | 1.002 | 0.9998 | 0.9998 |
| Australia | 0.5461 | 0.5427 | 1.8313 | 1.8425 |
| Austria | 0.06720 | 0.06720 | 14.882 | 14.922 |
| Bahrain | 2.6525 | 2.6525 | 0.3770 | 0.3770 |
| Belgium | 0.0229 | 0.0229 | 43.6272 | 43.7455 |
| Brazil | 0.5066 | 0.5079 | 1.9738 | 1.9688 |
| Britain | 1.4605 | 1.4559 | 0.6847 | 0.6869 |
| Canada | 0.6633 | 0.6610 | 1.5077 | 1.5129 |
| Chile | 0.001764 | 0.001762 | 566.76 | 567.46 |
| Colombia | 0.004446 | 0.004435 | 224.921 | 225.479 |
| Czech Republic | 0.02653 | 0.02636 | 37.692 | 37.933 |
| Denmark | 0.1238 | 0.1235 | 8.0764 | 8.0955 |
| Ecuador | 1.000 | 1.000 | 1.000 | 1.000 |
| Finland | 0.1555 | 0.1551 | 6.4302 | 6.4477 |
| France | 0.1410 | 0.1406 | 7.0941 | 7.1133 |
| Germany | 0.4728 | 0.4715 | 2.1152 | 2.1209 |
| Greece | 0.002714 | 0.002706 | 368.52 | 359.62 |
| Hong Kong | 0.1282 | 0.1282 | 7.7996 | 7.7997 |
| Hungary | 0.003487 | 0.003478 | 386.75 | 387.53 |
| India | 0.02151 | 0.02157 | 46.485 | 46.360 |
| Indonesia | 0.0001058 | 0.0001064 | 9450.00 | 9400.00 |
| Ireland | 1.1741 | 1.1708 | 0.8517 | 0.8541 |
| Israel | 0.2419 | 0.2410 | 4.1340 | 4.1490 |
| Italy | 0.0004775 | 0.0004763 | 2094.06 | 2099.73 |
| Japan | 0.008561 | 0.008485 | 116.81 | 117.85 |
| Jordan | 1.4085 | 1.4085 | 0.7100 | 0.7100 |
| Kuwait | 3.2626 | 3.2637 | 0.3065 | 0.3064 |
| Lebanon | 0.0006634 | 0.0006634 | 1507.50 | 1507.50 |
| Malaysia | 0.2632 | 0.2632 | 3.8000 | 3.8000 |
| Malta | 2.2717 | 2.2665 | 0.4402 | 0.4412 |
| Mexico | 0.1032 | 0.1033 | 9.6925 | 9.6780 |
| Netherlands | 0.4196 | 0.4184 | 2.3833 | 2.3898 |
| New Zealand | 0.4371 | 0.4385 | 2.2878 | 2.2805 |
| Norway | 0.1126 | 0.1121 | 8.8797 | 8.9200 |
| Pakistan | 0.1698 | 0.01702 | 58.900 | 58.750 |
| Peru | 0.2835 | 0.2834 | 3.5278 | 3.5280 |
| Philippines | 0.02033 | 0.02068 | 49.200 | 48.350 |
| Poland | 0.2420 | 0.2424 | 4.1325 | 4.1250 |
| | | | | *(continued)* |

Figure 24-2: Foreign Currency/U.S. Dollars Exchange Rate *(continued)*

| Portugal | 0.004612 | 0.004600 | 216.82 | 271.41 |
|---|---|---|---|---|
| Russia | 0.03523 | 0.03517 | 28.381 | 28.433 |
| Saudi Arabia | 0.2666 | 0.2666 | 3.7507 | 3.7507 |
| Singapore | 0.5740 | 0.5741 | 1.7412 | 1.7420 |
| Slovak Republic | 0.02120 | 0.02114 | 47.180 | 47.297 |
| South Africa | 0.1270 | 0.1268 | 7.8750 | 7.8850 |
| South Korea | 0.00078 | 0.00078 | 1276.50 | 1275.40 |
| Spain | 0.005557 | 0.005542 | 179.94 | 180.43 |
| Sweden | 0.1037 | 0.1036 | 9.6389 | 9.6496 |
| Switzerland | 0.7410 | 0.7450 | 1.3495 | 1.3422 |
| Taiwan | 0.03072 | 0.03072 | 32.550 | 32.550 |
| Thailand | 0.02326 | 0.02320 | 43.000 | 43.100 |
| Turkey | 0.0000014 | 0.0000014 | 677670.0 | 677670.0 |
| United Arab Emirates | 0.2723 | 0.2723 | 3.6728 | 3.6728 |
| Uruguay | 0.07973 | 0.07952 | 12.543 | 12.543 |
| Venezuela | 0.001429 | 0.001429 | 699.65 | 699.80 |
| Euro | 0.9247 | 0.9222 | 1.0814 | 1.0844 |

Four other governments call their currency the dollar—Australia, Canada, Hong Kong, and New Zealand. These dollars are not U.S. dollars; each is a separate currency. Several currencies share names, such as the franc, mark, peso, or pound. Throughout the world, each country prints its currency with that country's name.

http://cnnfn.cnn.com/ markets/currencies

Other currency converters are available at these Web sites and at many others on the Internet.

EXAMPLE A

The President of Worldwide Industries read in Friday's newspaper that the U.S. dollar had risen slightly against the Swiss franc, as shown in Figure 24-2. The Foreign Currency in Dollars column for Switzerland shows 0.7410 for Friday and 0.7450 for Thursday. Thus on Friday it took $0.7410 to buy one Swiss franc, whereas on Thursday it took $0.7450.

0.7450 (Thursday) − $0.7410 (Friday) = $0.004 less on Friday

EXAMPLE B

Again, refer to Figure 24-2. How many U.S. dollars did it take to buy 100 Swiss francs on Friday and on Thursday?

0.7410 × 100 = 74.10 U.S. dollars on Friday
0.7450 × 100 = 74.50 U.S. dollars on Thursday

In example B, because it cost fewer U.S. dollars to buy 100 Swiss francs on Friday (74.10) than it did on Thursday (74.50), the dollar was more valuable on Friday than on the day before. That is, it rose relative to the Swiss franc.

Because the exchange rate from foreign currencies to U.S. dollars is listed as a percent of $1.00, you usually can do the calculation mentally.

EXAMPLE C

How much would 100 Swiss francs be worth on Friday in U.S. dollars? In the *Dollars in Foreign Currency* column for Switzerland, Figure 24-2 shows that the exchange rate on Friday was 1.3495. Thus on Friday it would take 1.3495 Swiss francs to buy 1 U.S. dollar.
100 Swiss francs ÷ 1.3495 = 74.10 U.S. dollars (rounded to the nearest cent)

CONCEPT CHECK 24.1

a. Using the *Foreign Currency in Dollars* column from Figure 24-2, compute the amount of U.S. currency necessary on Friday to buy 1,000 units of each of the following foreign currencies listed below.

Austria's schilling: 0.06720 × 1,000 = $67.20 Denmark's krone: 0.1238 × 1,000 = $123.80
Germany's mark: 0.4728 × 1,000 = $472.80 Hong Kong's dollar: 0.1282 × 1,000 = $128.20

b. Using the *Dollars in Foreign Currency* column from Figure 24-2, compute the value in U.S. currency on Thursday of 1,000 units of each of the following foreign currencies (rounded to the nearest cent).

Belgium's franc: 1,000 ÷ 43.7455 = $22.86 Norway's krone: 1,000 ÷ 8.9200 = $112.11
Peru's new sol: 1,000 ÷ 3.5280 = $283.45 Saudi Arabia's riyal: 1,000 ÷ 3.7507 = $266.62

Computing the Effects of Exchange Rate Changes

One hazard of foreign trade is the uncertainty of future exchange rates between currencies. The relationship between the values of the U.S. dollar and a foreign currency can change between the time a contract is signed and the time payment is received. If a U.S. exporter agrees to accept foreign currency, a devaluation in the foreign currency could cause the exporter to lose money on the transaction.

LEARNING OBJECTIVES

Compute the effects of exchange rate changes on business transactions with other countries.

Worldwide Industries contracted to sell certain goods to a company in France. The French company contracted to pay 500,000 French francs for the shipment. At the time the contract was signed, the *Foreign Currency in Dollars* column in the financial section of the newspaper showed that 1 French franc was valued at $0.20 U.S. Worldwide Industries expected to receive $100,000 U.S.
500,000 francs × $0.20 U.S. = $100,000 U.S.

Between the date the contract was signed and the date on which payment was due, the French franc fell to a value of $0.18 U.S. How much did Worldwide Industries lose by having agreed to accept payment in French francs instead of U.S. dollars?
500,000 francs × $0.18 U.S. = $90,000 U.S.
$100,000 U.S. expected − $90,000 U.S. received = $10,000 lost

**http://pacific.
commerce.ubc.
ca/xr/today.html**

A list of the daily exchange rates can be found at this Web site.

EXAMPLE E

Worldwide Industries contracted to buy certain goods from another company in France. Worldwide Industries contracted to pay 1 million French francs for the shipment. On the day the contract was signed, the *Dollars in Foreign Currency* column in the newspaper showed that 5.2110 French francs would buy 1 U.S. dollar. Worldwide Industries expected to pay the equivalent of $191,902 U.S.

1,000,000 francs ÷ 5.2110 = $191,902 U.S. (rounded to the nearest dollar)

On the date payment was made, the *Dollars in Foreign Currency* column showed that 5.3100 French francs were required to buy 1 U.S. dollar. How many dollars did Worldwide Industries save by having agreed to pay in French francs?

1,000,000 francs ÷ 5.3100 = $188,324 U.S. (rounded to the nearest dollar)

$191,902 U.S. expected to pay − $188,324 actually paid = $3,578 saved

 CONCEPT CHECK 24.2

a. Global Company contracted to sell certain goods to a company in Germany. The German company contracted to pay 300,000 German marks for the shipment. At the time the contract was signed, the *Foreign Currency in Dollars* column in the newspaper showed that the German mark was worth $0.64.

 Between the date the contract was signed and the date on which payment was due, the German mark rose to a value of $0.68 U.S. How much did Global Company gain by having agreed to accept payment in German marks instead of U.S. dollars?

 300,000 marks × $0.64 = $192,000 expected
 300,000 marks × $0.68 = $204,000 received
 $204,000 − $192,000 = $12,000 gained

b. Global Company contracted to buy certain goods from another company in Germany. Global Company contracted to pay 600,000 German marks for the shipment. On the day the contract was signed, the *Dollars in Foreign Currency* column in the newspaper showed that 1.5558 German marks would buy 1 U.S. dollar. Global Company expected to pay the equivalent of $385,654 U.S. dollars.

 On the date payment was made, the *Dollars in Foreign Currency* column showed that the German mark had fallen to 1.5100. How many dollars did Global Company lose by having agreed to pay in German marks?

 600,000 marks ÷ 1.5558 = $385,654 (rounded to nearest dollar)
 600,000 marks ÷ 1.5100 = $397,351 (rounded to nearest dollar)
 $397,351 paid − $385,654 expected to pay = $11,697 loss

COMPLETE ASSIGNMENT 24.1.

Computing Duties on Imports

Compute duties on imports.

All items imported into the United States must go through the U.S. Customs Agency. Many imported items have a **duty** (charge or tax) imposed by the Customs Agency to protect U.S. manufacturers against foreign competition in domestic markets. Duties vary widely from item to item. A duty may be a set amount—such as $0.50 per item—or an **ad valorem duty**, which is a percent of the value of the item.

EXAMPLE F

Assume that a quartz analog wristwatch in a gold-plated case with a metal band has four duty rates imposed on importation: $0.40 per wristwatch + 6% of the value of the case + 14% of the value of the strap + 5.3% of the value of the battery. Anderson Jewelry Company imported three dozen quartz analog wristwatches from Switzerland. The value of the case was $16; the strap, $10; and the battery, $6. How much duty did the Anderson Jewelry Company pay for the three dozen wristwatches? (Round answer to nearest cent.)

Duty per wristwatch: $0.40
Ad valorem duty on case: $16 × 0.06 = 0.96
Ad valorem duty on strap: $10 × 0.14 = 1.40
Ad valorem duty on battery: $6 × 0.053 = $0.318
 Total duty per watch $3.078

$3.078 per watch × 36 watches = $110.81 total duty paid

EXAMPLE G

A computer printer costs $150 whether purchased from country A or country B. However, it has an ad valorem duty rate of 3.5% if purchased from country A and an ad valorem duty rate of 30% if purchased from country B. How much more would it cost a company to purchase the printer from country B than from country A?

Country A: $150 × 0.035 = $5.25
 $150 + $5.25 = $155.25 total cost

Country B: $150 × 0.30 = $45
 $150 + $45 = $195 total cost

$195 country B − $155.25 country A = $39.75 more

Foreign trade zones are domestic sites in the United States considered to be outside U.S. Customs territory. These foreign trade zones are used for import and export activities. No duty or federal excise taxes are charged on foreign goods moved into the zone until the goods or products made from them are moved into U.S. Customs territory. No duty is charged on imports that later are exported for sale, because they never entered U.S. Customs territory. Recently, there were more than 150 foreign trade zones in port communities in the United States. Operations in them include storage, repacking, inspection, exhibition, assembly, and manufacturing.

EXAMPLE H

A U.S. company located in a foreign trade zone imported $500,000 worth of goods. The duty rate on the goods is 5%. If 20% of the goods were moved into U.S. Customs territory for sale and 80% were exported for sale, how much money did the company save by being located in a foreign trade zone?

$500,000 × 5% duty = $25,000 duty if goods are sold in U.S. Customs territory
$25,000 × 80% exported = $20,000 saved

CONCEPT CHECK 24.3

a. Downtown Toy Store ordered from a foreign country 400 dolls on which an ad valorem duty of 4.5% is charged. Payment is to be made in U.S. dollars. The price of each doll is $23. What is the total cost to Downtown?

400 × $23 = $9,200 cost before duty

$9,200 × 0.045 duty = $414

$9,200 + $414 = $9,614 total cost to buyer

b. A company located in a foreign trade zone purchased $1 million worth of electronic equipment having an ad valorem duty of 4.1%. Thirty percent of the products were moved into U.S. Customs territory for sale, and 70% were repackaged and exported. How many dollars did the company save by being located in a foreign trade zone?

$1,000,000 × 70% = $700,000

$700,000 × 4.1% = $28,700 saved

CyberConnection

Visit the Bureau of Export Administration's Web site for answers to frequently asked questions (FAQs) about export duties and taxes, restrictions and embargoes, licenses, and paperwork.

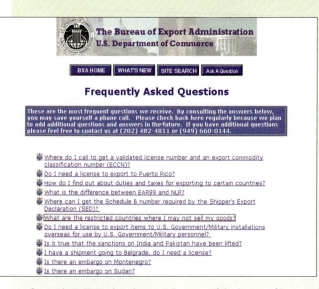

Source: http://www.bxa.doc.gov/. Courtesy of The Bureau of Export Administration.

Converting Between U.S. Weights and Measures and Metric Weights and Measures

LEARNING OBJECTIVES

Convert between U.S. weights and measures and metric weights and measures.

Some businesses, especially in the area of import—export activities, must convert U.S. customary units of weight and measure to the **metric system** of weights and measures used in most other countries. Figure 24-3 shows the conversion values for the U.S./metric units used most frequently in business.

Figure 24-3: U.S./Metric Unit Conversions

| To Convert U.S. to | Metric | Multiply by Number of Metric in U.S. | To Convert Metric to | U.S. | Multiply by Number of U.S. in Metric |
|---|---|---|---|---|---|
| Inches | Meters | 0.0254 | Meters | Inches | 39.37 |
| Feet | Meters | 0.305 | Meters | Feet | 3.281 |
| Yards | Meters | 0.914 | Meters | Yards | 1.09 |
| Miles | Kilometers | 1.609 | Kilometers | Miles | 0.621 |
| Ounces | Grams | 28.35 | Grams | Ounces | 0.035 |
| Pounds | Grams | 454 | Grams | Pounds | 0.0022 |
| Pounds | Kilograms | 0.454 | Kilograms | Pounds | 2.2 |
| Pints | Liters | 0.473 | Liters | Pints | 2.113 |
| Quarts | Liters | 0.946 | Liters | Quarts | 1.057 |
| Gallons | Liters | 3.785 | Liters | Gallons | 0.264 |

EXAMPLE I

Convert the following U.S. measures to metric measures.

a. Convert 20 inches to meters.
20 in. × 0.0254 = 0.508 m

b. Convert 20 feet to meters.
20 ft × 0.305 = 6.1 m

c. Convert 20 yards to meters.
20 yd × 0.914 = 18.28 m

d. Convert 20 miles to kilometers.
20 mi × 1.609 = 32.18 km

e. Convert 20 ounces to grams.
20 oz × 28.35 = 567 g

f. Convert 20 pounds to grams.
20 lb × 454 = 9,080 g

g. Convert 20 pounds to kilograms.
20 lb × 0.454 = 9.08 kg

h. Convert 20 pints to liters.
20 pt × 0.473 = 9.46 L

i. Convert 20 quarts to liters.
20 qt × 0.946 = 18.92 L

j. Convert 20 gallons to liters.
20 gal × 3.785 = 75.7 L

EXAMPLE J

Convert the following metric measures to U.S. measures.

a. Convert 20 meters to inches.
20 m × 39.37 = 787.4 in.

b. Convert 20 meters to feet.
20 m × 3.281 = 65.62 ft

c. Convert 20 meters to yards.
20 m × 1.09 = 21.8 yd

d. Convert 20 kilometers to miles.
20 km × 0.621 = 12.42 mi

e. Convert 20 grams to ounces.
20 g × 0.035 = 0.7 oz

f. Convert 20 grams to pounds.
20 g × 0.0022 = 0.044 lb

g. Convert 20 kilograms to pounds.
20 kg × 2.2 = 44 lb

h. Convert 20 liters to pints.
20 L × 2.113 = 42.26 pt

i. Convert 20 liters to quarts.
20 L × 1.057 = 21.14 qt

j. Convert 20 liters to gallons.
20 L × 0.264 = 5.28 gal

http:/

**www.worldwidemetric
.com/metcal.htm**

**www.sciencemadesimple
.com/conversions.html**

**www.wsdot.wa.gov/
metrics**

**www.allmath.com/
metric.asp**

It is important to know how to convert between U.S. and metric measure when you don't have access to the Internet. When you do, you can find conversion calculators at these sites and many others.

CONCEPT CHECK 24.4

Using Figure 24-3, make the following conversions:

a. Convert 28 inches to meters.
28 × 0.0254 = 0.7112

b. Convert 17 feet to meters.
17 × 0.305 = 5.185

c. Convert 3 meters to inches.
3 × 39.37 = 118.11

d. Convert 18 meters to feet.
18 × 3.281 = 59.058

e. Convert 3 ounces to grams.
3 × 28.35 = 85.05

f. Convert 7 pounds to grams.
7 × 454 = 3,178

g. Convert 36 grams to pounds.
36 × 0.0022 = 0.0792

h. Convert 18 kilograms to pounds.
18 × 2.2 = 39.6

i. Convert 8 pints to liters.
8 × 0.473 = 3.784

j. Convert 2 quarts to liters.
2 × 0.946 = 1.892

COMPLETE ASSIGNMENT 24.2.

Chapter Terms for Review

ad valorem duty

duty

export

Export Administration Regulations

foreign trade zones

import

metric system

Visit the Web

http://www.alamri.com—Click on "Doing Business in SA" for general information on customs duties on goods imported by Saudi Arabia is available here.

http://www.doc.gov/—Go to the Department of Commerce Web site for information on international trade.

http://www.islamic-banking.com, **http://www.islamic-finance.com/indexnew.htm**, and **http://islamic-finance.net**—Visit one of these Web sites for information on Islamic banking and finance.

http://www.imex-fx.com/—Read more on international monetary exchanges and other international financial topics here.

http://www.imf.org/—Information on international financial issues are found at the International Monetary Fund Web site.

http://www.travlang.com/money and **http://www.x-rates.com/calculator.html**—Foreign exchange converters can also be found at these Web sites.

CYBERCONNECTION

Sample Trades

While every transaction is unique with varying conditions and rates, the following are two sample trades indicative of the savings IMEX clients realize when compared to their banks. These examples are for purposes of illustration only and are based upon average savings of IMEX clients.

Sample Trade # 1

Assumption: Value of transaction = $100,000.00
Currency: German Mark (DM)

A. Average retail bank pricing for a trade of $100,000.00 in DM:
$100,000.00 x 1.6425 DM/$ = DM 164,250.00.

B. IMEX interbank rate pricing on same transaction:
$100,000.00 x 1.6750 DM/$ = DM 167,500.00

Savings in DM = DM 3250.00
Savings in $ = DM 3250.00 @ 1.6750 DM/$ = $1940.00.

Sample Trade# 2

Assumption: Value of transaction = $100,000.00
Currency: Canadian Dollar (C$)

A. Average retail bank pricing for a trade of $100,000.00. in C$:
$100,000.00. x 1.5020 C$/US$ = C$ 150,200.00

Source: http://www.imex-fx.com/. Copyright © 2001 by International Money Exchange, Inc. (IMEX); used with permission.

1. What are some of the differences a business would realize if it exchanged its foreign currency at its local bank than if it exchanged its foreign currency from Imex.fx?

2. Would you recommend the www.imex.fx.com web site to anyone who was interested in converting or exchanging foreign currencies into US dollars or interested in converting or exchanging US dollars into foreign currencies?

The Bottom Line

Summary of chapter learning objectives:

| Learning Objective | Summary | Example |
|---|---|---|
| **24.1** | Compute currency exchange rates | 1. Using the *Dollars in Foreign Currency* column in Figure 24-2, compute the value in U.S. dollars on Thursday of 5,000 units of each of the following foreign currencies. Round answers to the nearest cent.

 a. Czech Republic's koruna _____ b. Spain's peseta _____
 c. Uruguay's new peso _____ d. Thailand's baht _____

2. Using the *Foreign Currency in Dollars* column in Figure 24-2, compute the amount of U.S. dollars necessary on Thursday to buy 5,000 units of each of the following foreign currencies. Round answers to the nearest cent.

 a. Sweden's krona _____ b. Hungary's forint _____
 c. France's franc _____ d. Belgium's franc _____ |
| **24.2** | Compute the effects of exchange rate changes on business transactions with other countries | 3. A U.S. company has contracted to sell certain goods to a company in the Netherlands. The Dutch company has contracted to pay 700,000 guilders for the goods. At the time the contract was signed, the *Foreign Currency in Dollars* column in the newspaper showed that the Dutch guilder was worth $0.57. On the date payment was due, the guilder fell to a value of $0.53 U.S. How much did the U.S. company lose by having agreed to accept payment in guilders instead of U.S. dollars? |
| **24.3** | Compute duties on imports | 4. Broadway Department Store ordered from a foreign country 300 sets of dishes on which an ad valorem duty of 5.8% is charged. The price of each set of dishes is $72. Payment is to be made in U.S. currency. What is the total cost to Broadway? |
| **24.4** | Convert between U.S. weights and measures and metric weights and measures | 5. Using Figure 24-3, make the following conversions.
 a. Convert 100 inches to meters. b. Convert 1,000 meters to feet.
 c. Convert 6 miles to kilometers. d. Convert 100 grams to ounces.
 e. Convert 3 gallons to liters. f. Convert 7 liters to quarts. |

Answers: 1a. $131.80 **b.** $27.71 **c.** $397.60 **d.** $116.00; **2a.** $518.16 **b.** $12.90 **c.** $702.91 **d.** $114.30 **3.** $28,000 **4.** $22,852.80 **5a.** 2.54 m **b.** 3,281 ft **c.** 9.654 km **d.** 3.5 oz **e.** 11.355 L **f.** 7.399 qt

Self-Check Review Problems for Chapter 24

1. The Australian dollar is listed in the Friday *Foreign Currency in Dollars* column of Figure 24-2 as 0.5461. Compute the amount of U.S. dollars needed to buy 100 Australian dollars.

2. The Australian dollar is listed in the Friday *Foreign Currency in Dollars* column of Figure 24-2 as 0.5461. Compute the amount in Australian dollars you could purchase for $100 in U.S. currency.

3. The Australian dollar is listed in the Friday *Dollars in Foreign Currency* column of Figure 24-2 as 1.8313. Compute the value in U.S. dollars of 5,000 Australian dollars.

4. The Belgian franc is listed in the *Foreign Currency in Dollars* column of Figure 24-2 as 0.0229. Compute the amount of U.S. dollars needed to buy 100 Belgian francs.

5. Global Concerns, Inc., has contracted to sell certain goods to a company in Australia. The Australian company has contracted to pay 500,000 Australian dollars for the shipment. At the time the contract was signed, the *Foreign Currency in Dollars* column in the financial section of the morning paper showed that one Australian dollar was valued at $0.7570 U.S. Compute the value in U.S. dollars that Global Concerns expects to receive from the transaction.

6. In problem 5, suppose that, between the date the contract was signed and the date on which payment was received, the Australian dollar fell to a value of $0.72 in U.S. dollars. Compute the amount that Global Concerns lost by agreeing to accept payment in Australian dollars instead of U.S. dollars.

7. White's Manufacturing, Inc., has contracted to sell certain goods to a company in the Netherlands. The price agreed on for the purchase is 94,000 Netherlands guilders. On the date the contract was signed the *Foreign Currency in Dollars* column in the financial section of the local paper showed that the Netherlands guilder was valued at 0.5745. Compute the U.S. dollar value that White's Manufacturing expects to receive for the goods. (Round answer to the nearest U.S. dollar.)

8. The Italian lira is shown as 0.000628 and the Turkish lira is shown as 0.000028 in the *Foreign Currency in Dollars* column of the local paper. Compute how many more Turkish lira than Italian lira a U.S. citizen could buy for $100 U.S. (Round all calculations to the nearest lira.)

9. Princess Jewelry, Inc., has contracted to purchase 144 bracelets from a foreign manufacturer. The price of each bracelet is $32. An ad valorem duty of 20% is charged on each bracelet. Compute the duty Princess Jewelry will pay for the shipment.

10. ABC, Inc., plans to purchase 250 units of a certain component used in building computers in its U.S. factory. ABC can purchase the components from country X at a price of $45 each plus an ad valorem duty of 38%, or from country Y at a price of $52 plus an ad valorem duty of 3%. Compute the amount ABC will save by purchasing from country Y.

Assignment 24.1: Trading with Other Countries

Name

Date _____ Score _____

A (44 points) Solve the following problems. (1 point for each correct answer)

1. Using the *Foreign Currency in Dollars* column in Figure 24-2, find the amount of U.S. dollars needed to buy 100 units of each foreign currency listed, for Friday and for Thursday. Decide whether the U.S. dollar rose (+), fell (−), or stayed the same against that currency from Thursday to Friday.

| | Foreign Currency | Price of 100 Units Friday | Thursday | U.S. Dollar Rose, Fell, or Was Even |
|---|---|---|---|---|
| **a.** | Australian dollar | _____ | _____ | _____ |
| **b.** | Austrian schilling | _____ | _____ | _____ |
| **c.** | Belgian franc | _____ | _____ | _____ |
| **d.** | Brazilian real | _____ | _____ | _____ |
| **e.** | Canadian dollar | _____ | _____ | _____ |
| **f.** | German mark | _____ | _____ | _____ |
| **g.** | South African rand | _____ | _____ | _____ |
| **h.** | Swiss franc | _____ | _____ | _____ |

2. Using the *Dollars in Foreign Currency* column in Figure 24-2, determine the value in U.S. dollars of 1,000 units of each foreign currency listed for Thursday and for Friday. Round answers to the nearest cent.

| | Foreign Currency | Value of 1,000 units Friday | Thursday |
|---|---|---|---|
| **a.** | Argentinean peso | _____ | _____ |
| **b.** | British pound | _____ | _____ |
| **c.** | Danish krone | _____ | _____ |
| **d.** | Finnish mark | _____ | _____ |
| **e.** | Greek drachma | _____ | _____ |
| **f.** | Indian rupee | _____ | _____ |
| **g.** | Irish punt | _____ | _____ |
| **h.** | Italian lira | _____ | _____ |
| **i.** | Japanese yen | _____ | _____ |
| **j.** | Spanish peseta | _____ | _____ |

Score for A (44) _____

Assignment 24.1 *Continued*

B (56 points) Solve the following problems. Round pennies to the nearest dollar. (points for correct answers as marked)

3. Hadley Enterprises has contracted to sell certain goods to a company in Britain. The price agreed on for the goods is 80,000 British pounds. On the date the contract was signed, the *Foreign Currency in Dollars* column in the financial section of the local paper showed that the British pound was valued at $1.6554 U.S.

 a. How much in U.S. dollars does Hadley Enterprises expect to receive for the goods? (6 points) _____

 b. If the value of the British pound fell from 1.6554 to 1.550 on the date of payment, how much would Hadley Enterprises lose by having contracted in British pounds instead of U.S. dollars? (8 points) _____

 c. If the British pound rose to 1.7500 on the date of payment, how much would Hadley Enterprises gain by having contracted in British pounds instead of U.S. dollars? (8 points) _____

4. Miller Furniture Company imported chairs from a Danish firm. Each chair is valued at 890 Danish krone. What is the value of the chairs in U.S. dollars if the Danish krone is currently shown as 0.1694 in the *Foreign Currency in Dollars* column of the local paper? (8 points) _____

5. Oldtown Industries, Inc., is contracting to sell its product to a country whose currency is unstable and difficult to convert to U.S. currency. The value of the goods is $20,000 U.S. The currency of the country to which the goods will be shipped is currently valued at 0.0040 in the *Foreign Currency in Dollars* column of the local paper. Oldtown Industries is willing to accept the currency of a third country. The German mark is agreed on. The German mark is shown as 0.6428 in the *Foreign Currency in Dollars* column on the date the contract is signed.

 a. How many German marks does Oldtown Industries expect to receive? (Round the answer to the nearest mark.) (6 points) _____

 b. If the mark does not change before the date of payment, but the value of the currency of the receiving country falls from 0.0040 to 0.0003, how much did Oldtown Industries save by using the German mark? (10 points)

6. If the British pound is shown as 1.9000 and the Egyptian pound as 0.3700 in the *Foreign Currency in Dollars* column of the local paper, how many more Egyptian pounds than British pounds could a U.S. citizen buy for $1,000 U.S.? (Round the answer to the nearest pound.) (10 points) _____

Score for B (56)

Assignment 24.2: Duties and Metric Conversion

Name _____

Date _____ Score _____

A **(56 points) Solve the following problems. (points for correct answers as marked)**

1. Benjamin's Department Store ordered from a foreign country 150 music boxes on which an ad valorem duty rate of 3.2% is charged. Payment is to be made in US. dollars. The price of each music box is $18. (2 points for each correct answer)

 a. What is the price of the 150 music boxes before duty is added? _____

 b. What is the amount of duty charged on the shipment? _____

 c. What is the total cost to Benjamin's? _____

2. Gems International Company is purchasing from a foreign country one gross (144) of 20-inch gold necklaces at $75 each and six dozen 18-inch silver necklaces at $55 each. The ad valorem duty rate for gold and silver jewelry is 6.5%. What is the total cost of the shipment to the buyer? (8 points) _____

3. Sutter's Department Store is going to buy four gross (one gross = 144) of vases for the next Christmas season. It can buy porcelain vases or lead crystal vases for $45 each. The duty on porcelain vases is 9%. The duty on lead crystal vases is 5%. How much will Sutter's save in total cost by purchasing lead crystal instead of porcelain? (8 points) _____

4. Melody Piano Store can purchase pianos domestically for $1,360 each. It can purchase pianos from a foreign country for $1,300 plus 5.3% ad valorem duty.

 a. Melody Piano Store purchases the pianos with the lower total cost. Do they purchase from a domestic or a foreign manufacturer? (6 points) _____

 b. How much do they save on each piano? (2 points) _____

5. Broadway Office Equipment Company purchased the following equipment from a foreign country:

72 automatic typewriters at $150 each + 2.2% duty

24 addressing machines at $30 each + 4.2% duty

144 pencil sharpeners at $12 each + 5.3% duty

24 check-writing machines at $60 each, duty free

80 calculators at $24 each + $3.9% duty

a. What was the cost of the order before duty? (8 points) _____

b. What was the cost of the order after duty? (Round each calculation to the nearest cent.) (8 points) _____

c. If the 144 pencil sharpeners had been purchased at $12 each from a country with which trade was discouraged and the ad valorem duty rate was 50%, how much would the pencil sharpeners have cost? (4 points) _____

d. How much more duty would a buyer pay on the pencil sharpeners at the ad valorem rate of 50% than at an ad valorem duty rate of 6.3%? (2 points) _____

6. Adams Industries could purchase $30,000 worth of textiles from country A with an ad valorem duty rate of 2.5% or from country B with an ad valorem duty rate of 1.2%.

a. How much would the shipment cost if purchased from country A? (2 points) _____

b. How much would Adams Industries save by purchasing from country B? (2 points) _____

Score for A (56)

Assignment 24.2 *Continued*

B **(24 points) Solve the following problems. (points for correct answers as marked)**

7. The Allied Computer Company imports some computer components and manufactures other components and then assembles computers for sale within the United States or for export to foreign countries. The company is located in a district that has been designated by the International Trade Administration as a foreign trade zone. The company imported $250,000 worth of monitors having an ad valorem duty rate of 3.7%, $300,000 worth of power supplies having an ad valorem duty rate of 3.0%, and $500,000 worth of printers having an ad valorem duty rate of 3.7%. All products were finished and sold 1 year later.

 a. If all products were sold within U.S. Customs territories, how much duty—in U.S. dollars—did the company pay at the end of the year? (2 points) _____

 b. If 40% of the finished products were moved into U.S. Customs territories for sale and 60% were exported for sale in foreign countries, how many dollars of duty did the company pay at the end of the year? (8 points) _____

 c. If all products were exported for sale, how much duty did the company pay at the end of the year? (2 points) _____

8. The Allied Computer Company imported $260,000 worth of portable computers having an ad valorem duty rate of 3.9% and kept 20% of them for exhibition and company use on the premises.

 a. If the company repackaged and sold the remaining portable computers in U.S. Customs territories, how many dollars of duty did the company pay on the portable computers? (4 points) _____

 b. If the company repackaged and exported 50% of the portable computers for sale in foreign countries and moved the remaining 30% into U.S. customs territories for sale, how many dollars did the company pay in duty on the portable computers? (4 points) _____

9. A company imported $5 million worth of laptop computers having an ad valorem duty rate of 3.9%. The company repackaged and exported all the computers for resale. How many dollars did the company save by being located in a foreign trade zone? (4 points) _____

Score for B (24)

Assignment 24.2 *Continued*

C **(20 points) Solve the following problems using Figure 24-3. (1 point for each correct answer)**

10. Make the following conversions from U.S. measures to metric:

 a. Convert 15 inches to meters: _____

 b. Convert 15 feet to meters: _____

 c. Convert 15 yards to meters: _____

 d. Convert 15 miles to kilometers: _____

 e. Convert 25 ounces to grams: _____

 f. Convert 25 pounds to grams: _____

 g. Convert 25 pounds to kilograms: _____

 h. Convert 30 pints to liters: _____

 i. Convert 30 quarts to liters: _____

 j. Convert 30 gallons to liters: _____

11. Make the following conversions from metric to U.S. measures:

 a. Convert 15 meters to inches: _____

 b. Convert 15 meters to feet: _____

 c. Convert 15 meters to yards: _____

 d. Convert 15 kilometers to miles: _____

 e. Convert 25 grams to ounces: _____

 f. Convert 25 grams to pounds: _____

 g. Convert 25 kilograms to pounds: _____

 h. Convert 30 liters to pints: _____

 i. Convert 30 liters to quarts: _____

 j. Convert 30 liters to gallons: _____

Score for C (20)

Notes

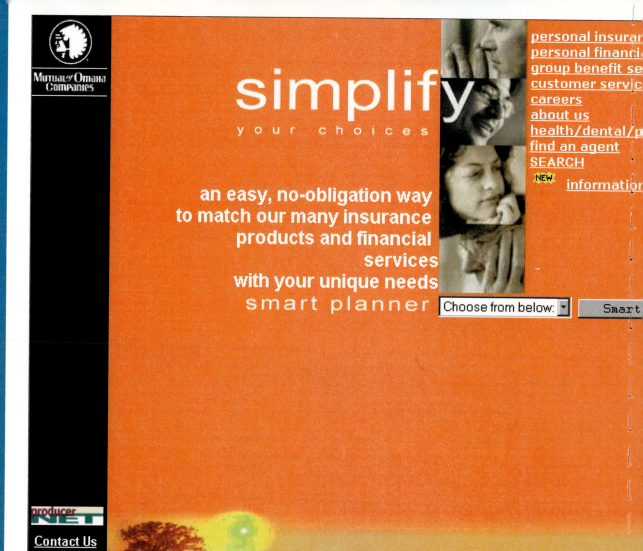

When someone asked Albert Einstein what he considered mankind's greatest invention to be, he replied without hesitation, "compound interest." To smart investors, compound interest is the secret of success. Simply put, it is interest that earns interest. Insurance companies, such as Mutual of Omaha, develop many types of life and retirement insurance plans based on the principles of compound interest. To learn about Mutual of Omaha, including investor information, professional opportunities, and small business insurance, go to its Web site at http://www.mutualofomaha.com.

Compound Interest and Present Value

25

LEARNING OBJECTIVES

By studying this chapter and completing all assignments you will be able to:

1 Compute future values and compound interest from future value tables.

2 Compute present values from future value tables.

3 Compute present values and compound interest from present value tables.

In a tax-deferred IRA, compound interest kicks into high gear. Investing $2,000 a year at 8% from age 35 until age 65 will yield $246,700. Compounding helps you work investment miracles—it's just a matter of time.

1. What is the fundamental difference between simple interest and compound interest?

2. If the number of compounding periods in 1 year increases, how does the future value change?

3. Suppose that you have a 5-year investment that pays simple interest. If you double the interest rate, will you double the future value?

4. Suppose that you have a 5-year investment that pays compound interest. If you double the interest rate, will you double the future value?

5. Suppose that the future value of an investment is given and that the present value is calculated by using compound interest. If the interest rate is doubled, will the present value stay the same, increase, or decrease?

Most Americans will buy at least one product that is financed over 1 or more years. The product will probably be large, such as a car or a home. The interest on the loan for the car or home is not the simple interest you studied in Chapter 14; it is *compounded* monthly. Most banks offer savings accounts and certificates of deposit (CDs) for which interest is compounded daily. Credit unions may pay interest that is compounded quarterly (four times a year). To evaluate the value of corporate bonds, an investor bases calculations on interest compounded semiannually (twice a year).

To understand even the simplest financial decisions in today's world, you must understand the fundamentals of compound interest, future values, and present values.

Computing Future Values and Compound Interest

LEARNING OBJECTIVES

Compute future values and compound interest from future value tables.

Simple interest is computed with the formula $I = P \times R \times T$, which you learned in Chapter 14. For example, the simple interest on $2,000 invested at 8% for 2 years is $I = P \times R \times T = \$2,000 \times 0.08 \times 2 = \320. The amount, or future value, of the investment is $A = P + I = \$2,000 + \$320 = \$2,320$.

Compound interest means that the computations of the simple interest formula are performed periodically during the term of the investment. The money from the previous interest computation is added to the principal before the next interest computation is performed. If an investment is *compounded annually for two years,* the simple interest is computed once at the end of each year. The simple interest earned in year 1 is added to the principal for the beginning of year 2. The total value of an investment is the principal plus all the compound interest, called the **future value** or the **compound amount**.

EXAMPLE A

Mary Simone invests $2,000 for 2 years in an account that pays 8% compounded annually. Compute the total compound interest and future value (compound amount).

| | |
|---|---|
| $2,000.00 | Original principal |
| × 0.08 | Interest rate |
| $160.0000 | First-year interest |
| + 2,000.00 | First-year principal |
| $2,160.00 | Second-year principal |
| × 0.08 | Interest rate |
| $172.8000 | Second-year interest |
| 2,160.00 | Second-year principal |
| $2,332.80 | Final compound amount (future value) |
| − 2,000.00 | Original principal |
| $332.80 | Total compound interest |

On the $2,000 investment in example A, the total amount of compound interest paid is $332.80, compared to $320 simple interest over the same 2 years.

The computations in example A are time-consuming and become more tedious with each compounding. Twice as many computations would be required for a 4-year investment. In actual practice, compound interest is computed using a **compound interest table**, a calculator with an exponent key, or a computer.

Table 25-1, on pages 548 and 549, is part of a future value table. The numbers in the table are called **future value factors** or **compound amount factors**. The columns (vertical)

represent interest rates, and the rows (horizontal) represent the number of times that interest is compounded. The following steps explain how to use Table 25-1 to find future values (compound amounts) and compound interest.

www.pine-grove.com/ pi08002.htm

www.finaid.org/ calculators/ compoundinterest. phtml

Use the interest calculator at either of these Web sites.

STEPS to Use the Future Value Table

1. Locate the factor in the proper row and column of Table 25-1.
2. Multiply the original principal by the factor. The product is the future value.
3. Subtract the original principal from the future value. The difference is the total amount of compound interest.

EXAMPLE B

Compute the future value and total amount of compound interest of a 2-year $2,000 investment at 8% compounded annually.

STEP 1
The interest rate is 8%. Interest is compounded twice—once each year for 2 years. Locate the intersection of the 8.00% column and row 2. The compound amount factor is 1.16640.

STEP 2
Future value = $2,000 × 1.16640 = $2,332.80

STEP3
Compound interest = $2,332.80 − $2,000 = $332.80

These results are identical to the results in example A.

EXAMPLE C

Ted Washington loans $1,000 to his son for 4 years at 5% compounded annually. Compute the future value and total compound interest.

STEP 1
The interest rate is 5%. Interest is computed four times, once each year for 4 years. The future value factor in the 5.00% column and row 4 is 1.21551.

STEP 2
Future value = $1,000 × 1.21551 = $1,215.51

STEP 3
Compound interest = $1,215.51 − $1,000 = $215.51

Future Value Formula

If you prefer to use a formula, the formula for future value is

Future value = Principal (Present value) × Future value factor (from Table 25-1)

Various Compounding Periods

In examples A, B, and C, the compounding was annual (i.e., done once each year). Compounding is also done daily (every day), monthly (every month), quarterly (every quarter), or semiannually (every half-year). The word **period** refers to the unit of time of the compounding. The period will be a day, a month, a quarter, a half-year, or a year. You can use Table 25-1 with some interest rates for all these compounding periods except 1 day. Daily compounding requires the use of a calculator with an exponent key.

To do monthly, quarterly, or semiannual compounding using Table 25-1, follow the same steps you used to do annual compounding. The only differences are that the column will be the **periodic interest rate** and that the row will be the **number of compounding periods**. Sometimes the periodic rate and the number of periods will be stated clearly.

© PHOTODISK, INC.

Much of the time, however, the interest rate will be stated as an annual rate and the time will be stated in years. When that happens, find the row and column as described in the following steps.

STEPS to Determine the Periodic Rate and the Number of Compounding Periods

i. Determine the number of compounding periods in 1 year (1 for annually, 2 for semi-annually, 4 for quarterly, 12 for monthly, etc.).

ii. Divide the stated rate by the number of periods in 1 year (from Step i). The quotient is the periodic rate, the correct column.

iii. Multiply the number of years by the number of periods in 1 year (from Step i). The product is the total number of compounding periods, the correct row.

EXAMPLE D

Find the periodic interest rate and the number of compounding periods in 2 years when 12% is compounded (a) semiannually (2 times per year), (b) quarterly (4 times per year), and (c) monthly (12 times per year). Then find the future value factors in Table 25-1.

Each term is for 2 years; each rate is 12%, but compounded differently:

| STEP i | STEP ii | STEP iii |
|---|---|---|
| **Periods per Year** | **Periodic Interest Rate** | **Compounding Periods** |
| a. 2 | 12% ÷ 2 = 6% | 2 years × 2 = 4 periods |
| b. 4 | 12% ÷ 4 = 3% | 2 years × 4 = 8 periods |
| c. 12 | 12% ÷ 12 = 1% | 2 years × 12 = 24 periods |

Future value factors from Table 25-1 are as follows.

| a. Semiannually | 6.00% column and row 4 | Factor = 1.26248 |
|---|---|---|
| b. Quarterly | 3.00% column and row 8 | Factor = 1.26677 |
| c. Monthly | 1.00% column and row 24 | Factor = 1.26973 |

To compute the future value and the compound interest, first determine the periodic rate and the number of compounding periods using Steps i, ii, and iii. Then do Steps 1, 2, and 3, as illustrated in examples B and C previously.

EXAMPLE E

Felix Tavis and his wife deposit $2,000 in her credit union, which pays interest of 8% compounded quarterly. Find the future value and the total compound interest after 2 years. (Use Table 25-1.)

STEP i There are 4 compounding periods in 1 year.

STEP ii Periodic interest rate = 8% ÷ 4 = 2% per period

STEP iii Number of periods = 2 years × 4 = 8 periods

STEP 1 Use Table 25-1, 2.00% column and row 8:
Future value factor = 1.17166.

STEP 2 Future value = $2,000.00 × 1.17166 = $2,343.32

STEP 3 Total compound interest = $2,343.32 − $2,000.00 = $343.32

Thus, $2,000 invested at 8% compounded quarterly will be worth $2,343.32 in 2 years.

CONCEPT CHECK 25.1

a. If $2,600 is invested for 2 years at 12% compounded quarterly, compute the future value of the investment.

Quarterly means 4 periods per year.
Periodic rate = 12% ÷ 4 = 3% per quarter
Number of periods = 2 years × 4 = 8 periods
The future value factor from row 8 of the 3.00% column in Table 25-1 is 1.26677.

Future value = $2,600 × 1.26677 = $3,293.602, or $3,293.60

b. If $3,200 is invested for 1 year at 9% compounded monthly, what is the compound interest on the investment?

Monthly means 12 periods per year.
Periodic rate = 9% ÷ 12 = 0.75% per month
Number of periods is 1 year × 12 = 12 periods
The future value factor from row 12 of the 0.75% column in Table 25-1 is 1.09381.

Future value = $3,200 × 1.09381 = $3,500.192, or $3,500.19
Compound interest = Future value − Present value (Principal)
 = $3,500.19 − $3,200 = $300.19

COMPLETE ASSIGNMENT 25.1.

Computing Present Values from Future Value Tables

The basic investment problem is to compute what a given sum of money invested today will be worth in the future. Example E was such a future value problem. There we found that $2,000 original principal (or present value) invested today at 8% compounded quarterly will have a future value of $2,343.32 in 2 years.

Some savers and investors want to compute future values; others want to compute present values. Consider the following present value problem.

LEARNING OBJECTIVES

Compute present values from future value tables.

EXAMPLE F

Roberta Mandell has a 12-year-old son and a 10-year-old daughter. Roberta inherits $100,000. Friends tell Roberta that she should plan to have $60,000 cash available for her son's education when he turns 18. She should also have $70,000 cash available for her daughter's education when she turns 18. Roberta wants to put enough money in an investment for each child so that in 6 and 8 years the two accounts will be worth $60,000 and $70,000, respectively. If Roberta can earn 10% compounded annually, how much money should she put into each investment today?

Roberta knows the future value of the investments—$60,000 and $70,000. What she wants to compute is the **present value**—the amounts that she needs to invest today for each child. We find the solution to this problem later, in example J.

Businesses make investments in the present to provide future revenues. Sometimes a business will estimate its future revenues and costs (future values). Then the business might use these numbers to compute the required amounts to invest initially (present values).

As given earlier, the formula for future value is

Future value = Present value × Future value factor (from Table 25-1)

Rewriting the formula to solve for present value gives

Present value = Future value ÷ Future value factor (from Table 25-1)

EXAMPLE G

How much money must be invested today to end up with $1,266.77 in 4 years? The interest rate is 6% compounded semiannually.

The $1,266.77 is the future value for which we want to find the present value. Interest is computed four times—once each year for 4 years. The future value in Table 25-1 in the 3.00% column and row 4 is 1.26677. Substitute these values into the formula to solve for present value.

Present value = Future value ÷ Future value factor (from Table 25-1)
= $1,266.77 ÷ 1.26677 = $1,000

Compare this result to that of example C, in which $1,000 was invested for 4 years at 3% compounded annually. The future value was $1,266.77.

EXAMPLE H

Edison Printing estimates that in 2 years it will cost $10,000 to repair its new printer. How much must Edison invest today to have $10,000 in 2 years, if the interest rate is 15% compounded monthly? How much interest will Edison Printing earn on its investment?

$10,000 is the future value for which Edison wants to know the present value.

STEP i There are 12 compounding periods in 1 year (monthly).

STEP ii Periodic rate = 15% ÷ 12 = 1.25%

STEP iii Number of compounding periods = 2 years × 12 = 24
The future value factor in the 1.25% column and row 24 of Table 25-1 is 1.34735.

Substitute these values into the formula to solve for present value:

Present value = Future value ÷ Future value factor (from Table 25-1)
= $10,000 ÷ 1.34735 = $7,421.976, or $7,421.98

If Edison Printing invests $7,421.98 today at 15% compounded monthly, it will have $10,000 at the end of 2 years.

The $10,000 is the sum of the amount invested plus the interest earned. To find the interest, subtract the amount invested from $10,000.

Interest = Future value − Present value = $10,000 − $7,421.98 = $2,578.02

CONCEPT CHECK 25.2

What principal (present value) invested for 3 years at 10% compounded semiannually will result in a total future value of $4,000? (Use Table 25-1.)

Semiannually means 2 periods per year.
Periodic rate = 10% ÷ 2 = 5% per half-year
Number of periods = 3 years × 2 = 6 periods
The future value factor from row 6 of the 5.00% column in Table 25-1 is 1.34010.

Present value = Future value ÷ Future value factor
= $4,000 ÷ 1.34010 = $2,984.852, or $2,984.85

Computing Present Values and Compound Interest from Present Value Tables

You may prefer to solve for present values by using **present value factors** rather than future value factors, as in the preceding formula. Table 25-2, on pages 550 and 551, is a table of present value factors. Use exactly the same procedure (Steps i, ii, and iii) to find present value factors as you used to find future value factors.

Compute present values and compound interest from present value tables.

Present Value Formula

If you use the present value factors (Table 25-2), you use a different formula, the present value formula,

> Present value = Future value × Present value factor (from Table 25-2)

EXAMPLE I

Rework example H using Table 25-2 and the present value formula. How much must Edison Printing invest today to have $10,000 in 2 years if the interest rate is 15% compounded monthly?

$10,000 is the future value, for which Edison wants to know the present value.

STEP i — There are 12 compounding periods in 1 year (monthly).

STEP ii — Periodic rate = 15% ÷ 12 = 1.25%

STEP iii — Number of compounding periods = 2 years × 12 = 24
The present value factor in the 1.25% column and row 24 of Table 25-2 is 0.74220.

Substitute these values into the present value formula.

Present value = Future value × Present value factor (from Table 25-2)
= $10,000 × 0.74220 = $7,422.00

The answer to example H was $7,421.98. The discrepancy between that result and $7,422.00 in example I is due to rounding. If the two tables had six or seven digits instead of just five, this discrepancy would disappear.

EXAMPLE J

Solve the present value problem from example F. If Roberta can earn 10% compounded annually, how much should she deposit today in investments for her son and daughter so that the investments will be worth $60,000 and $70,000 in 6 and 8 years, respectively?

| | | Son | Daughter |
|---|---|---|---|
| | Future value: | $60,000 | $70,000 |
| | Term: | 6 years | 8 years |
| | Rate: | 10% compounded annually | 10% compounded annually |
| STEP i | Periods per year: | 1 (annual) | 1 (annual) |
| STEP ii | Periodic rate: | 10% ÷ 1 = 10% | 10% ÷ 1 = 10% |
| STEP iii | Compounding periods: | 6 years × 1 = 6 | 8 years × 1 = 8 |
| | PV factor (Table 25-2): | 0.56447 | 0.46651 |
| | Present value: | $60,000 × 0.56447 = $33,868.20 | $70,000 × 0.46651 = $32,655.70 |

The reason for two formulas and two tables is historical, predating handheld calculators. Without a calculator, a multiplication problem is typically easier than a division problem with the same two numbers.

Theoretically, we need only one formula and one table. The second present value formula and the table of present value factors permit us to solve present value problems by using multiplication instead of division. Look at example H. To solve the problem requires that we divide $10,000 by 1.34735, which is time-consuming to do without a calculator. (The answer is $7,421.98.) Using Table 25-2, we can solve the same problem by multiplying $10,000 by 0.74220, a relatively easy calculation without a calculator. (The answer is $7,422; the difference is due to rounding in the creation of the table.)

Notes About the Future Value and Present Value Tables

The numbers in the future value table (Table 25-1) are actually just the future value of $1.00 at a specific interest rate and for a specific period of time. For example, suppose that you invest $1.00 for 2 years at 8% compounded annually. This is the same problem as example A, except that the principal is only $1.00 instead of $2,000.00

The calculations shown at the right have not been rounded off. The answer, which is $1.1664, is the future value of the $1.00 investment. Now, find row 2 and the 8.00% column of Table 25-1. The future value factor is 1.16640—exactly the same as $1.1644, without the dollar sign and with five decimal places.

| | |
|---|---|
| $1.00 | Original principal |
| × 0.08 | Interest rate |
| $0.0800 | First-year interest |
| + 1.00 | First-year principal |
| $1.0800 | Second-year principal |
| × 0.08 | Interest rate |
| $0.0864 | Second-year interest |
| + 1.08 | Second-year principal |
| $1.1664 | Final compound amount |

Each number in the present value table (Table 25-2) can be calculated directly from the corresponding number in the future value table. The corresponding numbers are *reciprocals* of each other. Recall that the reciprocal of a number is found by dividing the number into 1.

Look back at examples H and I, which showed two different ways to solve the same problem. In example H we used a future value factor, which was 1.34735. In example I we used a present value factor, which was 0.74220. Each factor is in row 24 and the 1.25% column of its respective table. With your calculator, divide 1 by 1.34735 to get 0.74219765, which rounded to five places is 0.74220. And dividing 1 by 0.74220 gives 1.34734573.

$1 \div 1.34735 = 0.74219765$, or 0.74220

$1 \div 0.74220 = 1.34734573$, or 1.34735

Examine your calculator. You may have a reciprocal key, labeled "1/x." If you have such a key, enter 1.34735 and press the 1/x key. The calculator will display 0.74219765. Press the 1/x key again and the calculator will display 1.34735, or perhaps 1.34735000.

CONCEPT CHECK 25.3

a. What principal (present value) invested for 3 years at 10% compounded semiannually will result in a total future value of $4,000? (Use Table 25-2.)

Semiannually means 2 periods per year.
Periodic rate = 10% ÷ 2 = 5% per half-year
Number of periods = 3 years × 2 = 6 periods
The present value factor from row 6 of the 5.00% column in Table 25-2 is 0.74622.

Present value = Future value × Present value factor
 = $4,000 × 0.74622 = $2,984.88

Note: The answers to Concept Checks 25.2 and 25.3 are essentially the same. If the future value table and the present value table had values with six decimals, the answers would both be $2,984.86.

b. Seven years ago, a woman invested money at 9% compounded annually. If the investment is now worth $6,000, how much compound interest did she earn in the 7 years? (Use Table 25-2.)

Annual means 1 period per year.
Periodic rate = 9% ÷ 1 = 9% per year
Number of periods is 7 years × 1 = 7 periods
The present value factor from row 7 of the 9.00% column in Table 25-2 is 0.54703.

Present value = Future value × Present value factor
 = $6,000 × 0.54703 = $3,282.18
Compound interest = Future value − Present value
 = $6,000 − $3,282.18 = $2,717.82

COMPLETE ASSIGNMENT 25.2.

Chapter Terms for Review

compound amount

compound amount factors

compound interest

compound interest table

future value

future value factors

number of compounding periods

period

periodic interest rate

present value

present value factors

Visit the Web

http://www.chicagofed.org—Visit the Chicago Fed Web site for information on currency, interest calculations, credit, and other financial topics.

http://islamic-finance.net/, **http://www.islamic-banking.com/**, and **http://www.islamic-finance.com/indexnew.htm**—Look up information on Islamic banking and finance at these Web sites.

CYBERCONNECTION

A 20-something $trategy
Use 'sacred' retirement savings and you will face taxes and penalties

July 24, 2000: 11:52 a.m. ET

NEW YORK (CNNfn)-Retirement may seem a long way off when you're in your 20s, and it may be tempting to dip into your long-term savings for short-term needs.

In response to a reader's question, Frank Armstrong III, a certified financial planner from Miami, and the president of Managed Account Services Inc., said you should dip into your nest egg only as a last resort.

I'm 26 years old. My spouse and I have approximately $17,000 in profit sharing and $9,000 in a 401(k) account. We also have an $8,000 credit card debt that we are paying $400 a month to pay off. I am thinking about leaving the company and cashing out the profit sharing account, paying off the credit card debt, and putting $2,000 into an IRA. Then we can put that $400 a month into a savings account or bump up the 401(k) allocation. Can you point me in the right direction?

You are very fortunate to have accumulated so much in your retirement accounts at such a young age. It would be a shame to blow it.

First, let's assume that you rolled over the entire account to an IRA when you changed jobs. Your $26,000 would grow to $412,440.42 by the time you are 55, or $664,239.42 at age 60, or $1,069,764.22 at age 65. We assume a conservative rate of 10 percent. That's a great example of the power of tax-free compounding over long time periods.

Leaving the account intact is exactly what the tax laws are trying to encourage. That way the rest of us won't have to support you on welfare when you reach retirement age. So, the government builds in some pretty strong penalties to try to prevent systematic dissipation of retirement savings when employees change jobs. After all, they gave you a big tax advantage when they let your employer deposit it into an account for you without paying any tax on it.

So, let's look at what happens if you withdraw some or all of it. Assume a $10,000 withdrawal. Right off the bat your employer is required to withhold 20 percent. You get a check for $8,000. But, it gets worse. When tax time rolls around next year, the $10,000 is taxed at ordinary income tax rates. If you are in the 28 percent bracket, that means $2,800 tax due. Then, of course, there is a 10 percent penalty tax of $1,000. Total tax: $3,800. Net to you only $6,200.

After you get the $8,000 distribution from your employer, if you changed your mind and decided to roll over the $10,000 within the 60-day rollover period, you would only have $8,000 in your pocket. (Remember the 20 percent withholding?) So, now you would have to come up with the additional $2,000 to make the rollover. If you couldn't, then the missing $2,000 is subject to the tax and penalty.

Assuming that you came up with the missing funds from another source, and that you rolled the entire amount over, when you file next year you can recover the $2,000. In the meantime you have made a $2,000 loan to Uncle Sam.

Retirement funds are "sacred money." You should consider them a source of last resort. I would tap into them only to avoid starvation or something equally horrible. It would be a shame to deplete them to pay off a credit card balance. As I have tried to illustrate above, the trade-off is pretty unfavorable.

But, consumer debt is ugly! Interest rates are obscenely high, and they are not even tax deductible. You should make every effort to get it under control. It is possible to consider a life on somewhat less than 100 percent of your net income. Even if you never want to retire, at some point in your life you will want to be in a "work optional" status.

To clean up your credit card debt, you might consider temporarily suspending contributions to the 401(k). Then you could pay the cards off. Later you can use the 401(k) "catch up" procedures to fully participate in the advantages of your retirement plans and make up for your lost contributions.

Remember, a tax advantage is a terrible thing to waste!

CYBERCONNECTION

SBA Online *Women's Business Center*

| | |
|---|---|
| About Us | |
| News | |
| Women's Bus. Centers | |
| Find Local Services | |
| Business Basics | |
| Networking | |
| Hot Topics | |
| Major Partners | |
| Español | |
| 繁體中文 | |
| 日本語 | |
| Русский | |
| Íslenska | |

← Home for this Area Previous Location →

Cost of Lease vs. Purchase

You can analyze the costs of the lease versus purchase problem through discounted cash flow analysis. This analysis compares the cost of each alternative by considering: the timing of the payments, tax benefits, the interest rate on a loan, the lease rate, and other financial arrangements.

To make the analysis you must first make certain assumptions about the economic life of the equipment, salvage value, and depreciation.

To evaluate a lease you must first find the net cash outlay (not cash flow) in each year of the lease term. You find these amounts by subtracting the tax savings from the lease payment. This calculation gives you the net cash outlay for each year of the lease.

Each year's net cash outlay must next be discounted to take into account the time value of money. This discounting gives you the present value of each of the amounts.

The present value of an amount of money is the sum you would have to invest today at a stated rate of interest to have that amount of money at a specified future date. Say someone offered to give you $100 five years from now. How much could you take today and be as well off?

Common sense tells you you could take less than $100, because you'd have the use of the money for the five year period. Naturally, how much less you could take depends on the interest rate you thought you could get if you invested the lesser amount. For example to have

Source: http://onlinewbc.gov/DOCS/Finance/Lease_cost.html.

Visit the Online Women's Business Center at the Small Business Administration's Web site to read about present value issues with regard to the leasing v. purchase decision in business. Many companies find it more cost effective to lease, rather than buy, real estate and equipment. http://onlinewbc.gov/DOCS/Finance/lease_cost.html

The Bottom Line

Summary of chapter learning objectives:

| Learning Objective | Summary | Example |
|---|---|---|
| **25.1** | Compute future values and compound interest from future value tables | 1. Compute the future value of $8,000 invested at 15% compounded monthly for 2 years.
 2. Compute the compound interest earned on $5,000 invested at 6% compounded quarterly for 5 years. |
| **25.2** | Compute present values from future value tables | 3. Compute the present value that has to be invested at 12% compounded semiannually for 7 years to result in $10,000. |
| **25.3** | Compute present values and compound interest from present value tables | 4. If $6,000 is the future value after 13 years at 9% compounded annually, compute the principal (present value).
 5. Compute the compound interest earned if the future value of an investment made 16 months ago at 6% compounded monthly is $7,000. |

Answers: 1. $10,778.80 **2.** $1,734.30 **3.** $4,423.02 **4.** $1,957.08 **5.** $536.90

1 Calculate the future value (compound amount) and compound interest. (Use Table 25-1.)

| Principal | Rate | Time | Future Value | Interest |
|---|---|---|---|---|
| $ 4,000 | 9% compounded monthly | 4 yr | a. _____ | b. _____ |
| $12,500 | 8% compounded quarterly | 7 yr | c. _____ | d. _____ |
| $25,000 | 10% compounded annually | 12 yr | e. _____ | f. _____ |
| $ 9,600 | 12% compounded semiannually | 11 yr | g. _____ | h. _____ |

2 Calculate the present value (principal) and compound interest. (Use Table 25-2.)

| Future Value | Rate | Time | Present Value | Interest |
|---|---|---|---|---|
| $30,000 | 9% compounded annually | 8 yr | a. _____ | b. _____ |
| $ 8,800 | 10% compounded semiannually | 13 yr | c. _____ | d. _____ |
| $14,000 | 12% compounded monthly | 3 yr | e. _____ | f. _____ |
| $55,500 | 6% compounded quarterly | 10 yr | g. _____ | h. _____ |

3 Dick Liebelt received a $10,000 bonus from his employer. He can invest it safely in his credit union at 5% compounded quarterly. What will be the value of the investment in 8 years?

4 Donna Rojas inherited $4,200. She invested it immediately in an investment fund paying 8% compounded semiannually. How much interest would Donna earn if she left principal and interest invested for 4 years?

5 Sandy Hopkins was planning to buy a new car in 3 years. She has some money today that she can invest for 3 years in an account that will pay 6% compounded quarterly. How much of it would she need to deposit today so that she will have $8,000 in her account in 3 years?

6 Doug Jurgensen will need to buy a $25,000 metal lathe in 2 years. He can deposit excess profits from this year in an investment that should pay 9% compounded monthly. If Doug earns the $25,000 in 2 years, how much did he earn in interest?

Assignment 25.1: Future Value (Compound Amount)

Name _____

Date _____ Score _____

A (28 points) Find the future value (compound amount) and the compound interest, as indicated, for each of the following investments. Round answers to the nearest cent. (2 points for each correct answer)

| | Principal | Rate | Term | Future Value | Compound Interest |
|---|---|---|---|---|---|
| **1.** | $7,000 | 15% compounded monthly | 3 years | _____ | _____ |
| **2.** | $750 | 8% compounded semiannually | 13 years | _____ | _____ |
| **3.** | $24,500 | 12% compounded quarterly | 10.5 years | _____ | _____ |
| **4.** | $8,400 | 10% compounded annually | 20 years | _____ | _____ |
| **5.** | $3,600 | 9% compounded monthly | 21 months | _____ | _____ |
| **6.** | $14,450 | 6% compounded quarterly | 4 years | _____ | _____ |
| **7.** | $2,800 | 12% compounded semiannually | 7 years | _____ | _____ |

Score for A (28) _____

B (32 points) Find the future value (compound amount) or the compound interest, as indicated, for each of the following investments or loans. Round answers to the nearest cent. (4 points for each correct answer)

8. Compute the future value (compound amount) of $4,500 invested for 10 years at 5% compounded quarterly.

 Future value _____

9. How much compound interest will you pay if you borrow $35,000 for 19 months at 18% compounded monthly?

 Compound interest _____

10. Calculate the compound amount (future value) on a loan of $6,250 at 9% compounded annually for 7 years.

 Compound amount _____

11. How much compound interest will you earn if you loan $7,900 for 16.5 years at 12% compounded semiannually?

 Compound interest _____

12. What total amount (principal and interest) must be repaid in $3\frac{1}{2}$ years on a loan of $17,000 at 6% compounded monthly?

 Compound amount _____

13. Determine the total compound interest that you will have to pay if you borrow $845 at 10% compounded semiannually and don't pay it back for 11 years.

 Compound interest _____

14. How much compound interest will you earn if you invest $10,000 for 13 years at 8% compounded annually?

 Compound interest _____

15. Compute the future value (compound amount) of $21,000 invested for 5.5 years at 6% compounded quarterly.

 Future value _____

Score for B (32)

Assignment 25.1 *Continued*

C (40 points) Business Applications. Find the compound amount (future value) or the compound interest, as indicated. Round answers to the nearest cent. (4 points for each correct answer)

16. Joyce Bodley thinks that she needs to borrow $2,400 for 3 years. She doesn't have a very good credit rating, so most finance companies want to charge her too much interest. She finally finds a lender that will loan her the money at 12% compounded monthly. How much interest will Joyce have to pay to this particular lender?

17. Mary Sousa receives a telephone call from a salesperson who describes "an incredible investment opportunity." The investment promises a return of 16% compounded semiannually for investments of $5,000 or more. One disadvantage is that no money will be paid out for a long time. Another disadvantage is that the investment is very risky. Mary doesn't think that she will need the money for six years, so she decides to invest $5,000. If the investment pays what it promises, how much interest will Mary earn in the 6 years?

18. Sean Ferguson wants to borrow money from his father to buy a car. Sean's father is trying to teach him how to manage money, so he agrees to loan him the money, but at 6% compounded quarterly. Sean borrows $8,800 and repays everything—principal plus all of the interest—in 2 years. How much does Sean pay back to his father?

19. Burton Hansen is trying to decide whether to invest money in a bank or in something a little riskier that will pay a higher return. One very simple investment promises to pay a minimum of 10% compounded annually for 5 years. How much will Burton earn after the 5 years if he invests $3,475 and the investment pays the minimum?

20. Marcia Juarez and her brother-in-law have a successful business with several employees. They decide to borrow $15,000 to pay their quarterly payments for payroll taxes and federal income tax. They get the money at 9% compounded monthly and repay all interest and principal after 9 months. How much do they repay?

21. Bonnie Bomar inherited $4,285. She wants to invest it in something relatively safe so that she can transfer all the money to her son's college fund in about 5 years. One investment brochure (called a prospectus) states that it will pay a return of 12% compounded quarterly. How much will she have total, principal plus interest, after 4 years and 9 months (4.75 years)?

22. To help his daughter and son-in-law purchase their first new car, Bill Chou loans them $20,000. They agree on an interest rate of 5% compounded annually, and Bill tells them that they can pay it all back, the $20,000 plus the interest, in 4 years. How much interest will Bill receive from them?

23. Walt Pierce owns and operates an art supply store in a suburban shopping center. Walt learns about an investment that claims to pay a return of 12% compounded semiannually if he leaves the money invested for $4\frac{1}{2}$ years. Walt decides to invest $8,250. Compute the amount of interest that he will earn in the $4\frac{1}{2}$ years.

24. Joe Littrell is a student at a community college. He borrowed $950 for 15 months at the rate of 12% compounded monthly. How much total, principal plus compound interest, must Joe repay at the end of the 15 months?

25. The County Employees Credit Union pays an interest rate of 8% compounded quarterly on savings accounts of $1,000 or more, with the requirement that the money be deposited for at least 6 months. How much interest will Marilyn Bunnell earn if she deposits $1,800 and leaves it in the credit union for 2 years?

Score for C (40)

Assignment 25.2: **Present Value**

Name _____

Date _____ Score _____

A **(28 points) Find the present value (principal) and the compound interest, as indicated, for each of the following investments. (Hint: Subtract the present value from the future value to find the compound interest.) Do problems 1–3 using Table 25-1; do problems 4–7 using Table 25-2. Round answers to the nearest cent. (2 points for each correct answer)**

| | Future Value | Rate | Term | Present Value | Compound Interest |
|---|---|---|---|---|---|
| **1.** | $3,900 | 6% compounded semiannually | 3 years | _____ | _____ |
| **2.** | $16,820 | 12% compounded quarterly | 8 years | _____ | _____ |
| **3.** | $35,000 | 5% compounded annually | 9 years | _____ | _____ |
| **4.** | $6,800 | 9% compounded monthly | 4 years | _____ | _____ |
| **5.** | $8,240 | 8% compounded quarterly | 5 years | _____ | _____ |
| **6.** | $75,000 | 10% compounded semiannually | 8 years | _____ | _____ |
| **7.** | $1,475 | 18% compounded monthly | 15 months | _____ | _____ |

Score for A (28)

B **(32 points) Find the present value (principal) or the compound interest, as indicated, for each of the following investments or loans. Use either Table 25-1 or Table 25-2. Round answers to the nearest cent. (4 points for each correct answer)**

8. Compute the present value (principal) if the future value 20 years from now is $25,000 and if the interest rate is 8% compounded semiannually.

 Present value _____

9. How much compound interest would you pay if you repay a total of $6,160 1 year and 3 months after borrowing the principal at 6% compounded monthly?

 Compound interest _____

10. Calculate the principal (present value) of a loan made 5 years ago at 12% compounded quarterly if the borrower repays a total of $8,225.

 Principal _____

11. Compute the amount that a company must invest (the present value) at 10% compounded annually if it wants to have $100,000 available (the future value) in 25 years.

 Present value _____

12. How much compound interest is earned on a 5.5-year investment that has a rate of return of 8% compounded quarterly and repays a total compound amount (future value) of $4,350?

 Compound interest _____

13. Determine the principal of a single deposit that is worth exactly $2,250 after 9 months at 12% compounded monthly.

 Principal _____

14. Calculate the amount of compound interest that has accrued on an investment that is now worth $15,000 after 14 years at 10% compounded semiannually.

 Compound interest _____

15. Compute the present value (principal) if the future value is $50,000 after 50 years at 6% compounded annually.

Present value _____

Score for B (32)

C **(40 points) Business Applications. Find the present value (principal) or the compound interest, as indicated. Use either Table 25-1 or Table 25-2. Round answers to the nearest cent. (4 points for each correct answer)**

16. Dick Shanley wants to buy a motorcycle. His mother will loan him some money at only 6% compounded quarterly. If Dick estimates that he will be able to repay a total of $5,200 in $2\frac{1}{2}$ years, how much should he borrow from his mother today?

17. Six years ago, Eleanor Baker invested money at 8% compounded annually. Today she received a check for $6,000 that represented her total payment of principal and interest. Compute the amount of the interest that she earned.

18. Karl Coke wants to have $4,500 available at the end of 4 years to buy a surround-sound audiovisual system for the new house he is planning to build. If he can invest money at 10% compounded semiannually, how much should he invest?

19. As part of their financial planning, Janice Garcia's grandparents made monetary gifts to each of their grandchildren. In addition, Janice's grandfather told her that, if she would save part of her gift for at least a year, he would pay her interest of 12% compounded monthly. Janice decided to save just enough so that she would have $1,000 at the end of 18 months, when she would be 16 years old. How much should she save?

20. Marilyn Whitehorse estimated that she would need $12,600 in $5\frac{1}{2}$ years to buy new equipment for her pottery shop. Having extra cash now, she invested the money in an extremely safe investment that advertised a return of 6% compounded semiannually. Marilyn invested just enough money to end up with the $12,600. How much of the $12,600 did Marilyn earn on her investment?

Table 25-1: Future Value (Compound Amount) Factors

| Period | 0.50% | 0.75% | 1.00% | 1.25% | 1.50% | 2.00% | 3.00% | 4.00% | 5.00% | 6.00% | 8.00% | 9.00% | 10.00% | 12.00% |
|---|---|---|---|---|---|---|---|---|---|---|---|---|---|---|
| 1 | 1.00500 | 1.00750 | 1.01000 | 1.01250 | 1.01500 | 1.02000 | 1.03000 | 1.04000 | 1.05000 | 1.06000 | 1.08000 | 1.09000 | 1.10000 | 1.12000 |
| 2 | 1.01003 | 1.01506 | 1.02010 | 1.02516 | 1.03023 | 1.04040 | 1.06090 | 1.08160 | 1.10250 | 1.12360 | 1.16640 | 1.18810 | 1.21000 | 1.25440 |
| 3 | 1.01508 | 1.02267 | 1.03030 | 1.03797 | 1.04568 | 1.06121 | 1.09273 | 1.12486 | 1.15763 | 1.19102 | 1.25971 | 1.29503 | 1.33100 | 1.40493 |
| 4 | 1.02015 | 1.03034 | 1.04060 | 1.05095 | 1.06136 | 1.08243 | 1.12551 | 1.16986 | 1.21551 | 1.26248 | 1.36049 | 1.41158 | 1.46410 | 1.57352 |
| 5 | 1.02525 | 1.03807 | 1.05101 | 1.06408 | 1.07728 | 1.10408 | 1.15927 | 1.21665 | 1.27628 | 1.33823 | 1.46933 | 1.53862 | 1.61051 | 1.76234 |
| 6 | 1.03038 | 1.04585 | 1.06152 | 1.07738 | 1.09344 | 1.12616 | 1.19405 | 1.26532 | 1.34010 | 1.41852 | 1.58687 | 1.67710 | 1.77156 | 1.97382 |
| 7 | 1.03553 | 1.05370 | 1.07214 | 1.09085 | 1.10984 | 1.14869 | 1.22987 | 1.31593 | 1.40710 | 1.50363 | 1.71382 | 1.82804 | 1.94872 | 2.21068 |
| 8 | 1.04071 | 1.06160 | 1.08286 | 1.10449 | 1.12649 | 1.17166 | 1.26677 | 1.36857 | 1.47746 | 1.59385 | 1.85093 | 1.99256 | 2.14359 | 2.47596 |
| 9 | 1.04591 | 1.06956 | 1.09369 | 1.11829 | 1.14339 | 1.19509 | 1.30477 | 1.42331 | 1.55133 | 1.68948 | 1.99900 | 2.17189 | 2.35795 | 2.77308 |
| 10 | 1.05114 | 1.07758 | 1.10462 | 1.13227 | 1.16054 | 1.21899 | 1.34392 | 1.48024 | 1.62889 | 1.79085 | 2.15892 | 2.36736 | 2.59374 | 3.10585 |
| 11 | 1.05640 | 1.08566 | 1.11567 | 1.14642 | 1.17795 | 1.24337 | 1.38423 | 1.53945 | 1.71034 | 1.89830 | 2.33164 | 2.58043 | 2.85312 | 3.47855 |
| 12 | 1.06168 | 1.09381 | 1.12683 | 1.16075 | 1.19562 | 1.26824 | 1.42576 | 1.60103 | 1.79586 | 2.01220 | 2.51817 | 2.81266 | 3.13843 | 3.89598 |
| 13 | 1.06699 | 1.10201 | 1.13809 | 1.17526 | 1.21355 | 1.29361 | 1.46853 | 1.66507 | 1.88565 | 2.13293 | 2.71962 | 3.06580 | 3.45227 | 4.36349 |
| 14 | 1.07232 | 1.11028 | 1.14947 | 1.18995 | 1.23176 | 1.31948 | 1.51259 | 1.73168 | 1.97993 | 2.26090 | 2.93719 | 3.34173 | 3.79750 | 4.88711 |
| 15 | 1.07768 | 1.11860 | 1.16097 | 1.20483 | 1.25023 | 1.34587 | 1.55797 | 1.80094 | 2.07893 | 2.39656 | 3.17217 | 3.64248 | 4.17725 | 5.47357 |
| 16 | 1.08307 | 1.12699 | 1.17258 | 1.21989 | 1.26899 | 1.37279 | 1.60471 | 1.87298 | 2.18287 | 2.54035 | 3.42594 | 3.97031 | 4.59497 | 6.13039 |
| 17 | 1.08849 | 1.13544 | 1.18430 | 1.23514 | 1.28802 | 1.40024 | 1.65285 | 1.94790 | 2.29202 | 2.69277 | 3.70002 | 4.32763 | 5.05447 | 6.86604 |
| 18 | 1.09393 | 1.14396 | 1.19615 | 1.25058 | 1.30734 | 1.42825 | 1.70243 | 2.02582 | 2.40662 | 2.85434 | 3.99602 | 4.71712 | 5.55992 | 7.68997 |
| 19 | 1.09940 | 1.15254 | 1.20811 | 1.26621 | 1.32695 | 1.45681 | 1.75351 | 2.10685 | 2.52695 | 3.02560 | 4.31570 | 5.14166 | 6.11591 | 8.61276 |
| 20 | 1.10490 | 1.16118 | 1.22019 | 1.28204 | 1.34686 | 1.48595 | 1.80611 | 2.19112 | 2.65330 | 3.20714 | 4.66096 | 5.60441 | 6.72750 | 9.64629 |
| 21 | 1.11042 | 1.16989 | 1.23239 | 1.29806 | 1.36706 | 1.51567 | 1.86029 | 2.27877 | 2.78596 | 3.39956 | 5.03383 | 6.10881 | 7.40025 | 10.80385 |
| 22 | 1.11597 | 1.17867 | 1.24472 | 1.31429 | 1.38756 | 1.54598 | 1.91610 | 2.36992 | 2.92526 | 3.60354 | 5.43654 | 6.65860 | 8.14027 | 12.10031 |
| 23 | 1.12155 | 1.18751 | 1.25716 | 1.33072 | 1.40838 | 1.57690 | 1.97359 | 2.46472 | 3.07152 | 3.81975 | 5.87146 | 7.25787 | 8.95430 | 13.55235 |
| 24 | 1.12716 | 1.19641 | 1.26973 | 1.34735 | 1.42950 | 1.60844 | 2.03279 | 2.56330 | 3.22510 | 4.04893 | 6.34118 | 7.91108 | 9.84973 | 15.17863 |
| 25 | 1.13280 | 1.20539 | 1.28243 | 1.36419 | 1.45095 | 1.64061 | 2.09378 | 2.66584 | 3.38635 | 4.29187 | 6.84848 | 8.62308 | 10.83471 | 17.00006 |

Table 25-1: Future Value (Compound Amount) Factors *(continued)*

| Period | 0.50% | 0.75% | 1.00% | 1.25% | 1.50% | 2.00% | 3.00% | 4.00% | 5.00% | 6.00% | 8.00% | 9.00% | 10.00% | 12.00% |
|---|---|---|---|---|---|---|---|---|---|---|---|---|---|---|
| 26 | 1.13846 | 1.21443 | 1.29526 | 1.38125 | 1.47271 | 1.67342 | 2.15659 | 2.77247 | 3.55567 | 4.54938 | 7.39635 | 9.39916 | 11.91818 | 19.04007 |
| 27 | 1.14415 | 1.22354 | 1.30821 | 1.39851 | 1.49480 | 1.70689 | 2.22129 | 2.88337 | 3.73346 | 4.82235 | 7.98806 | 10.24508 | 13.10999 | 21.32488 |
| 28 | 1.14987 | 1.23271 | 1.32129 | 1.41599 | 1.51722 | 1.74102 | 2.28793 | 2.99870 | 3.92013 | 5.11169 | 8.62711 | 11.16714 | 14.42099 | 23.88387 |
| 29 | 1.15562 | 1.24196 | 1.33450 | 1.43369 | 1.53998 | 1.77584 | 2.35657 | 3.11865 | 4.11614 | 5.41839 | 9.31727 | 12.17218 | 15.86309 | 26.74993 |
| 30 | 1.16140 | 1.25127 | 1.34785 | 1.45161 | 1.56308 | 1.81136 | 2.42726 | 3.24340 | 4.32194 | 5.74349 | 10.06266 | 13.26768 | 17.44940 | 29.95992 |
| 31 | 1.16721 | 1.26066 | 1.36133 | 1.46976 | 1.58653 | 1.84759 | 2.50008 | 3.37313 | 4.53804 | 6.08810 | 10.86767 | 14.46177 | 19.19434 | 33.55511 |
| 32 | 1.17304 | 1.27011 | 1.37494 | 1.48813 | 1.61032 | 1.88454 | 2.57508 | 3.50806 | 4.76494 | 6.45339 | 11.73708 | 15.76333 | 21.11378 | 37.58173 |
| 33 | 1.17891 | 1.27964 | 1.38869 | 1.50673 | 1.63448 | 1.92223 | 2.65234 | 3.64838 | 5.00319 | 6.84059 | 12.67605 | 17.18203 | 23.22515 | 42.09153 |
| 34 | 1.18480 | 1.28923 | 1.40258 | 1.52557 | 1.65900 | 1.96068 | 2.73191 | 3.79432 | 5.25335 | 7.25103 | 13.69013 | 18.72841 | 25.54767 | 47.14252 |
| 35 | 1.19073 | 1.29890 | 1.41660 | 1.54464 | 1.68388 | 1.99989 | 2.81386 | 3.94609 | 5.51602 | 7.68609 | 14.78534 | 20.41397 | 28.10244 | 52.79962 |
| 36 | 1.19668 | 1.30865 | 1.43077 | 1.56394 | 1.70914 | 2.03989 | 2.89828 | 4.10393 | 5.79182 | 8.14725 | 15.96817 | 22.25123 | 30.91268 | 59.13557 |
| 37 | 1.20266 | 1.31846 | 1.44508 | 1.58349 | 1.73478 | 2.08069 | 2.98523 | 4.26809 | 6.08141 | 8.63609 | 17.24563 | 24.25384 | 34.00395 | 66.23184 |
| 38 | 1.20868 | 1.32835 | 1.45953 | 1.60329 | 1.76080 | 2.12230 | 3.07478 | 4.43881 | 6.38548 | 9.15425 | 18.62528 | 26.43668 | 37.40434 | 74.17966 |
| 39 | 1.21472 | 1.33831 | 1.47412 | 1.62333 | 1.78721 | 2.16474 | 3.16703 | 4.61637 | 6.70475 | 9.70351 | 20.11530 | 28.81598 | 41.14478 | 83.08122 |
| 40 | 1.22079 | 1.34835 | 1.48886 | 1.64362 | 1.81402 | 2.20804 | 3.26204 | 4.80102 | 7.03999 | 10.28572 | 21.72452 | 31.40942 | 45.25926 | 93.05097 |
| 41 | 1.22690 | 1.35846 | 1.50375 | 1.66416 | 1.84123 | 2.25220 | 3.35990 | 4.99306 | 7.39199 | 10.90286 | 23.46248 | 34.23627 | 49.78518 | 104.21709 |
| 42 | 1.23303 | 1.36865 | 1.51879 | 1.68497 | 1.86885 | 2.29724 | 3.46070 | 5.19278 | 7.76159 | 11.55703 | 25.33948 | 37.31753 | 54.76370 | 116.72314 |
| 43 | 1.23920 | 1.37891 | 1.53398 | 1.70603 | 1.89688 | 2.34319 | 3.56452 | 5.40050 | 8.14967 | 12.25045 | 27.36664 | 40.67611 | 60.24007 | 130.72991 |
| 44 | 1.24539 | 1.38926 | 1.54932 | 1.72735 | 1.92533 | 2.39005 | 3.67145 | 5.61652 | 8.55715 | 12.98548 | 29.55597 | 44.33696 | 66.26408 | 146.41750 |
| 45 | 1.25162 | 1.39968 | 1.56481 | 1.74895 | 1.95421 | 2.43785 | 3.78160 | 5.84118 | 8.98501 | 13.76461 | 31.92045 | 48.32729 | 72.89048 | 163.98760 |
| 46 | 1.25788 | 1.41017 | 1.58046 | 1.77081 | 1.98353 | 2.48661 | 3.89504 | 6.07482 | 9.43426 | 14.59049 | 34.47409 | 52.67674 | 80.17953 | 183.66612 |
| 47 | 1.26417 | 1.42075 | 1.59626 | 1.79294 | 2.01328 | 2.53634 | 4.01190 | 6.31782 | 9.90597 | 15.46592 | 37.23201 | 57.41765 | 88.19749 | 205.70605 |
| 48 | 1.27049 | 1.43141 | 1.61223 | 1.81535 | 2.04348 | 2.58707 | 4.13225 | 6.57053 | 10.40127 | 16.39387 | 40.21057 | 62.58524 | 97.01723 | 230.39078 |
| 49 | 1.27684 | 1.44214 | 1.62835 | 1.83805 | 2.07413 | 2.63881 | 4.25622 | 6.83335 | 10.92133 | 17.37750 | 43.42742 | 68.21791 | 106.71896 | 258.03767 |
| 50 | 1.28323 | 1.45296 | 1.64463 | 1.86102 | 2.10524 | 2.69159 | 4.38391 | 7.10668 | 11.46740 | 18.42015 | 46.90161 | 74.35752 | 117.39085 | 289.00219 |

Table 25-2: Present Value Factors

| Period | 0.50% | 0.75% | 1.00% | 1.25% | 1.50% | 2.00% | 3.00% | 4.00% | 5.00% | 6.00% | 8.00% | 9.00% | 10.00% | 12.00% |
|--------|-------|-------|-------|-------|-------|-------|-------|-------|-------|-------|-------|-------|--------|--------|
| 1 | 0.99502 | 0.99256 | 0.99010 | 0.98765 | 0.98522 | 0.98039 | 0.97087 | 0.96154 | 0.95238 | 0.94340 | 0.92593 | 0.91743 | 0.90909 | 0.89286 |
| 2 | 0.99007 | 0.98517 | 0.98030 | 0.97546 | 0.97066 | 0.96117 | 0.94260 | 0.92456 | 0.90703 | 0.89000 | 0.85734 | 0.84168 | 0.82645 | 0.79719 |
| 3 | 0.98515 | 0.97783 | 0.97059 | 0.96342 | 0.95632 | 0.94232 | 0.91514 | 0.88900 | 0.86384 | 0.83962 | 0.79383 | 0.77218 | 0.75131 | 0.71178 |
| 4 | 0.98025 | 0.97055 | 0.96098 | 0.95152 | 0.94218 | 0.92385 | 0.88849 | 0.85480 | 0.82270 | 0.79209 | 0.73503 | 0.70843 | 0.68301 | 0.63552 |
| 5 | 0.97537 | 0.96333 | 0.95147 | 0.93978 | 0.92826 | 0.90573 | 0.86261 | 0.82193 | 0.78353 | 0.74726 | 0.68058 | 0.64993 | 0.62092 | 0.56743 |
| 6 | 0.97052 | 0.95616 | 0.94205 | 0.92817 | 0.91454 | 0.88797 | 0.83748 | 0.79031 | 0.74622 | 0.70496 | 0.63017 | 0.59627 | 0.56447 | 0.50663 |
| 7 | 0.96569 | 0.94904 | 0.93272 | 0.91672 | 0.90103 | 0.87056 | 0.81309 | 0.75992 | 0.71068 | 0.66506 | 0.58349 | 0.54703 | 0.51316 | 0.45235 |
| 8 | 0.96089 | 0.94198 | 0.92348 | 0.90540 | 0.88771 | 0.85349 | 0.78941 | 0.73069 | 0.67684 | 0.62741 | 0.54027 | 0.50187 | 0.46651 | 0.40388 |
| 9 | 0.95610 | 0.93496 | 0.91434 | 0.89422 | 0.87459 | 0.83676 | 0.76642 | 0.70259 | 0.64461 | 0.59190 | 0.50025 | 0.46043 | 0.42410 | 0.36061 |
| 10 | 0.95135 | 0.92800 | 0.90529 | 0.88318 | 0.86167 | 0.82035 | 0.74409 | 0.67556 | 0.61391 | 0.55839 | 0.46319 | 0.42241 | 0.38554 | 0.32197 |
| 11 | 0.94661 | 0.92109 | 0.89632 | 0.87228 | 0.84893 | 0.80426 | 0.72242 | 0.64958 | 0.58468 | 0.52679 | 0.42888 | 0.38753 | 0.35049 | 0.28748 |
| 12 | 0.94191 | 0.91424 | 0.88745 | 0.86151 | 0.83639 | 0.78849 | 0.70138 | 0.62460 | 0.55684 | 0.49697 | 0.39711 | 0.35553 | 0.31863 | 0.25668 |
| 13 | 0.93722 | 0.90743 | 0.87866 | 0.85087 | 0.82403 | 0.77303 | 0.68095 | 0.60057 | 0.53032 | 0.46884 | 0.36770 | 0.32618 | 0.28966 | 0.22917 |
| 14 | 0.93256 | 0.90068 | 0.86996 | 0.84037 | 0.81185 | 0.75788 | 0.66112 | 0.57748 | 0.50507 | 0.44230 | 0.34046 | 0.29925 | 0.26333 | 0.20462 |
| 15 | 0.92792 | 0.89397 | 0.86135 | 0.82999 | 0.79985 | 0.74301 | 0.64186 | 0.55526 | 0.48102 | 0.41727 | 0.31524 | 0.27454 | 0.23939 | 0.18270 |
| 16 | 0.92330 | 0.88732 | 0.85282 | 0.81975 | 0.78803 | 0.72845 | 0.62317 | 0.53391 | 0.45811 | 0.39365 | 0.29189 | 0.25187 | 0.21763 | 0.16312 |
| 17 | 0.91871 | 0.88071 | 0.84438 | 0.80963 | 0.77639 | 0.71416 | 0.60502 | 0.51337 | 0.43630 | 0.37136 | 0.27027 | 0.23107 | 0.19784 | 0.14564 |
| 18 | 0.91414 | 0.87416 | 0.83602 | 0.79963 | 0.76491 | 0.70016 | 0.58739 | 0.49363 | 0.41552 | 0.35034 | 0.25025 | 0.21199 | 0.17986 | 0.13004 |
| 19 | 0.90959 | 0.86765 | 0.82774 | 0.78976 | 0.75361 | 0.68643 | 0.57029 | 0.47464 | 0.39573 | 0.33051 | 0.23171 | 0.19449 | 0.16351 | 0.11611 |
| 20 | 0.90506 | 0.86119 | 0.81954 | 0.78001 | 0.74247 | 0.67297 | 0.55368 | 0.45639 | 0.37689 | 0.31180 | 0.21455 | 0.17843 | 0.14864 | 0.10367 |
| 21 | 0.90056 | 0.85478 | 0.81143 | 0.77038 | 0.73150 | 0.65978 | 0.53755 | 0.43883 | 0.35894 | 0.29416 | 0.19866 | 0.16370 | 0.13513 | 0.09256 |
| 22 | 0.89608 | 0.84842 | 0.80340 | 0.76087 | 0.72069 | 0.64684 | 0.52189 | 0.42196 | 0.34185 | 0.27751 | 0.18394 | 0.15018 | 0.12285 | 0.08264 |
| 23 | 0.89162 | 0.84210 | 0.79544 | 0.75147 | 0.71004 | 0.63416 | 0.50669 | 0.40573 | 0.32557 | 0.26180 | 0.17032 | 0.13778 | 0.11168 | 0.07379 |
| 24 | 0.88719 | 0.83583 | 0.78757 | 0.74220 | 0.69954 | 0.62172 | 0.49193 | 0.39012 | 0.31007 | 0.24698 | 0.15770 | 0.12640 | 0.10153 | 0.06588 |
| 25 | 0.88277 | 0.82961 | 0.77977 | 0.73303 | 0.68921 | 0.60953 | 0.47761 | 0.37512 | 0.29530 | 0.23300 | 0.14602 | 0.11597 | 0.09230 | 0.05882 |

Table 25-2: Present Value Factors *(continued)*

| Period | 0.50% | 0.75% | 1.00% | 1.25% | 1.50% | 2.00% | 3.00% | 4.00% | 5.00% | 6.00% | 8.00% | 9.00% | 10.00% | 12.00% |
|---|---|---|---|---|---|---|---|---|---|---|---|---|---|---|
| 26 | 0.87838 | 0.82343 | 0.77205 | 0.72398 | 0.67902 | 0.59758 | 0.46369 | 0.36069 | 0.28124 | 0.21981 | 0.13520 | 0.10639 | 0.08391 | 0.05252 |
| 27 | 0.87401 | 0.81730 | 0.76440 | 0.71505 | 0.66899 | 0.58586 | 0.45019 | 0.34682 | 0.26785 | 0.20737 | 0.12519 | 0.09761 | 0.07628 | 0.04689 |
| 28 | 0.86966 | 0.81122 | 0.75684 | 0.70622 | 0.65910 | 0.57437 | 0.43708 | 0.33348 | 0.25509 | 0.19563 | 0.11591 | 0.08955 | 0.06934 | 0.04187 |
| 29 | 0.86533 | 0.80518 | 0.74934 | 0.69750 | 0.64936 | 0.56311 | 0.42435 | 0.32065 | 0.24295 | 0.18456 | 0.10733 | 0.08215 | 0.06304 | 0.03738 |
| 30 | 0.86103 | .079919 | 0.74192 | 0.68889 | 0.63976 | 0.55207 | 0.41199 | 0.30832 | 0.23138 | 0.17411 | 0.09938 | 0.07537 | 0.05731 | 0.03338 |
| 31 | 0.85675 | 0.79324 | 0.73458 | 0.68038 | 0.63031 | 0.54125 | 0.39999 | 0.29646 | 0.22036 | 0.16425 | 0.09202 | 0.06915 | 0.05210 | 0.02980 |
| 32 | 0.85248 | 0.78733 | 0.72730 | 0.67198 | 0.62099 | 0.53063 | 0.38834 | 0.28506 | 0.20987 | 0.15496 | 0.08520 | 0.06344 | 0.04736 | 0.02661 |
| 33 | 0.84824 | 0.78147 | 0.72010 | 0.66369 | 0.61182 | 0.52023 | 0.37703 | 0.27409 | 0.19987 | 0.14619 | 0.07889 | 0.05820 | 0.04306 | 0.02376 |
| 34 | 0.84402 | 0.77565 | 0.71297 | 0.65549 | 0.60277 | 0.51003 | 0.36604 | 0.26355 | 0.19035 | 0.13791 | 0.07305 | 0.05339 | 0.03914 | 0.02121 |
| 35 | 0.83982 | 0.76988 | 0.70591 | 0.64740 | 0.59387 | 0.50003 | 0.35538 | 0.25342 | 0.18129 | 0.13011 | 0.06763 | 0.04899 | 0.03558 | 0.01894 |
| 36 | 0.83564 | 0.76415 | 0.69892 | 0.63941 | 0.58509 | 0.49022 | 0.34503 | 0.24367 | 0.17266 | 0.12274 | 0.06262 | 0.04494 | 0.03235 | 0.01691 |
| 37 | 0.83149 | 0.75846 | 0.69200 | 0.63152 | 0.57644 | 0.48061 | 0.33498 | 0.23430 | 0.16444 | 0.11579 | 0.05799 | 0.04123 | 0.02941 | 0.01510 |
| 38 | 0.82735 | 0.75281 | 0.68515 | 0.62372 | 0.56792 | 0.47119 | 0.32523 | 0.22529 | 0.15661 | 0.10924 | 0.05369 | 0.03783 | 0.02673 | 0.01348 |
| 39 | 0.82323 | 0.74721 | 0.67837 | 0.61602 | 0.55953 | 0.46195 | 0.31575 | 0.21662 | 0.14915 | 0.10306 | 0.04971 | 0.03470 | 0.02430 | 0.01204 |
| 40 | 0.81914 | 0.74165 | 0.67165 | 0.60841 | 0.55126 | 0.45289 | 0.30656 | 0.20829 | 0.14205 | 0.09722 | 0.04603 | 0.03184 | 0.02209 | 0.01075 |
| 41 | 0.81506 | 0.73613 | 0.66500 | 0.60090 | 0.54312 | 0.44401 | 0.29763 | 0.20028 | 0.13528 | 0.09172 | 0.04262 | 0.02921 | 0.02009 | 0.00960 |
| 42 | 0.81101 | 0.73065 | 0.65842 | 0.59348 | 0.53509 | 0.43530 | 0.28896 | 0.19257 | 0.12884 | 0.08653 | 0.03946 | 0.02680 | 0.01826 | 0.00857 |
| 43 | 0.80697 | 0.72521 | 0.65190 | 0.58616 | 0.52718 | 0.42677 | 0.28054 | 0.18517 | 0.12270 | 0.08163 | 0.03654 | 0.02458 | 0.01660 | 0.00765 |
| 44 | 0.80296 | 0.71981 | 0.64545 | 0.57892 | 0.51939 | 0.41840 | 0.27237 | 0.17805 | 0.11686 | 0.07701 | 0.03383 | 0.02255 | 0.01509 | 0.00683 |
| 45 | 0.79896 | 0.71445 | 0.63905 | 0.57177 | 0.51171 | 0.41020 | 0.26444 | 0.17120 | 0.11130 | 0.07265 | 0.03133 | 0.02069 | 0.01372 | 0.00610 |
| 46 | 0.79499 | 0.70913 | 0.63273 | 0.56471 | 0.50415 | 0.40215 | 0.25674 | 0.16461 | 0.10600 | 0.06854 | 0.02901 | 0.01898 | 0.01247 | 0.00544 |
| 47 | 0.79103 | 0.70385 | 0.62646 | 0.55774 | 0.49670 | 0.39427 | 0.24926 | 0.15828 | 0.10095 | 0.06466 | 0.02686 | 0.01742 | 0.01134 | 0.00486 |
| 48 | 0.78710 | 0.69861 | 0.62026 | 0.55086 | 0.48936 | 0.38654 | 0.24200 | 0.15219 | 0.09614 | 0.06100 | 0.02487 | 0.01598 | 0.01031 | 0.00434 |
| 49 | 0.78318 | 0.69341 | 0.61412 | 0.54406 | 0.48213 | 0.37896 | 0.23495 | 0.14634 | 0.09156 | 0.05755 | 0.02303 | 0.01466 | 0.00937 | 0.00388 |
| 50 | 0.77929 | 0.68825 | 0.60804 | 0.53734 | 0.47500 | 0.37153 | 0.22811 | 0.14071 | 0.08720 | 0.05429 | 0.02132 | 0.01345 | 0.00852 | 0.00346 |

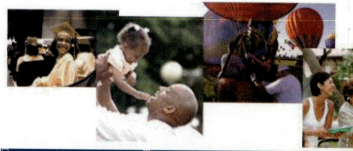

An annuity is a series of payments made regularly at monthly or other intervals. Insurance companies sell annuity policies so that retirees can be guaranteed a regular income after retirement. One insurance company specializing in annuities is The Best of America. Visit its site at http://www.bestofamerica.com/ to learn more about retirement planning, tools and tables, and annuities.

Annuities

26

LEARNING OBJECTIVES

By studying this chapter and completing all assignments you will be able to:

 1 Compute the future value of an annuity.

 2 Compute the regular payments of an annuity from the future value.

 3 Compute the present value of an annuity.

 4 Compute the regular payments of an annuity from the present value.

 5 Compute the loan payment required to amortize a loan.

 6 Create a loan amortization schedule.

Did you ever wonder why all state lotteries don't give winners all the money at once? The prize is often an annuity paid over time. Annuity investments can be a safe, popular supplement to retirement income. So you can invest in an annuity or buy a lottery ticket. The odds favor the former.

1. What determines whether financial payments form an annuity?
2. Is the amortization of a loan more closely associated with the future value of an annuity or with present value of an annuity?

John and Joan Popplewell just won their state's lottery and the prize was listed as $5,000,000. When they purchased the winning ticket, they had a choice of taking the prize over 20 years or taking one cash payment now. The $5,000,000 represents the sum of 20 annual payments of $250,000 each. The series of equal payments is called an **annuity**. Because they chose the single cash payment, they do not actually receive $5,000,000 in cash. The amount that they receive is the **present value of an annuity**.

In Chapter 23, we discuss corporate and government bonds. When a corporation issues $10,000,000 worth of 8%, 20-year bonds, the corporation is simply borrowing money from the public for 20 years. Each $1,000 bond pays 8% (or $80) each year. The $80 is paid out in two $40 payments every 6 months for 20 years. The series of $40 interest payments is an annuity. The amount that someone pays for the bond is the present value of the annuity. Some investors may worry that the corporation won't have $10,000,000 available in 20 years to repay the bonds. Therefore the corporation may decide to make 20 equal annual payments into a separate account managed by a neutral third party. At the end of the 20 years, the deposits plus accumulated interest will be worth the $10,000,000. This fund of deposits is called a **sinking fund**. Equal deposits into a sinking fund form an annuity. The total amount is the **future value of an annuity**.

Computing the Future Value of an Annuity

Compute the future value of an annuity.

An annuity is made up of a series of equal payments that occur at regular time intervals. The payments go into—or come out of—an interest-bearing account or investment. The constant interest rate is compounded at the same time the payments are made. (Perhaps obviously, the number of periods in an annuity is the same as the number of payments.)

We can illustrate an annuity by drawing a straight line, called a **time line**. On the time line, we insert equal marks and the payment dates and write in the payment amount.

EXAMPLE A

An annuity has four annual payments of $1,000, always on December 31. The date of the first $1,000 payment is December 31, 2002. Draw a time line showing the four years—2002, 2003, 2004, and 2005—and the four payments.

The annuity illustrated in Figure 26-1, with the payments occurring at the end of each period, is called an **ordinary annuity**. In this book, every annuity will have its payments at the end of each period. The date December 31, 2001, is the *beginning of the annuity*, and the date December 31, 2005, is the *end of the annuity*.

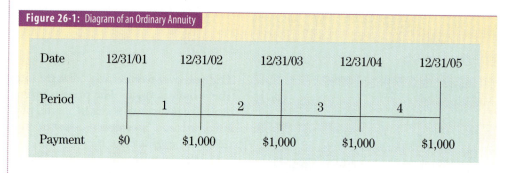

Figure 26-1: Diagram of an Ordinary Annuity

| Date | 12/31/01 | 12/31/02 | 12/31/03 | 12/31/04 | 12/31/05 |
|------|----------|----------|----------|----------|----------|
| Period | | 1 | 2 | 3 | 4 |
| Payment | $0 | $1,000 | $1,000 | $1,000 | $1,000 |

Again, the value of the annuity at the end of the annuity is called the *future value of the annuity*. In example A, it is the total value of all payments plus the compound interest

from the date of each payment until December 31, 2005. When a business or individual decides to deposit the same amount of money every year (or month or quarter) into an interest-bearing account for a specified amount of time, the future value of the annuity is the amount that will be in the account when the last deposit is made.

EXAMPLE B

In December, 2001, Barbara Luzardi accepted a job with a manufacturing company. Barbara decided to save $1,000 at the end of each year for 4 years. The company credit union allowed Barbara to open a savings account on December 31, 2001, but Barbara will not make any deposit until December 31, 2002. She also will make deposits on December 31 of 2003, 2004, and 2005. The credit union pays interest of 10% compounded annually. How much will be in the account after the last deposit? (*Hint:* Make a time line diagram and compute the future value of each of the four deposits.)

To find the future value of the annuity on December 31, 2005, first use Table 25-1 (see Chapter 25) to determine the future value of each of the four payments as of December 31, 2005. Then compute the total.

| Amount of Payment | Date of Payment | Years of Interest | Future Value on 12/31/05 |
|---|---|---|---|
| $1,000 | 12/31/02 | 3 | $1,000 × 1.33100 = $1,331 |
| $1,000 | 12/31/03 | 2 | $1,000 × 1.21000 = $1,210 |
| $1,000 | 12/31/04 | 1 | $1,000 × 1.10000 = $1,100 |
| $1,000 | 12/31/05 | 0 | $1,000 × 1.00000 = $1,000 |
| | | | Total = $4,641 |

Figure 26-2 illustrates how each of the four payments moves *forward* in time to December 31, 2005.

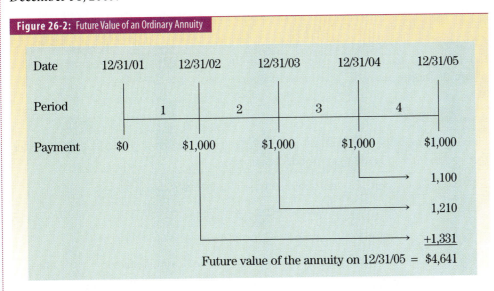

Figure 26-2: Future Value of an Ordinary Annuity

| Date | 12/31/01 | 12/31/02 | 12/31/03 | 12/31/04 | 12/31/05 |
|---|---|---|---|---|---|
| Period | | 1 | 2 | 3 | 4 |
| Payment | $0 | $1,000 | $1,000 | $1,000 | $1,000 |
| | | | | | 1,100 |
| | | | | | 1,210 |
| | | | | | +1,331 |

Future value of the annuity on 12/31/05 = $4,641

www.retireweb.com/ annuitycalc.html

www.nifcu.org/ calcannuity.html

Visit either of these Web sites to use an annuity calculator.

Annuity Tables

Annuity calculations are time-consuming, even with just four payments. With 20 or 30 payments, the calculations would be tiresome. Computers, financial calculators, and tables eliminate tedious computations. Table 26-1 is an abbreviated sample of a table of **annuity factors**. It is used the same way as Table 25-1 was used. As in Chapter 25, the columns indicate the periodic interest rate and the rows indicate the number of periods.

EXAMPLE C

Find the future value of an annuity of 4 annual payments of $1,000. Each payment is made at the end of the year, and 10% interest is compounded each year. Also find the total interest earned over the 4 years.

STEP 1 The annuity factor from Table 26-1 is 4.64100.

STEP 2 Future value of the annuity = $1,000 × 4.64100 = $4,641

STEP 3 Total of the payments = 4 × $1,000 = $4,000

STEP 4 Total interest = $4,641 − $4,000 = $641

Finding Annuity Amounts for Periods Other Than 1 Year

VIDEO

Annuities: Future Value/
Present Value

Payments may be made more often than once a year. The only additional requirement for an ordinary annuity is that the interest be compounded at the same time the payments are made—semiannually, quarterly, or monthly. We described the same method in Chapter 25, and also use steps i, ii, and iii in this chapter. However, in this chapter, the number computed in Step iii represents both the number of payments and the number of compounding periods.

EXAMPLE D

Find the future value of an annuity in which $200 is deposited at the end of each quarter for 3 years. Interest is 8% compounded quarterly.

STEP i There are 4 compounding periods in 1 year.

STEP ii Periodic interest rate = 8% ÷ 4 = 2% per period

STEP iii Number of payments = 3 years × 4 = 12 payments

STEP 1 Use Table 26-1, 2.00% column and row 12: annuity factor = 13.41209

STEP 2 Future value = $200 × 13.41209 = $2,682.418, or $2,682.42

CONCEPT CHECK 26.1

Assume that $1,800 is invested every 6 months for 7 years in an account that pays 8% compounded semiannually. Compute the future value of the investment. Then compute the total interest earned by the investment.

Semiannual means 2 periods per year.
Periodic rate = 8% ÷ 2 = 4% per period
Number of payments = 7 years × 2 = 14 payments
The future value annuity factor from row 14 of the 4.00% column in Table 26-1 is 18.29191.
Future value of the annuity = $1,800 × 18.29191 = $32,925.438, or $32,925.44

Total of all payments = $1,800 × 14 payments = $25,200
Total interest earned = Future value − Total payments = $32,925.44 − $25,200.00 = $7,725.44

Computing Regular Payments of an Annuity from the Future Value

In examples A–D, the amounts of the payments were known and the future values were unknown. If, however, the future value is known, then you can compute the amount of each payment.

STEPS **to Find the Size of the Payment in an Annuity, Given Its Future Value**

1. Locate the annuity factor in the correct row and column of Table 26-1.
2. Divide the future value by the annuity factor. The quotient is the amount of each payment in the annuity.

EXAMPLE E

Nate and Nan Roth want to have $25,000 in their credit union account when their son Danny starts college. They will make equal payments every month for 4 years. The credit union will pay 9% compounded monthly. What should their payment amount be?

The value of the annuity at the end, or the future value of the annuity, is $25,000. Use Table 26-1.

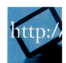

www. immediateannuity. com/

More on annuities is available at this Web site.

| | |
|---|---|
| STEP i | There are 12 compounding periods in 1 year. |
| STEP ii | Periodic interest rate = 9% ÷ 12 = 0.75% per period |
| STEP iii | Number of deposits = 4 years × 12 = 48 deposits |
| STEP 1 | Use Table 26-1, 0.75% column and row 48: annuity factor = 57.52071 |
| STEP 2 | Future value of the annuity = $25,000
Payment amount = $25,000 ÷ 57.52071 = $434.626, or $434.63 |

Sinking Funds

At the beginning of this chapter, we said that a $10,000,000 corporate bond issue may include a sinking fund feature. Sometimes a sinking fund means that the corporation will set aside an equal amount of money each year so that by the end of the 20 years,

© PHOTODISC, INC.

the corporation will have accumulated the $10,000,000. At other times, the sinking fund may be used by the corporation to buy back $500,000 worth of the bonds each year.

Although the term *sinking fund* may be most often associated with the repayment of a bond issue, its use isn't restricted to bonds. A corporation may set up a sinking fund to save money for an expensive piece of equipment that it knows it must replace in the future. The college fund set up by Nate and Nan Roth in example E is essentially a sinking fund.

EXAMPLE F

Macromedia Corporation is preparing a $10,000,000 bond issue. The company wants to make 20 equal payments in a sinking fund so that it will have a total of $10,000,000 available in 20 years to repay the bonds. What size should each of the payments be if the company can earn 6% per year on the payments?

| STEP i | There is 1 compounding period in 1 year. |
| STEP ii | Periodic interest rate = 6% ÷ 1 = 6% per period |
| STEP iii | Number of deposits = 20 years × 1 = 20 deposits |
| STEP 1 | Use Table 26-1, 6% column and row 20: annuity factor = 36.78559 |
| STEP 2 | Future value of the annuity = $10,000,000
Payment amount = $10,000,000 ÷ 36.78559 = $271,845.58 |

CONCEPT CHECK 26.2

Assume that an equal amount is invested every quarter for 9 years. After the last payment, the future value is $40,000. If the interest rate is 5% compounded quarterly, compute the size of each regular quarterly payment.

Quarterly means 4 periods per year.
Periodic rate = 5% ÷ 4 = 1.25% per period
The number of payments is 9 years × 4 = 36 payments
The future value annuity factor from row 36 of the 1.25% column in Table 26-1 is 45.11551.

Regular quarterly payment = $40,000 ÷ 45.11551 = $886.61305, or $886.61

COMPLETE ASSIGNMENT 26.1.

Computing the Present Value of an Annuity

LEARNING OBJECTIVES

Compute the present value of an annuity.

The annuity shown in Figure 26-3 begins December 31, 2001. Again, the value of the annuity on this date is called the *present value of the annuity.* For example, when a person deposits a large amount in a bank account and then makes a series of equal withdrawals from the account until it is empty, the series of withdrawals (the equal payments) is the annuity and the amount deposited is the present value. The interest earned equals the difference between the total amount withdrawn and the amount deposited.

In December 2001, Harry Hamilton inherited some money. He planned to donate part of the money immediately to the American Cancer Society and to make 4 equal donations of $1,000 each on December 31 of 2002, 2003, 2004, and 2005. To prepare for the 4 future payments, Harry went to his bank on December 31, 2001, and deposited money in a new account. The account paid 10% compounded annually. Harry would withdraw $1,000 each year; after the last withdrawal on December 31, 2005, the account would be empty.

How much must Harry deposit on December 31, 2001? (*Hint:* Make a time line diagram, and compute the present value of each of the 4 withdrawals.)

To find the present value of the annuity on December 31, 2001, first use Table 25-2 to find the present value of each of the 4 payments on December 31, 2001. Then compute the total.

VIDEO

Annuities: Future Value/
Present Value

| Amount of Payment | Date of Payment | Years of Interest | Present Value on 12/31/01 |
|---|---|---|---|
| $1,000 | 12/31/02 | 1 | $1,000 × 0.90909 = $ 909.09 |
| $1,000 | 12/31/03 | 2 | $1,000 × 0.82645 = 826.45 |
| $1,000 | 12/31/04 | 3 | $1,000 × 0.75131 = 751.31 |
| $1,000 | 12/31/05 | 4 | $1,000 × 0.68301 = $ 683.01 |
| | | | Present value of the annuity on 12/31/01 = $3,169.86 |

Figure 26-3 illustrates example G. The time line shows the equal withdrawals as each payment is moved from the future backward to the present (to December 31, 2001). In contrast, in Figure 26-2, each payment was projected forward into the future.

The method shown in Figure 26-3 aids instruction but is too time consuming to be practical. To get the same solution quickly, use Table 26-2.

STEPS to Use Table 26-2 to Compute Present Value and Total Interest Earned

1. Locate the annuity factor in the correct row and column of Table 26-2 on pages 580–581.
2. Multiply the payment amount by the annuity factor. The product is the present value of the annuity.
3. Multiply the payment amount by the number of payments. The product is the total of all payments.
4. Subtract the present value of the annuity from the total of all payments. The difference is the total interest earned.

For example G, the factor in the 10.00% column and row 4 of Table 26-2 is 3.16987 (Step 1), and $1,000 × 3.16987 = $3,169.87 (Step 2).

The application in example H may not sound time consuming, but even it would be tedious to do without Table 26-2. As the payments and compounding are quarterly, use Steps i, ii, and iii to find the periodic rate and the number of periods.

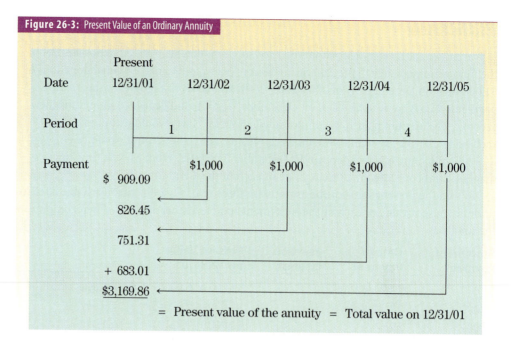

Figure 26-3: Present Value of an Ordinary Annuity

= Present value of the annuity = Total value on 12/31/01

EXAMPLE H

Mary Currie just won first prize in a fund-raising raffle. She has a choice: She can receive quarterly payments of $600 each for 5 years, with the first payment 3 months (one quarter) from now, or she can receive 1 lump sum today. Assuming an interest rate of 8% compounded quarterly, what lump sum today equals the future payments? (*Hint:* The series of $600 payments is an annuity, and the lump sum is the present value of the annuity.)

STEP i There are 4 compounding periods in 1 year.

STEP ii Periodic interest rate = 8% ÷ 4 = 2% per period

STEP iii Number of payments = 5 years × 4 = 20

STEP 1 Use Table 26-2, 2.00% column and row 20: annuity factor = 16.35143

STEP 2 Present value = $600.00 × 16.35143 = $9,810.858, or $9,810.86

CONCEPT CHECK 26.3

What present value (principal) must be invested today in an account to provide for 10 equal annual withdrawals (an annuity) of $6,000 each? The interest rate is 9% compounded annually.

Annual means 1 period per year.
Periodic rate = 9% ÷ 1 = 9% per year
Number of payments = 10 years × 1 = 10 payments
The present value annuity factor from row 10 of the 9.00% column of Table 26-2 is 6.41766.

Present value of the annuity = $6,000 × 6.41766 = $38,505.96

Computing Regular Payments of an Annuity from the Present Value

In examples G and H, the amounts of the payments were known and the present values were unknown. If, however, the present value is known, then you can compute the amount of the payments.

STEPS to Find the Size of the Payment in an Annuity, Given the Present Value

1. Locate the annuity factor in the correct row and column of Table 26-2.
2. Divide the present value by the annuity factor. The quotient is the amount of the payments in the annuity.

EXAMPLE I

Jim Schremp received a $10,000 bonus from his employer. Rather than spend it all at once, he decided to deposit it in a bank account that pays 6% compounded monthly. He will make equal monthly withdrawals for 2 years. After the last withdrawal, the account will be empty. How much will he withdraw each month?

The value of the annuity in the beginning (present value of the annuity) is $10,000. Use Table 26-2.

| | |
|---|---|
| STEP i | There are 12 compounding periods in 1 year. |
| STEP ii | Periodic interest rate = 6% ÷ 12 = 0.5% per period |
| STEP iii | Number of withdrawals = 2 years × 12 = 24 withdrawals |
| STEP 1 | Use Table 26-2, 0.5% column and row 24: annuity factor = 22.56287 |
| STEP 2 | Each withdrawal = $10,000 ÷ 22.56287 = $443.206, or $443.21 |

CONCEPT CHECK 26.4

Assume that $50,000 is deposited today (the present value) to provide for 44 equal quarterly withdrawals (an annuity) over the next 11 years. If the interest rate is 8% compounded quarterly, what is the size of each regular quarterly payment? What is the total interest earned during the term of the annuity?

Quarterly means 4 periods per year.
Periodic rate = 8% ÷ 4 = 2% per quarter
Number of payments = 11 years × 4 = 44 payments
The present value annuity factor from row 44 of the 2.00% column in Table 26-2 is 29.07996.

Regular quarterly payment = $50,000 ÷ 29.07996 = $1,719.3971, or $1,719.40

Total of all payments = $1,719.40 × 44 payments = $75,653.60
Total interest earned = Total payments − Present value
= $75,653.60 − $50,000.00 = $25,653.60

Computing the Payment to Amortize a Loan

LEARNING OBJECTIVES

Compute the loan payment required to amortize a loan.

Recall from your study of loan amortization in Chapter 17 that the borrower repays the loan by making equal monthly payments and that the interest is computed on the unpaid balance each month. Loan amortization creates an annuity because there is a series of equal periodic payments. Computing the interest each month makes it compound interest. The amount of the loan is the present value of the annuity.

Stated another way, in amortization, when the amount of the loan is known, the present value of the annuity is known. As in example J, you can use Table 26-2 to compute the amount of the monthly payments.

STEPS to Find the Size of the Payment to Amortize a Loan

1. Locate the annuity factor in the correct row and column of Table 26-2.
2. Divide the loan amount by the annuity factor. The quotient is the amount of the monthly loan payments.

EXAMPLE J

Nanda Cerrato wants to buy a new piano. Nanda pays $4,000 and also trades in her old piano. The balance is $2,400, and the piano dealer will amortize the $2,400 over 4 months at 12%. Find the size of the required monthly payments.

STEP i There are 12 compounding periods in 1 year.

STEP ii Periodic interest rate = 12% ÷ 12 = 1% per period

STEP iii Number of monthly payments = 4

STEP 1 Because the borrowing occurs at the *beginning* of the annuity, this is a present value problem and $2,400 is the present value of the annuity; use Table 26-2. In the 1.00% column and row 4, the annuity factor is 3.90197.

STEP 2 Size of each payment = $2,400 ÷ 3.90197 = $615.07392, or $615.07

CONCEPT CHECK 26.5

A bank loans $30,000 at an interest rate of 15% compounded monthly. Find the loan payment necessary to amortize the loan with monthly payments over 4 years.

Loan amortization involves an annuity. The amount borrowed is the present value of the annuity, and the monthly loan payment is the regular annuity payment.

Monthly means 12 periods per year.
Periodic rate = 15% ÷ 12 = 1.25% per period
Number of payments = 4 years × 12 = 48 payments
The present value annuity factor from row 48 of the 1.25% column of Table 26-2 is 35.93148.

Loan payment = $30,000 ÷ 35.93148 = $834.92247, or $834.92

Creating a Loan Amortization Schedule

Recall from Chapter 17 that the following procedure is used to create an amortization schedule.

LEARNING OBJECTIVES

Create a loan amortization schedule.

> **STEPS to Create an Amortization Schedule**
>
> For each row except the last:
> 1. Interest payment = Unpaid balance × Monthly interest rate
> 2. Principal payment = Monthly payment − Interest payment
> 3. New unpaid balance = Old unpaid balance − Principal payment
>
> For the last row:
> 1. Interest payment = Unpaid balance × Monthly interest rate
> 2. (Then ADD) Monthly payment = Unpaid balance + Interest payment
> 3. Principal payment = Unpaid balance

EXAMPLE K

Create an amortization schedule for the loan in example J: $2,400 to be amortized over 4 months with interest of 12% compounded monthly charged on the unpaid balance.

| Month | Beginning Unpaid Balance | STEP 1 (1%) Interest Payment | Annuity Payment | STEP 2 Principal Payment | STEP 3 New Balance |
|---|---|---|---|---|---|
| 1 | $2,400.00 | $24.00 | $615.07 | $591.07 | $1,808.93 |
| 2 | 1,808.93 | 18.09 | 615.07 | 596.98 | 1,211.95 |
| 3 | 1,211.95 | 12.12 | 615.07 | 602.95 | 609.00 |
| 4 | 609.00 | 6.09 | 615.09 | 609.00 | 0 |

In example J, note that each month's beginning unpaid balance is multiplied by the monthly interest rate (1%).

www.allstate.com/ Finance/pagerender. asp?Page=AFS.htm

Visit the Allstate Insurance Web site for information on different kinds of annuities.

CONCEPT CHECK 26.6

Amortize a $1,500 purchase over 3 months at an annual rate of 12%. First, use Table 26-2 to calculate the first two monthly payments. Then show the calculations to construct a 3-month amortization schedule.

The periodic interest rate is 12% ÷ 12 = 1%, and the number of periods is 3. The present value annuity factor from row 3 of the 1.00% column of Table 26-2 is 2.94099. The loan payment is $1,500 ÷ 2.94099 = $510.03.

| | Month 1 | | Month 2 | | Month 3 | |
|---|---|---|---|---|---|---|
| Unpaid balance: | Purchase price | $1,500.00 | From month 1 | $1,004.97 | From month 2 | $504.99 |
| Interest payment: | $1,500.00 × 0.01 = | $15.00 | $1,004.97 × 0.01 = | $10.05 | $504.99 × 0.01 = | $5.05 |
| Total payment: | From above | $510.03 | From above | $510.03 | $504.99 + $5.05 = | $510.04 |
| Principal payment: | $510.03 − $15.00 = | $495.03 | $510.03 − $10.05 = | $499.98 | (Unpaid balance) | $504.99 |
| New balance: | $1,500 − $495.03 = | $1,004.97 | $1,004.97 − $499.98 = | $504.99 | $504.99 − $504.99 = | $0.00 |

COMPLETE ASSIGNMENT 26.2.

Chapter Terms for Review

annuity

annuity factors

future value of an annuity

ordinary annuity

present value of an annuity

sinking fund

time line

Visit the Web

http://www.med.utah.edu/hr/ben/ben.htm—Read about a tax-deferred annuity at this Web site.

http://www.cigna.com—For more on annuities, go to the Cigna Web site.

http://www.timyounkin.com/heardarchive.html—Visit this Web site for FAQs on annuities and retirement funds.

http://www.aul.com/group/needs.html—If you are an employer considering providing a benefits package for your employees, a good start might be this Employer Benefit Checklist.

http://www.annuitywiz.com/html/annuity_risks.html—This site discusses the five principal risks of investing in annuities.

Other useful sites about annuities include
http://www.sectorupdates.com/annuity.asp
http://www.metlife.com/Lifeadvice/Money/Docs/annuitiesintro.html, and
http://www.teachmefinance.com/annuities.html.

CYBERCONNECTION

Source: http://www.retireearlyhomepage.com. Copyright 2001 by John P. Greaney. Used with permission.

Most people planning to retire, whether early or not, will have some savings to supplement Social Security—and many of these savings are in retirement accounts, deposited before taxes and earning money tax free. In general, the IRS prohibits withdrawing these funds without penalty before age $59\frac{1}{2}$ and requires that the saver begin making withdrawals by age $70\frac{1}{2}$. These general rules differ depending on the type of retirement account and there are legal ways to get around these rules.

To get an idea of the complexity of this, go from the Retire Early Home Page at http://retireearlyhomepage.com to the link labeled *"Can I withdraw money from my IRA before age 59-1/2?"* When a retiree takes money out of a tax-deferred retirement account, the IRS says that it should be in the form of an annuity—an equal amount each year. The retiree divides the amount in the account by an annuity factor to determine the annual withdrawal. However, the annuity factor is calculated differently than those discussed in this chapter. The annuity factor will differ depending on the retiree's age, and built into the factor are the probabilities that a retiree of any age will live an additional year.

The Web page tells you to input an IRS-accepted interest rate and gives you a link to find the rate. You can either use the link to find the appropriate rate, or use a rate of 5.5%. Next click on *Jump to Retire Early's JavaScript IRA Withdrawal Calculator.*

Suppose a person will retire at age 62, has a balance of $250,000 in her retirement account, and has a life expectancy of 90 years. Using an interest rate of 5.5%, complete the information in the Withdrawal calculator to find the annual payments that she can withdraw. Change the year of her life expectancy or her retirement age to see the changes in annual payments.

CYBERCONNECTION

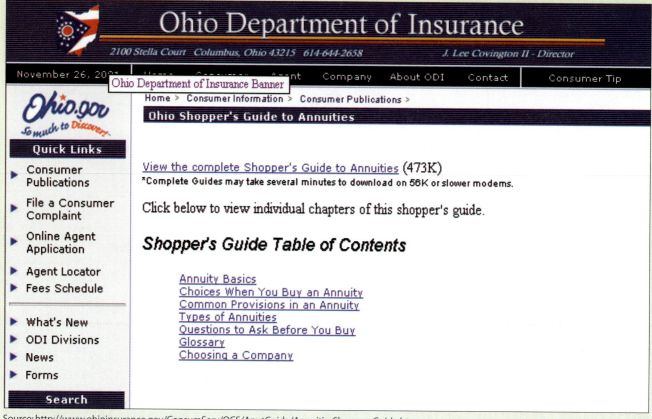

Ohio Department of Insurance

2100 Stella Court · Columbus, Ohio 43215 · 614-644-2658 J. Lee Covington II - Director

November 26, 2001 Home Consumer Agent Company About ODI Contact Consumer Tip

Ohio Department of Insurance Banner

Quick Links

► Consumer Publications

► File a Consumer Complaint

► Online Agent Application

► Agent Locator

► Fees Schedule

► What's New
► ODI Divisions
► News
► Forms

Search

Home > Consumer Information > Consumer Publications >

Ohio Shopper's Guide to Annuities

View the complete Shopper's Guide to Annuities (473K)

*Complete Guides may take several minutes to download on 56K or slower modems.

Click below to view individual chapters of this shopper's guide.

Shopper's Guide Table of Contents

Annuity Basics
Choices When You Buy an Annuity
Common Provisions in an Annuity
Types of Annuities
Questions to Ask Before You Buy
Glossary
Choosing a Company

Source: http://www.ohioinsurance.gov/ConsumServ/OCS/AnutGuide/AnnuitiesShoppersGuide.htm.

Learn more about annuities at the Ohio Department of Insurance at
http://www.ohioinsurance.gov/ConsumServ/OCS/AnutGuide/AnnuitiesShoppersGuide.htm.

The Bottom Line

Summary of chapter learning objectives:

| Learning Objective | Summary | Example |
|---|---|---|
| **26.1** | Compute the future value of an annuity | 1. Compute the future value of $800 invested every month for 3 years, with interest at 9% compounded monthly. |
| **26.2** | Compute the regular payments of an annuity from the future value | 2. Compute the regular annuity payment that is required to accumulate $6,000 after 17 quarterly payments at an interest rate of 8% compounded quarterly. |
| **26.3** | Compute the present value of an ordinary annuity | 3. Compute the present value of $1,200 withdrawn every half-year for $5\frac{1}{2}$ years, with interest at 10% compounded semiannually. |
| **26.4** | Compute the regular payments of an annuity from the present value | 4. An account starts at $4,000. Compute the regular annual withdrawal that is required to empty the account in 6 years if the interest is 10% compounded annually. |
| **26.5** | Compute the loan payment required to amortize a loan | 5. Compute the loan payment that is required to amortize a $10,000 loan in 15 monthly payments, with an interest rate of 12% compounded monthly. |
| **26.6** | Create a loan amortization schedule | 6. A $10,000 loan at a rate of 12% compounded monthly is amortized in 15 monthly payments of $721.24. Compute the entries for the first line of the amortization schedule. |

Answers: 1. $32,922.18 **2.** $299.82 **3.** $9,967.69 **4.** $918.43 **5.** $721.24 **6.** $10,000; $100.00; $721.24; $621.24, $9,378.76

1 For each annuity, find either the future value (compound amount) or the payment, as indicated. Then compute the compound interest earned by the annuity. (Use Table 26-1)

| Payment | Rate | Period | Time | Future Value | Interest |
|---------|------|--------|------|--------------|----------|
| $ 2,500 | 8% | quarterly | 6 yrs | a. _____ | b. _____ |
| $25,000 | 10% | annually | 30 yrs | c. _____ | d. _____ |

| Future Value | Rate | Period | Time | Payment | Interest |
|--------------|------|--------|------|---------|----------|
| $40,000 | 12% | semiannually | 10 yrs | e. _____ | f. _____ |
| $18,000 | 9% | monthly | 3 yrs | g. _____ | h. _____ |

2 For each annuity, find either the present value or the payment, as indicated. Then compute the compound interest earned by the annuity. (Use Table 26-2)

| Payment | Rate | Period | Time | Present Value | Interest |
|---------|------|--------|------|---------------|----------|
| $ 750 | 12% | monthly | 2 yrs | a. _____ | b. _____ |
| $1,875 | 6% | quarterly | 11 yrs | c. _____ | d. _____ |

| Present Value | Rate | Period | Time | Payment | Interest |
|---------------|------|--------|------|---------|----------|
| $26,000 | 10% | annually | 17 yrs | e. _____ | f. _____ |
| $80,000 | 8% | semiannually | 25 yrs | g. _____ | h. _____ |

3 Allan Abraham planned to save money for retirement. He put $350 every month in an investment that paid a return of 9% compounded monthly. How much would Allan have in his account after 4 years?

4 Med-West wanted to set up a sinking fund to have $15,000,000 in 20 years. The company would make annual payments that would pay a return of 6% per year. What size should the payments be?

5 Jeanne Fukuji received a payment of $50,000 from an insurance company. She put it in an account that would pay 8% compounded quarterly. Jeanne wanted to make equal quarterly withdrawals from the account for 10 years, when the account would be empty. What size payments can Jeanne make?

6 Steve Evans read about an investment opportunity on the internet. The Web site explained that Steve would receive payments of $450 every 6 months for 14 years. If the returns are based on 10% compounded semiannually, what is the present value of this investment opportunity?

Assignment 26.1: Annuities—Future Value

Name _____

Date _____ Score _____

A (28 points) For each of the following annuities, find the future value or the amount of the periodic payment. Round answers to the nearest cent. (4 points for each correct answer)

| | Payment Amount | Payment Periods | Interest Rate | Length of Annuity | Future Value |
|---|---|---|---|---|---|
| **1.** | $1,500 | quarterly | 12% compounded quarterly | 5 years | _____ |
| **2.** | _____ | monthly | 9% compounded monthly | 4 years | $18,000 |
| **3.** | $4,000 | annually | 12% compounded annually | 11 years | _____ |
| **4.** | _____ | semiannually | 8% compounded semiannually | 13 years | $50,000 |
| **5.** | $500 | monthly | 15% compounded monthly | 3 years | _____ |
| **6.** | _____ | quarterly | 5% compounded quarterly | 10 years | $30,000 |
| **7.** | $1,000 | semiannually | 10% compounded semiannually | 16 years | _____ |

Score for A (28)

Assignment 26.1 Continued

B **(32 points) For each of the following annuities, find the future value, the amount of the periodic payment, or the total amount of interest paid. Round answers to the nearest cent. (4 points for each correct answer)**

8. Calculate the future value of a 15-year annuity with payments of $2,000 each year and an interest rate of 6% compounded annually.

 Future value _____

9. How much total interest is earned on an annuity with payments of $300 per month for 4 years and an interest rate of 6% compounded monthly?

 Total interest earned _____

10. An annuity consists of quarterly payments of $1,200 each for 9 years at an interest rate of 5% compounded quarterly. Compute the future value of the annuity.

 Future value _____

11. An 8-year annuity has semiannual payments of $5,000 each and an interest rate of 10% compounded semiannually. What will be the total amount of interest earned?

 Total interest earned _____

12. A sinking fund has 7 annual payments, has an interest rate of 8% compounded annually, and has a future value of $15,000. Compute the amount of each annual payment.

 Payment amount _____

13. A 6-year annuity with quarterly payments and an interest rate of 8% compounded quarterly has a future value of $35,000. How much total interest does the annuity earn?

 Total interest earned _____

14. Calculate the amount of each monthly payment in a 1-year annuity that has a future value of $5,000 and an interest rate of 9% compounded monthly.

 Payment amount _____

Assignment 26.1 *Continued*

15. Determine the total interest earned by an annuity with semiannual payments for 18 years, an interest rate of 10% compounded semiannually, and a future value of $25,000.

 Total interest earned _____

 Score for B (32)

C **(4 points) In each of the following applications, find the future value of the annuity, the amount of the periodic payment, or the total amount of interest earned. Round answers to the nearest cent. (4 points for each correct answer)**

16. Linda Oxford decides to make quarterly deposits in her credit union account because it is guaranteeing a rate of 12% compounded quarterly for the next 3 years. How much will Linda have after making equal quarterly deposits of $1,750 for 3 years?

17. Delores Hensen is planning for her daughter's college education. An investment advisor recommends an investment whose prospectus claims it will return 10% compounded semiannually. If the investment does return 10% compounded semiannually how much must Delores invest each 6 months for 7 years if she wants to have a total of $40,000 after the fourteenth deposit?

18. Maxfield International is raising $25,000,000 by selling bonds that will mature in 20 years. Maxfield plans to make equal annual payments in a sinking fund to repay the bonds. If Maxfield can earn 6% per year, what amount should it deposit each year in order to have $25,000,000 at the end of 20 years?

19. Ruben Mendoza is quite certain he will need to replace some construction equipment in 2 years. He decides to set up a sinking fund now to help buy the equipment. Ruben estimates that he can deposit $1,100 each month for 2 years in a sinking fund that will pay 9% compounded monthly. How much will Ruben's sinking fund be worth after the last deposit?

20. Liz Schaer is planning to take her children on a trip to Asia in 4 years. She thinks that she should start saving now to try to accumulate $18,000 by the end of the 4 years. Assuming that Liz can find an investment that will pay 12% compounded quarterly, what amount must Liz deposit each quarter to have the necessary $18,000 at the end of the 4 years?

Assignment 26.1 *Continued*

21. Edward Chiu imports patio furniture from various countries. He prefers to have cash available when he goes on buying trips. Suppose that Edward makes 12 equal monthly deposits in a bank account that pays 6% compounded monthly. If he deposits enough each month to accumulate $50,000 by the end of the year, how much of the $50,000 will the bank have paid in interest?

———————

22. Jeanne Knowles will graduate from high school in a few months. She has found a part-time job and is trying to determine how much money she can save in 6 years. Calculate the future value after 6 years if Jeanne makes semiannual deposits of $600 each in an investment account that promises a return of 8% compounded semiannually.

———————

23. Medical Instrument Manufacturing, Inc., (MIMI) just sold $50,000,000 in bonds. The bonds will mature in 15 years. MIMI will make 30 equal semiannual payments into a sinking fund that will earn 8% compounded semiannually. If MIMI has the $50,000,000 after 15 years, what amount of the total was earned on the interest?

———————

24. Sharon Wilder sends money to her granddaughter, Jenny, every 3 months. Jenny deposits the money in a credit union account that pays her 6% compounded quarterly. If Jenny receives 6 payments of $925 each from her grandmother over 18 months and saves it all, how much total interest will Jenny have earned from the credit union by the time of her last deposit?

———————

25. This year, Doug McCombs charged all his family's Christmas gifts on a credit card, and the result was a minor financial disaster. Planning for next year, Doug decides to save money each month from January through November and put it into an account that will pay 9% compounded monthly. He plans to make 11 equal deposits, and he wants to have accumulated $2,000 once he makes the eleventh deposit. Calculate the size of each deposit.

———————

———————
Score for C (40)

Assignment 26.2: Annuities— Present Value

Name

Date _____ Score _____

A **(28 points)** For each of the following annuities, find the present value or the amount of the periodic payment. Round answers to the nearest cent. **(4 points for each correct answer)**

| | Payment Amount | Payment Periods | Interest Rate | Length of Annuity | Present Value |
|---|---|---|---|---|---|
| **1.** | $2,500 | semiannually | 10% compounded semiannually | 7 years | _____ |
| **2.** | _____ | quarterly | 8% compounded quarterly | 10 years | $25,000 |
| **3.** | $800 | monthly | 6% compounded monthly | 4 years | _____ |
| **4.** | _____ | annually | 6% compounded annually | 17 years | $50,000 |
| **5.** | $1,500 | quarterly | 12% compounded quarterly | 6 years | _____ |
| **6.** | _____ | semiannually | 6% compounded semiannually | 25 years | $100,000 |
| **7.** | $1,000 | monthly | 15% compounded monthly | 2 years | _____ |

Score for A (28)

B (32 points) For each of the following annuities, find the present value, the amount of the periodic payment, or the total amount of interest paid. Round answers to the nearest cent. (4 points for each correct answer)

8. An annuity consists of quarterly payments of $1,500 each for 11 years at an interest rate of 6% compounded quarterly. Determine the present value of the annuity.

Present value _____

9. Compute the amount of each payment in an annuity that has a present value of $10,000 after 9 years of semiannual payments at an interest rate of 16% compounded semiannually.

Payment amount _____

10. In a 15-year annuity, the annual payments are $2,500 each and the interest rate is 10% compounded annually. What is the present value of the annuity?

Present value _____

11. What is the total interest earned by an annuity that has a present value of $16,000 with monthly payments over a 2-year period at an interest rate of 9% compounded monthly?

Total interest earned _____

12. Calculate the size of the regular quarterly payments in a 3-year annuity that has a present value of $40,000 and an interest rate of 12% compounded quarterly.

Payment amount _____

13. An annuity has a present value of $75,000. Compute the total interest earned by the annuity if there are annual payments over 10 years at an interest rate of 12% compounded annually.

Total interest earned _____

14. Find the present value of a 6-year annuity with semiannual payments of $3,000 each, which earns interest at a rate of 10% compounded semiannually.

Present value _____

15. Compute the amount of the regular monthly payments in a 1-year annuity that has a present value of $20,000 and an interest rate of 12% compounded monthly.

Payment amount _____

Score for B (32)

C **(28 points) In each of the following applications, find the present value of the annuity, the amount of the periodic payment, or the total amount of interest earned. Round answers to the nearest cent. (4 points for each correct answer)**

16. Sandee Millet is making a budget for the next 2 years. She estimates that her rent will be about $250 per month. For calculations, Sandee considers her housing expense to be an annuity of 24 payments. If she uses an interest rate of 6% compounded monthly, what will be the present value of the annuity?

17. After their children moved away from home, Barbara Cain and her husband sold their large house and bought a smaller condominium. Barbara invested $25,000 of their after-tax profit in an annuity that would give them equal quarterly payments for 10 years. The fund will pay a return of 8% compounded quarterly. At the end of the 10 years, their annuity will be finished. What amount will they receive each quarter?

18. Ken Ortman is considering an investment that is somewhat like a bond. The investment is an annuity that would pay Ken $500 every 6 months for 20 years. He is trying to determine how much the investment is worth today. If he uses an interest rate of 10% compounded semiannually, what is the present value of the annuity?

19. Sammie Crass will receive a retirement bonus of $50,000. She has the option of either receiving the $50,000 now in one lump sum or having it invested and then receiving 10 equal annual annuity payments. If she selects payments over 10 years, the $50,000 is invested at a guaranteed rate of 9% compounded annually. Compute the amount of interest that Sammie would earn by choosing the payments over 10 years instead of the lump sum.

20. Nellie Van Calcar inherited money from her grandfather. Nellie's daughter is in her second year of college, and Nellie wants to give her $1,600 every quarter for 3 years. Nellie can invest the money for her daughter at 6% compounded quarterly. How much should she invest now to provide for all the quarterly withdrawals and have an empty account after the last withdrawal?

21. Kathy Shutter plans to buy a new car. She could either finance the car through the dealer or borrow the money from the bank. Either way, the amount borrowed would be amortized in equal payments over 3 years. If the bank's interest rate for new cars is 12% compounded monthly, compute Kathy's monthly payments for a bank loan of $16,000.

22. Don Hildebrand wanted to protect his home from fire and burglars, so he purchased a home security system. The total price including installation was $3,750. The alarm company convinced Don to amortize the cost over 30 months at an interest rate of 1.5% per month (which is 18% compounded monthly). Determine the amount of each of the equal monthly payments.

Score for C (28)

D **(12 points) Gary Robinson purchased some new equipment and furniture for his office. Instead of charging it on a credit card, which had a 21% interest rate, Gary arranged for financing through the office supply dealer. The total purchase amount was $5,248.95, and it was amortized over 4 months. The interest rate was 15% per year, or 1.25% per month. The first 3 monthly payments were each $1,353.50. Complete the first three lines of the amortization schedule below. Round answers to the nearest cent. (1 point for each correct answer)**

| | Month | Unpaid Balance | Interest Payment | Principal Payment | Total Payment | New Balance |
|---|---|---|---|---|---|---|
| **23.** | 1 | _____ | _____ | _____ | $1,353.50 | _____ |
| **24.** | 2 | _____ | _____ | _____ | $1,353.50 | _____ |
| **25.** | 3 | _____ | _____ | _____ | $1,353.50 | _____ |

Score for D (12)

Notes

Table 26-1: Future Value Annuity Factors

| Period | 0.50% | 0.75% | 1.00% | 1.25% | 1.50% | 2.00% | 3.00% | 4.00% | 5.00% | 6.00% | 8.00% | 9.00% | 10.00% | 12.00% |
|---|---|---|---|---|---|---|---|---|---|---|---|---|---|---|
| 1 | 1.00000 | 1.00000 | 1.00000 | 1.00000 | 1.00000 | 1.00000 | 1.00000 | 1.00000 | 1.00000 | 1.00000 | 1.00000 | 1.00000 | 1.00000 | 1.00000 |
| 2 | 2.00500 | 2.00750 | 2.01000 | 2.01250 | 2.01500 | 2.02000 | 2.03000 | 2.04000 | 2.05000 | 2.06000 | 2.08000 | 2.09000 | 2.10000 | 2.12000 |
| 3 | 3.01502 | 3.02256 | 3.03010 | 3.03766 | 3.04522 | 3.06040 | 3.09090 | 3.12160 | 3.15250 | 3.18360 | 3.24640 | 3.27810 | 3.31000 | 3.37440 |
| 4 | 4.03010 | 4.04523 | 4.06040 | 4.07563 | 4.09090 | 4.12161 | 4.18363 | 4.24646 | 4.31013 | 4.37462 | 4.50611 | 4.57313 | 4.64100 | 4.77933 |
| 5 | 5.05025 | 5.07556 | 5.10101 | 5.12657 | 5.15227 | 5.20404 | 5.30914 | 5.41632 | 5.52563 | 5.63709 | 5.86660 | 5.98471 | 6.10510 | 6.35285 |
| 6 | 6.07550 | 6.11363 | 6.15202 | 6.19065 | 6.22955 | 6.30812 | 6.46841 | 6.63298 | 6.80191 | 6.97532 | 7.33593 | 7.52333 | 7.71561 | 8.11519 |
| 7 | 7.10588 | 7.15948 | 7.21354 | 7.26804 | 7.32299 | 7.43428 | 7.66246 | 7.89829 | 8.14201 | 8.39384 | 8.92280 | 9.20043 | 9.48717 | 10.08901 |
| 8 | 8.14141 | 8.21318 | 8.28567 | 8.35889 | 8.43284 | 8.58297 | 8.89234 | 9.21423 | 9.54911 | 9.89747 | 10.63663 | 11.02847 | 11.43589 | 12.29969 |
| 9 | 9.18212 | 9.27478 | 9.36853 | 9.46337 | 9.55933 | 9.75463 | 10.15911 | 10.58280 | 11.02656 | 11.49132 | 12.48756 | 13.02104 | 13.57948 | 14.77566 |
| 10 | 10.22803 | 10.34434 | 10.46221 | 10.58167 | 10.70272 | 10.94972 | 11.46388 | 12.00611 | 12.57789 | 13.18079 | 14.48656 | 15.19293 | 15.93742 | 17.54874 |
| 11 | 11.27917 | 11.42192 | 11.56683 | 11.71394 | 11.86326 | 12.16872 | 12.80780 | 13.48635 | 14.20679 | 14.97164 | 16.64549 | 17.56029 | 18.53117 | 20.65458 |
| 12 | 12.33556 | 12.50759 | 12.68250 | 12.86036 | 13.04121 | 13.41209 | 14.19203 | 15.02581 | 15.91713 | 16.86994 | 18.97713 | 20.14072 | 21.38428 | 24.13313 |
| 13 | 13.39724 | 13.60139 | 13.80933 | 14.02112 | 14.23683 | 14.68033 | 15.61779 | 16.62684 | 17.71298 | 18.88214 | 21.49530 | 22.95338 | 24.52271 | 28.02911 |
| 14 | 14.46423 | 14.70340 | 14.94742 | 15.19638 | 15.45038 | 15.97394 | 17.08632 | 18.29191 | 19.59863 | 21.01507 | 24.21492 | 26.01919 | 27.97498 | 32.39260 |
| 15 | 15.53655 | 15.81368 | 16.09690 | 16.38633 | 16.68214 | 17.29342 | 18.59891 | 20.02359 | 21.57856 | 23.27597 | 27.15211 | 29.36092 | 31.77248 | 37.27971 |
| 16 | 16.61423 | 16.93228 | 17.25786 | 17.59116 | 17.93237 | 18.63929 | 20.15688 | 21.82453 | 23.65749 | 25.67253 | 30.32428 | 33.00340 | 35.94973 | 42.75328 |
| 17 | 17.69730 | 18.05927 | 18.43044 | 18.81105 | 19.20136 | 20.01207 | 21.76159 | 23.69751 | 25.84037 | 28.21288 | 33.75023 | 36.97370 | 40.54470 | 48.88367 |
| 18 | 18.78579 | 19.19472 | 19.61475 | 20.04619 | 20.48938 | 21.41231 | 23.41444 | 25.64541 | 28.13238 | 30.90565 | 37.45024 | 41.30134 | 45.59917 | 55.74971 |
| 19 | 19.87972 | 20.33868 | 20.81090 | 21.29677 | 21.79672 | 22.84056 | 25.11687 | 27.67123 | 30.53900 | 33.75999 | 41.44626 | 46.01846 | 51.15909 | 63.43968 |
| 20 | 20.97912 | 21.49122 | 22.01900 | 22.56298 | 23.12367 | 24.29737 | 26.87037 | 29.77808 | 33.06595 | 36.78559 | 45.76196 | 51.16012 | 57.27500 | 72.05244 |
| 21 | 22.08401 | 22.65240 | 23.23919 | 23.84502 | 24.47052 | 25.78332 | 28.67649 | 31.96920 | 35.71925 | 39.99273 | 50.42292 | 56.76453 | 64.00250 | 81.69874 |
| 22 | 23.19443 | 23.82230 | 24.47159 | 25.14308 | 25.83758 | 27.29898 | 30.53678 | 34.24797 | 38.50521 | 43.39229 | 55.45676 | 62.87334 | 71.40275 | 92.50258 |
| 23 | 24.31040 | 25.00096 | 25.71630 | 26.45737 | 27.22514 | 28.84496 | 32.45288 | 36.61789 | 41.43048 | 46.99583 | 60.89330 | 69.53194 | 79.54302 | 104.60289 |
| 24 | 25.43196 | 26.18847 | 26.97346 | 27.78808 | 28.63352 | 30.42186 | 34.42647 | 39.08260 | 44.50200 | 50.81558 | 66.76476 | 76.78981 | 88.49733 | 118.15524 |
| 25 | 26.55912 | 27.38488 | 28.24320 | 29.13544 | 30.06302 | 32.03030 | 36.45926 | 41.64591 | 47.72710 | 54.86451 | 73.10594 | 84.70090 | 98.34706 | 133.33387 |

Table 26-1: Future Value Annuity Factors *(continued)*

| Period | 0.50% | 0.75% | 1.00% | 1.25% | 1.50% | 2.00% | 3.00% | 4.00% | 5.00% | 6.00% | 8.00% | 9.00% | 10.00% | 12.00% |
|---|---|---|---|---|---|---|---|---|---|---|---|---|---|---|
| 26 | 27.69191 | 28.59027 | 29.52563 | 30.49963 | 31.51397 | 33.67091 | 38.55304 | 44.31174 | 51.11345 | 59.15638 | 79.95442 | 93.32398 | 109.18177 | 150.33393 |
| 27 | 28.83037 | 29.80470 | 30.82089 | 31.88087 | 32.98668 | 35.34432 | 40.70963 | 47.08421 | 54.66913 | 63.70577 | 87.35077 | 102.72313 | 121.09994 | 169.37401 |
| 28 | 29.97452 | 31.02823 | 32.12910 | 33.27938 | 34.48148 | 37.05121 | 42.93092 | 49.96758 | 58.40258 | 68.52811 | 95.33883 | 112.96822 | 134.20994 | 190.69889 |
| 29 | 31.12439 | 32.26094 | 33.45039 | 34.69538 | 35.99870 | 38.79223 | 45.21885 | 52.96629 | 62.32271 | 73.63980 | 103.96594 | 124.13536 | 148.63093 | 214.58275 |
| 30 | 32.28002 | 33.50290 | 34.78489 | 36.12907 | 37.53868 | 40.56808 | 47.57542 | 56.08494 | 66.43885 | 79.05819 | 113.28321 | 136.30754 | 164.49402 | 241.33268 |
| 31 | 33.44142 | 34.75417 | 36.13274 | 37.58068 | 39.10176 | 42.37944 | 50.00268 | 59.32834 | 70.76079 | 84.80168 | 123.34587 | 149.57522 | 181.94342 | 271.29261 |
| 32 | 34.60862 | 36.01483 | 37.49407 | 39.05044 | 40.68829 | 44.22703 | 52.50276 | 62.70147 | 75.29883 | 90.88978 | 134.21354 | 164.03699 | 201.13777 | 304.84772 |
| 33 | 35.78167 | 37.28494 | 38.86901 | 40.53857 | 42.29861 | 46.11157 | 55.07784 | 66.20953 | 80.06377 | 97.34316 | 145.95062 | 179.80032 | 222.25154 | 342.42945 |
| 34 | 36.96058 | 38.56458 | 40.25770 | 42.04530 | 43.93309 | 48.03380 | 57.73018 | 69.85791 | 85.06696 | 104.18375 | 158.62667 | 196.98234 | 245.47670 | 384.52098 |
| 35 | 38.14538 | 39.85381 | 41.66028 | 43.57087 | 45.59209 | 49.99448 | 60.46208 | 73.65222 | 90.32031 | 111.43478 | 172.31680 | 215.71075 | 271.02437 | 431.66350 |
| 36 | 39.33610 | 41.15272 | 43.07688 | 45.11551 | 47.27597 | 51.99437 | 63.27594 | 77.59831 | 95.83632 | 119.12087 | 187.10215 | 236.12472 | 299.12681 | 484.46312 |
| 37 | 40.53279 | 42.46136 | 44.50765 | 46.67945 | 48.98511 | 54.03425 | 66.17422 | 81.70225 | 101.62814 | 127.26812 | 203.07032 | 258.37595 | 330.03949 | 543.59869 |
| 38 | 41.73545 | 43.77982 | 45.95272 | 48.26294 | 50.71989 | 56.11494 | 69.15945 | 85.97034 | 107.70955 | 135.90421 | 220.31595 | 282.62978 | 364.04343 | 609.83053 |
| 39 | 42.94413 | 45.10817 | 47.41225 | 49.86623 | 52.48068 | 58.23724 | 72.23423 | 90.40915 | 114.09502 | 145.05846 | 238.94122 | 309.06646 | 401.44778 | 684.01020 |
| 40 | 44.15885 | 46.44648 | 48.88637 | 51.48956 | 54.26789 | 60.40198 | 75.40126 | 95.02552 | 120.79977 | 154.76197 | 259.05652 | 337.88245 | 442.59256 | 767.09142 |
| 41 | 45.37964 | 47.79483 | 50.37524 | 53.13318 | 56.08191 | 62.61002 | 78.66330 | 99.82654 | 127.83976 | 165.04768 | 280.78104 | 369.29187 | 487.85181 | 860.14239 |
| 42 | 46.60654 | 49.15329 | 51.87899 | 54.79734 | 57.92314 | 64.86222 | 82.02320 | 104.81960 | 135.23175 | 175.95054 | 304.24352 | 403.52813 | 537.63699 | 964.35948 |
| 43 | 47.83957 | 50.52194 | 53.39778 | 56.48231 | 59.79199 | 67.15947 | 85.48389 | 110.01238 | 142.99334 | 187.50758 | 329.58301 | 440.84566 | 592.40069 | 1081.08262 |
| 44 | 49.07877 | 51.90086 | 54.93176 | 58.18834 | 61.68887 | 69.50266 | 89.04841 | 115.41288 | 151.14301 | 199.75803 | 356.94965 | 481.52177 | 652.64076 | 1211.81253 |
| 45 | 50.32416 | 53.29011 | 56.48107 | 59.91569 | 63.61420 | 71.89271 | 92.71986 | 121.02939 | 159.70016 | 212.74351 | 386.50562 | 525.85873 | 718.90484 | 1358.23003 |
| 46 | 51.57578 | 54.68979 | 58.04589 | 61.66464 | 65.56841 | 74.33056 | 96.50146 | 126.87057 | 168.68516 | 226.50812 | 418.42607 | 574.18602 | 791.79532 | 1522.21764 |
| 47 | 52.83366 | 56.09996 | 59.62634 | 63.43545 | 67.55194 | 76.81718 | 100.39650 | 132.94539 | 178.11942 | 241.09861 | 452.90015 | 626.86276 | 871.97485 | 1705.88375 |
| 48 | 54.09783 | 57.52071 | 61.22261 | 65.22839 | 69.56522 | 79.35352 | 104.40840 | 139.26321 | 188.02539 | 256.56453 | 490.13216 | 684.28041 | 960.17234 | 1911.58980 |
| 49 | 55.36832 | 58.95212 | 62.83483 | 67.04374 | 71.60870 | 81.94059 | 108.54065 | 145.83373 | 198.42666 | 272.95840 | 530.34274 | 746.86565 | 1057.18957 | 2141.98058 |
| 50 | 56.64516 | 60.39426 | 64.46318 | 68.88179 | 73.68283 | 84.57940 | 112.79687 | 152.66708 | 209.34800 | 290.33590 | 573.77016 | 815.08356 | 1163.90853 | 2400.01825 |

Table 26-2: Present Value Annuity Factors

| Period | 0.50% | 0.75% | 1.00% | 1.25% | 1.50% | 2.00% | 3.00% | 4.00% | 5.00% | 6.00% | 8.00% | 9.00% | 10.00% | 12.00% |
|---|---|---|---|---|---|---|---|---|---|---|---|---|---|---|
| 1 | 0.99502 | 0.99256 | 0.99010 | 0.98765 | 0.98522 | 0.98039 | 0.97087 | 0.96154 | 0.95238 | 0.94340 | 0.92593 | 0.91743 | 0.90909 | 0.89286 |
| 2 | 1.98510 | 1.97772 | 1.97040 | 1.96312 | 1.95588 | 1.94156 | 1.91347 | 1.88609 | 1.85941 | 1.83339 | 1.78326 | 1.75911 | 1.73554 | 1.69005 |
| 3 | 2.97025 | 2.95556 | 2.94099 | 2.92653 | 2.91220 | 2.88388 | 2.82861 | 2.77509 | 2.72325 | 2.67301 | 2.57710 | 2.53129 | 2.48685 | 2.40183 |
| 4 | 3.95050 | 3.92611 | 3.90197 | 3.87806 | 3.85438 | 3.80773 | 3.71710 | 3.62990 | 3.54595 | 3.46511 | 3.31213 | 3.23972 | 3.16987 | 3.03735 |
| 5 | 4.92587 | 4.88944 | 4.85343 | 4.81784 | 4.78264 | 4.71346 | 4.57971 | 4.45182 | 4.32948 | 4.21236 | 3.99271 | 3.88965 | 3.79079 | 3.60478 |
| 6 | 5.89638 | 5.84560 | 5.79548 | 5.74601 | 5.69719 | 5.60143 | 5.41719 | 5.24214 | 5.07569 | 4.91732 | 4.62288 | 4.48592 | 4.35526 | 4.11141 |
| 7 | 6.86207 | 6.79464 | 6.72819 | 6.66273 | 6.59821 | 6.47199 | 6.23028 | 6.00205 | 5.78637 | 5.58238 | 5.20637 | 5.03295 | 4.86842 | 4.56376 |
| 8 | 7.82296 | 7.73661 | 7.65168 | 7.56812 | 7.48593 | 7.32548 | 7.01969 | 6.73274 | 6.46321 | 6.20979 | 5.74664 | 5.53482 | 5.33493 | 4.96764 |
| 9 | 8.77906 | 8.67158 | 8.56602 | 8.46234 | 8.36052 | 8.16224 | 7.78611 | 7.43533 | 7.10782 | 6.80169 | 6.24689 | 5.99525 | 5.75902 | 5.32825 |
| 10 | 9.73041 | 9.59958 | 9.47130 | 9.34553 | 9.22218 | 8.98259 | 8.53020 | 8.11090 | 7.72173 | 7.36009 | 6.71008 | 6.41766 | 6.14457 | 5.65022 |
| 11 | 10.67703 | 10.52067 | 10.36763 | 10.21780 | 10.07112 | 9.78685 | 9.25262 | 8.76048 | 8.30641 | 7.88687 | 7.13896 | 6.80519 | 6.49506 | 5.93770 |
| 12 | 11.61893 | 11.43491 | 11.25508 | 11.07931 | 10.90751 | 10.57534 | 9.95400 | 9.38507 | 8.86325 | 8.38384 | 7.53608 | 7.16073 | 6.81369 | 6.19437 |
| 13 | 12.55615 | 12.34235 | 12.13374 | 11.93018 | 11.73153 | 11.34837 | 10.63496 | 9.98565 | 9.39357 | 8.85268 | 7.90378 | 7.48690 | 7.10336 | 6.42355 |
| 14 | 13.48871 | 13.24302 | 13.00370 | 12.77055 | 12.54338 | 12.10625 | 11.29607 | 10.56312 | 9.89864 | 9.29498 | 8.24424 | 7.78615 | 7.36669 | 6.62817 |
| 15 | 14.41662 | 14.13699 | 13.86505 | 13.60055 | 13.34323 | 12.84926 | 11.93794 | 11.11839 | 10.37966 | 9.71225 | 8.55948 | 8.06069 | 7.60608 | 6.81086 |
| 16 | 15.33993 | 15.02431 | 14.71787 | 14.42029 | 14.13126 | 13.57771 | 12.56110 | 11.65230 | 10.83777 | 10.10590 | 8.85137 | 8.31256 | 7.82371 | 6.97399 |
| 17 | 16.25863 | 15.90502 | 15.56225 | 15.22992 | 14.90765 | 14.29187 | 13.16612 | 12.16567 | 11.27407 | 10.47726 | 9.12164 | 8.54363 | 8.02155 | 7.11963 |
| 18 | 17.17277 | 16.77918 | 16.39827 | 16.02955 | 15.67256 | 14.99203 | 13.75351 | 12.65930 | 11.68959 | 10.82760 | 9.37189 | 8.75563 | 8.20141 | 7.24967 |
| 19 | 18.08236 | 17.64683 | 17.22601 | 16.81931 | 16.42617 | 15.67846 | 14.32380 | 13.13394 | 12.08532 | 11.15812 | 9.60360 | 8.95011 | 8.36492 | 7.36578 |
| 20 | 18.98742 | 18.50802 | 18.04555 | 17.59932 | 17.16864 | 16.35143 | 14.87747 | 13.59033 | 12.46221 | 11.46992 | 9.81815 | 9.12855 | 8.51356 | 7.46944 |
| 21 | 19.88798 | 19.36280 | 18.85698 | 18.36969 | 17.90014 | 17.01121 | 15.41502 | 14.02916 | 12.82115 | 11.76408 | 10.01680 | 9.29224 | 8.64869 | 7.56200 |
| 22 | 20.78406 | 20.21121 | 19.66038 | 19.13056 | 18.62082 | 17.65805 | 15.93692 | 14.45112 | 13.16300 | 12.04158 | 10.20074 | 9.44243 | 8.77154 | 7.64465 |
| 23 | 21.67568 | 21.05331 | 20.45582 | 19.88204 | 19.33086 | 18.29220 | 16.44361 | 14.85684 | 13.48857 | 12.30338 | 10.37106 | 9.58021 | 8.88322 | 7.71843 |
| 24 | 22.56287 | 21.88915 | 21.24339 | 20.62423 | 20.03041 | 18.91393 | 16.93554 | 15.24696 | 13.79864 | 12.55036 | 10.52876 | 9.70661 | 8.98474 | 7.78432 |
| 25 | 23.44564 | 22.71876 | 22.02316 | 21.35727 | 20.71961 | 19.52346 | 17.41315 | 15.62208 | 14.09394 | 12.78336 | 10.67478 | 9.82258 | 9.07704 | 7.84314 |

Table 26-2: Present Value Annuity Factors *(continued)*

| Period | 0.50% | 0.75% | 1.00% | 1.25% | 1.50% | 2.00% | 3.00% | 4.00% | 5.00% | 6.00% | 8.00% | 9.00% | 10.00% | 12.00% |
|--------|-------|-------|-------|-------|-------|-------|-------|-------|-------|-------|-------|-------|--------|--------|
| 26 | 24.32402 | 23.54219 | 22.79520 | 22.08125 | 21.39863 | 20.12104 | 17.87684 | 15.98277 | 14.37519 | 13.00317 | 10.80998 | 9.92897 | 9.16095 | 7.89566 |
| 27 | 25.19803 | 24.35949 | 23.55961 | 22.79630 | 22.06762 | 20.70690 | 18.32703 | 16.32959 | 14.64303 | 13.21053 | 10.93516 | 10.02658 | 9.23722 | 7.94255 |
| 28 | 26.06769 | 25.17071 | 24.31644 | 23.50252 | 22.72672 | 21.28127 | 18.76411 | 16.66306 | 14.89813 | 13.40616 | 11.05108 | 10.11613 | 9.30657 | 7.98442 |
| 29 | 26.93302 | 25.97589 | 25.06579 | 24.20002 | 23.37608 | 21.84438 | 19.18845 | 16.98371 | 15.14107 | 13.59072 | 11.15841 | 10.19828 | 9.36961 | 8.02181 |
| 30 | 27.79405 | 26.77508 | 25.80771 | 24.88891 | 24.01584 | 22.39646 | 19.60044 | 17.29203 | 15.37245 | 13.76483 | 11.25778 | 10.27365 | 9.42691 | 8.05518 |
| 31 | 28.65080 | 27.56832 | 26.54229 | 25.56929 | 24.64615 | 22.93770 | 20.00043 | 17.58849 | 15.59281 | 13.92909 | 11.34980 | 10.34280 | 9.47901 | 8.08499 |
| 32 | 29.50328 | 28.35565 | 27.26959 | 26.24127 | 25.26714 | 23.46833 | 20.38877 | 17.87355 | 15.80268 | 14.08404 | 11.43500 | 10.40624 | 9.52638 | 8.11159 |
| 33 | 30.35153 | 29.13712 | 27.98969 | 26.90496 | 25.87895 | 23.98856 | 20.76579 | 18.14765 | 16.00255 | 14.23023 | 11.51389 | 10.46444 | 9.56943 | 8.13535 |
| 34 | 31.19555 | 29.91278 | 28.70267 | 27.56046 | 26.48173 | 24.49859 | 21.13184 | 18.41120 | 16.19290 | 14.36814 | 11.58693 | 10.51784 | 9.60857 | 8.15656 |
| 35 | 32.03537 | 30.68266 | 29.40858 | 28.20786 | 27.07559 | 24.99862 | 21.48722 | 18.66461 | 16.37419 | 14.49825 | 11.65457 | 10.56682 | 9.64416 | 8.17550 |
| 36 | 32.87102 | 31.44681 | 30.10751 | 28.84727 | 27.66068 | 25.48884 | 21.83225 | 18.90828 | 16.54685 | 14.62099 | 11.71719 | 10.61176 | 9.67651 | 8.19241 |
| 37 | 33.70250 | 32.20527 | 30.79951 | 29.47878 | 28.23713 | 25.96945 | 22.16724 | 19.14258 | 16.71129 | 14.73678 | 11.77518 | 10.65299 | 9.70592 | 8.20751 |
| 38 | 34.52985 | 32.95808 | 31.48466 | 30.10250 | 28.80505 | 26.44064 | 22.49246 | 19.36786 | 16.86789 | 14.84602 | 11.82887 | 10.69082 | 9.73265 | 8.22099 |
| 39 | 35.35309 | 33.70529 | 32.16303 | 30.71852 | 29.36458 | 26.90259 | 22.80822 | 19.58448 | 17.01704 | 14.94907 | 11.87858 | 10.72552 | 9.75696 | 8.23303 |
| 40 | 36.17223 | 34.44694 | 32.83469 | 31.32693 | 29.91585 | 27.35548 | 23.11477 | 19.79277 | 17.15909 | 15.04630 | 11.92461 | 10.75736 | 9.77905 | 8.24378 |
| 41 | 36.98729 | 35.18307 | 33.49969 | 31.92784 | 30.45896 | 27.79949 | 23.41240 | 19.99305 | 17.29437 | 15.13802 | 11.96723 | 10.78657 | 9.79914 | 8.25337 |
| 42 | 37.79830 | 35.91371 | 34.15811 | 32.52132 | 30.99405 | 28.23479 | 23.70136 | 20.18563 | 17.42321 | 15.22454 | 12.00670 | 10.81337 | 9.81740 | 8.26194 |
| 43 | 38.60527 | 36.63892 | 34.81001 | 33.10748 | 31.52123 | 28.66156 | 23.98190 | 20.37079 | 17.54591 | 15.30617 | 12.04324 | 10.83795 | 9.83400 | 8.26959 |
| 44 | 39.40823 | 37.35873 | 35.45545 | 33.68640 | 32.04062 | 29.07996 | 24.25427 | 20.54884 | 17.66277 | 15.38318 | 12.07707 | 10.86051 | 9.84909 | 8.27642 |
| 45 | 40.20720 | 38.07318 | 36.09451 | 34.25817 | 32.55234 | 29.49016 | 24.51871 | 20.72004 | 17.77407 | 15.45583 | 12.10840 | 10.88120 | 9.86281 | 8.28252 |
| 46 | 41.00219 | 38.78231 | 36.72724 | 34.82288 | 33.05649 | 29.89231 | 24.77545 | 20.88465 | 17.88007 | 15.52437 | 12.13741 | 10.90018 | 9.87528 | 8.28796 |
| 47 | 41.79322 | 39.48617 | 37.35370 | 35.38062 | 33.55319 | 30.28658 | 25.02471 | 21.04294 | 17.98102 | 15.58903 | 12.16427 | 10.91760 | 9.88662 | 8.29282 |
| 48 | 42.58032 | 40.18478 | 37.97396 | 35.93148 | 34.04255 | 30.67312 | 25.26671 | 21.19513 | 18.07716 | 15.65003 | 12.18914 | 10.93358 | 9.89693 | 8.29716 |
| 49 | 43.36350 | 40.87820 | 38.58808 | 36.47554 | 34.52468 | 31.05208 | 25.50166 | 21.34147 | 18.16872 | 15.70757 | 12.21216 | 10.94823 | 9.90630 | 8.30104 |
| 50 | 44.14279 | 41.56645 | 39.19612 | 37.01288 | 34.99969 | 31.42361 | 25.72976 | 21.48218 | 18.25593 | 15.76186 | 12.23348 | 10.96168 | 9.91481 | 8.30450 |

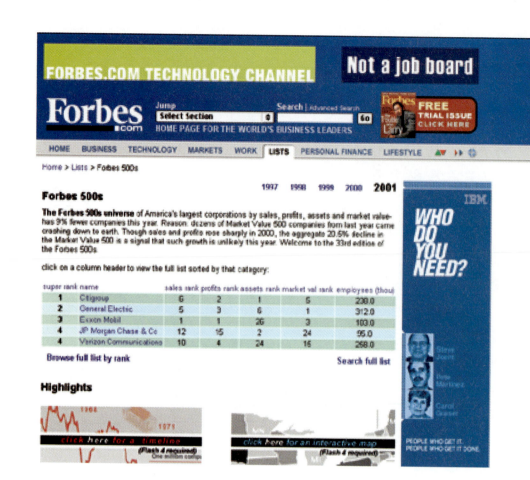

Computing and using accurate statistical information are vital for business success. Statistical data are gathered by businesses, government, and other organizations. In addition, many business periodicals collect and provide data on short- and long-term trends. One business magazine that presents updated statistics in each issue is Forbes. To learn about Forbes and the diverse array of business statistics it publishes, visit www.forbes.com.

Business Statistics

27

LEARNING OBJECTIVES

By studying this chapter and completing all assignments you will be able to:

 1 Compute the mean.

 5 Construct histograms.

 2 Determine the median.

 6 Construct bar graphs.

 3 Determine the mode.

 7 Construct line graphs.

 4 Construct frequency tables.

 8 Construct pie charts.

According to the U.S. Office of Management and Budget, the number of hours Americans spend filling out federal forms per year is 784,862,000. Another statistic estimates that if you were guilty of every crime shown on TV in 1 week, you would go to jail for 1,600 years.

1. How are the mean, median and mode similar? Different?
2. When would the median be more appropriate than the mean? Explain or give an example.
3. Explain why summarizing data into a frequency distribution may be useful.
4. Explain why creating a histogram from a frequency distribution may be useful.
5. What is the difference between a comparative bar chart and a component bar graph?

McDonald's has sold billions of hamburgers. Housing prices are higher in New York City than in St. Louis. The United States has a trade deficit, which means that the country has been importing more goods than it has been exporting. More families buy television sets during December than during any other single month of the year. These examples are based on collections of information about businesses. The information is called **business statistics**. The word **statistics** also refers to a field of study that includes the collection, organization, analysis, and presentation of data. Businesses use statistics for two primary purposes: (1) to summarize and report the performance of the business and (2) to analyze their options in making business decisions.

Individuals and groups who want information about business performance include upper management, the board of directors, investors, and government agencies. Once statistics have been reported, individuals and groups use the statistics to make business decisions. For example, depending on the amount of profits, the board of directors decides how much dividend to pay the shareholders. Likewise, after hearing about current profits and projected profits, investors decide whether to purchase or sell shares of the company's stock. After studying sales figures for its products and those of competitors, management makes decisions about which markets to enter, what products to emphasize, and how to advertise.

If a McDonald's analyst wants to report data on sales of hamburgers, she could list the number of hamburgers sold at every restaurant. But McDonald's has so many restaurants that there would be too many numbers to be meaningful. To make the data meaningful, the analyst can make some summary calculations and/or organize the data in tables. To make presentation of the data more meaningful and easier to interpret, she may draw charts, diagrams, or graphs.

Statistical Averages: Computing the Mean

LEARNING OBJECTIVES 1

Compute the mean.

The objective in reporting statistics is to summarize the data in a simple, yet meaningful manner. One way to simplify data is to compute an **average**. An average is a single number that is supposed to be "typical" or "representative" of the group. The most common way to find an average is to add all the data values and divide by the number of values. In statistics, this particular average is called the *mean*. When the mean isn't typical or representative of an entire group of data, another average might be more representative. We also discuss two other averages: the median and the mode.

The **mean** of a group of values is computed by dividing the sum of the group of values by the number of values in the group.

EXAMPLE A

Find the mean salary of five employees whose actual salaries are $42,500, $45,400, $48,000, $53,600, and $55,500.

Sum = $42,500 + $45,400 + $48,000 + $53,600 + $55,500 = $245,000
Mean = $245,000 ÷ 5 = $49,000

CONCEPT CHECK 27.1

Find the mean for the following set of numbers: 14, 11, 12, 13, 10, 16, 15, 12, 13, 11, 12, 15, 17, 13, 14, 15, 12, 18, 12

There are 19 numbers. The mean equals their sum divided by 19.

Sum = 255
Mean = 255 ÷ 19 = 13.421, or 13.4 rounded to one decimal place

Determining the Median

The **median** of a group of numbers is determined by arranging the numbers in numerical order and finding the middle number. The median is useful when one value in the group is much larger or much smaller than the rest of the numbers.

EXAMPLE B

Find the median salary of five employees whose salaries are $42,500, $45,400, $48,000, $53,600, and $145,500.

The salaries are already in numerical order; the median is $48,000 because it is the middle number of the five numbers arranged in order.

In example B, the mean is $335,000 ÷ 5 = $67,000, but $67,000 isn't representative of the salaries of the five employees. The mean is large because one employee (perhaps the owner) has a very large salary compared to the rest of the group. The median salary, $48,000, is more typical of the group.

If the number of values is even, the median will be halfway between the two middle values. (It is the mean of the middle two values.)

EXAMPLE C

Find the median salary of six employees whose salaries are $42,500, $45,400, $48,000, $53,600, $55,500, and $145,500.

The median is halfway between the middle two numbers, $48,000 and $53,600. It is ($48,000 + $53,600) ÷ 2, or $101,600 ÷ 2 = $50,800.

2
LEARNING OBJECTIVES

Determine the median.

http://

www.imacc.org/ standards/examples. html

This Web site presents some exercises in statistics.

CONCEPT CHECK 27.2

Find the median for the following set of numbers: 14, 11, 12, 13, 10, 16, 15, 12, 13, 11, 12, 15, 17, 13, 14, 15, 12, 18, 12

The median is the middle number, after all the numbers have been arranged by order of size: 10, 11, 11, 12, 12, 12, 12, 12, 13, 13, 13, 14, 14, 15, 15, 15, 16, 17, 18

The median is the tenth number, or 13.

Determining the Mode

Determine the mode.

The **mode** of a group of numbers is the number that occurs most often. None of examples A, B, and C, has a mode because each number occurs only once. The mode is useful when the word *average* implies "most typical" or "happening most often." Retail businesses keep track of the items that sell most frequently so that they can avoid shortages of those items.

http://cne.gmu.edu/ modules/dau/stat/

This Web site has a statistics tutorial for more practice. You'll want to read the introductory material and then select the index to reach pages on the mean, median, and mode.

EXAMPLE D

Find the mode shoe size of 12 pairs of Reebok running shoes, sizes $6, 6, 7\frac{1}{2}, 7\frac{1}{2}, 8, 9\frac{1}{2}, 10,$ $10, 10, 10, 10,$ and $10\frac{1}{2}$.

The mode is size 10, because 10 occurs most frequently.

Note that in example D neither the mean nor the median makes any sense. The mean is $105 \div 12 = 8.75$, or $8\frac{3}{4}$. The median is halfway between sizes $9\frac{1}{2}$ and 10, which would be 9.75, or $9\frac{3}{4}$. The store owner could not buy any shoes in either size $8\frac{3}{4}$ or size $9\frac{3}{4}$ because those sizes don't exist. However, the store owner does want to stock enough shoes in size 10.

 CONCEPT CHECK 27.3

Find the mode for the following set of numbers: 14, 11, 12, 13, 10, 16, 15, 12, 13, 11, 12, 15, 17, 13, 14, 15, 12, 18, 12

The mode is the number that occurs most often. It is easiest to find if you arrange the numbers by size first:

10, 11, 11, 12, 12, 12, 12, 12, 13, 13, 13, 14, 14, 15, 15, 15, 16, 17, 18

There are five 12s, so the mode is 12.

Constructing Frequency Tables

Construct frequency tables.

The data in examples A–D are sometimes called **ungrouped data** because the numbers are listed individually. Business applications, such as sales results for all McDonald's restaurants, often involve hundreds or thousands of numbers. Interpreting data that are literally pages of raw numbers is impossible. To make sense of such data, we organize the individual values into groups called **classes of data** or *data classes*. The number of values in each class, called the **frequency** of the class, is summarized in a table called a **frequency table**.

STEPS to Develop a Frequency Table

1. Determine the classes of data, and list the classes in one column.
2. Tally the data by making one mark for each data item in the column next to the appropriate class.
3. Count the tally marks for each class and write the number in the column next to the tally marks.

EXAMPLE E

Listed are the salaries of 25 part-time and full-time employees of a department store. Make a frequency table with five classes: $40,000 up to but not including $45,000, $45,000 up to $50,000, and so on.

| | | | | |
|---|---|---|---|---|
| $42,500 | $41,300 | $43,500 | $62,400 | $47,500 |
| 45,400 | 54,600 | 41,000 | 44,400 | 59,100 |
| 48,000 | 52,000 | 57,500 | 49,500 | 44,000 |
| 53,600 | 46,200 | 53,500 | 51,800 | 56,400 |
| 55,500 | 46,000 | 45,200 | 46,000 | 60,800 |

The frequency table for these salaries appears in Figure 27-1.

Figure 27-1: Frequency Table

| Class | Tally | Frequency (F) |
|---|---|---|
| $40,000 up to $45,000 | ٮᴴᴸI | 6 |
| $45,000 up to $50,000 | ᴴᴸ III | 8 |
| $50,000 up to $55,000 | ᴴᴸ | 5 |
| $55,000 up to $60,000 | IIII | 4 |
| $60,000 up to $65,000 | II | 2 |
| Total | | 25 |

Computing the Mean of Large Data Sets

When a data set contains many numbers, as in example E, a computer spreadsheet can be used to compute the mean. If you use a calculator, be sure to check your work. One way to do so is to add all the numbers twice; one way to add them twice, but in different order, is the following.

STEPS to Compute the Mean for a Large Data Set

1. Add all the numbers in each column.
2. Add all the numbers in each row.
3. Compute the grand total by adding all the column totals.
4. Check the grand total by adding all the row totals.
5. Divide the grand total by the number of values to get the mean.

EXAMPLE F

Compute the mean of the 25 salaries in example E.

| | | | | | |
|---|---|---|---|---|---|
| 42,500 | 41,300 | 43,500 | 62,400 | 47,500 | 237,200 |
| 45,400 | 54,600 | 41,000 | 44,400 | 59,100 | 244,500 |
| 48,000 | 52,000 | 57,500 | 49,500 | 44,000 | 251,000 |
| 53,600 | 46,200 | 53,500 | 51,800 | 56,400 | 261,500 |
| 55,500 | 46,000 | 45,200 | 46,000 | 60,800 | 253,500 |
| 245,000 | 240,100 | 240,700 | 254,100 | 267,800 | 1,247,700 |

The sum of the row totals and the sum of the column totals are both $1,247,700.

Mean = $1,247,700 ÷ 25 = $49,908

CONCEPT CHECK 27.4

Make a frequency table for the following set of data. Use the classes 1,500 up to 2,000, 2,000 up to 2,500, and so on.

| | | | | |
|---|---|---|---|---|
| 2,550 | 2,575 | 3,410 | 2,650 | 3,140 |
| 3,480 | 3,400 | 2,860 | 3,810 | 3,680 |
| 2,475 | 3,280 | 2,940 | 2,480 | 3,325 |
| 1,975 | 4,270 | 3,520 | 2,440 | 2,325 |
| 3,710 | 3,300 | 2,290 | 4,140 | 3,990 |
| 2,570 | 2,150 | 2,840 | 4,325 | 2,720 |

COMPLETE ASSIGNMENT 27.1.

| Class | Tally | Frequency |
|---|---|---|
| 1,500 up to 2,000 | I | 1 |
| 2,000 up to 2,500 | HHL I | 6 |
| 2,500 up to 3,000 | HHL III | 8 |
| 3,000 up to 3,500 | HHL II | 7 |
| 3,500 up to 4,000 | HHL | 5 |
| 4,000 up to 4,500 | III | 3 |
| Total | | 30 |

Charts and Graphs: Constructing Histograms

Construct histograms.

In business, statistical information is first summarized clearly in tables. For presentation, the results are then often displayed in charts or graphs. Popular graphs include the histogram, the bar graph, and the pie chart (circle graph). Histograms, bar graphs, and line graphs all have a rectangular shape. Labels are placed at the left (the vertical axis) and bottom (the horizontal axis).

A **histogram** is a diagram that presents the **grouped data** from a frequency table. The classes are positioned adjacent to each other along the horizontal axis, and the frequencies are written along the vertical axis. Figure 27-2 shows the histogram for the frequency table in Figure 27-1. The numbers on the horizontal axis increase from left to right. The numbers on the vertical axis increase from bottom to top.

Figure 27-2: Histogram

http://exploringdata.
cqu.edu.au/histsum2.
htm.

For a quick quiz on histograms and more on statistical charts, go to this Web site.

CONCEPT CHECK 27.5

Construct a histogram from the frequency table below.

| Class | Frequency |
|---|---|
| 1,500 up to 2,000 | 1 |
| 2,000 up to 2,500 | 6 |
| 2,500 up to 3,000 | 8 |
| 3,000 up to 3,500 | 7 |
| 3,500 up to 4,000 | 5 |
| 4,000 up to 4,500 | 3 |
| Total | 30 |

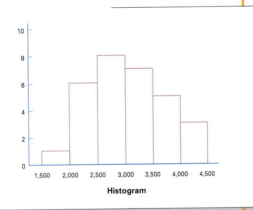

Constructing Bar Graphs

A **bar graph**, or bar chart, resembles the histogram except that there may not be a numeric scale along the horizontal axis and the bars normally do not touch each other. Figure 27-3 shows the sales revenue, cost of goods sold, operating expenses, and net profit for each of the first six months of the current year. Figure 27-3 also shows the net profit amounts for the first six months of last year. Figure 27-3 data are used in all of the remaining charts and graphs in this chapter. Figure 27-4 is a bar graph that displays only the revenue for the current six months.

6
LEARNING OBJECTIVES

Construct bar graphs.

Figure 27-3: Revenues, Expenses, and Net Profit

| Month | Sales Revenue | Cost of Goods Sold | Operating Expenses | Net Profit | Net Profit Last Year |
|---|---|---|---|---|---|
| Jan. | $315,000 | $155,000 | $125,000 | $35,000 | $30,000 |
| Feb. | 300,000 | 145,000 | 145,000 | 10,000 | 35,000 |
| March | 360,000 | 180,000 | 135,000 | 45,000 | 30,000 |
| April | 350,000 | 185,000 | 135,000 | 30,000 | 40,000 |
| May | 375,000 | 200,000 | 130,000 | 45,000 | 35,000 |
| June | 400,000 | 210,000 | 165,000 | 25,000 | 20,000 |

Comparative Bar Graph

Two bar graphs can be drawn together on one grid to make a **comparative bar graph**, which compares statistical data. In Figure 27-5, bars for last year's and this year's monthly net profit for the first 6 months are drawn next to each other. Each month has one pair of bars. Each bar in a pair is shaded differently to help the person reading the graph tell the bars apart.

VIDEO

How to Create Bar Graphs/ Line Graphs/ Pie Charts

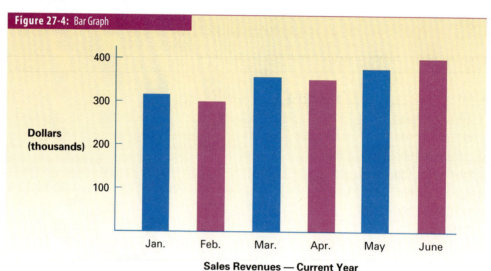

Figure 27-4: Bar Graph

Sales Revenues — Current Year

Figure 27-5: Comparative Bar Graph

Net Profits — Current Year and Last Year

Component Bar Graph

A bar graph constructed to show how certain data are composed of various parts is a **component bar graph**. Figure 27-6 shows how the current sales revenue is composed of cost of goods sold, operating expenses, and net profit. As in the comparative bar graph, the component parts are colored or shaded differently to permit easier reading.

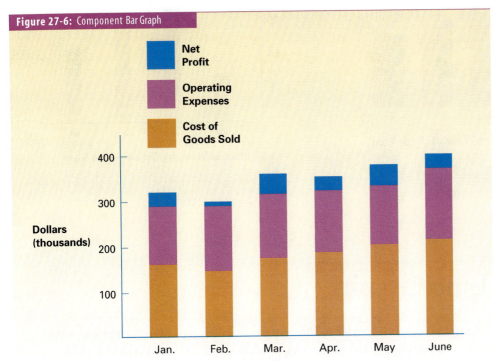

Figure 27-6: Component Bar Graph

Sales Revenues—Current Year

CONCEPT CHECK 27.6

A real estate firm has three offices, all of which sell some homes. The Shopping Mall Office sells homes almost exclusively; last year it sold 150 homes. The Downtown Office handles mostly commercial property, but it sold 60 homes last year. The Mountain Office primarily manages various resort properties, but it did sell 30 homes. Following are the numbers of homes sold in each quarter of last year. The first quarter is January through March; the second quarter is April through June; the third quarter is July through September; and the fourth quarter is October through December.

Home Sales Last Year

| Quarter | 1st | 2nd | 3rd | 4th |
|---|---|---|---|---|
| Shopping Mall Office | 20 | 60 | 40 | 30 |
| Downtown Office | 5 | 20 | 25 | 10 |
| Mountain Office | 10 | 3 | 5 | 12 |
| Total sales last year | 35 | 83 | 70 | 52 |
| Total sales prior year | 30 | 75 | 65 | 55 |

a. Construct a bar graph showing total home sales for each quarter last year. Make the vertical scale from 0 to 100, and mark the four quarters on the horizontal scale.

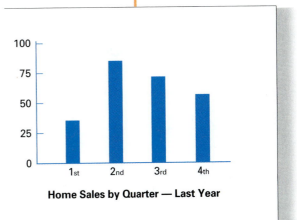

Home Sales by Quarter — Last Year

b. Construct a comparative bar graph showing quarterly home sales for last year and the prior year.

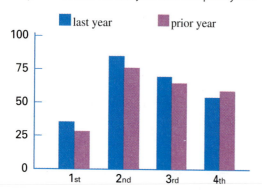

Home Sales Last Year vs. Prior Year

c. Construct a component bar graph showing quarterly home sales for each office last year.

Home Sales by Office—Last Year

Constructing Line Graphs

Construct line graphs.

A **line graph** can also be used to show data over a period of time. As with bar graphs, the time element is plotted on the horizontal axis and the other variable is plotted on the vertical axis. Often, the first time interval is graphed directly on the vertical axis. In Figure 27-7, January is shown on the vertical axis. More than one line graph can be drawn on a single grid. The same information presented in Figure 27-5 is graphed in Figure 27-7. Examine both figures and determine how they differ.

VIDEO

How to Create Bar Graphs/ Line Graphs/ Pie Charts

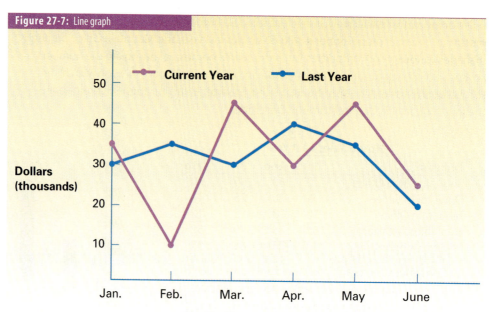

Figure 27-7: Line graph

Net Profits — Current Year and Last Year

CONCEPT CHECK 27.7

The real estate firm's quarterly home sales for last year and the prior year are as follows.

Quarterly Home Sales

| Quarter | 1st | 2nd | 3rd | 4th |
| --- | --- | --- | --- | --- |
| Sales last year | 35 | 83 | 70 | 52 |
| Sales prior year | 30 | 75 | 65 | 55 |

Construct two line graphs on the same grid showing quarterly home sales for last year and for the prior year. Make the vertical scale from 0 to 100, and mark the four quarters on the horizontal scale.

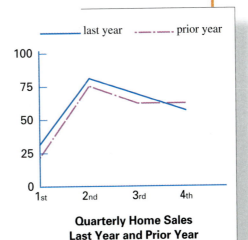

Quarterly Home Sales
Last Year and Prior Year

CyberConnection

Visit the Bureau of Labor Statistics' Web site for statistics on employment, production, and other aspects of the economy. You'll see line graphs like those you've learned about in this chapter, and the one shown here.

Do you understand how to read and interpret them?

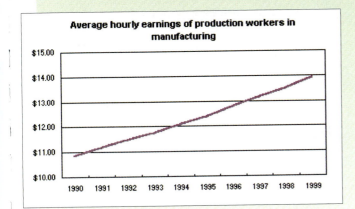

Source: http://www.bls.gov. Copyright © 2001. Courtesy of The Bureau of Labor Statistics.

Constructing Pie Charts

A **pie chart**, sometimes called a circle graph, resembles a component bar graph because it shows how one quantity is composed of different parts. In a pie chart, however, the parts normally are written as percents. Figure 27-8 shows another version of the March data from Figure 27-3. The pie chart shown in Figure 27-9 indicates how sales revenue for March is composed of cost of goods sold, operating expenses, and net profit.

LEARNING OBJECTIVES

Construct pie charts.

Before the graph is drawn, the data are changed into percents, as shown in Figure 27-8. The size of each part of the circle can be reasonably estimated by using the fractional equivalents of the percents. In Figure 27-9, cost of goods sold is 50%, or $\frac{1}{2}$, of the circle. Operating expenses make up 37.5%, or $\frac{3}{8}$, of the circle. The remaining $\frac{1}{8}$ represents net profit.

Figure 27-8: March Sales Revenue

| Item | Amount | Percent |
|------|--------|---------|
| Cost of goods sold | $180,000 | 50.0% |
| Operating expenses | 135,000 | 37.5% |
| Net profit | 45,000 | 12.5% |
| Sales revenue | $360,000 | 100.0% |

$180,000 \div $360,000 = 50.0\%$
$135,000 \div $360,000 = 37.5\%$
$45,000 \div $360,000 = 12.5\%$

VIDEO

How to Create Bar Graphs/ Line Graphs/ Pie Charts

Figure 27-9: Pie Chart

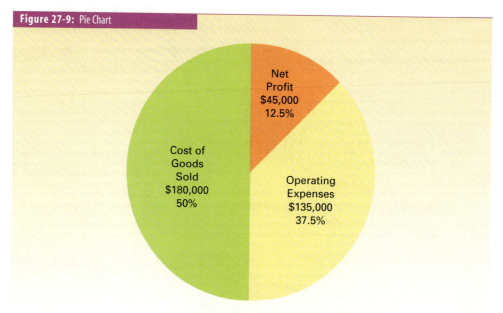

Net Profit
$45,000
12.5%

Cost of Goods Sold
$180,000
50%

Operating Expenses
$135,000
37.5%

March Sales Revenue

CONCEPT CHECK 27.8

The total home sales by three real estate offices for the past year are shown. Calculate the percent of total sales for each office, and make a pie chart showing each office's share of the sales.

| Office | Homes | Percent |
|--------|-------|---------|
| Shopping Mall Office | 150 | $150 \div 240 = 62.5\%$ |
| Downtown Office | 60 | $60 \div 240 = 25\%$ |
| Mountain Office | 30 | $30 \div 240 = 12.5\%$ |
| Total | 240 | $240 \div 240 = 100\%$ |

Downtown
60 homes
25%

Mountain
30 homes
12.5%

Shopping Mall
150 homes
62.5%

Home Sales Last Year

25% is $\frac{1}{4}$ of the circle; 12.5% is half of another quarter, or $\frac{1}{8}$; 62.5% is the remaining eighth plus the remaining half, or $\frac{5}{8}$.

COMPLETE ASSIGNMENT 27.2.

Chapter Terms for Review

average
bar graph
business statistics
classes of data
comparative bar graph
component bar graph
frequency
frequency table
grouped data

histogram
line graph
mean
median
mode
pie chart
statistics
ungrouped data

Visit the Web

http://www.ups.edu/community/tofu/lev1f/conframe.htm—Get more practice in statistics, including exercises and problems using pie charts, data tables, and more at this Web site.

http://www.cbot.com—The Chicago Board of Trade deals with statistical data in trading. Click on "charts" on the board's Web site and select the chart you want to see.

CYBERCONNECTION

The U.S. government's Preliminary Report on Manufacturers' Shipments, Inventories, and Orders in December of 2000 stated that "new orders for manufactured goods in December increased $4.1 billion or 1.1 percent to $382.5 billion, the Department of Commerce, Bureau of the Census reported today. This follows a 1.9 percent November increase." The data for the year were graphed as follows.

Explain what you see on this bar graph. Were new orders increasing or decreasing by the end of the year? What might be the reason for your answer? Do you see a pattern throughout the year?

Compare that chart with the following chart showing data on new orders for durable goods. Consider only the year 2000. Do you see any similarities or differences? If so, what are they, and why might they exist?

Source: http://www.whitehouse.gov/fsbr/production.html.

The Bottom Line

Summary of chapter learning objectives:

| Learning Objective | Summary | Example |
|---|---|---|
| **27.1** | Compute the mean | 1. Determine the mean (rounded to one decimal place) for these seven values: 24, 26, 17, 9, 21, 24, and 15. |
| **27.2** | Determine the median | 2. Determine the median for these seven values: 24, 26, 17, 9, 21, 24, and 15. |
| **27.3** | Determine the mode | 3. Determine the mode for these seven values: 24, 26, 17, 9, 21, 24, and 15. |
| **27.4** | Construct a frequency table | 4. Use the following set of data to construct a frequency table. Use the classes 0 up to 100, 100 up to 200, and so on. |
| **27.5** | Construct histograms | 5. Construct a histogram from the following frequency table. |

For objective 27.4:

| | | | | |
|---|---|---|---|---|
| 50 | 427 | 134 | 254 | 75 |
| 8 | 134 | 228 | 317 | 284 |
| 347 | 289 | 129 | 180 | 125 |
| 197 | 27 | 430 | 246 | 308 |
| 310 | 330 | 297 | 141 | 182 |

For objective 27.5:

| Class | Frequency |
|---|---|
| 0 up to 100 | 4 |
| 100 up to 200 | 8 |
| 200 up to 300 | 6 |
| 300 up to 400 | 5 |
| 400 up to 500 | 2 |
| Total | 25 |

Summary of chapter learning objectives:

| Learning Objective | Summary | Example |
|---|---|---|
| **27.6** | Construct bar graphs | 6. The monthly car sales for one car salesperson for May, June, July, and August of this year are arranged as follows by type of vehicle. The total sales for these 4 months of last year are also given. |

| Vehicle Type | May | June | July | Aug. |
|---|---|---|---|---|
| Two-door coupe | 6 | 4 | 8 | 6 |
| Four-door sedan | 12 | 8 | 6 | 7 |
| Sport utility vehicle | 3 | 10 | 5 | 11 |
| Pick-up truck | 3 | 7 | 6 | 4 |
| Totals this year | 24 | 29 | 25 | 28 |
| Totals last year | 17 | 21 | 20 | 24 |

Construct a bar graph showing the four monthly totals for this year. Make a vertical scale from 0 to 30, and mark the horizontal scale May, June, July, and August.

7. Construct a comparative bar graph showing the four monthly totals for this year and last year.

8. Construct a component bar graph showing car sales by model for May through August of this year.

| **27.7** | Construct line graphs | 9. The monthly car sales for May, June, July, and August of this year and last year for one salesperson are as follows. |

| Period | May | June | July | Aug. |
|---|---|---|---|---|
| This year | 24 | 29 | 25 | 28 |
| Last year | 17 | 21 | 20 | 24 |

On one grid, construct line graphs showing sales for these 4 months during this year and last year.

Summary of chapter learning objectives:

| Learning Objective | Summary | Example |
|---|---|---|
| **27.8** | Construct a pie chart | 10. In May of this year, one car salesperson sold the following numbers of cars, arranged by type of vehicle. |

| Vehicle Type | Sales | Percent |
|---|---|---|
| Two-door coupe | 6 | |
| Four-door sedan | 12 | |
| Sport Utility Vehicle | 3 | |
| Pick-up Truck | 3 | |
| Totals | 24 | |

Calculate the percent for each model, and make a pie chart showing each model.

10. 25%, 50%, 12.5%, 12.5%

| 4. Class | Tally | Frequency |
|---|---|---|
| 0 up to 100 | IIII | 4 |
| 100 up to 200 | THL III | 8 |
| 200 up to 300 | THL I | 6 |
| 300 up to 400 | THL | 5 |
| 400 up to 500 | II | 2 |
| Total | | 25 |

Answers: 1. 19.4 2. 21 3. 24

Self-Check Review Problems for Chapter 27

1 For the data 65, 53, 77, 88, 82, 66, 52, 59, 62, 47, 68, 78, 59, 45, and 59, find (a) the mean, (b) the median, and (c) the mode.

2 Use the data given to complete the following frequency distribution:

86, 67, 65, 57, 72
61, 77, 53, 85, 67
69, 83, 79, 68, 51
59, 62 , 88, 64, 80

| Class | Tally | Frequency |
|---|---|---|
| 50 up to 60 | _____ | a. _____ |
| 60 up to 70 | _____ | b. _____ |
| 70 up to 80 | _____ | c. _____ |
| 80 up to 90 | _____ | d. _____ |

3 Use the frequency distributions from problem 2(a)–(d) to create the appropriate histogram. (Each vertical bar should represent one part of the problem.

4 Kevin and Al Bianchini own two meat markets, Bianchini's and Foodville. In a typical week, both stores sell approximately 2,000 lb of red meat, pork and poultry. Red meat includes beef, lamb, and veal. Typical amounts are as follows.

| Location | Red Meat | Pork | Poultry |
|---|---|---|---|
| Bianchini's | 750 | 500 | 750 |
| Foodville | 1000 | 400 | 600 |

a. Draw a comparative bar graph showing the sales of the two markets (two vertical bars for each type of meat).

b. Draw a component bar graph showing the sales of the two markets. Make one vertical bar for each store, and each bar should show the amount of each meat sold in that store.

c. Draw a pie chart for the meat sales for Bianchini's market only.

Assignment 27.1: **Statistical Averages**

Name _____

Date _____ Score _____

A **(52 points) Solve the following problems. (points for correct answers as marked**

1. A department store has three local locations: Mason Plaza, Corbin Center, and Balbo Mall. The store gives every applicant for any type of managerial job a test of basic business skills. Listed below are the scores from the tests given to applicants at the three locations last week.

| Mason Plaza | Corbin Center | Balbo Mall |
|:---:|:---:|:---:|
| 59 | 46 | 65 |
| 62 | 57 | 44 |
| 47 | 89 | 53 |
| 68 | 55 | 77 |
| 88 | 46 | 58 |
| 78 | 74 | 77 |
| 59 | 64 | 82 |
| 45 | 89 | 66 |
| 59 | 46 | 52 |
| 87 | 55 | |
| | 60 | |

Sum _____ _____ _____

b. Combine all the scores into one frequency distribution with the classes as shown. (2 points for each correct line)

| Class | Tally | Frequency |
|---|---|---|
| 40 up to 50 | _____ | _____ |
| 50 up to 60 | _____ | _____ |
| 60 up to 70 | _____ | _____ |
| 70 up to 80 | _____ | _____ |
| 80 up to 90 | _____ | _____ |

a. Find the mean, median, and mode for each location. (3 points for each correct answer)

| | Mason | Corbin | Balbo |
|---|---|---|---|
| | _____ | _____ | _____ |
| Mean | _____ | _____ | _____ |
| Median | _____ | _____ | _____ |
| Mode | _____ | _____ | _____ |

2. Monique Bidondo operates a popular coffee cart from which she also sells sandwiches. She has the opportunity to open another cart in the inner patio of a complex of office buildings, but she won't be allowed to sell sandwiches. Perform a statistical analysis on Monique's sales receipts for nonsandwich items for the first 15 work days of April and October. (3 points for each correct answer)

| April | | | October | | |
|---|---|---|---|---|---|
| $430 | $470 | $450 | $200 | $320 | $430 |
| 240 | 350 | 240 | 340 | 240 | 310 |
| 280 | 260 | 340 | 280 | 230 | 360 |
| 320 | 360 | 370 | 320 | 370 | 420 |
| 310 | 190 | 250 | 220 | 250 | 180 |

a. Find the mean for April. _____

b. Find the mean for October. _____

c. Find the median for April. _____

d. Find the median for October. _____

e. Find the combined mean for all 30 days. (Hint: Add the two sums and divide by 30.)

Assignment 27.1 *Continued*

Score for A (52)

B **(48 points) Solve the following problems. (points for correct answers as marked**

3. Valley National Bank has several branches. Bank management wants to compare the ages of retail bank customers at two specific branches—the Financial District Branch, downtown, and the University Branch, in a residential area between the local university and a retirement community. The bank's analyst randomly selects 30 retail bank customers from each bank and writes down their ages. The following two tables show the results.

| Financial District Branch | | | | | |
|---|---|---|---|---|---|
| 43 | 30 | 43 | 51 | 60 | ____ |
| 68 | 32 | 72 | 52 | 27 | ____ |
| 28 | 73 | 43 | 19 | 64 | ____ |
| 70 | 35 | 56 | 55 | 48 | ____ |
| 63 | 24 | 47 | 44 | 34 | ____ |
| 52 | 59 | 66 | 57 | 58 | ____ |

| University Branch | | | | | |
|---|---|---|---|---|---|
| 74 | 82 | 46 | 19 | 20 | ____ |
| 21 | 36 | 73 | 57 | 61 | ____ |
| 54 | 17 | 18 | 75 | 84 | ____ |
| 76 | 77 | 24 | 19 | 68 | ____ |
| 27 | 21 | 75 | 34 | 18 | ____ |
| 81 | 64 | 22 | 60 | 70 | ____ |

____ ____ ____ ____ ____ ____ ____ ____ ____ ____ ____ ____

a. Compute the mean age of the group of customers from the Financial District Branch. (8 points) _____

b. Compute the mean age of the group of customers from the University Branch. (8 points) _____

c. Make two frequency tables of customer ages, one for the Financial District Branch and one for the University Branch. For each table, use frequency classes 10 up to 20, 20 up to 30,…,80 up to 90. (2 points for each correct row in each table)

| Financial District Branch | | | | University Branch | | |
|---|---|---|---|---|---|---|
| Class | Tally | Frequency | | Class | Tally | Frequency |

Score for B (48)

602 | Part 8 Advanced Applications

Assignment 27.2: **Graphs and Charts**

Name

Date Score

A **(18 points) Complete the following problem as directed. (9 points for each correct graph**

1. After doing the initial research in problem 3 of Assignment 27.1, Valley National Bank's analyst randomly selects 100 customers from the Financial District Branch and 100 customers from the University Branch. She finds the age of each customer and summarizes the data in the following two frequency tables.

| **Financial District Branch** | | | **University Branch** | |
|---|---|---|---|---|
| **Class** | **Frequency** | | **Class** | **Frequency** |
| 10 up to 20 | 6 | | 10 up to 20 | 18 |
| 20 up to 30 | 10 | | 20 up to 30 | 17 |
| 30 up to 40 | 15 | | 30 up to 40 | 11 |
| 40 up to 50 | 17 | | 40 up to 50 | 8 |
| 50 up to 60 | 21 | | 50 up to 60 | 6 |
| 60 up to 70 | 16 | | 60 up to 70 | 8 |
| 70 up to 80 | 13 | | 70 up to 80 | 17 |
| 80 up to 90 | 2 | | 80 up to 90 | 15 |
| Total | 100 | | Total | 100 |

a. Draw a histogram for the Financial District Branch. Label each axis, and write a title under the graph.

b. Draw a histogram for the University Branch. Label each axis, and write a title under the graph.

B **(54 points) Complete the following problems as directed. (18 points for each correct graph**

2. Carrie Ortega owns two printing/copying businesses: Ortega Printing and Carrie's Copies. Ortega Printing is near City Hall and does most of its work for corporations. Carrie's Copies is in a residential district and does primarily printing and copies for individuals and small businesses. The following table shows sales revenues for the two shops for the last 6 months of the year.

| Shop | July | August | September | October | November | December |
|------|------|--------|-----------|---------|----------|----------|
| Ortega Printing | $240,000 | $280,000 | $220,000 | $230,000 | $270,000 | $250,000 |
| Carrie's Copies | 130,000 | 110,000 | 100,000 | 160,000 | 170,000 | 150,000 |

a. Make a comparative bar graph showing the monthly sales revenue for each shop. Label each axis, and write a title under the graph. Shade the bars for each shop differently.

b. On the same grid, make line graphs showing the monthly sales revenue for each shop. Label each axis, and write a title under the graph. Use a solid line for Ortega Printing and a dashed line for Carrie's Copies.

3. Statewide Insurance Agency records the totals of residential (as opposed to commercial) insurance policy premiums billed each month. The results for the first 6 months of the year are shown classified by automobile insurance, homeowner's insurance, and life insurance. Construct a component bar graph showing the premiums for each insurance type each month. Label each axis, and write a title under the graph. Shade the three types of insurance differently.

| Insurance Type | January | February | March | April | May | June |
|---|---|---|---|---|---|---|
| Auto | $100,000 | $100,000 | $ 80,000 | $ 80,000 | $100,000 | $ 80,000 |
| Home | 180,000 | 160,000 | 200,000 | 220,000 | 180,000 | 240,000 |
| Life | 80,000 | 60,000 | 80,000 | 100,000 | 60,000 | 60,000 |
| Total | $360,000 | $320,000 | $360,000 | $400,000 | $340,000 | $380,000 |

Score for B (54)

C **(28 points) Complete the following problem. (points for correct answers as marked**

4. Compact Disc Depot is a large discount store selling audio and video products. For its internal analysis, the store classifies all music sales as Rock, Folk, Classical, or Jazz. Every music sale is included in one of these four categories. In October, the store recorded the sales shown.

 a. Compute the percent of the total and the fraction of the total represented by each category of music.
 (2 points for each correct percent, 1 point for each correct fraction)

| Music Type | Sales | Percent | Fraction |
|---|---|---|---|
| Rock | $276,000 | _____ | _____ |
| Folk | 138,000 | _____ | _____ |
| Classical | 69,000 | _____ | _____ |
| Jazz | 69,000 | _____ | _____ |
| Total | _____ | 100.0% | $\frac{8}{8}$, or 1 |

b. Complete the pie chart to approximate the percent of total October sales for each category of music. Label each section with the category and percent and write a title under the graph. (8 points)

c. The percents of music sales at Compact Disc Depot for September are shown. Complete the pie chart to approximate the percent of total September sales for each category of music. Label each section with the category and percent and write a title under the graph. (*Hint:* 37.5% is $\frac{3}{8}$; 12.5% is $\frac{1}{8}$; 30% is somewhere between 25% and 37.5%; 20% is between 12.5% and 25%.) (8 points)

| Music Type | Percent |
|------------|---------|
| Rock | 37.5% |
| Folk | 30.0% |
| Classical | 12.5% |
| Jazz | 20.0% |
| | 100.0% |

Score for C (28)

Notes

 AT&T

Mon., Feb. 26, 2001

ABOUT AT&T

FOR HOME

FOR BUSINESS

DIRECTORIES

ACCOUNT MANAGEMENT

EMPLOYMENT

AS ADVERTISED

◀ **Tell Me More!**

AT&T connects you with wireless, long distance and local phone, the Internet, cable TV, and beyond...

AT&T WorldNet Service and Long Distance in One Package

OFFER

Search [] go

Advanced Search
Site Map
A to Z Index

Enter AT&T Keyword or Search Terms

Stock Info: AT&T 21.24 +0.29 Feb 26, 11:22 am ET
 AWE 20.98 1.13 Feb 26, 11:15 am ET

▸ **AT&T NEWSROOM**

- AT&T Wireless Services Expands to the Big Island of Hawaii
- AT&T And MentorNet Launch Program To Help Close The Digital Divide
- AT&T Celebrates Grand Opening Of Orlando Internet Data Center

▸ **WHAT'S NEW**

- Search available toll-free numbers for your small business
- Get help selecting telecommunications services
- Get the latest on AT&T services in our FREE consumer newsletter!

About AT&T | For Home | For Business | Directories

Source: http://www.att.com. Copyright © 2001 by AT&T; used with permission.

Corporations need to know how changes in currency values will affect their bottom line and how much to charge to make a profit. These kinds of information are determined mathematically, often through the use of equations.

A company such as AT&T uses equations for calculations in its development labs, in pricing services, and to answer financial questions about its worldwide operations. Learn about AT&T's research, services, investment opportunities, and career choices at its Web site at http://www.att.com.

Business Applications of Equations

28

LEARNING OBJECTIVES

By studying this chapter and completing all assignments you will be able to:

 1 LEARNING OBJECTIVES
Solve addition/subtraction and multiplication/division equations.

 2 LEARNING OBJECTIVES
Solve complex equations.

3 LEARNING OBJECTIVES
Solve for variables in inventory equations.

 4 LEARNING OBJECTIVES
Solve for variables in simple interest equations.

 5 LEARNING OBJECTIVES
Solve for variables in total cost equations.

A contractor determines amounts of materials to order. A teacher determines grades. A business determines profits. Complex equations are used to calculate a multinational corporation's sales projections. Simple equations are used to estimate the cost of changing a light bulb. Equations are just formulas for finding mathematical answers.

1. What is meant by the phrase *solution of an equation*?
2. In an algebraic equation, what are three different notations that you can use to indicate multiplication?
3. Assume that you already know how much money someone borrowed, how long she had the money, and how many dollars she paid in simple interest. Explain how you would calculate the simple interest rate.
4. Assume that you already know a company's cost of goods sold, its beginning inventory, and its ending inventory. How would you calculate its purchases?

In most chapters of this book, we presented formulas that you could use to help you solve problems efficiently. Writing $I = P \times R \times T$ is easier and clearer than writing "Interest is equal to Principal times Rate times Time." Writing $BI + P - EI = CGS$ is easier and clearer than writing "Beginning inventory plus Purchases minus Ending inventory equals Cost of goods sold."

Recall that the basic formula for simple interest, $I = P \times R \times T$, is used to find Interest when you know the Principal, Rate, and Time. In Chapter 15, you also learned to find the Principal when the Interest, Rate, and Time were given. There, we used the formula: $P = I/(R \times T)$. In this chapter, we will find the solution by using only the original formula, $I = P \times R \times T$, and algebra. **Algebra** is essentially arithmetic with some letters instead of all numbers. In algebra, the formula $I = P \times R \times T$ is called an **equation**. When you find the answer, you have obtained the **solution** to the equation. That is, you have found the number that makes the equation true. You can always *check* your answers by substituting the solution back into the original equation.

In previous chapters, we usually used the common symbol \times to indicate multiplication. However, in algebra, multiplication is also indicated with other notation. In Chapter 14, we mentioned that $I = P \times R \times T$ was often written as $I = PRT$. By placing two letters adjacent to each other, you can indicate multiplication. This notation is most common in algebra when a number and a letter are multiplied together. The number is written first, and no multiplication symbol is used. In the equation of example B, we write $3y = 21$ instead of $3 \times y = 21$. The 3 is called the **coefficient** of y. This notation is not used when only numbers are multiplied together, because 583 could mean 5×83 or 58×3 or simply 583. Sometimes people use a dot between two numbers or between two letters to indicate multiplication. We could write $5 \cdot 83$ instead of 5×83. In this chapter, we use all three notations to indicate multiplication, choosing the simplest and clearest in each particular situation.

Some equations in business have variables represented by more than one letter. In $BI + P - EI = CGS$, BI means Beginning inventory, not B times I, and CGS means Cost of goods sold, not C times G times S. Usually, the formula is familiar and you will know its meaning from the context. To eliminate confusion, we could use just one letter to represent a variable. Thus we might write $B + P - E = C$ instead of $BI + P - EI = CGS$. Sometimes it will also be useful to write parentheses around one or two of the factors to emphasize the multiplication. This often occurs when one of the factors is a decimal or a fraction. For example, $(0.08) \cdot \left(\frac{60}{360}\right)$ might be easier to read and understand than $0.08 \cdot \frac{60}{360}$.

Solving Addition/Subtraction and Multiplication/Division Equations

LEARNING OBJECTIVES

Solve addition/subtraction and multiplication/division equations.

Equations can be classified by the kind of arithmetic they perform. $BI + P - EI = CGS$ is an *addition/subtraction equation* because the primary operations are addition and subtraction. $I = PRT$ is a *multiplication/division equation* because the primary operation is multiplication. Most equations are more complex and have addition/subtraction operations as well as multiplication/division operations.

Examples A and B illustrate the two basic types of equations—example A is an addition/subtraction equation, and example B is a multiplication/division equation. Each equation is written in two forms: (1) a sentence with words and (2) an equation with letters and numbers. The two forms are equivalent and have the same solution.

The phrase *solve the equation* means to find the number that makes the equation or sentence true. One important way to solve an equation is by trial and error. To do so you substitute various values into the equation until you find the number that makes it true.

EXAMPLE A

Write the following sentence as an equation: Five added to what number gives eight? Then solve the equation by trial and error.

Sentence: Five added to what number gives eight?

Equation: $x + 5 = 8$

Solution: Try 0: $0 + 5 = 5$, which is too small. Try 5: $5 + 5 = 10$, which is too large. The solution must be between 0 and 5. Eventually we find that 3 works: $3 + 5$ does equal 8.

EXAMPLE B

Write the following equation in words: $3y = 21$. Then find the solution by trial and error.

Equation: $3y = 21$ (Remember, $3y$ means $3 \times y$.)

Sentence: Three times what number gives twenty-one?

Solution: Try 1: $3 \times 1 = 3$, which is too small. Try 5: $3 \times 5 = 15$, which is still too small. Try 10: $3 \times 10 = 30$, which is too large. The solution is between 5 and 10. Eventually we find that 7 works: 3×7 equals 21.

Note: The reason that the *trial and error method* is so important is that many important problems are easy to understand but are absolutely impossible to solve by using algebra. For example, recall annuities. You want to have $12,000 at the end of 4 years, and you can deposit $200 per month for the entire 4 years. What interest rate is required? Use row 48 in Table 26-1. When you try the 0.75% column, the future value is too small. When you try the 1% column, the future value is too large. Because the compounding is monthly and the periodic rate is between 0.75% and 1%, the interest rate that is the answer lies between 11% and 12%. It is impossible to solve for the answer using algebra. A computer, and some financial calculators, can find the answer, but the computer and financial calculator are programmed to use the trial and error method.

Two Basic Principles of Algebra

You can solve example A more easily by using a fundamental principle of algebra. Example A presents an addition equation—the only operation is to *add* 5. Solve the equation by doing the opposite operation to both sides; that is, *subtract* 5 from both sides. Subtracting 5 from 8 gives the solution: $x = 3$.

Rule 1a: **Solve an addition equation by doing the reverse—subtraction—to both sides of the equation.**

Rule 1b: **Solve a subtraction equation by doing the reverse—addition—to both sides of the equation.**

Example B uses another fundamental principle of algebra. It gives a multiplication equation—the only operation is to *multiply* by 3. Solve the equation by doing the opposite operation to both sides; *divide* both sides by 3. Dividing 21 by 3 gives the solution: $y = 7$.

Rule 2a: **Solve a multiplication equation by doing the reverse—division—to both sides of the equation.**

Rule 2b: **Solve a division equation by doing the reverse—multiplication—to both sides of the equation.**

a. Solve for x: $\qquad\qquad\qquad x - 7 = 38$
 Add 7 to each side: $x - 7 + 7 = 38 + 7$
 Simplify: $\qquad\qquad\qquad\qquad x = 45$

 Check: $x - 7 = 45 - 7 = 38$. It checks.

b. Solve for y: $\qquad\qquad\qquad y + 5 = 8$
 Subtract 5 from each side: $y + 5 - 5 = 8 - 5$
 Simplify: $\qquad\qquad\qquad\qquad y = 3$

 Check: $y + 5 = 3 + 5 = 8$. It checks.

c. Solve for x: $\qquad\qquad\qquad 5x = 45$

 Divide each side by 5: $\quad \dfrac{5x}{5} = \dfrac{45}{5}$

 Simplify: $\qquad\qquad\qquad\qquad x = 9$

 Check: $5x = 5 \cdot 9 = 45$. It checks.

d. Solve for y: $\qquad\qquad\qquad \dfrac{y}{6} = 3$

 Multiply each side by 6: $\quad \dfrac{y}{6} \cdot 6 = 3 \cdot 6$

 Simplify: $\qquad\qquad\qquad\qquad y = 18$

 Check: $\dfrac{y}{6} = \dfrac{18}{6} = 3$. (OK)

Solving Complex Equations

LEARNING OBJECTIVES

Solve complex equations.

www.gcse.com/Maths/ algmen.htm

Practice exercises are available at this Web site.

This book contains many diverse equations. They include basic addition/subtraction, basic multiplication/division, and some complex combinations of the two basic equations.

Compare these two equations:

(1) $6 + 3x = 48$ \qquad and \qquad (2) $3 \cdot (x + 6) = 48$

Both equations have multiplication and addition operations, and both have the same numbers. But the equations are different. The order of operations is different. Equation (1) says to multiply "something" by 3 and then add the product to 6. Equation (2) says to add 6 to "something" and then multiply the sum by 3.

(1) $6 + 3x = 48$

 The product $3x$ is added to 6 to give 48. Equation (1) is an addition equation, so we subtract 6 from each side to get $3x = 42$—a multiplication equation. We divide each side by 3 to get

$x = 14$ \quad The solution to equation (1) is 14.

Check: $6 + 3x = 6 + 3 \cdot 14 = 6 + 42 = 48$ \quad (OK)

(2) $3 \cdot (x + 6) = 48$

 3 times the sum gives 48. Equation (2) is a multiplication equation, so we divide each side by 3 to get $x + 6 = 16$—an addition equation. We subtract 6 from each side to get

$x = 10$ \quad The solution to equation (2) is 10.

Check: $3 \cdot (x + 6) = 3 \cdot (10 + 6) = 3 \cdot 16 = 48$ \quad (OK)

Order of Operations

Equations (1) and (2) are called **complex equations** because they have both addition/subtraction and multiplication/division components. They therefore illustrate yet another basic rule of algebra.

Rule: \quad **First perform multiplication and division. Then perform addition and subtraction. To change the order, use parentheses.**

Look at the check of equation (1), where $6 + 3 \cdot 14$. The $3 \cdot 14$ is performed before the addition, even though the addition symbol comes first. Now look at the check of equation (2), where $3 \cdot (10 + 6)$. The multiplication symbol is written first, and multiplication is supposed to be done before addition. But the parentheses tell us to perform the addition before the multiplication.

Note: In the check of equation (2), we wrote $3 \cdot (x + 6)$ and $3 \cdot (10 + 6)$. We can eliminate the dot and still have the same meaning: $3(x + 6)$ means "3 times $(x + 6)$" and $3(10 + 6)$ means "3 times $(10 + 6)$." The equation $3 \cdot (x + 6) = 48$ is normally written without the dot, as $3(x + 6) = 48$.

http://mthwww.uwc.
edu/wwwmahes/
courses/math/
mm110nts.htm

You'll find information and practice on algebraic word problems at this Web site.

EXAMPLE C

Solve the equation $4y - 5 = 55$

| | |
|---|---|
| $4y - 5 = 55$ | This is a subtraction equation, so add 5. |
| $4y - 5 + 5 = 55 + 5$ | Add 5 to each side and simplify. |
| $4y = 60$ | This is a multiplication equation, so divide by 4. |
| $\dfrac{4y}{4} = \dfrac{60}{4}$ | Divide each side by 4 and simplify. |
| $y = 15$ | The solution |

Check the solution in the original equation:

$4y - 5 = 4 \cdot 15 - 5 = 60 - 5 = 55$ It checks.

EXAMPLE D

Solve the equation $2 = \dfrac{(x + 9)}{7}$

| | |
|---|---|
| $2 = \dfrac{(x + 9)}{7}$ | This is a division equation, so multiply by 7. |
| $2 \cdot 7 = \dfrac{(x + 9)}{7} \cdot 7$ | Multiply each side by 7 and simplify. |
| $14 = x + 9$ | This is an addition equation, so subtract 9. |
| $14 - 9 = x + 9 - 9$ | Subtract 9 from each side and simplify. |
| $5 = x$ | The solution |

Check the solution in the original equation:

$\dfrac{(x + 9)}{7} = \dfrac{(5 + 9)}{7} = \dfrac{14}{7} = 2.$ It checks.

CONCEPT CHECK 28.2

a. Solve for x: $5x - 7 = 38$

Add 7 to each side: $5x - 7 + 7 = 38 + 7$

Simplify: $5x = 45$

Divide each side by 5: $\dfrac{5x}{5} = \dfrac{45}{5}$

Simplify: $x = 9$

Check: $5x - 7 = 5 \cdot 9 - 7 = 45 - 7 = 38$ (OK)

b. Solve for y: $\dfrac{y}{6} + 5 = 8$

Subtract 5 from each side: $\dfrac{y}{6} + 5 - 5 = 8 - 5$

Simplify: $\dfrac{y}{6} = 3$

Multiply each side by 6: $\dfrac{y}{6} \cdot 6 = 3 \cdot 6$

Simplify: $y = 18$

Check: $\dfrac{y}{6} + 5 = \dfrac{18}{6} + 5 = 3 + 5 = 8$ (OK)

Solving Inventory Equations

LEARNING OBJECTIVES

Solve for variables in inventory equations.

In Chapter 19, when you knew beginning inventory, purchases, and ending inventory, you could use the formula BI + P − EI = CGS to solve for the cost of goods sold. We can rewrite the formula as BI + P = EI + CGS. Then, using only one letter to represent a variable, we can write the basic inventory equation as $B + P = E + C$, which we use in this chapter. If we know any three of the variables, we can use algebra to solve for the fourth.

EXAMPLE E

For ending inventory = \$27.3, cost of goods sold = \$107, and purchases = \$95, use the formula $B + P = E + C$ and algebra to solve for beginning inventory. (All numbers are in thousands.)

| | |
|---|---|
| $B + P = E + C$ | The basic inventory equation |
| $B + \$95 = \$27.3 + \$107$ | Substitute each value for its letter. |
| $B + 95 = 134.3$ | Simplify: 27.3 + 107 = 134.3 |
| | This is an addition equation. |
| $B = 134.3 - 95$ | Subtract 95 from each side. |
| $B = 39.3$ | Beginning inventory is \$39,300. |

Check the solution in the original equation:
$E + C = 27.3 + 107 = 134.3$
$B + P = 39.3 + 95 = 134.3$ It checks.

EXAMPLE F

For ending inventory = \$47.21, cost of goods sold = \$92.62, and beginning inventory = \$43.3, use the formula $B + P = E + C$ and algebra to solve for purchases. (All numbers are in thousands.)

| | |
|---|---|
| $B + P = E + C$ | The basic inventory equation |
| $\$43.3 + P = \$47.21 + \$92.62$ | Substitute each value for its letter. |
| $43.3 + P = 139.83$ | Simplify: 47.21 + 92.62 = 139.83 |
| | This is an addition equation |
| $P = 139.83 - 43.3$ | Subtract 43.3 from each side. |
| $P = 96.53$ | Purchases are \$96,530. |

Check the solution in the original equation:
$E + C = 47.21 + 92.62 = 139.83$
$B + P = 43.3 + 96.53 = 139.83$ It checks.

CONCEPT CHECK 28.3

Find Beginning inventory when purchases = \$56,000, ending inventory = \$24,000, and cost of goods sold = \$70,000. Use the formula $B + P = E + C$ and algebra.

| | |
|---|---|
| Write the basic inventory equation: | $B + P = E + C$ |
| Substitute values for letters: | $B + \$56,000 = \$24,000 + \$70,000$ |
| Simplify: | $B + 56,000 = 94,000$ |
| Subtract \$56,000 from each side: | $B + 56,000 - 56,000 = 94,000 - 56,000$ |
| Simplify: | $B = 38,000$, or Beginning inventory is \$38,000 |

Check: $B + P = \$38{,}000 + \$56{,}000 = \$94{,}000$
$E + C = \$24{,}000 + \$70{,}000 = \$94{,}000$ (OK)

Solving Simple Interest Equations

The basic formula for simple interest is $I = PRT$. In chapter 14, you used the formula to solve for Interest when Principal, Rate, and Time were given. In chapter 15, Interest was given, and you solved for one of the other three variables, P or R or T. The method in chapter 15 was to use one of four formulas. Here we use only the basic *simple interest equation*, $I = PRT$; substitute in the given values; and use algebra to solve for the missing value. Remember, to solve the equation we must write T in years, and in this chapter we assume that every year is a 360-day year.

LEARNING OBJECTIVES

Solve for variables in simple interest equations.

EXAMPLE G

Interest = \$10, Rate = 12%, and Time = 60 days (i.e., $\frac{60}{360}$ year). Use the formula $I = PRT$ and algebra to solve for P.

$I = PRT$ — The basic simple interest equation

$\$10 = P \cdot (0.12) \cdot \left(\dfrac{60}{360} \right)$ — Substitute each value for its letter.

$10 = P \cdot (0.02)$ — Simplify: $(0.12) \cdot \left(\dfrac{60}{360} \right) = 0.02$.

$\dfrac{10}{0.02} = \dfrac{P(0.02)}{0.02}$ — Divide each side by 0.02 and simplify.

$500 = P$ — Principal is \$500.

Check the solution in the original equation:

$PRT = (\$500) \cdot (0.12) \cdot \left(\dfrac{60}{360} \right) = \10 It checks.

EXAMPLE H

Interest = \$45, Principal = \$1,200, and Rate = 15%. Use the formula $I = PRT$ and algebra to solve for T, and state the time of the loan in days.

$I = PRT$ — The basic simple interest equation

$\$45 = (\$1{,}200) \cdot (0.15) \cdot T$ — Substitute each value for its letter.

$45 = 180 \cdot T$ — Simplify: $(1{,}200) \cdot (0.15) = 180$.

$\dfrac{45}{180} = \dfrac{180T}{180}$ — Divide each side by 180 and simplify.

$0.25 = T$ — T is 0.25 year, which is 90 days.
$0.25 \cdot 360$ days = 90 days.

Check the solution in the original equation:

$PRT = (\$1{,}200) \cdot (0.15) \cdot \left(\dfrac{90}{360} \right) = \45 It checks.

CONCEPT CHECK 28.4

Use the basic formula for simple interest and algebra to find the Rate when Interest = $26, Principal = $1,600, and Time = 90 days. Use a 360-day year.

| | |
|---|---|
| Write the basic interest equation: | $I = PRT$ |
| Substitute numbers for letters: | $\$26 = \$1,600 \cdot R \cdot \dfrac{90}{360}$ |
| Simplify: | $26 = 400 \cdot R$ |
| Divide each side by 400: | $\dfrac{26}{400} = \dfrac{400R}{400}$ |
| Simplify: | $0.065 = R$, or Rate is 6.5% |

Check: $PRT = (\$1,600) \cdot (0.065) \cdot \left(\dfrac{90}{360}\right) = \26 (OK)

Solving Total Cost Equations

Solve for variables in total cost equations.

Bill and Carrie Walters had been studying furniture design and construction, and they decided to open a small business manufacturing tables in their garage. They purchased the necessary power equipment—saws, drill press, planer, sander, and the like. The Walters must have the equipment, regardless of how many tables they make. Therefore the cost of the equipment is a **fixed cost (FC)**. Fixed costs do not vary with the quantity of tables that the Walters build.

To manufacture a table, the Walters must purchase materials such as wood and hardware. And to actually make the table, they need to use labor. The costs for materials and labor vary with the number of tables they build; if no tables are built, no materials and labor are used. Each cost for materials and labor is called a **variable cost (VC)**. Variable costs are stated in terms of dollars per item manufactured. The total variable cost is the product of VC times **Quantity produced (Q)**, or VC · Q.

Total cost (TC) is the sum of all of these costs. The basic **total cost equation** is TC = VC · Q + FC. As mentioned earlier, equations are less confusing with only one letter per variable. Using this format, we can write either $T = VQ + F$, without the multiplication dot, or $T = V \cdot Q + F$, with the dot. Use the equation that seems clearest to you. Example I uses the original equation, and example J uses the last one.

Summary: FC = F = Fixed cost VC = V = Variable cost per item
Q = Quantity produced TC = T = Total cost

TC = VC · Q + FC or $T = V \cdot Q + F$ or $T = VQ + F$

© CORBIS

EXAMPLE I

Bill and Carrie Walters' fixed costs for building tables total $42,000. The variable costs for material and labor total $160 for each table. What will be the total cost if the Walters build 100 tables?

TC = Total cost VC = $160 per table
FC = $42,000 Q = 100 tables produced

TC = VC · Q + FC = $160 · 100 + $42,000 = $16,000 + $42,000 = $58,000
The total cost for the Walters to build 100 tables is $58,000.

The total cost equation is an example of a complex equation—it has both a multiplication/division component and an addition/subtraction component. If you know three of the four variables, use algebra to solve for the fourth variable.

EXAMPLE J

The Walters' fixed costs for building tables (example I) total $42,000. The variable costs for material and labor total $160 for each table. How many tables can the Walters build for a total cost of $62,000?

T = $62,000 V = $160 per table
F = $42,000 Q = Quantity produced

| | |
|---|---|
| $T = V \cdot Q + F$ | The equation for total cost |
| $\$62,000 = \$160 \cdot Q + \$42,000$ | Substitute each value for the letter |
| $62,000 - 42,000 = 160 \cdot Q + 42,000 - 42,000$ | Subtract 42,000 from each side |
| $20,000 = 160 \cdot Q$ | Simplify: $62,000 - 42,000 = 20,000$ |
| $\dfrac{20,000}{160} = \dfrac{160 \cdot Q}{160}$ | Divide each side by 160 |
| $125 = Q$ | They can build 125 tables for $62,000 |

Check the solution in the original equation:
$V \cdot Q + F = \$160 \cdot 125 + \$42,000$
$= \$20,000 + \$42,000 = \$62,000$ It checks.

CONCEPT CHECK 28.5

Use the basic equation for total cost to find the quantity when the total cost is $48,000, the variable cost is $4 per item, and the fixed cost is $18,000.

| | |
|---|---|
| Write the total cost equation: | $T = VQ + F$ |
| Substitute numbers for letters: | $\$48,000 = \$4 \cdot Q + \$18,000$ |
| Subtract 18,000 from each side: | $48,000 - 18,000 = 4 \cdot Q + 18,000 - 18,000$ |
| Simplify: | $30,000 = 4 \cdot Q$ |
| Divide each side by 4: | $\dfrac{30,000}{4} = \dfrac{4 \cdot Q}{4}$ |
| Simplify: | $7,500 = Q$, or the quantity is 7,500 items |

Check: $VQ + F = \$4 \cdot 7,500 + \$18,000 = \$30,000 + \$18,000 = \$48,000$ (OK)

COMPLETE ASSIGNMENTS 28.1 and 28.2.

Try this Web site for practice problems. The additional challenge site might be fun!

Source: http://www.figurethis.org/challenges/math_index.htm.
Copyright © 2001 by Widmeyer Communications; used with permission.

Chapter Terms for Review

algebra
coefficient
complex equations
equation
fixed cost (FC)

quantity produced
solution
total cost
total cost equation
variable cost (VC)

Visit the Web

http://campus.northpark.edu/wicksBook/LinearAlgebra/—Equations are used in many areas of business. Visit this Web site for examples of problems applicable to business.

http://mathforum.com/algpow, http://www2.hawaii.edu/suremath/intro_algebra. html, http://www.algebra-online.com/quickmath.php3—These also are excellent sites for help with algebra and the solution of problems.

CyberConnection

USA TODAY Tech Report

01/26/99- Updated 08:02 PM ET

Algorithms solve business problems

Remember algorithms from math class? You might want to get reacquainted. New high-octane algorithms, often running on desktop computers, are changing entire industries and solving problems once thought unsolvable.

Algorithms are step-by-step equations for getting something done. A recipe for a cake is an algorithm. The equations have been around since the Greeks, who used algorithms to build the Parthenon.

But this is a new breed of algorithms. They have been constructed by mathematicians with the power of today's computers in mind. They take advantage of mountains of digital information never before available but now gleaned from bar-code readings, bills, transactions and other databases. Algorithms have become a critical force in the global marketplace because they can often find ways to run a business faster, cheaper or better.

"It's like taking raw information and spinning money out of it," says Robert Cross, whose company, DFI Aeronomics, sells algorithm-based systems

Source: http://www.usatoday.com/life/cyber/tech/ctb873.htm. Copyright © 1999, USA TODAY. Reprinted with permission.

Read the article at http://www.usatoday.com/life/cyber/tech/ctb873.htm and learn how algorithms (step-by-step equations for getting something done) are becoming much more important today due to powerful computers that do the computations. An algorithm like the one used by The Gap sounds straightforward—management wants to optimize the number of employees working at every store during the Christmas rush. But as you read on, you begin to get an idea about how complicated such a simple-sounding algorithm really is.

You need information about a huge number of variables, such as "time of the day, day of the week, each store's customer traffic at each time and day, employee availability at each time and day, whether or not there is a sale going on, employee salaries, union restrictions about hours worked, regular budget and overtime budget." Then equations and inequalities must be written that express the relationships among all of these variables. The final solution will be done by a computer, but the algorithm may be written to find the best solution, or one that is simply acceptable. Finding the best solution could take too long, or could even be impossible.

CYBERCONNECTION

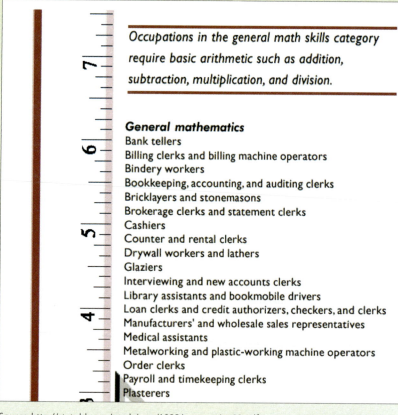

Occupations in the general math skills category require basic arithmetic such as addition, subtraction, multiplication, and division.

General mathematics
Bank tellers
Billing clerks and billing machine operators
Bindery workers
Bookkeeping, accounting, and auditing clerks
Bricklayers and stonemasons
Brokerage clerks and statement clerks
Cashiers
Counter and rental clerks
Drywall workers and lathers
Glaziers
Interviewing and new accounts clerks
Library assistants and bookmobile drivers
Loan clerks and credit authorizers, checkers, and clerks
Manufacturers' and wholesale sales representatives
Medical assistants
Metalworking and plastic-working machine operators
Order clerks
Payroll and timekeeping clerks
Plasterers

Other careers that require math skills
Math skills are useful in a number of other occupations. For example, most jobs in the financial industry use math skills. Bank tellers must have strong math skills to be both accurate and efficient. Accountants need proficiency in math to calculate and analyze numbers. Air traffic controllers need to understand maps and geometry when directing planes. Managers of all kinds use math skills; for example, hotel managers and assistants must be able to estimate costs for items the hotel needs to order, such as food and drinks.

Preparing for careers in math
The accompanying lists show occupations that require different levels of math skills: Advanced, applied, practical, or general. Occupations in the advanced or theoretical math skills category require an understanding of more complex math concepts such as calculus and linear algebra. Occupations in the applied math skills category include those in which

Source: http://stats.bls.gov/opub/ooq/1999/summer/art03.pdf

The Occupational Outlook Quarterly Online at the Occupational Outlook Handbook Web site (http://stats.bls.gov/oco/) discusses the need for a well-rounded education in mathematics, as well as the sciences and social sciences. It describes the value of different kinds of mathematical expertise and lists many occupations that require math skills, including cashiers, billing clerks, payroll clerks, and even construction workers.

This page is from the summer 1999 issue; visit this site or search through other issues for similar information.

The Bottom Line

Summary of chapter learning objectives:

| Learning Objective | Summary | Example |
|---|---|---|
| **28.1** | Solve addition/subtraction and multiplication/division equations | 1. Solve the following equation for *x* and check your answer: $x - 11 = 33$

2. Solve the following equation for *x* and check your answer: $\dfrac{x}{4} = 9$ |
| **28.2** | Solve complex equations | 3. Solve the following equation for *x* and check your answer: $7x + 11 = 53$ |
| **28.3** | Solve for variables in inventory equations | 4. Beginning inventory = $111,700, Purchases = $132,200, and Cost of goods sold = $151,900. Solve for Ending inventory, using BI + P = EI + CGS or $B + P = E + C$. Check your answer. |
| **28.4** | Solve for variables in simple interest equations | 5. Interest = $42, Rate = 9%, and Time = 60 days. Solve for Principal, using $I = PRT$. Use a 360-day year, and check your answer. |
| **28.5** | Solve for variables in total cost equations | 6. Fixed cost = $37,500, Quantity = 25,000 items, and Total cost = $100,000. Find the Variable cost per item, using TC = VC · Q + FC or $T = V \cdot Q + F$. Check your answer. |

Answers: 1. $x = 44$ 2. $x = 36$ 3. $x = 6$ 4. $E = \$92,000$ 5. $P = \$2,800$ 6. $V = \$2.50$

1 Solve each of the following equations for the missing unknown.

a. $x + 7 = 15$

b. $x - 13 = 10$

c. $26 = 9 + y$

d. $21 = y - 7$

e. $3x = 39$

f. $84 = 6x$

g. $\dfrac{x}{3} = 5$

h. $7 = \dfrac{y}{7}$

i. $4x - 6 = 38$

j. $25 + 6y = 79$

k. $19 = 8y - 37$

l. $99 = 27 + 18x$

m. $\dfrac{9 + x}{2} = 5$

n. $\dfrac{2x - 5}{3} = 7$

o. $13 = \dfrac{y - 4}{5}$

p. $35 = \dfrac{13 + 3y}{2}$

q. $11 = 7 + \dfrac{2x}{3}$

r. $\dfrac{3y}{5} - 7 = 5$

s. $12 = 2 + \dfrac{5y - 20}{6}$

t. $\dfrac{3x + 4}{4} - 2 = 8$

2 Melody Chan borrowed $12,000 from her parents for 3 years at a modest simple interest rate. When she repaid the principal, she paid her parents $1,440 as the full amount of her simple interest. What was the simple interest rate that her parents charged Melody?

3 Gomez Granite and Marble Company had a Beginning inventory of $575,000, a Cost of goods sold of $544,000 and made Purchases of $467,000. What was its ending inventory?

4 Lipinski Clothing Company makes a woman's dress that has fixed costs of $88,600. The total cost to make 2,550 dresses is $269,850. What is the variable cost per dress?

Assignment 28.1: Introduction to Algebra

Name

Date Score

A (24 points) Use algebra to solve each of the following basic equations. (3 points for each correct answer)

1. $6y = 72$

2. $x - 9 = 21$

3. $38 = 11 + x$

4. $\dfrac{x}{5} = 7$

5. $y + 15 = 61$

6. $11 = \dfrac{y}{3}$

7. $78 = 13x$

8. $6 = x - 15$

Score for A (24)

B (40 points) Use algebra to solve each of the following complex equations. (5 points for each correct answer)

9. $4y - 5 = 43$

10. $5(4 + x) = 65$ or $5 \cdot (4 + x) = 65$

11. $4 = \dfrac{x - 11}{6}$

12. $34 = 9 + 5y$

13. $39 = (x - 4)3$ or $39 = (x - 4) \cdot 3$

14. $\dfrac{x}{7} - 2 = 3$

15. $17 = 13 + \dfrac{x}{5}$

16. $\dfrac{3y - 7}{4} = 5$

Score for B (40)

C **(36 points) Use algebra to solve each of the following business equations. In the simple interest applications, use a 360-day year. (6 points for each correct answer)**

17. If $I = \$450$, $R = 6\%$, and $T = \frac{90}{360}$ year, solve $I = PRT$ for P.

18. If $B = \$47,000$, $P = \$22,000$, and $E = \$42,000$, solve $B + P = E + C$ for C.

19. If $T = \$106,000$, $Q = 50$ items, and $F = \$42,000$, solve $T = V \cdot Q + F$ for V.

20. If $T = \$150,000$, $V = \$80$, and $F = \$90,000$, solve $T = V \cdot Q + F$ for Q.

21. If $P = \$89,000$, $E = \$72,000$, and $C = \$61,000$, solve $B + P = E + C$ for B.

22. If $I = \$640$, $P = \$16,000$, and $T = \frac{180}{360}$ year, solve $I = PRT$ for R.

Score for C (36)

Assignment 28.2: Algebra in Business Equations

LEARNING OBJECTIVES

Name _____

Date _____ Score _____

A **(24 points) Use the equation** $B + P = E + C$ **(or BI + P = EI + CGS) and algebra to solve each of the following inventory problems. (6 points for each correct answer)**

1. Patel Plumbing Supply had cost of goods sold of $92,400, purchases of $105,900, and beginning inventory of $122,500. Compute the ending inventory. _____

2. Lopez Stamp Exchange has an ending inventory of $135,000, a beginning inventory $126,300, and purchases of $155,000. Compute cost of goods sold. _____

3. Hong Kong Clothing had a beginning inventory of $352,000, an ending inventory of $364,000, and cost of goods sold of $344,000. Compute the amount of purchases. _____

4. Ohio Valley Furniture has an ending inventory of $285,000, cost of goods sold of $246,000, and purchases of $262,000. What was its beginning inventory? _____

Score for A (24)

B **(24 points) Use the equation** $I = PRT$ **and algebra to solve each of the following simple interest rate problems. Use a 360-day year. (6 points for each correct answer)**

5. Shen Xiu borrowed money from a finance company for 60 days at 15% simple interest. If Shen paid $37.50 interest, how much did he borrow? _____

6. Mary Piros borrowed $1,200 for 90 days. She paid $21 interest. Compute the simple interest rate. _____

7. JoAnn Evans invested $3,000 at 10% simple interest. If JoAnn earned $225 in interest, for how long was the money invested? _____

8. Rich Koppes loaned $800 to his son Jeff for 90 days at 5% simple interest. How much interest (in dollars) should Jeff pay? _____

Score for B (24)

C **(24 points) Use the equation $T = V \cdot Q + F$ (or TC = VC \cdot Q + FC) and algebra to solve each of the following total cost problems. (6 points for each correct answer)**

9. Adirondack Lawn Furniture makes cedar chairs. Fixed costs are $126,000. Variable costs are $88 per chair. What is the total cost of manufacturing 3,500 chairs? _____

10. Juan's Skateboard City makes high-quality skateboards. Juan's fixed costs are $4,500. If the total cost to make 400 skateboards is $21,700, what are the variable costs per board? _____

11. Marilyn Smith manufactures leather sandals. Her variable costs are $42 per pair. When Marilyn produced 350 pairs of sandals, her total costs were $43,400. What were her fixed costs? _____

12. Prescott Manufacturing, Inc., makes barbeques. The fixed costs are $42,000, and the variable costs are $31.50 per barbeque. How many barbeques can be produced for a total cost of $79,800? _____

Score for C (24)

Assignment 28.2 *Continued*

D **(28 points) Use the correct equation and algebra to solve each of the following business applications. (7 points for each correct answer)**

13. Major League Collectibles has an ending inventory of $151,800 and a beginning inventory of $158,300. If cost of goods sold is $206,100, compute purchases. _____

14. Southwestern Solar Shield makes awnings to protect windows from the sun. Fixed costs are $231,700, and variable costs are $145 per awning. What number of awnings will result in total costs of $368,000? _____

15. Penny Wong borrowed $4,500 for $2\frac{1}{2}$ years. If Penny paid $900 in simple interest, what was the simple interest rate? _____

16. The total cost to manufacture 500 high-quality glass coffee tables is $642,000. If the fixed costs are $280,000, what are the variable costs per table? _____

Score for D (28)

hp UNITED STATES

select another count

hp gets
in the groove

hp
i n v e n t

products
&
services

drivers

hp's internet
software platform

highlights

→ hp's storage strategy delivers groundbreaking efficiency

→ hp integrates inktomi software with servers

→ learn more about PDAs

→ hp taps mobile communications exec to head embedded and personal systems division

→ hp replaces PDU to resolve potential safety issue

solutions

→ home & home office

→ small & medium business

→ large & corporate business

→ government

→ education

→ IT professionals

→ developers

→ more solutions...

→ comp

→ jobs

→ inves

→ how

The language of computers is the language of business. Computers store and transmit data using simple binary numbers. It is important that users of business computers understand the basic components and functions of numerical operations in computers.

Hewlett-Packard Company is a leading manufacturer of computers, printers, and other related equipment. Visit HP at www.hp.com.

Business Math in Computer Operations

29

LEARNING OBJECTIVES

By studying this chapter and completing all assignments you will be able to:

1 Solve simple problems, using the binary (power of 2) system.

2 Compute computer hard disk and CD-ROM storage capacities.

3 Compute computer modem communication speeds.

U.S. business has moved into the "Information Age." Computer-based transactions dominate business operations and communication. Incredibly rapid advances are being made in the areas of computer and information technology. Computer storage capacity and data transmission speeds are increasing rapidly. The future of business will be a future of technology.

1. What is the *binary number system* and why is it important to computers?
2. How does the binary number system differ from the *decimal system*?
3. What is meant by the terms *kilobytes*, *megabytes*, *gigabytes*, and *terabytes*?
4. What is a *bit*, and what is a *byte*?
5. Do all computers operate at the same speed?

Solving Simple Problems Using the Binary Number System

LEARNING OBJECTIVES

Solve simple problems, using the binary (power of 2) system.

Bit (b) is short for *binary digit,* the smallest unit of information on a computer. A bit holds only one of two values: 0 or 1. A computer can read these values as "on" or "off" or "charged" or "uncharged." All information stored, processed, and communicated by computers is reduced to bits.

Bits are based on the binary numbering system. Mathematical operations that are possible in the decimal system (addition, subtraction, multiplication, and division) are also possible in the binary system. We use the decimal system in everyday life, and it seems natural to us. For computers, the binary system is natural because computers operate with charged and uncharged electrical units, which computer scientists reduce to 1s and 0s.

In the decimal system, each digit in a number represents a power (or multiple) of 10, progressing from right to left. Ten options are available with the digits 0, 1, 2, 3, 4, 5, 6, 7, 8, and 9.

| 1,000s | 100s | 10s | 1s |
|--------|------|-----|-----|
| (10^3) | (10^2) | (10^1) | (10^0) |

For example, to write the number 3,142, you would select three 1,000s, one 100, four 10s, and two 1s.

In the binary system, each digit represents a power (or multiple) of 2. Two options available with the digits 0 and 1.

| 128s | 64s | 32s | 16s | 8s | 4s | 2s | 1s |
|------|-----|-----|-----|-----|-----|-----|-----|
| (2^7) | (2^6) | (2^5) | (2^4) | (2^3) | (2^2) | (2^1) | (2^0) |

For example, to write the binary equivalent of 202, you would select one 128, one 64, zero 32s, zero 16s, one 8, zero 4s, one 2, and zero 1s. The binary equivalent of 202 is 11001010.

EXAMPLE A

To add 17 and 34 in the decimal system, you write the numbers in two rows, aligning the digits in columns from right to left, and then add. In the binary system you use six rows.

Decimal Addition

| 10s | 1s |
|-----|-----|
| 1 | 7 |
| + 3 | 4 |
| 5 | 1 |

Equivalent Binary Addition

| 32s | 16s | 8s | 4s | 2s | 1s |
|-----|-----|-----|-----|-----|-----|
| | 1 | 0 | 0 | 0 | 1 |
| + 1 | 0 | 0 | 0 | 1 | 0 |
| 1 | 1 | 0 | 0 | 1 | 1 |

To add binary numbers in which 1 and 1 need to be added, you carry over to the next higher digit, as you would in the decimal system. Add 15 and 37.

| 10s | 1s |
|---|---|
| 1 ← carry over | |
| 1 | 5 |
| + 3 | 7 |
| 5 | 2 |

| 32s | 16s | 8s | 4s | 2s | 1s |
|---|---|---|---|---|---|
| | 1 | 1 | 1 | | 1 ← carry over |
| | | 1 | 1 | 1 | 1 |
| + 1 | 0 | 0 | 1 | 0 | 1 |
| 1 | 1 | 0 | 1 | 0 | 0 |

In the 1s column, 1 and 1 are 2. Write 0 for that column and carry 1 to the 2s column. That puts 1 and 1 in the 2s column, so write 0 and carry 1 to the 4s column. That puts three 1s in the 4s column. Carry two of them to the 8s column (two 4s become one 8s digit) and write 1 in the 4s column. In the 8s column, write 0 and carry 1 to the 16s column. Write 1 in the 16s column and 1 in the 32s column. The answer is 110100, which is 52.

EXAMPLE C

To subtract 36 from 54, you write the numbers as in examples A and B and then subtract.

| Decimal Subtraction | Equivalent Binary Subtraction | | | | | |
|---|---|---|---|---|---|---|
| | 32s | 16s | 8s | 4s | 2s | 1s |
| 54 | 1 | 1 | 0 | 1 | 1 | 0 |
| − 36 | − 1 | 0 | 0 | 1 | 0 | 0 |
| 18 | | 1 | 0 | 0 | 1 | 0 |

EXAMPLE D

To subtract binary numbers in which 1 needs to be subtracted from 0, you borrow from the next higher digit, as you would in the decimal system. Subtract 12 from 26:

| 10s | 1s |
|---|---|
| 2 | 6 |
| −1 | 2 |
| 1 | 4 |

| 16s | 8s | 4s | 2s | 1s |
|---|---|---|---|---|
| 1 | 1 | 0 | 1 | 0 |
| − | 1 | 1 | 0 | 0 |
| 1 | 1 | 1 | | 0 |

Do the subtractions in the 1s and 2s columns. In the 4s column, borrow 1 from the 8s column. One digit from the 8s column is the same as two 4s. Thus the answer in the 4s column is 1. For the 8s column, borrow 1 from the 16s column. One digit from the 16s column is the same as two 8s. The digit from the 8s column has already been lent, so we borrow one 16, or two 8s. Subtract 1 from that 2, and the answer in the 8s column is 1.

EXAMPLE E

To multiply 27 by 5 in binary numbers, write the numbers as you did for addition and subtraction. Multiply, being careful to align the numbers correctly. Add the columns as in the addition problems.

| | 256 | 128 | 64 | 32 | 16 | 8 | 4 | 2 | 1 |
|---|---|---|---|---|---|---|---|---|---|
| 27 | | | | | 1 | 1 | 0 | 1 | 1 |
| × 5 | | | | × | | | 1 | 0 | 1 |
| 135 | | | | | 1 | 1 | 0 | 1 | 1 |
| | | | 1 | 1 | 0 | 1 | 1 | | |
| | | 1 | 0 | 0 | 0 | 0 | 1 | 1 | 1 |

EXAMPLE F

For larger binary numbers, extend the binary digits to the left as needed.

$$121 \times 9 = 1{,}089$$

| | 1,024 | 512 | 256 | 128 | 64 | 32 | 16 | 8 | 4 | 2 | 1 |
|---|---|---|---|---|---|---|---|---|---|---|---|
| | | | | | 1 | 1 | 1 | 1 | 0 | 0 | 1 |
| × | | | | | | | | 1 | 0 | 0 | 1 |
| | | | | | 1 | 1 | 1 | 1 | 0 | 0 | 1 |
| | | 1 | 1 | 1 | 1 | 0 | 0 | 1 | | | |
| | 1 | 0 | 0 | 0 | 1 | 0 | 0 | 0 | 0 | 0 | 1 |

EXAMPLE G

To divide binary numbers, write the binary digits in order and complete the division process as you do with decimals.

$$\begin{array}{r} 41 \\ 4\overline{)164} \end{array}$$

| | 128 | 64 | 32 | 16 | 8 | 4 | 2 | 1 |
|---|---|---|---|---|---|---|---|---|
| (quotient) | | | 1 | 0 | 1 | 0 | 0 | 1 |
| 100) | 1 | 0 | 1 | 0 | 0 | 1 | 0 | 0 |
| | 1 | 0 | 0 | | | | | |
| | | | | 1 | 0 | 0 | | |
| | | | | 1 | 0 | 0 | | |
| | | | | | | 1 | 0 | 0 |
| | | | | | | 1 | 0 | 0 |

CONCEPT CHECK 29.1

a. Write the binary equivalent of 53.

 1 1 0 1 0 1

b. Add the binary numbers 10110 and 101.

```
   1 0 1 1 0
+      1 0 1
   1 1 0 1 1
```

c. Subtract the binary number 101 from the binary number 1110.

```
   1 1 1 0
 −   1 0 1
   1 0 0 1
```

d. Multiply the binary number 1101 and 1010.

```
       1 1 0 1
     × 1 0 1 0
     1 1 0 1 0
   1 1 0 1 0
 1 0 0 0 0 0 1 0
```

e. Divide the binary number 10100101 by the binary number 101.

```
            1 0 0 0 0 1
  1 0 1)1 0 1 0 0 1 0 1
        1 0 1
            0 0 1 0 1
```

COMPLETE ASSIGNMENT 29.1.

Computing Computer Memory and Storage Capacity

Bits become meaningful when combined in a series. A series of eight consecutive bits is called a **byte (B)**. A byte is a computer unit capable of holding a single character. A computer's memory and storage capacities are measured in bytes. The most common measures are the following.

2 LEARNING OBJECTIVES

Compute computer hard disk and CD-ROM storage capacities.

| Term | | Verbal Number | Approximate Bytes | Actual Bytes |
|------|---|---------------|-------------------|--------------|
| Kilobyte | (KB) | Thousand | 1,000 | 1,024 |
| Megabyte | (MB) | Million | 1,000,000 | 1,048,576 |
| Gigabyte | (GB) | Billion | 1,000,000,000 | 1,073,741,824 |
| Terabyte | (TB) | Trillion | 1,000,000,000,000 | 1,099,511,627,776 |

A computer's **memory** is a *space* where programs and documents are stored. All programs and documents are stored in bits and bytes.

Data, including a computer's programs and documents, are usually stored on a magnetic disk called a *hard disk* (or hard disk space). Microsoft Windows, a common computer program, requires approximately 120 MB of hard disk space. A complete office program that includes a word processing, spreadsheet, data base, and graphics presentation program requires approximately 180 MB of space.

A compact disk (CD) holds 600 MB, or 600 million bytes, or 600 million characters of information. In written communications, the average word length is 5 letters (or characters).

EXAMPLE H

About how many words can be stored on a CD that has a capacity of 600 MB?

600 MB ÷ 5 (characters per word) = 120,000,000 words

Many computers now come with hard disk capacity in gigabytes. Laptop computers frequently offer a 1 GB hard drive, and desktop PCs offer from 1 to 9 GB of hard disk capacity. Common data storage capacities are

1.0 gigabyte, or approximately 1,000,000,000 characters
1.2 gigabytes, or approximately 1,200,000,000 characters
3.2 gigabytes, or approximately 3,200,000,000 characters
4.3 gigabytes, or approximately 4,300,000,000 characters
6.4 gigabytes, or approximately 6,400,000,000 characters
9.0 gigabytes, or approximately 9,000,000,000 characters

EXAMPLE I

About how many business words can be stored on a 1.2 GB hard disk that has 400 MB of programs stored?

1.2 GB − 0.4 GB = 0.8 GB
0.8 GB ÷ 5 = 160,000,000 words

The amount of text contained in business letters varies widely. For this chapter, we assume that the average business letter, with letterhead, date, inside address, salutation, message, and complimentary close, contains the equivalent of 200 words.

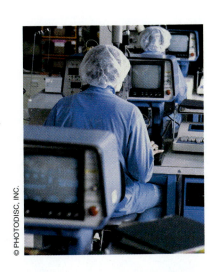

Pages of data are generally either double-spaced or single-spaced. Double-spaced pages usually have about 250 words; single-spaced pages usually have about 500 words.

EXAMPLE J

About how many business letters could be stored on a hard disk that has 4.3 GB of free space?

4,300,000,000 bytes ÷ 5 bytes per word = 860,000,000 words per disk
860,000,000 words ÷ 200 (words per business letter) = 4,300,000 letters

EXAMPLE K

About how many pages of single-spaced business text could be stored on a computer hard disk that has 6.2 GB of free space?

6,200,000,000 bytes ÷ 5 = 1,240,000,000 words
1,240,000,000 words ÷ 500 = 2,480,000 pages

A portable disk called a **floppy disk** can also be used to store data. A $3\frac{1}{2}$-inch floppy disk holds 1.44 MB, or approximately 1,440,000 characters. Newer Zip drive disks hold 100 (or more) MB, or approximately 100 million characters.

EXAMPLE L

About how many pages of double-spaced business text (250 words per page) could be stored on one 1.44 MB floppy disk?

1,440,000 bytes ÷ 5 bytes per word = 288,000 words
288,000 words ÷ 250 = 1,152 pages

CONCEPT CHECK 29.2

a. How many gigabytes are on a hard disk that has a total storage capacity of 640,000,000 words?

640,000,000 words × 5 bytes per word = 3,200,000,000 bytes
3,200,000,000 bytes ÷ 1,000,000,000 = 3.2 GB

b. About how many business letters could be stored on a 1.0 GB disk?

1.0 GB = 1,000,000,000 characters
1,000,000,000 characters ÷ 5 characters per word = 200,000,000 words
200,000,000 words ÷ 200 words per letter = 1,000,000 letters

c. About how many single-spaced pages of data could be stored on a 6.4 GB disk?

6.4 GB = 6,400,000,000 characters
6,400,000,000 characters ÷ 5 characters per word = 1,280,000,000 words
1,280,000,000 words ÷ 500 words per page = 2,560,000 pages

Computing Computer Modem Communication Speeds

A **modem** is a device that enables a computer to transmit data over telephone lines. Most modems for home PCs run at 28,800 to 56,600 **bits per second (bps)**. Note that computer communication speed is usually measured in bits (b), whereas computer hard disk space for storage is usually measured in bytes (B); one byte (1 B) = eight bits (8 b).

3
LEARNING OBJECTIVES

Compute computer modem communication speeds.

EXAMPLE M

How many bytes are transmitted in 1 minute by a modem transmitting at 28,800 bps?

28,800 bps × 60 seconds = 1,728,000 bits per minute
1,728,000 bits ÷ 8 bits per byte = 216,000 bytes

or

28,800 bps ÷ 8 bits per byte = 3,600 bytes per second
3,600 bytes × 60 seconds = 216,000 bytes

EXAMPLE N

How many 5-character words are transmitted in 5 minutes by a modem running at 33,600 bps?

33,600 bps ÷ 8 bits per byte = 4,200 bytes per second
4,200 bytes ÷ 5 characters per word = 840 words per second
840 words × 60 seconds × 5 minutes = 252,000 words

In the future, regular telephone lines for computers will be replaced by special integrated services digital network (ISDN) lines and wireless satellite dish systems. ISDN lines commonly transmit 1.5 megabits per second (Mbps); the new wireless satellite dishes commonly transmit 7.5 Mbps. At 7,500,000 bps, the satellite dishes send data more than 500 times faster than the common telephone line, which transmits at 28,800 bps.

In the future, more and more business communication will be via the **Internet**, a global web that currently connects more than 80 million computers in over 120 countries. Data transmission speed is increasing rapidly. During the next decade speeds are expected to be measured in terabits (one trillion bits) per second.

EXAMPLE O

Max's Office Supply uses the Internet to send detailed product specifications to its customers. If the length of one product specification is 200,000 characters (bytes), how long would it take to transmit it using a 28,800-bps modem, a 1.5-Mbps ISDN line, and a 7.5-Mbps satellite dish system?

200,000 × 8 bits per byte = 1,600,000 bits

For a 28,800 bps modem,

1,600,000 bits ÷ 28,800 bps = 55.56 seconds

For a 1.5 Mbps ISDN line,

1,600,000 bits ÷ 1.5 Mbps (1,500,000) = 1.07 seconds

For a 7.5 Mbps satellite dish system,

1,600,000 bits ÷ 7.5 Mbps (7,500,000) = 0.21 second

CONCEPT CHECK 29.3

a. How many bytes are transmitted in 1 minute by a modem transmitting at 33,600 bps?

33,600 bps × 60 seconds = 2,016,000 bits per minute
2,016,000 bits ÷ 8 bits per byte = 252,000 bytes per minute
or
33,600 bits ÷ 8 bits per byte = 4,200 bytes per second
4,200 bytes × 60 seconds = 252,000 bytes per minute

b. About how many business words are transmitted in 1 hour by a modem running at 28,800 bps?

28,800 bps ÷ 8 bits per byte = 3,600 bytes
3,600 bytes ÷ 5 bytes per word = 720 words per second
720 words × 60 seconds = 43,200 words per minute
43,200 words × 60 minutes = 2,592,000 words per hour

COMPLETE ASSIGNMENT 29.2

Chapter Terms for Review

bit (b) hard disk
bits per second (bps) Internet
byte (B) memory (computer)
floppy disk modem

Visit the Web

http://cap.ic.gc.ca/nb/stmartins/techbits/jargon.htm—This Web site explains in simple terms the meaning of various computer jargon.

http://www.computerworld.com/, http://www.freep.com/tech/, http://www.cddataguys.com/, http://www.eshop.com/, and **http://www.nectech.com/storage/, http://www.storage.ibm.com/**—Check some of these sites for more specific information, news, and other materials on computer issues.

CYBERCONNECTION

Going up: Broadband prices may be increasing
The price of high-speed Internet access could be headed higher.

By Corey Grice, News.com
February 26, 2001 6:17 AM PT

After rates have remained unchanged at about $40 a month for the past few years, industry analysts and executives say a confluence of several events may mean rate increases are on the way.

Among the warning signs:

Many DSL (digital subscriber line) companies are seeking buyers or are in bankruptcy court—including Flashcom, PSINet and others—or are financially weakened, such as Covad Communications. Less competition opens the door for survivors to bump up rates because consumers have fewer alternatives.

A plunge in advertising revenue is forcing providers to find other sources of revenue.

Demand remains high for broadband services.

Consumers have grown accustomed to regular increases in rates for utilities such as cable TV, natural gas and electricity. This may embolden broadband companies to up their own rates.

The largest ISP, AOL Time Warner, is considering increasing its monthly dial-up fee of $21.95, which could spill over to other dial-up providers and broadband companies.

"Raising prices is a natural reaction. I think it's to be expected," said Jeff Sadler, an ISP stock analyst at FAC/Equities. "If you're not making money on advertising and you've got objectives to meet, raising prices is one of the ways to do it. Who is going to step in to counter it?"

Charles Ardai, chief executive at Juno Online Services, echoes that sentiment. "The cost of rolling out the (broadband) network is too high relative to the revenue it produces. If the revenue is not meeting their costs, I do think you'll see a trend toward higher prices," he said.

Juno provides primarily free dial-up access, but also offers broadband access through several partners for $49.95.

One major broadband provider already bumped up its rates, and there is speculation that even dial-up Internet leader America Online will raise its rates this year.

. . .

Microsoft

The company's top executives have said they are considering price hikes, and now some analysts believe a dial-up rate hike is likely.

"We are becoming skeptical that AOL Time Warner can fully reach its goal of $11 billion EBITDA (earnings before interest, taxes, depreciation and amortization) in 2001 without a price increase at AOL," ABN AMRO equity analyst Arthur Newman wrote in a recent research report. "If a price increase of $2 per subscriber were to be instituted around midyear, and all the Time Warner sectors perform at optimistic levels, the $11 billion could be reached, but it might be a stretch."

Making the economics work

As AOL considers a price hike, analysts point out that its dial-up service is cheaper to maintain—additional evidence that price pressure could spill over to the broadband sector, where the cost of delivering service is higher.

The equipment necessary to install a broadband connection is expensive, meaning providers generate roughly the same amount of profit margin from a high-speed customer as dial-up ISPs collect for their $20-a-month services.

"The profit margin for a DSL customer is about the same dollar contribution as from a dial-up customer," Sadler said.

"But on a percentage basis it's a smaller profit margin.

In other words, the capital investment costs of provisioning service for a broadband customer are higher, but the actual profit is roughly the same as offering dial-up Net access, analysts say.

. . .

Competition will keep prices competitive

Not everyone is convinced, however, that consumers who enjoy high-speed downloading and Web surfing need worry about imminent increases.

Gary Hancock, a senior analyst for communications infrastructure at Current Analysis, a market analysis firm, argues that DSL line-sharing laws, which allow competitors to carry traffic over consumers' existing phone wires without incurring the cost of leasing a second phone line, will lead to lower-cost broadband connections. Hancock also believes there is enough competition yet to ensure further pricing battles.

"I think prices are likely to fall," Hancock said. "If they lower their price, they're likely to get more people. The lower you make the price point, the more subscribers you'll get."

. . .

Even if prices are to rise, analysts say consumers will come to tolerate small rate hikes rather than endure the hassles of switching providers, e-mail addresses and other tangles associated with adopting an alternative provider. [According to Emily Meehan, a senior Internet industry analyst at the Yankee Group, a market research firm,] "Incremental increases are acceptable. Large ones must be accompanied by some additional value."

Notes

The Bottom Line

Summary of chapter learning objectives:

| Learning Objective | Summary | Example |
|---|---|---|
| **29.1** | Solve simple problems, using the binary (power of 2) system | 1. Subtract 32 from 58, using the digital and binary systems.
 2. Multiply 32 × 58, using the digital and binary systems. |
| **29.2** | Compute hard disk and CD-ROM storage capacities | 3. Stored on Warren Jameson's 6.2 GB hard disk are software programs that take 1.4 GB of space. About how many business words can he put on the remaining disk space? |
| **29.3** | Compute computer modem communication speeds | 4. Consumes Junior College is installing a new network system. It is considering two different modems with transmitting speeds of 128 Kbps and 1.6 Mbps. How much time would the faster modem save if the college transmitted 100 documents, each containing 800 five-character words? |

Answers: 1. 26; 11010 2. 1,856; 1110100000000 3. 960,000,000 4. 23 sec

1 Write and add the binary equivalent of:
 a. $37 + 7 = 44$
 b. $29 + 16 = 45$

5 How many business words (average of 5 letters) can be stored
 a. on a floppy disk of 1.44 MB?
 b. on a 1.2 GB hard disk that has 600 MB of programs stored on it?
 c. on a hard disk that has 3.2 GB of free space?

2 Write and subtract the binary equivalent of:
 a. $17 - 5 = 12$
 b. $128 - 63 = 65$

6 How many bytes are transmitted in 1 minute by a modem transmitting at 56,600 bps?

3 Write and multiply the binary equivalent of:
 a. $7 \times 5 = 35$
 b. $32 \times 2 = 64$

7 How many 5-letter business words are transmitted in 1 minute by a modem running at
 a. 30,000 bps?
 b. 14,400 bps?

4 Write and divide the binary equivalent of:
 a. $18 \div 3 = 6$
 b. $100 \div 25 = 4$

8 If a letter contains 200 words, how many such letters can be transmitted in 1 minute by a modem running at 1.5 million bps?

Assignment 29.1: Math in Computer Operations

LEARNING
OBJECTIVES

Name _____

Date _____ Score _____

A (32 points) Solve the following binary system addition problems and then translate each into its decimal system equivalent. (2 points for each correct answer)

1. 100100
 + 10011

2. 110100
 + 10011

3. 111100
 + 101011

4. 110000
 + 11000

5. 111111
 + 11111

6. 11011
 + 1011

7. 101010
 + 10010

8. 11111
 + 1001

Score for A (32)

B (24 points) Solve the following binary system subtraction problems, and then translate each into its decimal system equivalent. (2 points for each correct answer)

9. 11011
 − 1011

10. 101010
 − 10010

11. 11011
 − 1111

12. 10000
 − 111

13. 110000
 − 11000

14. 100111
 − 10111

Score for B (24)

C (24 points) Solve the following binary system multiplication problems, and then translate each into its decimal system equivalent. (2 points for each correct answer)

15. 1111
 × 100

16. 1010
 × 11

17. 11000
 × 1101

18. 1110
 × 111

19. 1100
 × 101

20. 110000
 × 11000

Score for C (24)

D (20 points) Solve the following binary system division problems, and then translate each into its decimal system equivalent. (2 points for each correct answer)

21. 101)1111

22. 110)1100110

23. 11)11011

24. 110010)11001000

25. 111)111111

Score for D (20)

Assignment 29.2: Math in Computer Operations

Name _____

Date _____ Score _____

A **(50 points) Solve the following disk capacity problems. (5 points for each correct answer)**

1. Rupert Goings has an older computer with a 1.2 GB hard drive. If he installed software that uses 240 MB and then produced documents that use 180 MB, what capacity would remain for adding software or documents? _____

2. Anne Boris wants to send document files to a friend to load onto her computer. The file sizes are 122 KB, 207 KB, 4 MB, 11 KB, and 107 KB. How many floppy disks will Anne need? _____

3. How many floppy disks would be needed to download a software program that requires 1.1 GB of space? _____

4. Zane's 2.3 GB hard disk stores software programs that take 1.1 GB of space. About how many business words can be put onto the remaining disk space? _____

5. Refer to problem 4. If the average business document contained 300 words, how many documents could be stored on the remaining disk space? _____

6. How many gigabytes of disk space would be required to store 1 million business documents that average 700 words each? _____

7. Nancy Borderman has a computer with 3.5 GB of hard disk storage space. She has used 85% of it. She plans to create new documents that will require approximately 800 MB of space. What percent of the new documents can she put onto her hard disk? _____

8. Refer to problem 7. How many 100 MB Iomega disks would be required to store the data remaining after the hard disk was filled? _____

9. The Metcalf Company has an intranet system with seven computers on line. Its central server has 7.3 GB capacity, 30% of which is taken up by software programs. How many single-spaced pages of approximately 500 words each could be stored in the remaining space? _____

10. How many floppy disks would be required to copy the contents of a 2.3 GB hard disk that is 93% used?

B **(50 points) Solve the following computer communication problems. (10 points for each correct answer)**

11. An Internet service provider announces that it can give its customers the capacity to download Web pages at up to 7.5 Mbps. How many times faster is this than a regular telephone modem downloading at 28,800 bps? _____

12. T1 telephone lines transmit Internet data at 1.5 Mbps. How many times faster is this than a phone line that transmits at 54,000 bps? _____

13. A new version of an Internet browser required 54 MB of hard disk space. It was offered free to those computer users who wanted to download the program. How many minutes would it take to download the program at each of the following transmission speeds? (Round all computations to the nearest whole number.) (2 points for each subpart.)

 a. 33,000 bps _____

 b. 56,000 bps _____

 c. 128,000 bps _____

 d. 1,500,000 bps _____

 e. 9,000,000 bps _____

14. Paton Company is considering installing a network system with a modem transmission speed of either 128 KB or 1.5 MB. What would be the difference in time required to transmit 75 business documents, each containing 700 five-character words? _____

15. How many graphic illustrations are transmitted in 3 minutes by a modem running at 56,000 bps if each illustration contains 7,000 bytes? _____

Notes

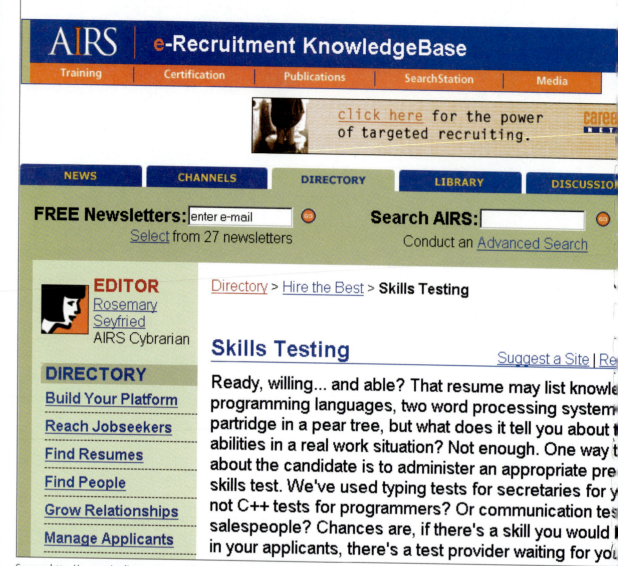

Founded in 1997, AIRS began life as a training firm, offering seminars to help recruiters and human resource professionals find passive candidates hidden among the 250 million people who use the Internet. Once controversial, AIRS active search techniques are now mainstream recruitment methods.

The firm's Web site is intended for management personnel who can link to sites providing consulting and/or testing for candidates for jobs. Visit this robust site at http://www.airsdirectory.com/, look around, and try some of its many useful links. Some can give you an idea of what to expect on employment tests and some offer brief sample tests.

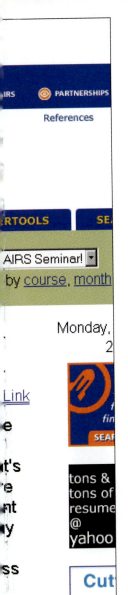

Math in Employment Tests

30

LEARNING OBJECTIVES

By studying this chapter and completing all assignments you will be able to:

1
LEARNING OBJECTIVES
Solve rate, time, and distance problems.

2
LEARNING OBJECTIVES
Solve proportion problems.

3
LEARNING OBJECTIVES
Solve time and work problems.

4
LEARNING OBJECTIVES
Solve weight and measure problems.

5
LEARNING OBJECTIVES
Solve percentage problems.

6
LEARNING OBJECTIVES
Solve relationship problems.

Taking employment tests is stressful for many applicants for business and government jobs. Stress can be reduced by confidence, which can be built up through practice and familiarity with the kinds of mathematical problems that often appear on the tests.

1. Most employment tests include math problems. Why?
2. What is the benefit of practicing the kind of problems commonly used in employment tests?
3. Why do employers regularly use tests to help them determine whether to hire a prospective employee?
4. Are most "rate, time, and distance" problems solved with common sense?
5. Why is it important to read a problem thoroughly and think about what is being asked before starting to solve it?

A knowledge of business math is important in applying for both business and government jobs. This chapter reviews the types of problems commonly found on employment tests. Previous chapters in this book have introduced each type of problem solution.

Solving Rate, Time, and Distance Problems

LEARNING OBJECTIVES

Solve rate, time and distance problems.

In all rate, time, and distance problems, the formula is simple:

> Rate (or Speed) × Time = Distance

If you know any two factors, you can easily find the third:

Rate × Time = Distance
Distance ÷ Time = Rate
Distance ÷ Rate = Time

EXAMPLE A

A train leaves a station traveling at 40 mph. How far has the train traveled 6 hours later?

 Rate × Time = Distance
40 mph × 6 hr = 240 mi

EXAMPLE B

A train traveled 550 miles in 11 hours. How fast was the train going?

Distance ÷ Time = Rate
 550 mi ÷ 11 hr = 50 mph

EXAMPLE C

A train traveled 300 miles at 60 mph. How many hours did the journey require?

 Distance ÷ Rate = Time
300 mi ÷ 60 mph = 5 hr

EXAMPLE D

www.tapdance.com

This Web site provides some information on what to expect in a business math test.

Train X leaves station X traveling 40 mph on a 400-mile trip to station Y. Train Y leaves station Y traveling 60 mph over the same 400 miles toward station X. How many hours will the trains have traveled when they pass each other?

 Distance = 400 mi
 Total rate = 40 mph (X) + 60 mph (Y) = 100 mph
 Distance ÷ Rate = Time
400 mi ÷ 100 mph = 4 hr

EXAMPLE E

How far will train X in example D have traveled when the two trains pass each other?

 Rate × Time = Distance
40 mph × 4 hr = 160 mi

Ellen Ross walks 4 miles per hour. Her brother walks 6 miles per hour. He leaves home an hour after Ellen, going in the same direction. How far ahead of Ellen will her brother be when Ellen has walked 32 miles?

Determine the time required for Ellen to walk 32 miles:

Distance ÷ Rate = Time

32 mi ÷ 4 mph = 8 hr

Determine the distance her brother walks:

Rate × Time = Distance

6 mph × (8 − 1 = 7 hr) = 42 mi

Subtract:

42 mi − 32 mi = 10 mi ahead

EXAMPLE G

Margie Conklin types 50 words per minute. Sharon Levi types 75 words per minute. They each typed a page with 300 words. Margie started 1 minute before Sharon. Who finished first?

Margie's time: (300 words ÷ 50 wpm) − 1 min head start = 5 min

Sharon's time: 300 words ÷ 75 wpm = 4 min

Sharon finished first: 4 min vs. 5 min

CONCEPT CHECK 30.1

a. Sam Jones and Mary McDonnell start traveling toward each other from 360 miles apart. Sam is traveling at 40 miles per hour, Mary at 50 miles per hour. How much time will elapse before they meet?

Distance = 360 mi

Total rate = 40 mph + 50 mph = 90 mph

360 mi ÷ 90 mph = 4 hr

b. Julie Smith types 50 words per minute. Brad Faxon types 75 words per minute. They each typed a page with 400 words. Julie started 2 minutes before Brad. Who finished first?

Julie's time: (400 words ÷ 50 wpm) − 2 min head start = 6 min

Brad's time: 400 words ÷ 75 wpm = $5\frac{1}{3}$ min

Brad finished first: $5\frac{1}{3}$ min vs. 6 min

Solving Proportion Problems

Almost all employment tests include proportion problems. The *unit method* is a simple and fast way to solve proportion problems. To use this method, find a single basic unit of 1 in the problem and then proceed to the answer. These problems may also be solved by proportionate shares (see example J).

LEARNING OBJECTIVES

Solve proportion problems.

EXAMPLE H

To sort 360 letters, three clerks require 4 hours. How many letters can seven clerks sort in 2 hours?

$$3 \times 4 = 12 \text{ clerk hours to sort 360 letters}$$
$$360 \div 12 = 30 \text{ letters per clerk hour (1 unit)}$$
$$7 \text{ clerks} \times 2 \text{ hours} = 14 \text{ clerk hours}$$
$$14 \times 30 = 420 \text{ letters}$$

EXAMPLE I

The number of pennies in a cash box was twice the number of nickels. There were five times as many nickels as there were dimes. All the coins totaled $36. How many pennies were there?

Group the coins:

10 pennies + 5 nickels + 1 dime = 45 cents

Determine the number of single units:

$36 ÷ $0.45 = 80 units

There are 10 pennies in each of the 80 units; therefore, multiply to find the answer:

80 units × 10 pennies = 800 pennies

EXAMPLE J

X, Y, and Z invest in a business as follows: X, $150; Y, $250; Z, $400. Later, the three divide $1,200 profit in proportion to their investments. How much does each receive?

Total investment = $150 + $250 + $400 = $800

X's share: $\dfrac{150}{800} \times \$1{,}200 = \$225$

Y's share: $\dfrac{250}{800} \times \$1{,}200 = \$375$

Z's share: $\dfrac{400}{800} \times \$1{,}200 = \$600$

CONCEPT CHECK 30.2

a. To wash 36 cars, three men require 6 hours. How many cars can nine men wash in 4 hours?

$$3 \times 6 = 18 \text{ man hours to wash 36 cars}$$
$$36 \div 18 = 2 \text{ cars per man hour (1 unit)}$$
$$9 \text{ men} \times 4 \text{ hours} = 36 \text{ man hours}$$
$$36 \times 2 = 72 \text{ cars}$$

b. A, B, C, and D invest in a building together, as follows: A, $4,000; B, $6,000; C, $10,000; D, $12,000. Later, they sell the building for a profit of $8,000. They divide the profit in proportion to their investments. How much profit does each receive?

Total investment = $4,000 + $6,000 + $10,000 + $12,000 = $32,000

A's share: $\dfrac{4{,}000}{32{,}000} \times \$8{,}000 = \$1{,}000$ 　　B's share: $\dfrac{6{,}000}{32{,}000} \times \$8{,}000 = \$1{,}500$

C's share: $\dfrac{10{,}000}{32{,}000} \times \$8{,}000 = \$2{,}500$ 　　D's share: $\dfrac{12{,}000}{32{,}000} \times \$8{,}000 = \$3{,}000$

www. mycareereducation. org/

The Michigan Department of Career Development provides information and assistance to job seekers. Visit its Web site and check your own state's Web site to see if you can find similar help.

Solving Time and Work Problems

Time and work problems are another form of problem frequently included in employment tests. The first and most important step in solving time and work problems is to find the *fraction* of the job that can be completed in one unit of time (for example a day, an hour, or a minute). Once you have found this amount, divide the *denominator* by the *numerator* to get your answer.

EXAMPLE K

X can do a job in 5 days. Y can do the job in 10 days. How long will it take the two of them working together to do the job?

X takes 5 days: 1 day $= \frac{1}{5}$ of job.

Y takes 10 days: 1 day $= \frac{1}{10}$ of job.

X and Y together do $\frac{1}{5} + \frac{1}{10}$, or $\frac{3}{10}$, of the job in 1 day.

Therefore they can finish the job in $3\frac{1}{3}$ days $\left(10 \div 3 = 3\frac{1}{3}\right)$.

EXAMPLE L

X and Y together can do a job in 4 hours. X can do the job alone in 16 hours. How long would it take Y to do the job alone?

X and Y take 4 hours: 1 hour $= \frac{1}{4}$ of job.

X alone takes 16 hours: 1 hour $= \frac{1}{16}$ of job.

Y alone can do $\frac{1}{4} - \frac{1}{16}$, or $\frac{3}{16}$, of the job in 1 hour.

Therefore Y can complete the job alone in $5\frac{1}{3}$ hours $\left(16 \div 3 = 5\frac{1}{3}\right)$.

© PHOTODISC, INC.

✓ CONCEPT CHECK 30.3

a. A can do a job in 6 minutes. B can do the job in 4 minutes. C can do the job in 4 minutes. How long will it take the three of them working together to do the job?

A takes 6 minutes: 1 minute $= \frac{1}{6}$ of job.

B takes 4 minutes: 1 minute $= \frac{1}{4}$ of job.

C takes 4 minutes: 1 minute $= \frac{1}{4}$ of job.

A, B, and C together do $\frac{1}{6} + \frac{1}{4} + \frac{1}{4}$, or $\frac{2}{3}$, of the job in 1 minute.

Therefore they can finish the job in $1\frac{1}{2}$ minutes $\left(3 \div 2 = 1\frac{1}{2}\right)$.

b. Carl Samuelson and Ron Delaney together can paint a house in 6 days. Carl can paint the house alone in 10 days. How long would it take Ron to do the job alone?

Carl and Ron take 6 days: 1 day $= \frac{1}{6}$ of job.

Carl alone takes 10 days: 1 day $= \frac{1}{10}$ of job.

Ron alone can do $\frac{1}{6} - \frac{1}{10}$, or $\frac{1}{15}$, of the job in 1 day.

Therefore Ron can complete the job in 15 days (15 ÷ 1 = 15).

Solving Weight and Measure Problems

LEARNING OBJECTIVES

Solve weight and measure problems.

The traditional measures shown in Figure 30.1 are commonly included in employment tests. You should memorize them.

Test problems generally involve changing from smaller units to larger ones and vice versa; adding, subtracting, multiplying, and dividing measures; and reasoning to logically apply units of measure. We used Figure 30.1 in working the following examples.

EXAMPLE M

How many hours equal 9,000 seconds?

9,000 ÷ 60 (sec in 1 min) = 150 min

 150 ÷ 60 (min in 1 hr) = 2 hr 30 min or

 2 hr 30 min $= 2\frac{30}{60} = 2\frac{1}{2}$ hr

60 sec = 1 min; 60 min = 1 hr

60 sec × 60 min = 3,600 sec = 1 hr

9,000 sec ÷ 3,600 sec $= 2\frac{1}{2}$ hr

EXAMPLE N

How many inches are there in 6 yards?

 6 yd × 3 ft = 18 ft

18 ft × 12 in. = 216 in.

EXAMPLE O

Add 2 yards, 1 foot, 9 inches and 3 yards, 2 feet, 6 inches.

| Yards | Feet | Inches |
|-------|------|--------|
| 2 | 1 | 9 |
| 3 | 2 | 6 |
| 1 | 1 | |
| 5 | 3 | 15 |
| | −3 | −12 |
| 6 | 1 | 3 |

In the final step, smaller units are changed to the next larger units, where possible.

Weight

| | |
|---|---|
| 16 ounces (oz) | = 1 pound (lb) |
| 2,000 pounds | = 1 ton |

Length

| | |
|---|---|
| 12 inches (in.) | = 1 foot (ft) |
| 3 feet | = 1 yard (yd) |
| 5,280 feet | = 1 mile (mi) |
| 1,760 yards | = 1 mile |

Time

| | |
|---|---|
| 60 seconds (sec) | = 1 minute (min) |
| 60 minutes | = 1 hour (hr) |
| 24 hours | = 1 day (da) |
| 7 days | = 1 week (wk) |
| $4\frac{1}{3}$ weeks | = 1 month (mo) |
| 52 weeks | = 1 year (yr) |
| 12 months | = 1 year |

Capacity

| | |
|---|---|
| 2 cups | = 1 pint (pt) |
| 2 pints | = 1 quart (qt) |
| 4 quarts | = 1 gallon (gal) |
| 16 ounces (1 lb) | = 1 pint |

Area

| | |
|---|---|
| 144 square inches | = 1 square foot (sq ft) |
| 9 square feet | = 1 square yard (sq yd) |
| 43,560 square feet | = 1 acre (a) |
| 640 acres | = 1 square mile (sq mi) |

Volume

| | |
|---|---|
| 1,728 cubic inches | = 1 cubic foot (cu ft) |
| 27 cubic feet | = 1 cubic yard (cu yd) |
| 1 cubic foot | = $7\frac{1}{2}$ gallons of water |

EXAMPLE P

Subtract 3 yards 10 inches from 15 yards 2 feet 7 inches.

| Yards | Feet | Inches |
|---|---|---|
| | 1 | 19 |
| 15 | 2̶ | 7̶ |
| −3 | − | −10 |
| 12 | 1 | 9 |

Because 10 inches cannot be subtracted from 7 inches, 1 foot (12 inches) is taken from the Feet column and added to the 7 to make 19 inches, from which 10 inches can be subtracted.

EXAMPLE Q

How many square yards of linoleum are used for a floor 18 feet wide and 27 feet long?

18 ft × 27 ft = 486 sq ft

486 sq ft ÷ 9 = 54 sq yd

EXAMPLE R

If a 14 foot 9 inch rod were cut into 3 equal pieces, how long would each piece be?

14 ft × 12 in. = 168 in.

168 in. + 9 in. = 177 in.

177 in. ÷ 3 = 59 in.

59 in. ÷ 12 in. = 4 ft 11 in.

www.ase-solutions.co. uk/html/practice/ index.htm

Visitors to the United Kingdom's ASE Web site are encouraged to try the practice verbal and numerical skills tests before their employment interviews and testing.

www.brainbench.com/ xml/bb/common/ testcenter/betatests. xml#

Visit this Web site for sample tests for accounting, finance, health care, hospitality, purchasing, computers, x-ray technology, and more.

EXAMPLE S

Find the gallons in a tank 18 feet long by 6 feet wide with 4 feet of water.

Length × Width × Height = Volume

18 ft × 6 ft × 4 ft = 432 cu ft

$432 \text{ cu ft} \times 7\frac{1}{2} \text{ gal per cu ft} = 3{,}240 \text{ gallons}$

CONCEPT CHECK 30.4

a. Subtract 1 yard 2 feet 11 inches from 4 yards 2 feet 6 inches

| Yards | Feet | Inches |
|-------|------|--------|
| | 4 | |
| 3 | ~~3~~ | 18 |
| 4 | ~~2~~ | ~~6~~ |
| −1 | −2 | −11 |
| 2 | 2 | 7 |

b. A contractor dug a swimming pool 36 feet long and 20 feet wide and filled it to a depth of 6 feet. What was the volume of water in the pool? (Answer in gallons.)

Find the cubic feet of water:
36 ft × 20 ft × 6 ft = 4,320 cu ft of water

Find the volume of 4,320 cu ft of water:

$4{,}320 \text{ cu ft} \times 7\frac{1}{2} \text{ gal per cu ft} = 32{,}400 \text{ gal}$

c. How many minutes are there in 24 hours?

24 hr × 60 min per hr = 1,440 min

d. How many square yards of carpet does it take to carpet a room 24 feet long and 18 feet wide?

24 ft × 18 ft = 432 sq ft
432 sq ft ÷ 9 sq ft per sq yd = 48 sq yd

e. A carpenter saws a board 20 feet 10 inches long into two equal pieces. How long is each piece?

20 ft ÷ 2 = 10 ft
10 in. ÷ 2 = 5 in.
Each piece is 10 ft 5 in. long.

Solving Percentage Problems

LEARNING OBJECTIVES

Solve percentage problems.

A percent is a fractional expression whose denominator is 100. Percent may be expressed by using a percent sign (%) or a decimal point (.). Fifteen percent, for example, is 15% or 0.15.

EXAMPLE T

A worker earning $400 a week saves 12% of her earnings. How much does she save each week?

Base × Rate = Percentage
$400 × 12% = $48 saved each week

EXAMPLE U

If a customer's 15% discount on a purchase amounted to $30, what was the total amount of the sale?

Percentage ÷ Rate = Base
$30 ÷ 0.15 = $200 total sale

EXAMPLE V

Of 120 employees in an organization, 90 attended the company picnic. What percent of the employees attended the picnic?

Percentage ÷ Base = Rate
90 ÷ 120 = 75% attended

EXAMPLE W

A company earned a profit of $8,000 in 1997 and $6,000 in 1998. What was the rate of decrease?

Find the dollar amount of decrease:

$8,000 − $6,000 = $2,000 decrease

Apply the formula:

Amount of decrease ÷ Original amount = Rate of decrease
$2,000 decrease ÷ $8,000 original amount = 25% rate of decrease

EXAMPLE X

A company had a profit of $9,000 in 1997 and $10,500 in 1998. What was the rate of increase?

Find the dollar amount of increase:

$10,500 (1998) − $9,000 (1997) = $1,500 increase

Apply the formula:

Amount of increase ÷ Original amount = Rate of increase
$1,500 increase ÷ $9,000 original amount = 16.67% rate of increase

 CONCEPT CHECK 30.5

a. A customer was given a 15% discount on furniture costing $180. What was the amount of discount?

Base × Rate = Percentage
$180 × 15% = $27

b. Another customer was given a 10% discount of $25 on a dining room set. What was the original cost of the dining room set?

Percentage ÷ Rate = Base
$25 ÷ 10% = $250

c. A third customer was given $30 off on a $200 couch. What was the rate of discount?

Percentage ÷ Base = Rate
$30 ÷ $200 = 15%

d. Three hundred customers purchased season tickets to ball games last year; 250 purchased them this year. What is the rate of decrease?

Find the amount of decrease:
300 − 250 = 50

Apply the formula:
50 decrease ÷ 300 original amount = 16.67% rate of decrease

e. A salesman sold 14 cars last year and 16 this year. What is the rate of increase?

Find the amount of increase:
16 − 14 = 2

Apply the formula:
2 ÷ 14 = 14.29% rate of increase

Solving Relationship Problems

6
LEARNING
OBJECTIVES

Solve relationship problems.

Relationships in a series of numbers may be found by comparing the first three or four terms in the series.

EXAMPLE Y

Complete the series: 3, 6, 9, 12, 15, _____, _____
Add 3 to the preceding number. The last two terms are 18 and 21.

A series might combine two or more steps.

Complete the series: 4, 8, 6, 10, 8, 12, _____, _____
Alternately add 4 and subtract 2. The last two terms are 10 and 14.

A series might be progressive.

Complete the series: 1, 3, 6, 10, _____, _____
The difference between successive numbers increases by 1. The last two terms are 15 and 21.

Complete the series: 1, 2, 6, 24, _____, _____
Multiply consecutively times 2, times 3, times 4, and so on. The last two terms are 120 and 720.

Complete the series: 1, 4, 2, 8, 4, 16, _____, _____
Alternately multiply by 4 and divide by 2. The last two terms are 8 and 32.

Number relationships may also be visualized from a verbal description.

EXAMPLE Z

A person walked 2 miles south, then 3 miles east, then 2 miles north, then 4 miles west. How far was he from his starting point?

1 mile

✓ CONCEPT CHECK 30.6

a. Complete the series: 4, 8, 12, 16, _____, _____

 Add 4; the last two terms are 20 and 24.

b. Complete the series: 50, 47, 44, 41, _____, _____

 Subtract 3; the last two terms are 38 and 35.

c. Complete the series: 5, 10, 7, 12, 9, 14, _____, _____

 Alternately add 5 and subtract 3; the last two terms are 11 and 16.

d. Complete the series: 2, 4, 8, 16, _____, _____

 Multiply by 2; the last two terms are 32 and 64.

e. Complete the series: 2, 8, 4, 16, 8, 32, _____, _____

 Alternately multiply by 4 and divide by 2; the last two terms are 16 and 64.

f. A person drove 4 miles east, then back 3 miles west, then back 2 miles east, then back 1 mile west. How far was she from her starting point?

 2 miles

COMPLETE ASSIGNMENTS 30.1, 30.2, and 30.3.

Visit the Web

www.brgarrison.com—This Web site features various types of employment tests and permits downloading of examples for some commonly used tests.

www.metronet.lib.mi.us/WEST/tclcareer.html—This Web site offers career information, links to the "Occupational Outlook Handbook," and tests to determine job suitability.

www.CareerPerfect.com—Online career planning, tests, interview questions, job search techniques, sample resumés, and job search assistance are presented on this Web site.

http://collegegradjobs.about.com/careers/collegegradjobs/—Here is a major resource for college graduates looking for jobs and careers. It provides interview schedules, test preparation, and worldwide job opportunities.

www.pse-net.com—This major resource for government jobs provides listings for local, state, and federal government jobs in the United States; job interview questions; and civil service tests.

http://www.employeeselect.com/—For fun, take the personality test at this Web site.

http://www.psionline.com/bstsample.htm, **http://www.asktheheadhunter.com/gv000802.htm**, and **http://amby.com/educate/math.html**—Additional information on job seeking and testing may be found at these Web sites.

http://works.state.mo.us/tips/a13.htm

The State of Missouri provides information on what a person should expect when preparing to take employment tests.

CYBERCONNECTION

COMPETENCY MANAGEMENT, INC

Testing

Bottom Line Payoff From Personnel Selection Programs

Can you afford costly hiring mistakes?

Does a 30% reduction in unnecessary turnover interest you?

Now more than ever, turnover is a primary concern of many business organizations. Excessive turnover has significant impact on workforce productivity because it leads to morale problems, team dysfunction, and results in high recruitment and retraining costs. A Competency Management Incorporated (CMI) designed pre-employment selection program can directly reduce turnover related to personnel selection.

What are the Benefits of CMI Basic Skills Testing Program? A time series study assessing the effectiveness of a CMI pre-employment process over time indicated:

 o A 57% reduction in turnover rates from 1991-1994 (pre-CMI) through 1997 (post-CMI).
 o A return on investment analysis indicated that this reduction resulted in an estimated annual savings of
 $720,000 to two positions alone. Other clients have realized over one million dollars savings in salary cost

[Return to Contents] [Return to Home]

Basic Arithmetic Test
Measures individual's ability to accurately solve basic addition, subtraction, multiplication and division problems in relation to:

- Handling cash
- Processing claims
- Reviewing numeric documents

Test used for positions such as:

- Secretaries
- Administrative Support Staff
- Entry Level Technicians
- Cashiers
- Building Services

[Return to Contents] [Return to Home]

Practical Arithmetic Test
Measures individual's ability to accurately solve arithmetic problems using whole numbers, fractions, decimals and percentages in relation to:

Source: http://www.competencymanagement.com/testing.shtml. Used with permission, Competency Management, Inc., www.competencymanagement. com, © 2001, Competency Management Inc.

At the Competency Management, Inc., Web site, you can find what you need to know for basic mathematics testing for employment.

 Study this chapter thoroughly, complete the assignments, and practice the review problems. Then visit this site, and the others highlighted in this chapter, as you prepare for your first employment test. Do you feel ready?

Summary of chapter learning objectives:

| Learning Objective | Summary | Example |
|---|---|---|
| **30.1** | Solve rate, time, and distance problems | 1. Plane A leaves San Francisco for Honolulu, a distance of 2,600 miles, at a speed of 500 miles an hour. Plane B leaves for the same flight a half hour later, flying at a speed of 600 miles an hour. How far will plane A have flown when plane B overtakes it? |
| **30.2** | Solve proportion problems | 2. To stuff 600 envelopes, four clerks require 3 hours. How many envelopes can six clerks stuff in 3 hours? |
| **30.3** | Solve time and work problems | 3. X can do a job in 6 days. Y can do the same job in 5 days, and Z in 4 days. If X, Y, and Z work together on the job, how long will it take them to finish? |
| **30.4** | Solve weight and measure problems | 4. A tank is 15 feet by 18 feet and is filled to a depth of 5 feet with water. What is the volume of the water in the tank? (Give your answer in gallons.) |
| **30.5** | Solve percentage problems | 5. A company employs 480 workers. Half of them drive to work. Of those who drive, 15% drive black cars. How many black cars are in the company parking lot?
6. A store has 2,000 customers with store credit cards. Of these customers, 420 attended a special sale. What percent of credit card customers attended the sale? |
| **30.6** | Solve relationship problems | 7. Complete the series: 12, 24, 36, 48, _____, _____
8. Complete the series: 2, 5, 3, 6, _____, _____
9. Complete the series: 3, 12, 6, 24, _____, _____
10. Complete the series: 1, 3, 6, 10, _____, _____ |

Answers: 1. 1,500 miles 2. 900 3. $1\frac{23}{60}$ days, or 1.38 days 4. 10,125 gal 5. 36 6. 21% 7. 60, 72 8. 4, 7 9. 12, 48 10. 15, 21

1 a. Carl Weber and Joe Krafcik painted a wall together in 5 hours. Earlier Carl had painted the wall alone in 12 hours. How long would it take Joe alone to paint the wall?

b. Carl painted a house in 36 hours. Previously, Joe had painted the house in 60 hours. How long would it take the two of them working together to paint the house?

c. Carl determined that 1 gallon of paint would cover 164 square feet. He painted three surfaces. One measured 30 ft × 36 ft; one 18 ft × 30 ft; and one 42 ft × 20 ft. How many gallons of paint did Carl use?

2 a. If a field measuring 54 feet and 9 inches were divided into three equal lengths, how much would each be?

b. What is the total of 18 yards, 2 feet, 9 inches and 3 yards, 2 feet, 9 inches?

c. How many gallons would it take to fill a tank that measures 18 inches by 24 inches by 60 inches?

d. If a field measures 3 miles by 4 miles, how much would planting cost if it cost $12 to plant each acre?

3 If a thrifty person saved $2,200 last year and this amount was 20% of the person's earnings, what were the person's total earnings for the year?

4 Talmage Corporation earned $200,000 in 2001 and $140,000 in 2002. What was the rate of decrease?

5 a. Complete the series: 12, 6, 3, _____, _____

b. Complete the series: 12, 80 24, 40, 48, 20, _____, _____

6 John Slocum sold 160 cars 2 years ago, 180 cars last year, and 220 cars this year. What was this year's percent gain over 2 years ago?

7 Add: 3 miles, 800 yards, 2 feet, 7 inches and 3 miles, 1,100 yard, 1 foot, 6 inches.

8 a. Earle Pitts drives the 400 miles from Dallas to his home in 10 hours. John Chi does the drive in 8 hours. How much faster is John driving than Earle?

b. The local train leaves Cody at 6:00 A.M. and travels at the rate of 50 mph. The express train leaves Cody at 9:00 A.M. and travels in the same direction at 100 mph. At what time would the trains be side by side?

c. A local train leaves Los Angeles for San Francisco traveling at 40 mph for the 400 mile trip. A direct train leaves San Francisco for Los Angeles traveling at 60 mph. How many hours will the trains have traveled when they pass each other?

9 a. Wanda Lopez inputs data to her computer at 100 words per minute. Allan Clark inputs at 60 words per minute. They each inputted a 600-word document. Allan started 5 minutes before Wanda. Who finished first?

b. To fold 400 napkins three people require 2 hours. How many napkins can the three people fold in 6 hours?

10 a. A cash box contained $72. The number of pennies is twice the number of nickels. There were five times as many nickels as there were dimes. How many pennies were there?

b. A person walked 4 miles south, 6 miles east, then 6 miles north, then 1 mile west. Continuing west then south how far was he from his starting point?

Assignment 30.1: Math Problems on Employment Tests

Name _____

Date _____ Score _____

A **(68 points) Solve the following problems. (4 points for each correct answer)**

1. Dan Kovacks can wash 3 cars per hour. Al Jameson can wash 2 cars per hour. Working together, how many cars can they wash in 8 hours? _____

2. Jane can do a job in 8 hours. Martha takes 12 hours to do the same job. How long will it take Jane and Martha working together to do the job? _____

3. Team A can do a project in 10 days. Team B requires 15 days to complete the same project. How long will it take both teams working together to finish the project? _____

4. Betty Kusack and Theresa Peña together can do a job in 20 hours. Working alone, Betty can do the job in 60 hours. How long would it take Theresa, working alone, to do the job? _____

5. One bricklayer can lay 450 bricks per day. Another can lay 540 bricks per day. Working together, how many days would it take them to lay 6,435 bricks? _____

6. If Chantall Jefferson can do a job in 3 days, Colleen O'Hara in 3 days, and Claire Most in 4 days, how long would the job take if Chantall, Colleen, and Claire worked together? _____

7. A plane leaves Los Angeles for New York and travels at 600 mph. At the same instant, a plane leaves New York for Los Angeles and travels 480 mph. If the total distance between the two cities is 3,600 miles, how much time will elapse before the planes meet? _____

8. Refer to problem 7. If the Los Angeles to New York plane departed at 8:30 A.M., how many miles would it have traveled by 2:30 P.M.? (*Ignore time zones.*) _____

9. Refer to problem 7. Suppose that both planes landed in Chicago, which is 1,200 miles from New York and on a straight line between the two cities. What time would the Los Angeles to New York plane arrive in Chicago if it departed from Los Angeles at 12 noon? (*Ignore time zones.*) _____

10. Faye Dunston and Franco Girelli start toward each other from 240 miles apart. Faye leaves 1 hour before Franco. Faye travels at 30 miles per hour, Franco at 40 miles per hour. How many miles will Franco have traveled when they meet? _____

11. Al Franck walks at 5 miles per hour; Michael Ovitz walks at 4 miles per hour. If Michael starts to walk in a certain direction 1 hour before Al, how far ahead of Al will he be when Al has walked 15 miles? _____

Assignment 30.1 *Continued*

12. Two cars started toward each other from 375 miles apart. The speed of one car was 40 mph. It met the other car after 5 hours. What was the speed of the other car? _____

13. If four postal clerks require 60 minutes to sort 1,200 letters, how many letters can ten clerks sort in 8 hours? _____

14. Three different bakers produce the following numbers of loaves of bread per hour: baker A, 250; baker B, 300; baker C, 350. If they all work the same number of hours and produce a combined total of 12,600 loaves, how many of these loaves does baker B produce? _____

15. Refer to problem 14. How many hours does each baker work? _____

16. Paul Eisner, Dylan Connor, and Martina Ybarria invest $25,000, $50,000, and $75,000, respectively, in a business. Later, they sell the business for $96,000 and divide the proceeds in proportion to their original investment. How much does Martina get? _____

17. A cash box had an equal number of dimes and quarters. It had twice as many pennies and three times as many nickels. The total cash was $416. How many nickels were there? _____

Score for A (68)

B **(32 points) Solve the following problems. (4 points for each correct answer)**

18. Subtract 5 hours 20 minutes 30 seconds from 10 hours 23 minutes. _____

19. Add 5 yards 2 feet 9 inches, 4 yards 1 foot 7 inches, and 1 yard 6 inches. _____

20. A room is 24 feet long and 15 feet wide. How much will it cost to cover the floor with carpet costing $36 a square yard if an extra 4 square yards must be purchased for matching? _____

21. A swimming pool is 60 feet long and 30 feet wide. It is 3 feet deep at one end and slopes evenly to a depth of 9 feet at the other end. How many gallons of water will be required to fill it to 1 foot from the top. (*Hint:* $(3 + 9) \div 2 = 6$ ft average depth of pool.) _____

22. A customer received a 30% discount on a $200 item, a 20% discount on a $150 item, and a 15% discount on an $80 item. What was the total dollar amount of the discount? _____

23. At a company employing 140 people, 40% of the employees took the bus to work, and 5% lived close enough to walk. The others drove cars. How many employees drove cars to work? _____

24. Complete the following series: 4, 12, 8, 16, 12, _____, _____

25. Complete the following series: 2, 8, 4, 16, 8, _____, _____

Score for B (32) _____

Assignment 30.2: Math Problems on Employment Tests

Name _____

Date _____ Score _____

A **(55 points) Solve the following problems. (5 points for each correct answer)**

1. United Airlines flight A flies 2,000 miles from Minneapolis to New York in 5 hours. United Airlines flight B flies 2,000 miles from Minneapolis to New York in 4 hours. If United flight A starts 1 hour before United flight B, how far ahead of flight B will it be when flight B has flown 1,500 miles? _____

2. A car leaves Cleveland for Albuquerque and travels at 55 mph. Simultaneously, a second car leaves Albuquerque for Cleveland and travels at 50 mph. If the total distance is 1,575 miles, how much time will go by before the two cars meet? _____

3. If the Cleveland to Albuquerque car (from problem 2) left at 9 a.m., how far from Albuquerque would it be at 6 p.m.? _____

4. Mozelle Williams and Melvin Kast start toward each other from 150 miles apart. Mozelle leaves 1 hour before Melvin. Mozelle travels at 30 mph, Melvin at 20 mph. How many miles will Melvin have traveled when they meet? _____

5. The Swenson nursery plants 14 trees per hour. The Johnson nursery plants 12 trees per hour. Working as a team, how many trees can the two nurseries plant in 22 hours? _____

6. One manufacturer can produce 800 VCRs per day; another can produce 960 per day. Working together, how many days would it take the two manufacturers to produce 39,600 VCRs? _____

7. One computer printer prints at 1,060 words per minute. A second brand prints at 510 words per minute. A third unit prints at 1,430 words per minute. How many words do the three computers produce in 1 hour?

8. Alberta Emery can sew a dress in 3 days. Allison Taylor requires only 2 days. If they combine efforts to fill one order to sew 30 dresses, how long will they take to fill the order? _____

9. If Warren Stevens can do a job in 4 days, Ted Burnsted in 6 days, Sam Chang in 2 days, and Jane Reilly in 5 days, how long would it take to complete the job if all four worked together? _____

10. Refer to problem 9. How long would it take Warren and Ted to do the job? _____

11. Refer to problems 9 and 10. How much less time would it take Sam and Jane than Warren and Ted to complete the job? _____

Score for A (55)

B **(45 points) Solve the following problems. (4 points for each correct answer)**

12. How many minutes are there in $1\frac{1}{2}$ days?

13. How many inches are there in 32 yards?

14. How many gallons of water would it take to fill a swimming pool that was 10 feet wide, 30 feet long, and 5 feet deep? _____

15. Refer to problem 14. If water cost $0.07 per gallon, how much would it cost to fill the pool?

16. Subtract 1 yard 2 feet 9 inches from 3 yards 2 feet 2 inches. _____

17. A room measures 21 feet long and 15 feet wide. How much will it cost to carpet the room with carpet that costs $18.50 per square yard?

18. If a Jacuzzi that is 5 feet long, 4 feet wide, and 3 feet deep is filled with water, how many gallons of water does it contain? _____

19. Complete the following series: 6, 10, 14, 18, 22,

20. Complete the following series: 3, 4, 8, 9, 18, 19,

Score for B (45)

Assignment 30.3: Practice Employment Test

Name _____

Date _____ Score _____

A **(100 points) Solve the following problems and fill in the blank with the letter of the correct answer.**
(2$\frac{1}{2}$ points for each correct answer)

Answers

1. Kay Mosier typed 50 letters per day for 7 days; Joe Plante typed 50 letters per day for 5 days. What is the total number of letters typed by both?

 a. 350; **b.** 600; **c.** 100; **d.** 550

 1. _____

2. An automobile traveled for 3 hours 20 minutes at an average speed of 45 mph. How many miles did it travel?

 a. 180; **b.** 135; **c.** 157; **d.** 150

 2. _____

3. A sales representative received commissions of $39.50 in March, $49.20 in April, $18.00 in May, and $97.70 in June. What was the average monthly commission?

 a. $49.20; **b.** $51.10; **c.** $204.40; **d.** $40.00

 3. _____

4. If a person receives a 30% discount on a purchase of $96.80, how much will that person pay?

 a. $29.04; **b.** $93.90; **c.** $32.27; **d.** $67.76

 4. _____

5. Of 465 students in school, 93 went to a ball game. What percent of the students did *not* go to the game?

 a. 20%; **b.** 40%; **c.** 60%; **d.** 80%

 5. _____

6. Two cars are traveling in the same direction, one at 50 mph, one at 55 mph. If the slower car started an hour earlier, how many hours will it take the faster car to catch up to it?

 a. 11; **b.** 9; **c.** 10; **d.** 5.5

 6. _____

7. A homeowner has enough lawn seed to plant $\frac{2}{3}$ of a backyard that is 90 ft by 20 ft. How many square feet will go unplanted?

 a. 330; **b.** 180; **c.** 900; **d.** 600

 7. _____

8. Julio Martinez saves twice as much as Ken Shaw. Ken saves twice as much as Phyllis Catchings. If Julio saves a total of $1,500, how much does Phyllis save?

 a. $1,500; **b.** $600; **c.** $375; **d.** $500

 8. _____

9. If $15,300 is divided among Ibn Rashid, Harold Pierce, and Monica Snead in the proportions 3, 5, and 9, respectively, how much will Ibn receive?

 a. $2,700; **b.** $5,900; **c.** $5,100; **d.** $900

 9. _____

Answers

10. Two trains were 780 miles apart. They were headed directly toward each other. One traveled at 30 mph. The other traveled at 35 mph. How many hours did it take for the trains to meet?

 a. 26; **b.** 13; **c.** 18; **d.** 12

10. _____

11. At $14.25 per sq yd, how much would it cost to carpet a room 18 ft by 27 ft?

 a. $2,308.50; **b.** $692.55; **c.** $769.50; **d.** $6,925.50

11. _____

12. Two partners, Joseph Alcara and Merriam Trask, own a restaurant. They sell a $\frac{1}{3}$ interest to Shelley Cantrell. If the part of the restaurant that Joseph and Merriam still own is worth a total of $15,000, how much was the original value?

 a. $22,500; **b.** $30,000; **c.** $45,000; **d.** $7,500

12. _____

13. A bus left San Diego at 1:30 and traveled 50 mph. A train left San Diego at 3:30 traveling in the same direction at 70 mph. At what time will the train catch up with the bus?

 a. 6:30; **b.** 9:00; **c.** 11:18; **d.** 8:30

13. _____

14. If plane X averages 800 mph and plane Y averages 400 mph, how many hours will plane X travel before it overtakes plane Y if plane Y has a 2 hour and 30 minute head start?

 a. $1\frac{1}{4}$; **b.** $2\frac{1}{2}$; **c.** 5; **d.** $7\frac{1}{2}$

14. _____

15. How long will it take a car to travel 1,100 miles at an average speed of 55 mph?

 a. 20 hr; **b.** $18\frac{2}{11}$ hr; **c.** 55 hr; **d.** $16\frac{2}{3}$ hr

15. _____

16. Two teenagers who were 60 miles apart walked toward each other. They met in 4 hours. One teenager averaged 7 mph. How fast did the other travel?

 a. 7 mph; **b.** $8\frac{4}{7}$ mph; **c.** $6\frac{3}{7}$ mph; **d.** 8 mph

16. _____

17. Two cars started toward each other from 400 miles apart. They met in 5 hours. Car X averaged 45 mph. How many mph did car Y average?

 a. 45; **b.** 55; **c.** 35; **d.** 37

17. _____

18. A bus averaging 45 mph leaves New York at 9:30 a.m. How many miles will it have traveled at 4:45 p.m.?

 a. 281.25; **b.** 326.25; **c.** 236.25; **d.** 282.5

18. _____

19. A submarine travels at a rate of 12 mph under water and 24 mph on top. In a 100-mile trip, it travels 20 miles below and the rest on top. How many hours does the 100 mile trip take?

 a. $4\frac{1}{6}$; **b.** $8\frac{1}{2}$; **c.** 5; **d.** $7\frac{1}{24}$

19. _____

20. How long will it take a train averaging 60 mph to cover its entire route of 400 miles if it loses 45 minutes travel time in stops?

 a. 7 hr 25 min; **b.** 6 hr 40 min; **c.** 5 hr 55 min; **d.** 9 hr

21. If a hiker travels 18 miles in 4 hours, how many miles will be covered in 7.5 days walking 8 hours per day?

 a. 240; **b.** 262.5; **c.** 320; **d.** 270

22. Thai Le can do a job in 3 days. Anh Huynh can do the same job in 2 days. How many days would it take them to do the job together?

 a. 1.8; **b.** 1.5; **c.** 1; **d.** 1.2

23. Aren Chomski, Warner Howe, and Brown Ledbetter invested $1,000, $1,500, and $3,500, respectively, in a business partnership. If the annual profit of $1,500 is divided among them in proportion to their investments, how much will Aren receive?

 a. $300; **b.** $250; **c.** $1,000; **d.** $375

24. Three salespeople, Melissa Lever, Tom Borders, and Rose Tanaka sold a combined total of $8,400. Melissa sold $3,360; Tom and Rose split the remainder. If a $300 bonus was divided among the three in proportion to their sales, how much did Rose receive?

 a. $90; **b.** $180; **c.** $50.40; **d.** $75.60

25. From Los Angeles to Dallas, a plane takes 3 hours, 25 minutes. A train takes 11 hours, 10 minutes. How many hours are saved by taking the plane?

 a. $7\frac{3}{4}$; **b.** $17\frac{1}{4}$; **c.** $31\frac{3}{4}$; **d.** $19\frac{3}{4}$

26. To 7 gallons 3 quarts 1 pint of liquid are added 2 gallons 4 quarts. How many pint jars would the total quantity fill?

 a. 87; **b.** 55; **c.** 17; **d.** 173

27. Boat Y and boat Z start traveling toward each other from 600 miles apart. Y is traveling at 35 mph, Z at 40 mph. How many hours will elapse before they meet?

 a. 7; **b.** 8; **c.** 9; **d.** 10

28. Refer to problem 27. Y and Z start traveling toward each other from 600 miles apart. Y is traveling at 35 mph, Z at 40 mph. How many miles will Y travel before they meet?

 a. 400; **b.** 320; **c.** 350; **d.** 280

Answers

20. ____

21. ____

22. ____

23. ____

24. ____

25. ____

26. ____

27. ____

28. ____

29. Departments X, Y, and Z had sales of $1,100, $1,900, and $2,500, respectively. A $700 advertising charge was allocated proportionately. How much is Department X's share of the advertising charge?

 a. $140; **b.** $350; **c.** $110; **d.** $116.67

Answers

29. ____

30. In problem 29, what would Department Z's expense be if the advertising charge was increased to $2,200?

 a. $100; **b.** $900; **c.** $1,000; **d.** $484

30. ____

31. If Alain Junev and Parc Lafontaine together do a job in 6 hours and Alain alone does the job in 10 hours, how long does it take Parc alone to do the job?

 a. 12 hr; **b.** 20 hr; **c.** 15 hr; **d.** 9 hr

31. ____

32. A company had expenses of $15,500 in 1994 and $18,600 in 1995. What was the percent increase?

 a. 15%; **b.** $16\frac{2}{3}$%; **c.** 25%; **d.** 20%

32. ____

33. This year, Sheridan Company had last year's expenses of $24,000 increased by 20%. If $\frac{1}{8}$ of this year's expenses were charged to overhead, what would be the amount?

 a. $2,400 **b.** $3,600; **c.** $4,800; **d.** $3,000

33. ____

34. In an agent's 8% commission for selling a product amounted to $1,200, what was the total amount of the sale?

 a. $12,000; **b.** $9,600; **c.** $8,600; **d.** $15,000

34. ____

35. How many days will be required for 5 people to build 5 machines if 5 people can build 20 machines in 8 days?

 a. 5; **b.** 4; **c.** 2; **d.** 1.6

35. ____

36. Mae Miles drove 231 miles in 1 day. If this distance is 40% more than she drove the day before, how many miles did she drive the day before?

 a. 165; **b.** 162.7; **c.** 323.4; **d.** 57.75

36. ____

37. How many square feet does one wall of a 24 ft by 24 ft room with a 12 ft high ceiling contain?

 a. 48; **b.** 6,912; **c.** 288; **d.** 64

37. ____

Assignment 30.3 *Continued*

38. Add: 2.001 + 3.14 + 280.3 + 0.7

 a. 512.5; **b.** 286.141; **c.** 287.007; **d.** 285.611

39. A secretary earns $1,400 per month, spends 90% of what is left after deductions of 22%, and saves the rest. How many months will it take the secretary to save $1,528.80?

 a. 11; **b.** 12; **c.** 13; **d.** 14

40. A $300 lamp is sold at a 40% discount; 15% of the sales price goes for advertising. What is the advertising cost?

 a. $180; **b.** $60; **c.** $27; **d.** $18

Score for A (100)

Appendix A
Answers to Odd-Numbered Problems

Chapter 1

Assignment 1.1
1. 347
3. 377
5. 491
7. 639
9. 337
11. 1,823
13. 2,437
15. 1,626
17. 1,589
19. 2,362
21. 1,897.20
23. 1,286.33
25. 1,395.30
27. 1,904.78
29. 7,269.37
31. 175.93
33. 132.44
35. 265.86
37. 296.36
39. 224.25

Assignment 1.2
1. 61
3. 47
5. 5
7. 76
9. 50
11. 9
13. 59
15. 29
17. 14
19. 411; 584
21. (399); 92
23. 283; 616
25. $73.98
27. $60.82
29. $38.61
31. $4,642.81
33. $8,216.01
35. $3,151.61
37. $6,983.78
39. $48.80
41. $1,790,906.69

Assignment 1.3
1. 24
3. 520
5. 90
7. 240
9. 72
11. 144
13. 48
15. 80
17. 36
19. 88
21. 28
23. 136
25. 72,576
27. 317,327,062
29. 1,080,000
31. 4,184,998
33. 548,784
35. 2,266,875
37. 184,200
39. 166,050
41. 34,400
43. 52,640
45. 9,800
47. 1,000
49. 60,264
51. 144.00
53. 366.08
55. 1,787.50
57. 2,352
59. 3,234
61. 26,400

Assignment 1.4
1. 12
3. 12
5. 42
7. 4
9. 13
11. 30
13. 13
15. 28
17. 52
19. 17
21. 7 (41)
23. 587

25. 390
27. 90 (5)
29. 38 (500)
31. 22 (16)
33. 612
35. 178 (28)
37. 184 (137)
39. 1,000 (7)
41. 20 (118)
43. 517 (597)
45. 1,111 (49)
47. $2.20
49. 1 (49)
51. 1,112 (36)
53. 260 (49)
55. 2,000,148 (24)
57. 45
59. 105 (9)

Assignment 1.5
1. 400,000
3. 2,400,000
5. 5,400,000
7. 30,000
9. 2,000,000
11. 640,000
13. 7,000,000
15. 1,000,000
17. 4,000
19. 4,000
21. 270,000; 259,602
23. 10,000,000; 9,822,780
25. 200,000; 200,532
27. 60; 51
29. 200; 208

Chapter 2

Assignment 2.1
1. 26
3. 7
5. 2
7. 90
9. 23
11. 15
13. 9

15. 114

17. 251

19. 500

21. $12.30

23. 2

25. 11

27. $400

29. $310

31. $28

33. $62.50

35. $74.00

37. 20

39. 385

41. 11

43. 7

45. 3

47. 16

49. 21

51. 50

53. 5

55. 15

57. (a) 30, 25

 (b) 36, 31

 (c) 66, 60

 (d) 510, 460

59. (a) 25; 5

 (b) 9, 3

 (c) 100, 20

 (d) 100, 50

Assignment 2.2

1. $0.72

3. $7.12

5. 103 lb

7. $1.50

9. 900 mi

11. $9.95

13. $79.92

15. $14.85

17. $23.70

19. $760

21. $801

23. $799.60

25. $240

27. $55.79

29. $89.40

31. $6 + 4 + 2 = 17 - 5$

33. $9 - 3 - 1 = 2 + 3$

35. $20 + 1 + 2 = 16 + 7$

37. $12 + 3 - 3 = 7 + 5$

39. $64 - 32 - 8 = 8 + 16$

Chapter 3

Assignment 3.1

1. 31,680

3. 1,728

5. 1,200

7. 26 yd; 0 ft; 6 in.

9. 40 lb; 2 oz

11. 9 gal; 2 qt; 0 pt

13. $544,500

15. $63

17. $6 per sq ft

19. $76.80

21. 250 cu ft

23. 15 hr, or 900 min

Assignment 3.2

1. $1,512

3. $7,776

5. Room A

7. 4 ft

9. 24 mi

Chapter 4

Assignment 4.1

1. 0.613

3. 0.42

5. 860.00098

7. fifteen and seventy-three hundredths

9. four hundred ninety-two and three tenths

11. fifty-eight and nine hundred nine thousandths

13. two thousand four hundred three and seven hundredths

15. 48.5 mi

17. 374.3 yd

19. 6.4 in.

21. $0.07

23. $3.11

25. $72.63

27. 0.005 qt

29. 5.041 in.

31. 0.100 lb

33. $0.16

35. $2.10

37. $0.66

39. 22.2363

41. 104.4996

43. 50.801

45. 249.202

47. 0.459

49. 16.932

51. 27.63

53. 2.2309

55. 0.099

57. 2.485

Assignment 4.2

1. $798.00; $798.00

3. $289.46000; $289.46

5. 90.0564; 90.0564

7. 22.8562608; 22.8562608

9. $1.85; $1.85

11. $45.25; $45.25

13. 6.122; 6.12

15. 4.25

17. 972

19. 0.00571

21. $1,448.20

23. $480.00

25. $2.13

27. (d) 0.004

29. (d) 280

31. (c) 0.48

33. (a) 27,000

35. (b) 1,200

37. (a) 0.004

39. (d) 140

41. (b) 7

Assignment 4.3

1. 5.75 ft

3. 15.45 mi

5. 529 oz

7. $235

9. $125

11. $0.08

13. 7.8 gal

Chapter 5

Assignment 5.1

1. $1\frac{6}{7}$

3. $2\frac{1}{4}$

5. $1\frac{5}{6}$

7. $\frac{37}{10}$

9. $\frac{37}{8}$

11. $\frac{33}{5}$

13. $\frac{3}{5}$

15. $\frac{2}{3}$

17. $\frac{2}{3}$

19. $\frac{3}{4}$

21. $\frac{15}{18}$

23. $\frac{21}{24}$

25. $\frac{44}{48}$

27. $\frac{18}{45}$

29. $\frac{10}{10} = 1$

31. $3\frac{17}{12} = 4\frac{5}{12}$

33. $7\frac{12}{6} = 9$

35. $10\frac{73}{45} = 11\frac{28}{45}$

37. $2\frac{10}{12} = 2\frac{5}{6}$

39. $\frac{1}{12}$

41. $2\frac{13}{20}$

43. $3\frac{7}{20}$

45. $8\frac{13}{30}$ gallons

47. $\frac{7}{8}$ in.

Assignment 5.2

1. $\frac{2}{9}$

3. $\frac{5}{8}$

5. $2\frac{1}{3}$

7. $6\frac{3}{4}$

9. $\frac{5}{6}$

11. $1\frac{1}{5}$

13. $3\frac{1}{3}$

15. $4\frac{1}{6}$

17. $18\frac{1}{3}$ cu yd

19. $1\frac{1}{2}$ qt

21. $14\frac{2}{3}$ times

Chapter 6

Assignment 6.1

1. 0.21
3. 0.0367
5. 400%
7. 12%
9. 125%
11. 3.245
13. 52%
15. 8.75%
17. 500%
19. 0.001
21. 0.24
23. 13.19
25. 0.384
27. $0.28
29. $5
31. 1.6
33. 60
35. 1.12
37. 25%
39. 100%
41. $0.60
43. 225%
45. $44
47. $4,000
49. 88
51. 400
53. 40%
55. $21.00
57. 140%
59. 25

Assignment 6.2

1. 300
3. 104
5. $7,760
7. 542
9. $60
11. 10%
13. 25%
15. 20%
17. +74; +16.9%
19. +230; +12.7%
21. +1,715; +16.0%
23. (189); (17.4%)
25. (608); (21.7%)
27. +$116.47; +19.6%
29. (14.40); (5.9%)
31. +22.74; +15%
33. (806.61); (16.5%)
35. (716.61); (17.6%)

Assignment 6.3

1. 270
3. 5,500
5. 55%
7. 240,000
9. $720
11. $3,000
13. $49,920
15. 15%
17. $50,000
19. 100%

Assignment 6.4

1. (a) 2,100; 28%; $3,360
 (b) 1,800; 24%; 2,880
 (c) 2,400; 32%; 3,840
 (d) 1,200; 16%; 1,920
3. $7,000; $4,000; $6,000; $8,000
5. $3,040; $4,160; $5,440; $3,360

Chapter 7

Assignment 7.1

1. 425.00
 4,412.50
 2,669.50
 1,679.50
 1,614.20
 1,542.20
 4,827.95
 4,701.53
 3,851.53

3,201.53

3. 1076.82

1076.82

1076.82

764.02

5. 886.03

3046.03

3046.03

2962.87

7. $1923.83

9. 4,020.90

3,872.70

4,849.50

4,035.50

$3,750.33

11. 1,678.36

1,541.95

1,177.77

2,302.62

$1,527.52

Assignment 7.2

1. 802.50

752.90

678.71

904.21

791.89

758.56

746.56

678.79

466.79

328.79

422.79

3. (a) $802.85

(b) $1,630.27

(c) $951.41

(d) $942.32

(e) $877.38

Assignment 7.3

1. Central Heating, Inc.

Reconciliation of Bank Statement, November 30

| | | |
|---|---|---|
| Checkbook balance | | $ 896.91 |
| Minus unrecorded bank charges: | | |
| Service charge | $ 9.50 | |
| | | $ 887.41 |
| Plus bank interest credit | | $ 12.00 |
| Adjusted checkbook balance | | $ 899.41 |
| Bank balance on statement | | $1,132.46 |
| Minus outstanding checks: | | |
| No. 148 | $ 26.62 | |
| No. 156 | 100.16 | |
| No. 161 | 77.17 | |
| No. 165 | $ 29.10 | $ 233.05 |
| Adjusted bank balance | | $ 899.41 |

3. Dawson & Sons, Inc.

Reconciliation of Bank Statement, July 31, 19—

| | | |
|---|---|---|
| Checkbook balance | | $17,940.44 |
| Minus unrecorded bank charges: | | |
| Service charge | $ 18.00 | |
| Automatic transfer— insurance | $1,765.00 | |
| Returned check charged to account | $1,500.00 | $ 3,283.00 |
| | | $14,657.44 |
| Plus bank interest credit | | $ 83.20 |
| Adjusted checkbook balance | | $14,740.64 |
| | | |
| Bank balance on statement | | $17,464.54 |
| Plus deposit not recorded by bank | | $ 2,004.35 |
| | | $19,468.89 |
| Minus outstanding checks: | | |
| No. 730 | $ 78.60 | |
| No. 749 | 2,765.23 | |
| No. 753 | 556.25 | |
| No. 757 | 271.84 | |
| No. 761 | 97.73 | |
| No. 768 | 958.60 | 4,728.25 |
| Adjusted bank balance | | $14,740.64 |

Chapter 8

Assignment 8.1

1. $360.00; $108.00; $18.00; $486.00

320.00; 320.00

400.00; 120.00; 40.00; 560.00

360.00; 67.50; 427.50

352.00; 352.00

280.00; 280.00

320.00; 84.00; 404.00

360.00; 13.50; 373.50

352.00; 105.60; 17.60; 475.20

352.00; 352.00

380.00; 114.00; 38.00; 532.00

400.00; 60.00; 460.00

$4,236.00; $672.60; $113.60; $5,022.20

3. $2,728.38

5. $612.54

7. $23.05

$23.00

$0.05

9. $29.28

$29.00

$0.28

Assignment 8.2

1. $424.00; $348.56

421.40; 329.16

398.80; 302.45

600.00; 489.10

650.00; 519.27

649.70; 537.00

392.00; 307.02

516.00; 403.53

500.00; 406.75

$4,549.90; $3,642.84

3. $379.87

350.02

350.02

399.14

352.49

410.53

362.00

368.50

425.81

422.41

430.46

418.92

445.77

$5,115.94

5. (a) $22,528.40

(b) $1,396.75

(c) $326.67

(d) $2,500.95

(e) $4,224.37

7. (a) $19,500; $7,000

(b) $56

(c) $378

(d) $434

Chapter 9

Assignment 9.1

1. (a) $21,500

(b) $34,800

(c) $10,950

(d) $7,692

(e) $18,688

3. (a) $3,500

(b) $525

5. (a) $41,900

(b) $6,285

Assignment 9.2

1. $26

3. Tanakas: $4,495

5. $20,260

Chapter 10

Assignment 10.1
1. $0.41; $6.30; $3.70
 $0.21; $3.28; $1.72
 $0.99; $15.11; $4.89
 $0.56; $8.53; $2.00
 $0.06; $0.91; $0.09
 $1.62; $24.77; $5.23
 $0.67; $10.27; $9.73
 $1.98; $30.33; $19.67
 $0.92; $14.13; $0.87
 $1.22; $18.71; $0.29
3. $88.29
5. King Carpets

Assignment 10.2
1. $940,000,000; $412,500,000;
 $318,000,000
3. $1.40; $0.95
5. $2,565
7. $248
9. (a) 1.7% (0.017)
 1.7% (0.017)
 1.35% (0.0135)
 1.5% (0.015)
 (b) 17 mills
 17 mills
 13.5 mills
 15 mills
11. $1,392

Chapter 11

Assignment 11.1
1. $3,780; $3,780
3. $1,710; $3,310
5. $2,600; $3,800
7. $2,680
9. $3,310
11. $4,400
13. $852; $27,305.75
15. $504; $7,611.35
17. $224; $5,176.00
19. $597; $5,767.00
21. $228; $3,296.80

Assignment 11.2
1. $5,400
3. $3,750
5. $3,620
7. $1,256.85
9. $562
11. $9,000

Chapter 12

Assignment 12.1
1. $375; $875
3. $1,190; $2,210
5. 66%; $1,980
7. $880; $330; —; $990
9. 70%; 85%; —; $1,071
11. 70%; 80%; 95%; 46.8%
13. $466

Assignment 12.2
1. Apr. 2; Apr. 22; $15.55; $502.75
3. June 5; July 5; $12.70; $495.30
5. Feb. 1; 99%; $477.97
7. $507.61; $340.39

Chapter 13

Assignment 13.1
1. $657.68
3. $455.48
5. $280.47
7. $225; $975
9. $960; $2,160
11. $210; $1,050
13. $2,500; $5,000
15. 130%; $1,000
17. 200%; $55
19. 140%; $600
21. 250%; $420
23. $1,479; $3,944
25. $570; 50%

Assignment 13.2
1. $253.53
3. $1,535
5. $37.49
7. $28.80; $35.20
9. $220.80; $147.20
11. $1,330; $570
13. $412; $412
15. 70%; $1,220
17. 25%; $256
19. 60%; $4,100
21. 75%; $816
23. $99.20; $148.80
25. $72.96; 60%

Chapter 14

Assignment 14.1
1. $28

3. $48
5. $247.50
7. $2,240
9. $850.50
11. $17.55; $17.31; $0.24
13. $50; $49.32; $0.68
15. $49.25; $48.58; $0.67
17. $65; $64.11; $0.89
19. $8.50; $8.50; $0.00
21. $14; $14.12; −$0.12
23. $6.25; $6.10; +$0.15
25. $60; $60.32; −$0.32

Assignment 14.2
1. $7.50; $757.50
3. $90; $3,690
5. $1,600; $76,600
7. $565.07; $25,565.07
9. $67.81; $5,067.81

Chapter 15

Assignment 15.1
1. 98
3. 189
5. 122
7. March 4, 2003
9. November 5, 2003
11. Jan. 6, 2004; $120; $8,120
13. Apr. 12, 2004; $350; $24,350
15. 105; $21.88; $866.88
17. 200; $425.75; $9,675.75
19. $180.00
21. $9,600.00
23. 9%
25. 135 days
27. $21.15
29. 8.5%
31. 240 days
33. $1,278.01

Assignment 15.2
1. $95; August 16, 2002
3. 6%; December 22, 2003
5. $1,800; October 4, 2004
7. 150 days; April 28, 2003
9. 5%; July 11, 2003

Chapter 16

Assignment 16.1
1. $60.00; 20; $19.33; $40.67
3. $525.00; 20; $74.41; $450.59

5. $92.00; 30; $67.37; $24.63

7. $650; 20; $118.07; $531.93

9. 31; $12,000.00; $99.88; $7,000.00;
$6,900.12; $5,099.88
30; 5,099.88; 41.08; 5,140.96;
5,099.88; 0.00

Assignment 16.2

1. $77; $6,523; 7.08%

3. $28.50; $2,821.50; 12.12%

5. Nov. 12; 63; $96.25; $4,903.75

7. $70; $4,270; Aug. 17; 47; $55.75;
$4,214.25

9. $132.30; $6,612.30; Jan. 22; 82;
$195.80; $6,416.50

Assignment 16.3

1. Sept. 18; 39; $31.41; $2,068.59

3. Feb. 16; 58; $66.74; $3,433.26

5. $165.70; $5,765.70; Sept. 7; 84;
$172.50; $5,593.20

7. $29.59; $1,829.59; June 28; 39;
$23.46; $1,806.13

Chapter 17

Assignment 17.1

1. (a) 1%

 (b) 1.25%

 (c) $1\frac{1}{3}\%$

 (d) $\frac{2}{3}\%$

 (e) 0.5%

 (f) $1\frac{2}{3}\%$

 (g) 2%

 (h) 1.5%

 (i) 0.75%

 (j) 1.75%

 (k) $\frac{5}{6}\%$

3. $25.41; $1,432.57

5. $42.61; $1,444.67

7. $22.10; $861.24

9. $704.20; $8.29; $1211.55

11. $833.58; $9.59; $1,351.31

Assignment 17.2

1. (a) $60

 (b) $1,250

 (c) 14.4%

3. $2,000; $15.00; $500; $515.00;
$1,500

5. $1,000; 7.50; 500; 507.50; 500

7. (a) $37.50

 (b) $1,250

 (c) 9%

9. 16.8%

Assignment 17.3

1. 0.25944; $908.04

3. 0.04707; $583.67

5. 0.02076; $158.81

7. $45.00; $1,483.26; $4,516.74

9. 3,022.36; 22.67; 1,505.59; 1,516.77

11. $45.00; $1,155.00; $4,845.00

13. 3,681.34; 27.61; 1,172.39; 2,508.95

15. (a) $562.50

 (b) $387.75

 (c) $74,612.25

Chapter 18

Assignment 18.1

1. (a) $960

 (b) $220

 (c) $1,650

 (d) $1,430

3. (a) $1,875

 (b) $1,250

 (c) $220

 (d) $3,233

5. (a) $53,340

 (b) $50,000

 (c) $6,000

 (d) $3,440

 (e) $56,000

Assignment 18.2

1. (a) $3,724

 (b) $2,793

3. $170,000

5. (a) $165,000

 (b) $55,000

 (c) $180,000

 (d) $120,000

7. $360,000

Assignment 18.3

1. $19.30; $289.50
$8.26; $1,982.40
$27.04; $1,081.60
$4.91; $1,964.00

$16.83; $1,683.00
$53.86; $4,039.50

3. $3,840

5. (a) $9,050

 (b) $9,500

 (c) $6,545

7. (a) $574

 (b) $2,524

Chapter 19

Assignment 19.1

1. $765.60; $655.20; $368.00; $744.00;
$1,785.00; $486.00; $4,803.80

3. (a) $22,950

 (b) $22,200

 (c) $21,700

5. (a) 20; 12; 62; 45; 23; 8; 73; 48

 (b) Yes; −3

Assignment 19.2

1. A: $120,000; $75,000

 B: $120,000; $82,000

 C: $72,000; $86,000

 D: $96,000; $90,000

 E: $47,000; $15,000

3. (a) $46,716

 (b) 6.12

5. (a) $30,123

 (b) $50,205

7. $160,000

9. (a) $200,000; $4,000

 (b) $182,000; $86,000

 (c) $255,500; $188,500

 (d) $275,591; $168,409

 (e) $24,000; $13,500

 (f) $160,000; $208,000

 (g) $360,000; $60,000

 (h) $313,043; $126,957

 (i) $112,500; $12,500

 (j) $100,000; $30,000

Chapter 20

Assignment 20.1

1. (a) $565; $2,260; $3,740

 (b) $800; $3,200; $3,200

 (c) $700; $2,100; $6,300

 (d) $2,240; $29,120; $5,480

3. (a) $14,000

 (b) $18,000

5. (a) $4,000.00; $3,062.50
 (b) $3,715.20; $343.68
 (c) $2,000.00; $1,000.00
 (d) $1,920.00; $1,228.80
 (e) $2,362.50; $1,328.91
 (f) $7,695.00; $6,232.95
7. $1540.39
9. $9,600; $6,400; $3,200
11. straight line; $31,885

Assignment 20.2
1. (a) $2,275
 (b) $8,640
3. $40,000
5. $15,670

Chapter 21

Assignment 21.1
1. 14.24%; 16.15%; $5,000; 2.38%
16.56%; 13.46%; 75,000; 42.86%
19.87%; 21.15%; 25,000; 9.09%
50.66%; 50.77%; 105,000; 15.91%
18.54%; 21.54%; —; 0.00%
7.95%; 7.69%; 20,000; 20.00%
10.60%; 13.85%; (20,000); −11.11%
23.18%; 20.77%; 80,000; 29.36%
15.56%; 14.62%; 45,000; 23.68%
49.34%; 49.23%; 105,000; 16.41%
100.00%; 100.00%; $210,000; 16.15%
5.96%; 4.77%; $28,000; 45.16%
2.98%; 2.69%; 10,000; 28.57%
0.99%; 1.54%; (5,000); −25.00%
9.93%; 9.00%; 33,000; 28.21%
19.87%; 23.85%; (10,000); −3.23%
11.92%; 16.15%; (30,000); −14.29%
31.79%; 40.00%; (40,000); −7.69%
41.72%; 49.00%; (7,000); −1.10%
26.49%; 30.77%; —; 0.00%
21.85%; 16.92%; 110,000; 50.00%
9.93%; 3.31%; 107,000; 248.84%
58.28%; 51.00%; 217,000; 32.73%
100.00%; 100.00%; $210,000; 16.15%
3. 10.6%; 8.9%; $10,000; 29.4%
6.9%; 6.5%; 3,700; 14.9%
8.1%; 8.6%; 1,300; 4.0%
25.6%; 24.0%; 15,000; 16.4%
13.9%; 13.7%; 5,500; 10.6%
3.5%; 3.5%; 1,300; 9.8%
10.3%; 10.2%; 4,200; 10.9%
48.2%; 52.6%; —; 0.00%

15.9%; 13.2%; 15,800; 31.6%
74.4%; 76.0%; 20,000; 6.9%
100.0%; 100.0%; 35,000; 9.2%
3.0%; 2.9%; 1,600; 14.5%
2.2%; 1.8%; 2,200; 31.4%
0.3%; 0.3%; 200; 20.0%
5.5%; 5.0%; 4,000; 21.1%
16.9%; 15.8%; $10,000; 16.7%
6.3%; 7.9%; (4,000); −13.3%
23.1%; 23.7%; 6,000; 6.7%
28.7%; 28.7%; 10,000; 9.2%
43.4%; 47.4%; —; 0.0%
21.2%; 22.1%; 4,000; 4.8%
6.7%; 1.8%; 21,000; 300.0%
71.3%; 71.3%; 25,000; 9.2%
100.0%; 100.0%; $35,000; 9.2%

Assignment 21.2
1. 103.66%; 103.57%; $125,000; 17.24%
3.66%; 3.57%; 5,000; 20.00%
100.00%; 100.00%; $120,000; 17.14%
25.61%; 27.14%; 20,000; 10.53%
53.66%; 55.17%; 50,000; 12.82%
79.27%; 82.86%; 70,000; 12.07%
30.49%; 30.00%; 40,000; 19.05%
48.78%; 52.86%; 30,000; 8.11%
51.22%; 47.14%; 90,000; 27.27%
15.24%; 14.00%; 27,000; 27.55%
10.00%; 11.14%; 4,000; 5.13%
1.83%; 2.86%; (5,000); −25.00%
0.49%; 0.50%; 500; 14.29%
0.43%; 0.40%; 700; 25.00%
0.17%; 0.13%; 500; 55.56%
0.85%; 0.71%; 2,000; 40.00%
0.26%; 0.26%; 300; 16.67%
29.27%; 30.00%; 30,000; 14.29%
21.95%; 17.14%; 60,000; 50.00%
3.66%; 2.86%; 10,000; 50.00%
18.29%; 14.29%; 50,000; 50.00%
3. 102%; 102%; 11,500; 16%
2%; 2%; 500; 33%
100%; 100%; 11,000; 15%
30%; 24%; 7,000; 40%
46%; 51%; 1,500; 47%
76%; 76%; 8,500; 16%
31%; 34%; 1,500; 6%
45%; 42%; 7,000; 23%
55%; 58%; 4,000; 10%
14%; 14%; 1,300; 13%
9%; 10%; 200; 3%
1%; 3%; (1,060); −57%

0%; 0%; (140); −41%
1%; 0%; 550; 275%
0%; 0%; 190; 119%
1%; 1%; 450; 85%
0%; 0%; (30); −14%
26%; 28%; 1,460; 7%
29%; 30%; 2,540; 12%
3%; 2%; 940; 56%
26%; 28%; 1,600; 8%

Assignment 21.3
1. $3,200; 3.1%
25,400; 22.5%
39,000; 29.3%
$67,600; 19.3%
$(6,500); −14.0%
21,000; 18.3%
$14,500; 9.0%
$82,100; 16.1%
$5,800; 20.1%
8,200; 7.2%
$14,000; 9.9%
$(31,000); −16.3%
$(17,000); −5.1%
$99,100; 55.2%
$82,100; 16.1%
$(85,000); −10.2%
$3,000; 2.3%
(41,000); −6.7%
$(38,000); −5.1%
39,000; 29.3%
$(77,000); −12.6%
$(8,000); −3.6%
$2,800; 3.6%
$(1,600); −5.6%
$1,200; 1.1%
$(9,200); −7.9%
3. $9,000; 56.3%
4,000; 50.0%
15,000; 48.4%
$28,000; 50.9%
$(4,000); −9.3%
4,000; 36.4%
$0; 0.0%
$28,000; 25.7%
$(1,000); −18.2%
3,500; 58.3%
$2,500; 21.7%
$(8,000); −21.1%
$(5,000); −11.1%
$33,500; 56.3%

$28,000; 25.7%
$85,000; 70.8%
$3,500; 12.7%
69,500; 82.2%
$73,000; 65.2%
15,000; 48.4%
$58,000; 71.6%
$27,000; 69.2%
$9,500; 44.2%
5,750; 79.3%
$15,250; 53.0%
$11,750; 114.6%

5. (a) Bear Country Boutique, 38.3% vs. 23.7%

(b) Banner Boutique, 5.93:1 vs. 2.7:1

Chapter 22

Assignment 22.1

1. (a) $52,400
(b) $10,499.25
3. 7.9%
5. 20.41%
7. (a) $4; 6.90%
(b) $6; 7.5%
(c) $2; 4.49%
(d) $4.50; 4.59%
(e) $3.25; 5.6%

Assignment 22.2

1. 6,000 shares; 32,000 shares
3. (a) $102,400
(b) $12,000
(c) $2.00
(b) $409,600
(e) $457,600
(f) $4.80
5. $9,750; $0.16
7. 2000 $38,000 ÷ 25,000 = $1.52; 0
2001 $42,000 ÷ 25,000 = $1.68;
$14,000 ÷ 50,000 = $0.28
2002 $40,000 ÷ 25,000 = $1.60;
$65,000 ÷ 50,000 = $1.30
25,000 shares × $20 par × 8% =
$40,000 per yr for preferred;
2000 $76,000 × 50% = $38,000;
$40,000 − $38,000 = $2,000
2001 $112,000 × 50% = $56,000
2002 $130,000 × 50% = $65,000
9. $550
11. (a) 12
(b) 15

Chapter 23

Assignment 23.1

1. $1,155
3. $5,400,000
5. (a) $4,000.00; $300.00
(b) 2,940.00; 236.25
(c) 7,740.00; 911.25
(d) 6,540.00; 562.50
21,220.00; 2,010.00
7. (a) 45; $2,080.00
(b) 100; $2,912.50
9. (a) 8.65%
(b) 7.64%
(c) 10.73%
(d) 7.32%

Assignment 23.2

1. (a) 9.24%
(b) 10.63%
3. (a) $3,838.33
(b) 10.33%
B. (a) $400.00; $40.00; $4,900.00;
9.00%
(b) 23.00; −6.00; 2,045.00; 11.0%
(c) 1,100.00; +540.00; 17,300.00;
9.5%
(d) 770.00; −233.33; 7,350.00;
7.3%
(e) 210.00; +20.00; 2,850.00; 8.1%
(f) 65.00; −6.00; 1,022.50; 5.8%

Chapter 24

Assignment 24.1

1. (a) $54.61; $54.27; −
(b) $6.72; $6.72; even
(c) $2.29; $2.29; even
(d) $50.66; $50.79; +
(e) $66.33; $66.10; −
(f) $47.28; $47.15; −
(g) $12.70; $12.68; −
(h) $74.10; $74.50; +
3. (a) $132,432
(b) $8,432
(c) $7,568
5. (a) 31,114
(b) $18,500

Assignment 24.2

1. (a) $2,700
(b) $86.40
(c) $2,786.40

3. $1,036.80
5. (a) $16,608
(b) $17,042.30
(c) $2,592
(d) $755.14
7. (a) $36,750
(b) $14,700
(c) 0
9. $195,000
11. (a) 590.55 in.
(b) 49.215 ft
(c) 16.35 yd
(d) 9.315 mi
(e) .875 oz
(f) .055 lb
(g) 55 lb
(h) 63.39 pt
(i) 31.71 qt
(j) 7.92 gal

Chapter 25

Assignment 25.1

1. $10,947.58; $3,947.58
3. $84,787.15; $60,287.15
5. $4,211.60; $611.60
7. $6,330.52; $3,530.52
9. $11,443.25
11. $46,140.66
13. $1,626.84
15. $29,138.76
17. $7,590.85
19. $2,121.52
21. $7,513.79
23. $5,688.21
25. $308.99

Assignment 25.2

1. $3,266.19; $633.81
3. $22,561.29; $12,438.71
5. $5,545.27; $2,694.73
7. $1,179.78; $295.22
9. $444.02
11. $9,229.60
13. $2,057.26
15. $2,714.42
17. $2,218.97
19. $836.02
21. $3,366.63
23. $4,884.71
25. $42.88

Chapter 26

Assignment 26.1
1. $40,305.56
3. $82,618.32
5. $22,557.76
7. $75,298.83
9. $1,829.35
11. $38,287.45
13. $7,388.24
15. $15,609.04
17. $2,040.96
19. $28,807.32
21. $1,360.16
23. $23,254,852.40
25. $175.10

Assignment 26.2
1. $24,746.60
3. $34,064.26
5. $25,403.31
7. $20,624.23
9. $1,067.02
11. $1,543.04
13. $57,738.20
15. $1,776.98
17. $913.89
19. $27,910.00
21. $531.43
23. $5,248.95; $65.61; $1,287.89; $3,961.06
25. $2,657.07; $33.21; $1,320.29; $1,336.78

Chapter 27

Assignment 27.1
1. **(a)** Mean: 65.2; 61.9; 63.8
 Median: 60.5; 57; 65
 Mode: 59; 46; 77
 (b) 6; 9; 6; 4; 5
3. **(a)** 49.1
 (b) 49.1
 (c) Financial District Branch
 10 up to 20: 1
 20 up to 30: 3
 30 up to 40: 4
 40 up to 50: 6
 50 up to 60: 8
 60 up to 70: 5
 70 up to 80: 3
 80 up to 90: 0
 Total: 30

University Branch
10 up to 20: 5
20 up to 30: 6
30 up to 40: 2
40 up to 50: 1
50 up to 60: 2
60 up to 70: 4
70 up to 80: 7
80 up to 90: 3
Total: 30

Assignment 27.2
1. **(a)**

Financial District Customer Age

(b)

University Branch Customer Age

3.

Residential Insurance Premiums by Insurance Type

Chapter 28

Assignment 28.1
1. $y = 12$
3. $27 = x$
5. $y = 46$
7. $6 = x$
9. $y = 12$
11. $35 = x$
13. $17 = x$
15. $20 = x$
17. $30,000 = P$
19. $1,280 = V$, or $V = \$1,280$ per item
21. $B = \$44,000$

Assignment 28.2
1. $136,000
3. $356,000
5. $1,500
7. 270 days
9. $434,000
11. $28,700

13. $199,600
15. 8%

Chapter 29

Assignment 29.1
1. 110111; 55
3. 1100111; 103
5. 1011110; 94
7. 111100; 60
9. 10000; 16
11. 1100; 12
13. 11000; 24
15. 111100; 60
17. 100111000; 312
19. 111100; 60
21. 11; 3
23. 1001; 9
25. 1001; 9

Assignment 29.2
1. 780 MB
3. 8
5. 800,000
7. 65.6%
9. 2,044,000
11. 260.4
13. **(a)** 218 minutes
 (b) 129 minutes
 (c) 56 minutes
 (d) 5 minutes
 (e) 1 minute
15. 180

Chapter 30

Assignment 30.1
1. 81
3. 6 days
5. $6\frac{1}{2}$ days
7. $3\frac{1}{3}$ hr, or 3 hr 20 min
9. 4:00 P.M.
11. 1 mi
13. 24,000 letters
15. 14 hr
17. 24,000 nickels
19. 11 yd 1 ft 10 in.
21. 67,500 gallons
23. 77
25. 32, 16

Assignment 30.2

1. 100 mi
3. 1,080 mi
5. 572 trees
7. 180,000 words
9. $\frac{60}{67}$ days
11. 1.7 days less
13. 1,152 in.
15. $787.50
17. $647.50
19. 26

Assignment 30.3

1. b
3. b
5. d
7. d
9. a
11. c
13. d
15. a
17. c
19. c
21. d
23. b
25. a
27. b
29. a
31. c
33. b
35. c
37. c
39. d

Appendix B
Solutions to Self-Check Review Problems

Chapter 1
1. 38
2. 127, 67, 240, 204; 140, 229, 121, 148, 638
3. 2,696
4. 51 (3)
5. 1,274
6. 21 (33)
7. 168,400
8. 81 (2)
9. 32 (12)
10. 2,400
11. 5
12. 32
13. 10,000
14. 222, 313, 205; 1774, 1034; 740
15. 41,216
16. 705,408
17. 28 (4)
18. 640
19. 30
20. 1,600,000

Chapter 2
1. 30
2. 42
3. 96
4. 2
5. $31,256
6. $43,244
7. 427 miles
8. $400
9. $225.75
10. $250
11. 23 hours
12. 19 hours
13. 12 hours
14. 8
15. 156
16. 3
17. 3
18. 13
19. 20
20. $2 and $0.01

Chapter 3
1. 7 yd, 2 ft, 3 in.
2. 10 gal, 1 qt, 0 pt
3. 7 sq yd, 2 sq ft, 122 sq in.
4. 9 yd, 2 ft, 10 in.
5. 5 gal, 1 qt, 0 pt
6. 2,250 lb
7. 60 boxes
8. Vikings
9. 16 oz
10. 600 cu yd
11. 25 cu ft
12. 190 yd
13. 36 cu ft
14. 320 cows
15. 45 gallons

Chapter 4
1. 206.014
2. five thousand twenty-five and three thousand eighty-four ten-thousandths
3. 5.9
4. $14.33
5. 743.64475
6. 18.237
7. 3.737
8. 178.7712
9. 1.797726
10. 259.51
11. 3.23
12. 0.74
13. 7,421
14. 1.29
15. 500
16. $0.80
17. 2,956.98
18. 206,017.59
19. 302.5
20. 11.2037037

Chapter 5
1. $\dfrac{23}{6}$
2. $6\dfrac{1}{4}$
3. $\dfrac{7}{10}$
4. $\dfrac{36}{56}$
5. $1\dfrac{11}{30}$
6. $2\dfrac{1}{24}$
7. $6\dfrac{11}{20}$
8. $\dfrac{4}{15}$
9. $1\dfrac{17}{18}$
10. $2\dfrac{38}{45}$
11. $\dfrac{2}{15}$
12. $1\dfrac{1}{20}$
13. $2\dfrac{1}{3}$
14. $2\dfrac{1}{4}$
15. $2\dfrac{1}{10}$
16. $\dfrac{8}{9}$
17. $6\dfrac{3}{4}$
18. $23\dfrac{15}{16}$
19. $6\dfrac{1}{8}$
20. $10\dfrac{22}{37}$

Chapter 6
1. 87.5%
2. 0.161
3. 0.25%

4. 2.5 or 2.50

5. 60%

6. 0.0005

7. 6

8. 90

9. 160

10. 60

11. 70%

12. 250%

13. $12,500

14. $9,375

15. 100%

16. 50%

17. 1,160

18. 325%

19. $7,600

20. 65%

Chapter 7

1 a. B

 b. D

 c. A

 d. D

 e. C

 f. C

 g. D

 h. D

2.

| Checkbook balance | | $12,583.40 |
|---|---|---|
| Add: Interest | $ 52.50 | |
| Error | 3.00 | 55.50 |
| Total | | $12,638.90 |
| Less: SC | $ 20.00 | |
| NSF ck. | 150.00 | 170.00 |
| Adjusted Balance | | $12,468.90 |

* * *

| Bank statement balance | | $10,961.65 |
|---|---|---|
| Add: Deposit in transit | | 1,850.15 |
| Total | | $12,811.80 |
| Less outstanding checks: | | |
| No. 602 | $ 35.80 | |
| No. 610 | 212.00 | |
| No. 612 | 95.10 | 342.90 |
| Adjusted Balance | | $12,468.90 |

Chapter 8

1. a. $712.50

 b. Social Security, $44.18; Medicare, $10.33

 c. $104.47

 d. $553.52

2. a. Percentage method, $62.17; wage-bracket method, $63

 b. Percentage method, $62.73; wage-bracket method, $63

3. Jan. $1,260.35; Feb. $1,198.35; Mar. $888.35

4. Social Security, $3,844; Medicare, 899; Federal income tax, 7,800; Total $12,543

5. Social Security, $117.80; Medicare, $94.25; Total $212.05

6. $614.08; $532.00; $464.40

Chapter 9

1. $26,500 ($15,800 + $8,600 + $1,500 + $600); state bond interest and inheritance are not taxable

2. $14,000 (5 × $2,800)

3. Standard deduction; the standard deduction for a married couple filing jointly, one aged 65+ is $8,400, which exceeds their itemized deductions.

4. $25,950 ($38,000 − $5,600 exemptions − $6,450 standard deduction)

5. $12,500 ($6,577.50 + ($65,000 − $43,850) × 28%) = $12,499.50, rounded to $12,500

6. $80,750 ($22,250 + ($250,000 − $100,000) × 39%)

Chapter 10

1. Company A has a lower total cost
Company A's cost:
$65,000 × 0.06 = $3,900 sales tax
$65,000 + $3,900 + $1,200 = $70,100
Company B's cost:
$73,500 × 0.9 = $66,150 net of discount
$66,150 × 0.06 = $3,969 sales tax
$66,150 + $3,969 = $70,119

2. a. 1.5%
$650,000,000 × 80% = $520,000,000 assessed value
$7,800,000 + $520,000,000 = 0.015, or 1.5%

 b. $2,100 ($350,000 × 80% = $280,000 × 1.5% = $4,200 ÷ 2 = $2,100); $1,161 ($215,000 × 80% = $172,000 ×

90% = $154,800 × 1.5% = $2,322 ÷ 2 = $1,161)

3. a. $443.50 ($324 ÷ 12 = $27 + $860 = $887 ÷ 2 = $443.50)

 b. $289 ($887 × $\frac{4}{12}$ = $295.67)

Chapter 11

1. a. $2,880.00

 b. $6,380.00

2. a. $3,802.50

 b. $6,602.50

3. a. $2,601.00

 b. $7,101.00

4. a. $6,926.00

 b $6,926.00

5. $5,500.00

6. $2,250.00

7. $7,000.00

8. $3,000.00

9. $1,300.00

10. $6,800.00

11. $4,600.00

12. $4,900.00

13. $8,525.00

14. $29,860.00

15. $2,960.00

16. $22,106.121

Chapter 12

1. a. $120.00

 b. $360.00

2. a. $480.00

 b. $144.00

 c. $576.00

3. a. 65%

 b. $416.00

4. a. 70%

 b. 90%

 c. $1,008.00

5. a. 60%

 b. 80%

 c. 90%

 d. 56.8%

6. a. Aug. 4

 b. Aug. 24

 c. 15.17

 d. $859.38

7. a. Nov. 3

 b. Nov. 13

c. 97%
d. $1,782.86
8. a. $10,204.08
b. $6,335.92

Chapter 13

1. a. $40.30
b. $207.38
c. $1,270
d. $522.63
2. a. $250
b. $750;
3. a. $23.40
b. $59.40
4. a. 140%
b. $123.20
5. a. 160%
b. $672
6. a. 200%
b. $360
7. a. 140%
b. $60
8. a. $240
b. 100%
9. a. $400
b. 25%
10. a. $13
b. $39
11. a. $50.40
b. $117.60
12. a. 60%
b. $744
13. a. 25%
b. $132
14. a. 40%
b. $2,400
15. a. 75%
b. $48
16. a. $240
b. 50%
17. a. $400
b. 20%
18. a. $12
b. 60%
c. 37.5%

Chapter 14

1. a. $44.10
b. $43.50
c. $0.60

2. a. $114.58
b. $113.01
c. $1.57
3. a. $46.75
b 45.91
c. $0.84
4. a. $124.00
b. $124.62
c. 0.62
5. a. $140.00
b. $133.33
c. $6.67
6. a. $60.00
b. $59.00
c. $1.00

Chapter 15

1. a. Oct. 15, 2002
b. $2,550.00
2. a. Feb. 9, 2004
b. $12,225.00
3. a. 92 days
b. $1,428.62
4. a. 122 days
b. $9,059.94
5. a. Jan. 22, 2003
b. $25,345.21
6. a. Sep. 16, 2002
b. $4,815.06
7. a. 61 days
b. $992.28
8. a. 122 days
b. $16,665.70
9. $4,200.00
10. 7%
11. 219 days
12. $15,140.00

Chapter 16

1. a. $32.00
b. 20 days
c $8.59
d. $23.41
2. a. $135.00
b. $4,365
c. 9.28%
3. a. 31 days
b. $26,000
c. $176.66
d. $12,823.34

e. $13,176.66
f. 30 days
g. $13,176.66
h. $86.64
i. $13,263.30
j. $13,176.66
k. $0
4. a. February 8
b. 60 days
c. $61.15
d. $3,038.85
5. a. $77.85
b. $3,037.85
c. September 12
d. 59 days
e. $73.66
f. $2,964.19

Chapter 17

1. a. 6%
b. 9%
c. 15%
d. 8%
2. a. 1%
b. $\frac{1}{3}$%
c. 1.75%
d. 1.5%
3. a. $26.72
b. $2,387.35
4. a $37.50
b. $1,037.50
c. $2,000.00
d. $2,000.00
e. $25.00
f. $1,025.00
g. $1,000.00
h. $1,000.00
i. $12.50
j. $1,012.50
5. 15%
6. $1,025.10
7. a. $37.50
b. $987.60
c. $2,012.40
d $2,012.40
e. $25.16
f. $999.94
g. $1,012.46
h. $1,012.46
i. $12.66

j. $1,025.12

k. $1,012.46

Chapter 18

1. Jim's company, $5,300; Joshua's company, $0

2. $313.20

3. $2,695

4. $29,250

5. $90,000

6. $4,389

7. $3,255

8. $730

Chapter 19

1. 80

2.

| | Inventory Value | Cost of Goods Sold |
|---|---|---|
| **a.** Average cost: | $86,385.00 | $352,115.00 |
| **b.** FIFO cost: | $87,562.50 | $350,937.50 |
| **c.** LIFO cost: | $83,125.00 | $355,375.00 |

3. $76,000

4. a. Average inventory: $38,600

b. Cost of goods sold: $271,800

c. Turnover: 7.04

Chapter 20

1. a. 12.5%

b. 25%

c. 25%

d. 40%

2. Year 1, 4/10; year 2, 3/10; year 3, 2/10; year 4, 1/10

3. Declining Balance and MACRS

4. a. $9,000

b. $71,000

c. $3,000

d $0.90

e $2,124

5. a. $9,000

b. $13,500

6. a. $9,280

b. $16,620

7. a. $1,040

b. $1,487

Chapter 21

1. a $285,000

b. $382,000

c. Assets, 14.8%; Liabilities, −1.5%; Equity, 34.0%

2. Gross Profit, $139,650; 30.3%

3. 16.3%

4. a. 16.67%

b. −25.00%

c. Cannot divide by 0

d. 0.00%

e. −100.00%

5. a. 1.65:1

b. 1.08:1

c. 1.05

d. 17.72%

e. 10.79%

f. 0.56:1

Chapter 22

1. a. 365,100 shares

b. $68.125

c. Boeing, $68.14 − $2.18 = $65.96; Chevron, $82.50 + $3.50 = $86

d. $72.50 − $63 = $9.50

e. $82.50 ÷ 16 = $5.16

2. a. $32,050.00

b. $18,986.25

c. $8,280

3. a. $660.10 gain

b. $660.10 ÷ $8,119.95 = 8.1%

4. a. $1.80 × 4 = $720
$720 ÷ $8,119.95 = 8.9%

b. $720 + $660.10 = $1,380.10 ÷ $8,119.95 = 17.0%

5. 300 × 4 = 1,200 shares

6. 300 × $20 × 8% = $480 pref. divd.
1,200 × $0.60 = $720 common divd.
$720 − $480 = $240 more

7. 6,000 × 50 × 7% = $21,000 ÷ 6,000 = $3.50/sh pref
$65,000 − $21,000 = $44,000 ÷ 50,000 sh = $0.88/sh. common

8. $21,000 × 2 = $42,000 ÷ 6,000 = $7/sh pref.
$75,000 − $42,000 = $33,000 ÷ 50,000 = $0.66/sh common

Chapter 23

1. a. $12,360 ($12,000 × 103%)

b. Semiannually

c. $510 ($12,000 × 8.5% × $\frac{1}{2}$)

d. $119 ($12,000 × 8.5% × 42 days ÷ 360)

e. $12,479 ($12,360 + $119)

f. Premium (103 = 3% above face value)

g. $360 ($12,000 × 3%) or ($12,360 − $12,000)

h. 2014

i. 8.25% ($1,020 annual interest ÷ $12,360)

j. 8.13%; $360 premium ÷ 12 yrs = $30 amortization; $1,020 − $30 = $990 annual interest adjusted for amortization; ($12,000 + $12,360) ÷ 2 = $12,180 average principal invested; $990 ÷ $12,180 = 8.13% yield to maturity

2. a. 150 (6 × 25)

b. $7,837.50 (150 sh. × $52.25)

c. $2,317.50 gain; $7,837.50 − $5,520 ($6,000 × 92%)

d. No, the market value of the bonds ($6,000) is greater than the market value of the stocks (150 shares at $38 = $5,700)

3. $10,500,000
$50,000,000 × 9.5% = $4,750,000 interest annually
$50,000,000 × 8% = $4,000,000 interest annually $ 750,000
per year × 14 yr (20 − 6) = $10,500,000

Chapter 24

1. $54.61 U.S.

2. 183.12 Australian dollars

3. $2,730.30 U.S.

4. $2.29 U.S.

5. $378,500 U.S.

6. $18,500 U.S.

7. $54,003 U.S.

8. 3,412,193 more Turkish lira

9. $921.60

10. $2,135

Chapter 25

1. a. $5,725.64

b. $1,725.64

c. $21,762.75

d. $9,262.75

e. $78,460.75

f. $53,460.75

g. $34,593.98

h. $24,993.98

2. a. $15,056.10

b. $14,943.90

c. $2,474.88

d. $6,325.09

e. $9,784.88

f. $4,215.12

g. $30,319.30

h. $24,680.70

3. $14,881.30

4. $1,547.99

5. $6,691.12

6. $4,104.25

Chapter 26

1. a. $76,054.65

b. $16,054.65

c. $4,112,350.50

d. $3,362,350.50

e. $1,087.38

f. $18,252.40

g. $449.55

h. $2,316.38

2. a. $15,932.54

b. $2,067.46

c. $60,076.16

d. $22,423.84

e. $3,241.27

f. $29,101.59

g. $3,724.02

h. $106,201.00

3. $20,132.25

4. $407,768.37

5. $1,827.79

6. $6,704.16

Chapter 27

1. a. Mean = 64

b. Median = 62

c. Mode = 59

2. a. 50 up to 60: 4

b. 60 up to 70: 8

c. 70 up to 80: 3

d. 80 up to 90: 5

3.

4. a.

b.

c.

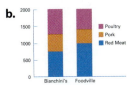

Bianchini's

Chapter 28

1. a. $x = 8$

b. $x = 23$

c. $17 = y$

d. $28 = y$

e. $x = 13$

f. $14 = x$

g. $x = 15$

h. $49 = y$

i. $x = 11$

j. $y = 9$

k. $7 = y$

l. $4 = x$

m. $x = 1$

n. $x = 13$

o. $69 = y$

p. $19 = y$

q. $6 = x$

r. $y = 20$

s. $16 = y$

t. $x = 12$

2. Rate = 4%

3. Ending inventory = $498,000

4. Variable cost = $71.08

Chapter 29

1. a. 100101 + 111 = 101100

b. 11101 + 10000 = 101101

2. a. 10001 − 101 = 1100

b. 10000000 − 111111 = 1000001

3. a. 111 × 101 = 100011

b. 100000 × 10 = 1000000

4. a. 10010 ÷ 11 = 110

b. 1100100 ÷ 11001 = 100

5. a. 288,000

b. 120,000,000

c. 3,200,000

6. 424,500

7. a. 45,000

b. 216,000

8. 11,250

Chapter 30

1. a. $8\frac{4}{7}$ hr

b. $22\frac{1}{2}$ hr

c. 15 gal

2. a. 18 ft 3 in

b. 22 yds, 2ft, 6 in.

c. 112.5 gal

d. $92,160

3. $11,000

4. 30%

5. a. $1\frac{1}{2}, \frac{3}{4}$

b. 96, 10

6. 37.5%

7. 7 mi, 141 yd, 1 ft, 1 in.

8. a. 10 mph

b. 12:00 noon

c. 4 hr

9. a. Allan, by 1 min

b. 1,200

10. a. 1,600 pennies

b. 7 mi

Appendix C
Solutions to CyberConnections

Chapter 1

Page 6

(a) $206.81
(b) 21,787,762 shares

Page 9

For highway driving, the Navigator would go the farthest. For city driving, the Metro would go the farthest.

Page 17

166 shares

Chapter 2

Page 36

One van: 600 mi per wk ÷ 17 mi per gal = 35.29 gal per wk
 35.29 gal × $1.331 per gal = $46.97
Other van: 250 mi per wk ÷ 17 mi per gal = 14.71 gal per wk
 14.71 gal × $1.331 per gal = $19.58
Gasoline cost per week for both vans: $46.97 + $19.58 = $66.55

Page 43

Costs were higher in the Midwest. The total rate increase was 53 percent.

Chapter 3

Page 62

1. 12 cartridges × 2 oz = 24 oz
 24 oz ÷ 16 oz per pt = 24 ÷ 16 = 1.5 pt
2. 12 × 1.5 = 18 oz

 $18 ÷ 24 = \dfrac{3}{4}$ $\dfrac{3}{4} \times 12 \text{ mo} = 9 \text{ mo}$

Chapter 4

Page 76

| | |
|---|---|
| 1999 | 377.743789 |
| 1980 | 126.661949 |
| Diff. | 251.081840 billion dollars more total personal income than in 1980 |

The increase is in <u>total</u> personal income for the entire state, not per capita income. As both population and salaries were higher in 1999, the total personal income also had to be higher.

Page 85

Suggested Answers: "Trading" in decimals means that the buyer and seller can compromise their transaction to the nearest one cent. The seller might be "asking" to sell at $52.34, the buyer might be "bidding" to buy at $52.32, and they might agree to consummate the trade at $52.33. Before decimal trading, they would have compromised at the nearest $1/16$ of a dollar (on the New York Stock Exchange and $1/32$ of dollar on the NASDAQ). The seller would "ask" to get 52\frac{3}{8}$ for one share; the seller would "bid" 52\frac{1}{4}$ for one share. They could compromise only at 52\frac{3}{16}$. "Quoting" stock market prices in decimals, as used in this article, means that the SEC was requiring the stock markets to publish the prices of stocks in decimals even though the actual trades at the stock market were still being made in fractions.

According to SEC Chairman Levitt, decimal trading will benefit investors by allowing greater competition and making it easier to compare prices. It will also bring about consistency with foreign markets, which now trade in decimals. Finally, and perhaps most important, trading in decimals means less money for the brokerage firms and therefore more money for the investors. The reason is that, as you can see from the preceding example, using decimals meant a $0.02 margin between the buyer and seller, whereas in fractions it was a $0.125 margin. This margin is one place where a brokerage firm makes profits.

Chapter 5

Page 108

No. You would only have 2 baths and 2 half-baths. In contrast, if you physically added $1\frac{1}{2}$ cups of sugar to $1\frac{1}{2}$ cups of sugar in one bowl, you would have 3 cups of sugar. You can't physically combine two half baths that are in two separate locations!

Chapter 9

Page 198

28% now, or $11,760; 25% in the plan, or $10,500; savings of 3%, or $11,760 − $10,500, or $1,260.

Chapter 10

Page 216

For a family of 4:

| | |
|---|---:|
| If you earned | $ 31,000 |
| Subtract the poverty level amount | − 18,588 |
| You would have to pay tax on purchases of this amount and over | $ 12,412 |
| | × .15 |
| Total tax | $1,861.80 |
| Income tax for year 2000 | $ 4,654 |

Chapter 11

Page 233

Agency A: 6% × $200,000 + $500 = $12,000 + $500 = $12,500
Agency B: 6.25% × $200,000 = $12,500

At a price of $200,000 the two agencies charge exactly the same amount. If the price is less than $200,000, $190,000 for example, then Agency A charges more. Agency A would charge $11,900 and Agency B would charge $11,875. If the price is more than $200,000, $210,000 for example, then Agency B charges more. Agency A would charge $13,100 and Agency B would charge $13,125.

Chapter 12

Page 254

30 × $35.31 = $1,059.30

| | | |
|---|---|---:|
| First $500: | 0% discount | $500.00 |
| Second $500: | 20% × $500 = $100 discount | 400.00 |
| More than $1,000: | 30% × $59.30 = $17.79 discount | 41.51 |
| Total | | $941.51 |

The amount on the invoice will be $941.51 plus freight, if any. If paid within 10 days, the remittance will be 98% × $941.51 = $922.68 plus freight, if any.

Chapter 13

Page 272

It cannot be referring to markup based on selling price. Suppose an item has a cost of $10. If there is a 100% markup on cost, then the markup amount will be $10. This would result in a selling price of $10 + $10 = $20.

Suppose that a different item has a selling price of $10. If there is a 100% markup on selling price, the markup amount will be $10. But now the cost would be $10 − $10 = $0. A cost of $0 does not make good business sense!

Chapter 19

Page 401

3 months + 1 month = 4-month cycle. 12 ÷ 4 = 3 turnovers
1 month + 1 month = 2-month cycle. 12 ÷ 2 = 6 turnovers

Chapter 20

Page 426

Dollar depreciation: $25,140 − $17,620 = $7,520
Percent depreciation: $7,520 ÷ $25,140 = 0.2991, or 30%
Total depreciation: 55,000 × $0.40 = $22,000
Current book value: $25,140 − $22,000 = $3,140

Chapter 27

Page 596

In general, new orders are increasing from January to June, with April as an exception. Then, new orders decrease from June to December. There are, of course, some recoveries when there were sharp decreases in July and October, but the recoveries were smaller each time.

By the end of the year, new orders were decreasing. The only time since July that there were increases was when new orders actually went negative.

Perhaps the new orders in June and July are for the Christmas retail season. Then, as Christmas gets closer, there is less and less need for more new orders. It is impossible for anyone to make such a generalization based upon just one graph of one year of data and without knowing a significant amount of economics. It would be important, for example, to know exactly the types of goods that are included in Manufacturers' New Orders and what percentage make up each type.

In general, orders increase from January to July and then decrease from July to December. Also, after each two consecutive months of increases, there is a sharp decrease in orders.

First, the scales on the two graphs are different. The fluctuations in the bottom graph are actually much greater.

Second, similar to the top graph, the new orders for durable goods increase in general from January to July and decrease July to December. There is the same general pattern of a monthly decrease following two months of increase. The reason for the similar patterns might be that new orders for durable goods are included as one part of manufacturers' new orders, so that a strong pattern might be reflected. Furthermore, the changes are not so extreme in the top graph, perhaps because there are other manufacturers' new orders that would tend to offset the more extreme changes viewed in durable goods.

Glossary

A

Account purchase. A detailed statement from the commission merchant to the principal.

Account sales. A detailed statement of the amount of the sales and the various deductions sent by the commission merchant to the consignor.

Accounts receivable. Amounts owed to a business for services performed or goods delivered.

Accrued interest. Interest earned from the last payment date.

Accumulated depreciation. The total of all the depreciation recognized up to a specified time.

Acid test ratio. Used to determine the amount of assets that can be quickly turned into cash to pay current liabilities; acid test ratio = total of cash plus receivables ÷ total current liabilities.

Ad valorem duty. Duty that is a percent of the value of the item.

Addends. Any of a set of numbers to be added.

Additional death benefit (ADB). Benefits, available with some life insurance policies, that allow the insured to purchase, at a low rate per thousand dollars of coverage, additional insurance up to the full face value of the policy. In case of death of the insured by accident, both the full value of the policy and the ADB would be paid to the beneficiaries of the insured. If death occurs other than by accident, the full value of the policy is paid but no ADB is paid. Sometimes referred to as accidental death benefit.

Additional discount. Extra trade discount based on purchase volume.

Adjusted bank statement balance. The dollar amount obtained by adding to or subtracting from the bank statement balance checkbook activities not yet known to the bank. This amount should equal the adjusted checkbook balance.

Adjusted checkbook balance. The dollar amount obtained by adding to or subtracting from the checkbook balance those activities appearing on the bank statement that do not yet appear in the checkbook. This amount should equal the adjusted bank statement balance.

Adjusted gross income (AGI). Gross income minus certain income adjustments.

Algebra. A form of mathematics in which letters as well as numbers are used to represent values.

Amortization payment factor. A number which, when multiplied by the per $1,000 loan amount, calculates the amount of each loan payment.

Amortization schedule. A schedule of payments; the schedule shows the amount of interest and the amount of principal in each payment.

Amortization. The process by which a loan's monthly payments are always equal in dollar amount while the interest amount, which is calculated on the unpaid balance, always varies.

Amount credited. The total amount paid plus the amount of cash discount.

Amount of an annuity. The accumulated cash value at the end of the annuity term (at the end of the last period). The amount includes the sum of all payments plus the accumulated compound interest on all payments.

Amount of decrease. The rate of decrease times the base amount.

Amount of increase. The rate of increase times the base amount.

Annual discount amortization. The spreading of the discount amount evenly over the life of the bonds; annual interest ÷ total cost = average annual yield.

Annual percentage rate (APR). The annual equivalent interest rate charged.

Annual premium amortization. The spreading of the premium amount evenly over the life of the bonds.

Annuity. A series of equal payments that are made at regular equal time intervals.

Annuity factors. Present value factors or future value factors.

Annuity insurance. Life insurance that pays a certain sum of money to the insured every year after the insured reaches a specified age or until the insured's death.

Assessed valuation. A property value determined by a representative of the local or state government.

Assets. Things of value owned by a business or a person.

Auto collision insurance. Insurance that protects the vehicle of the insured against collision damage.

Auto comprehensive insurance. Insurance that protects the vehicle of the insured against fire, water, theft, vandalism, falling objects, and other damage not caused by collision.

Auto liability and property damage insurance. Insurance that protects the insured against claims resulting from personal injuries and property damage.

Automatic teller machine (ATM). A computerized electronic machine, many of which are located outside of banks and in numerous other locations, that allows customers to perform various banking functions, such as checking balances, making deposits, and withdrawing funds.

Average. A single number that represents a group of numbers. The process of determining that single number.

Average annual yield. The average annual income rate from an investment; annual interest ÷ total cost = average annual yield.

Average cost method. A method of valuing inventory that is based on the assumption that the costs of all items on hand are averaged and shared evenly among all units.

Average daily balance. The sum of each day's balance divided by the number of days in the month. Payments are usually included; new purchases may or may not be included.

Average inventory. The inventory average calculated by summing each inventory valuation (determined by physical inventory) and divided by the number of physical inventories over a specified period of time; average annual inventory = (beginning inventory value + ending inventory value) ÷ 2.

Average principal. The average unpaid balance of a note or loan.

Average principal invested. The maturity value plus the cost price then divided by 2.

B

Balance sheet. The financial statement of what is owned (assets), what is owed (liabilities), and the difference between the two (net worth) on a specific date.

Bank charge. A fee for services performed by the bank.

Bank discount. The decrease in value of a discounted note.

Bank statement. A formal accounting by a bank of the adding and subtracting activities that have occurred in one bank account over a stated period of time (usually a month).

Bar graph. A diagram similar to a histogram except that there may not be a number scale along the horizontal axis and that the bars normally do not touch each other.

Base (B). The whole quantity, or 100%, of an amount.

Base value. The amount in the first year, month, or other period of time. A value to which another value is compared.

Basic depreciation rate. A rate of depreciation determined by dividing 100% by the estimated total years of useful life of the item.

Bearer bonds. Bonds with interest coupons attached; title to such bonds is not

recorded in the name of a person but passes from bearer (holder) to bearer with delivery of the bonds.

Beginning inventory (BI). The cost of inventory on hand at the beginning of a time period.

Beneficiary. A person, a company, or an organization that benefits from an insurance policy.

Bit (b). An abbreviation of *binary digit*. It is the smallest unit of information on a computer.

Bits per second (bps). A measurement of the speed of transmission of electronic data.

Board of directors. A group of people elected by the stockholders to oversee the running of a corporation.

Bond ratings. Based on experience and research of bond-rating firms, bond ratings give buyers some indication of how safe their bond investments are.

Bonds. Long-term notes issued by a corporation or government entity as a means of borrowing money.

Book value. The original cost of an asset minus accumulated depreciation.

Broker. A person who performs services of buying and/or selling for a commission.

Business statistics. The collection, tabulation, and/or presentation of number information about business operations.

Byte (B). A series of eight consecutive bits. It is a computer unit capable of holding a single character.

C

Callable bonds. Bonds that have a provision for the issuer of the bonds to repurchase or call the bonds in at specified dates if the board of directors authorizes the retirement (payoff) of the bonds before their maturity date.

Cancel. "Divide out" common factors that occur in both the numerator and denominator.

Cancellation. Process of dividing out common factors.

Capital stock. The general term applied to the shares of ownership in a corporation.

Cash discount. A reduction in an invoice amount available to the buyer for paying all or part of the amount due within a stated period of time.

Cash surrender value. The amount of cash that a company will pay the insured on the surrender, or "cashing-in," of an insurance policy.

Charges. The commission and any other sales expenses, such as transportation, advertising, storage, and insurance.

Charter. The basic rules under which a corporation operates.

Check. A written order directing the bank to pay a certain sum to a designated party.

Checkbook. Checks and check stubs to record deposits, withdrawals, check numbers, dates of transactions, other additions or subtractions, and the account balance.

Check register. A place for recording important information about each transaction.

Child tax credit. Taxpayers with dependent children under age 17 can receive a credit of $500 per qualifying child. The credit phases out at higher income levels.

Classes of data. Groups of organized individual values.

Coefficient. The constant value in an algebraic expression which is the product of a variable and a constant (as 2x).

Coinsurance clause. An insurance policy clause specifying that, if a property is not insured up to a certain percentage of its value, the owner is the bearer of part of the insurance and will not be covered for the full amount of damages.

Commercial paper. Documentation of a promise to repay a loan or pay for merchandise.

Commission. Payment to an employee or to an agent for performing or helping to perform a business transaction or service. A fee charged by the broker for buying or selling stock for a client.

Commission merchant. A person who performs services of buying and/or selling for a commission.

Common denominator. A denominator that is shared by two or more fractions. The product of the denominators of two or more fractions is always a common denominator.

Common stock. The usual type of stock issued by a corporation. Most common stock gives it owners one vote for each share owned.

Comparative bar graph. Two bar graphs can be drawn together on one grid to compare statistical data.

Complement method. A method for finding the net price.

Complement rate. A rate equal to 100% minus the discount rate; used with the complement method in determining trade or cash discounts.

Complex equation. An equation that has both a multiplication/division component and an addition/subtraction component.

Component bar graph. A bar graph that is constructed to show how certain data is composed of various parts.

Compound amount. The original principal multiplied by the compound amount factor.

Compound amount factors. Number factors used to shorten the process of calculating future values (compound amounts).

Compound interest. The accumulation of simple interest applied several times during the term.

Compound interest table. A table used to calculate compound interest.

Consignee. The party to whom a consignment shipment is sent.

Consignment. Goods from a producer to a commission merchant for sale at the best possible price.

Consignor. The party who sends a consignment.

Constant divisors. Divisor is usually the last number entered in and is usually the first number entered in multiplication problems.

Constant multipliers. A calculator feature that provides that a number, once entered, can be automatically repeated as a multiplier by entering the new multiplicand and depressing the equals key.

Convertible bonds. Bonds that have a provision to be converted into a designated number of shares or a designated value of the corporation's stock at the request of the bondholder.

Convertible preferred stock. Preferred stock that gives the owner the option of converting that preferred share into a stated number of common shares.

Corporate bonds. Long-term notes, such as convertible bonds and callable bonds, issued by a company for the purpose of raising funds. Are typically bought and sold on the open market, much like stocks.

Corporation. A body that is granted a charter by a state legally recognizing it as an entity with rights, privileges, and liabilities distinct from those of its owners.

Cost-based markup. The difference between the price and the seller's cost of an item for sale which is calculated as a percent of that cost.

Cost of goods sold. The seller's cost of items (goods) that have been sold during a certain time.

Coupon bonds. Bonds with interest coupons attached; title to such bonds is not recorded in the name of a person but passes from bearer (holder) to bearer with delivery.

Credit. A deposit to a bank account.

Credit balance. A negative difference.

Credit card. Credit extended by a third party.

Credit rating. The reputation of a buyer.

Cross-checking. Adding columns vertically and then adding these totals horizontally.

Cumulative preferred stock. Preferred stock for which all dividends in arrears must be paid before current period dividends can be paid.

Current yield. The average annual yield (income rate) from an investment.

D

Debenture bonds. Bonds that have the backing of the corporation but do not specify any of the corporation's tangible assets.

Decimal equivalent. The presentation of a nondecimal number in decimal form.

Decimal point. The period between two numerals.

Decimal representation. A number containing a decimal part that represents a coin.

Declare a dividend. The authorized distribution of earnings to stockholders; this is done by the board of directors.

Declining-balance (DB) method. A depreciation method based on the assumption that a greater use (and greater productivity) in the earlier years of an asset's life;

thus depreciation is greatest in the first year and less in each succeeding year.

Declining-balance depreciation rate. A multiple of the basic depreciation rate, such as two (double-declining-balance) or 1.5 (150%-declining-balance).

Deductible. A cost that must be paid by the employee before any cost is paid by the insurance company.

Deductible clause. An insurance policy clause that stipulates that the insured will pay the first portion of collision damage and that the insurance company will pay the remainder up to the value of the insured vehicle.

Denominator. In a fraction, the number below the line.

Dependency exemptions. Reductions to taxable income for each of one or more dependents.

Deposit slip. A written form that lists cash and checks being deposited in a bank account and cash received from the amount being deposited.

Depreciation. The decrease in the value of an asset through use.

Derived. Developed.

Difference. The result of subtracting the subtrahend from the minuend.

Discount. A fee charged when someone buys the note before maturity. The difference between a bond's par value and its market value when the market value is less. When bonds are sold at a discount, the yield rate will be higher than the stated (face) rate.

Discount amount. The decrease in value of a discounted note.

Discount broker. A stockbroker who buys and sells stock according to the clients' instructions.

Discount date. The last day on which a cash discount may be taken. The day on which a note is discounted (sold).

Discount method. Useful when you want to know both the net price and the actual amount of the trade discount.

Discount period. A certain number of days after the invoice date, during which a buyer may receive a cash discount. The time between a note's discount date and its maturity date.

Discount rate. The percent used for calculating a trade or cash discount. The interest percent charged by the buyer of a discounted note.

Discounting. A lending method in which the loan fee is subtracted from the face value of the note at the time the loan is made.

Discounting a note. Selling a note before its maturity date.

Dividend. The number being divided.

Dividend in arrears. An unpaid amount of dividends on cumulative preferred stock that is due but will be paid to the holder at a future date.

Divisor. The number used to divide another number.

Dollar markup. The total of operating expenses and net profit. Markup expressed as an amount rather than as a percent.

Double-declining-balance. A method that determines a depreciation amount for the first year that is approximately twice the straight-line rate.

Down payment. A partial payment made at the time of a purchase with the balance due at a later time.

Due date. The final day by which time an invoice is to be paid. After that day the buyer may be charged interest. Also the date by which a loan is to be repaid.

Duty. Charge or tax imposed on imported items by the Customs Agency to protect U.S. manufacturers against foreign competition in domestic markets.

E

Effective interest rate. The actual annual rate of interest.

Electronic fund transfers (EFTs). Money that is transmitted electronically, primarily via computers and automatic teller machines.

Employee's earnings record. Summary by quarter of the employee's gross earnings, deductions, and net pay.

Employer's Quarterly Federal Tax Return. A tax report, filed on Form 941 every three months by all employers, that provides the IRS with details about the number of employees, total wages paid, income and FICA taxes withheld, and other figures that determine whether a tax balance is due from the company.

Ending inventory. The cost of the inventory on hand at the end of a time period.

Endowment insurance. Insurance payable upon the insured's death if it occurs within a specified period, and an endowment of the same amount as the policy, payable if the insured is alive at the end of that period.

Equalization. Adjustment of assessments.

Equation. A sentence consisting of numbers and/or letters that represent numbers, divided into two sections by an equals sign (=).

Equivalent single discount rate. A single trade discount rate that can be used in place of two or more trade discount rates to determine the same discount amount.

Estimated physical life. The amount of usefulness that an owner expects to get from an item before it is physically worn out.

Estimated service life. The amount of usefulness that an owner expects to get from an item before it will need to be replaced owing to obsolescence.

Exact interest method. The calculation of interest based on the assumption that a year is 365 (or 366) days long.

Excise tax. A tax assessed on each unit, such as is levied on the sale of gasoline, cigarettes, and alcoholic beverages.

Exemption. A reduction of adjusted gross income that is allowed for each taxpayer, qualifying spouse, and qualifying dependent.

Export. Domestic goods (produced in the United States) for sale in foreign countries.

Export Administration Regulations. International Trade Administration rules and regulations to which all international trade by U.S. businesses is subject.

Extend credit. To give a buyer immediate possession or immediate service with payment due in the future.

Extension. When taking an inventory, the dollar amount derived by multiplying the quantity of an item by its unit price or average cost.

F

Face value. The dollar amount written on a note or bond; it is the same as the amount borrowed, or the principal (P).

Factors. Term used in multiplication to mean *numbers*.

Federal Insurance Contributions Act (FICA). Provides for a federal system of old-age, survivors, disability, and hospital insurance.

Federal Unemployment Tax Act (FUTA). Law that requires employers to pay the IRS an annual tax of 6.2% on the first $7,000 paid to each employee. The federal government uses the money to help fund State Employment Security Agencies, which administer unemployment insurance and job service programs.

Filing status. One of five conditions, including single, married, and married filing separate return, that a taxpayer qualifies for on Form 1040 that will determine such factors as tax rates and allowable deductions.

Finance charge. The fee that the seller charges for the privilege of buying on credit.

Financial statements. Statements presenting financial information about a company; two of these statements are the balance sheet and the income statement.

Finished goods. Those items that have completed the manufacturing process.

First-in, first-out (FIFO) costing method. A method of valuing inventory that assumes that costs for units used or sold are charged according to the order in which the units were manufactured or purchased.

Fixed cost. A cost that remains the same even though the number of units manufactured varies.

Floppy disk. A portable disk used for storage of electronic data.

Foreign trade zones. Domestic sites in the United States considered to be outside U.S. Customs territory.

Form 1040. One of the basic income tax return forms filed by taxpayers.

Form W-2. The wage and tax statement sent to each employee by the employer at the end of the year.

Form W-4. The form used to inform the government of a person's marital status and to claim withholding allowances.

Formulas. Equations written with letters.

For-profit corporation. Business that operates for a profit and whose owners are stockholders.

Fractions. Number expressions of one or more equal parts of whole units.

Fractional equivalent. The presentation of the value of a nonfraction in fraction form.

Frequency. The number of values in each class of data.

Frequency table. A schedule of information presenting a tally of data frequency.

Full service broker. A stockbroker who not only buys and sells upon request for clients but also provides investment advice and manages some of the investments.

Future value. The estimated future amount of money being loaned or invested today or over time.

Future value factors. Number factors used to calculate future values.

Future value of an annuity. A value that will accumulate at a certain time in the future from the investment of annuity payments.

G

Government bonds. Long-term notes, such as treasury bonds and municipal bonds, issued by a government entity for the purpose of raising funds. Are typically bought and sold on the open market, much like stocks.

Graduated commission rates. A system of rates by which graduated commissions increase as the level of sales increase.

Gross cost. The prime cost and all charges paid by the principal.

Gross income. All income received from any source except for that specifically excluded.

Gross proceeds. The price that a commission merchant gets for a consignment; also, the full sales price before any allowances, returns, or other adjustments are considered.

Gross profit. The term used by accountants to describe the difference between the sales price and the cost of the sold goods.

Gross profit method. A method of estimating inventory without a physical count or perpetual inventory system.

Group insurance. Health insurance coverage extended to a group of people. The cost for each person's coverage is less expensive than it would be under an individual policy.

Grouped data. Individual values organized into groups.

H

Hard disk. A magnetic disk used for storage of electronic data, including programs and documents.

Health maintenance organization (HMO). Group health insurance coverage with limited options as a means of keeping health insurance costs lower than that of regular group policies.

HI. Hospital Insurance (medicare).

Higher terms. A fraction in which both the numerator and denominator have been multiplied by the same number.

High-risk driver. A driver with a record of numerous citations or accidents.

Histogram. A diagram that presents the grouped data from a frequency table.

I

Import. A good produced in a foreign country and brought into the United States.

Improper fraction. One whole unit or more. The numerator is greater than or equal to the denominator.

In arrears. A term used to describe unpaid dividends on cumulative preferred stock.

Income statement. The financial statement that shows the revenues, the expenses, and the net income for a certain period of time.

Insured. For life insurance, the person whose life is being insured; for other types of insurance, the person who receives the benefit of the insurance.

Interest. A fee, usually charged for the use of money.

Interest-bearing note. A note that has a maturity value greater than its face value.

Interest dollars. The interest stated as an amount of money rather than as a percent.

Interest period. The period of time between the loan date and the repayment date.

Internet. A global web that connects more than 80 million computers in over 120 countries.

Inventory sheet. A form used for recording information when taking a physical inventory.

Inventory turnover. The number of times the average inventory is converted into sales during the year.

Inventory turnover at cost. Cost of goods sold divided by average inventory for the same period computed at cost prices.

Inventory turnover at retail. Net sales divided by average inventory for the same period computed at retail prices.

Invoice. A document from a seller requesting payment from the buyer; the supplier's bill.

Invoice date. The date stated on an invoice; the beginning of the discount period.

Itemized deductions. Potential reductions to income allowed for certain payments made during the tax year.

J

Junk bond. A bond with a low rating and thus a higher risk.

L

Last-in, first-out (LIFO) costing method. A method of valuing inventory based on the assumption that the inventory on hand at the end of a period of time is composed of the units received first.

Late charge. A fee for invoice payment after a deadline.

Least common denominator. The lowest shared multiple of two or more denominators.

Levy. A government charge or fee.

Liabilities. The sum total of all that a business owes at any point in time; debt.

Limited-payment life insurance. A certain premium to be paid every year for a certain number of years specified at the time of insuring, or until the death of the insured, should that occur during the specified period. The policy is payable on the death of the insured, although there may be some options available at the end of the payment period.

Line graph. Presentation of data on a graph in which the data points are connected by a line.

Liquid. A term used to describe an investment that can be readily sold for, or "converted to" cash.

List price. The price amount listed in the catalog.

Loan initiation fee. A fee charged by a bank or other lender to cover the cost of processing the loan application.

Loan payment schedules. Loan schedules showing each period's beginning unpaid principal balance, its total payment, and the amount of interest and of principal in each payment.

Loan value. The amount that an insured may borrow on a policy from the insurance company.

Long-term credit. Loans that are for longer than 1 year.

Lower of cost or market value (LCM). An inventory valuation method by which the lower amount of either the market value or the cost value is chosen.

Lower terms. A fraction that has been reduced by a common divisor.

Lowest terms. A fraction that cannot be reduced by any common divisor.

Low-risk driver. A driver with a long-standing, clear driving record.

M

Market value. The dollar amount required to replace the inventory as of the inventory date.

Markup. The difference between price and a seller's cost of an item for sale. In dollars it is the amount added to the cost of the goods in order to have a gross profit high enough to cover operating expenses and to make a net profit.

Markup percent based on cost. The percent that is calculated by dividing the desired amount of dollar markup by the cost.

Markup percent. A percent that is used to compute the amount of dollar markup by multiplication. It could be a percent that multiplies the cost to find the dollar markup; or, it could be a percent that multiplies the selling price to find that dollar markup.

Markup percent based on selling price. The percent that is calculated by dividing the desired amount of dollar markup by the selling price.

Markup rate. Markup percent.

Maturity date. The final day of a note (or bond) on which the borrower (the maker of the note or bond) pays the face value and any interest due to the holder of the note or bond. The due date.

Maturity value (MV). For an interest-bearing note (or bond), it is the sum of the face value (principal) and the interest dollars: $MV = P + I$. For a non–interest-bearing note, it is the same as the face value.

Mean. An "average" that is found by dividing the sum of the group of values by the number of values in the group.

Median. An "average" that is the middle number.

Memory (computer). A temporary work space where programs and documents are stored while the computer is running a program. It holds information only while the computer is on.

Merchandise. Goods that a company offers for sale.

Metric system. A decimal system of weights and measures based on meters, grams, and liters. It is used in most European and Asian countries.

Mill. One tenth of one cent, or $0.001; a tax rate may be expressed in mills.

Minuend. Number from which subtraction is being made.

Mixed decimal. A number containing a decimal point and both a whole-number part and a decimal part.

Mixed number. A number that represents more than one whole unit by combining a whole number and a proper fraction.

Mode. An "average" that is represented by the number that occurs most often.

Modem. A device that enables a computer to transmit data over telephone lines.

Modified Accelerated Cost Recovery System (MACRS). The accelerated depreciation method required by the IRS.

Modified complement method. A method for finding the net price when two or more trade discount rates are involved.

Multiplicand. The factor that is multiplied.

Multiplier. The factor that indicates how many times to multiply.

Municipal bonds. Bonds issued by states, cities, school districts, and other public entities.

N

Negotiable note. A note that can be sold by the lender.

Negotiable promissory note. A promissory note that may be sold to a third party.

Net price. The price that a distributor will charge a customer after any trade discounts have been subtracted from the list price.

Net proceeds. The amount sent to the consignor as a result of consignment sales; gross proceeds minus charges.

Net profit. The return for the owner after all expenses have been subtracted from sales.

Net purchase amount. The price of the merchandise actually purchased, including allowances for returns and excluding handling and other costs.

Net revenue. Total revenue less any returns and allowances; frequently called net sales.

Net sales. Total sales for the time period minus sales returned and adjustments made during the same time.

Net worth. The difference between what a business owns (its assets) and what it owes (its liabilities). Also known as owners' or stockholders' equity.

No-fault insurance. Insurance coverage under which the driver of each vehicle involved in an injury accident submits a claim to his or her own insurance company to cover medical costs for injuries to the driver and passengers in that person's own vehicle. The insurance does not cover damage to either vehicle involved in an accident.

No-par stock. Stock issued without par.

Non–interest-bearing note. A note having a maturity value equal to its face value.

Nonparticipating preferred stock. Preferred stock which has a priority right limited to a specified percentage.

Number of compounding periods. The number of time periods of equal length over which interest is calculated.

Number sentence. Numbers used in sentence form.

Number sentence equations. Number sentences in which both sides of the equals sign contain calculations.

Numeric sentence. A mathematical or logical statement, such as an equation, expressed in numbers and symbols.

Numerator. In a fraction, the number above the line.

O

OASDI. Old-Age, Survivors, and Disability Insurance.

Obsolescence. Becoming out-of-date.

Odd lot. Any number of stock shares less than 100.

Odd lot differential. A small extra charge by the broker for buying or selling an odd lot.

Of. "Multiply," particularly when "of" is preceded by the Rate and followed by the Base.

150%-declining-balance. A method that determines a depreciation amount for the first year that is approximately one and one-half the straight-line rate.

Operating expenses. The costs of operating a business.

Ordinary annuity. An annuity in which payment is made at the *end* of each time interval.

Ordinary interest method. The calculation of interest based on the assumption that a year is 360 days long.

Original cost. The cost of building or buying an asset and getting it into use.

Outstanding check. One that has been written but hasn't yet cleared the bank and been charged to the customer's account.

Outstanding deposit. A credit that hasn't yet been recorded by the bank.

Overhead costs. General costs not directly related to sales merchandise.

Over-the-counter. A term that means the securities traded are not listed on a stock or bond exchange.

P

Par. An arbitrary monetary figure frequently assigned to capital stock.

Par value. The face value of a bond; the amount that will be paid to the holder on redemption but not necessarily the market value.

Participating preferred stock. Preferred stock that provides for possible dividends greater than a stated percentage.

Partnership. A business that is not incorporated and has more than one owner.

Payee. Party to whom a check is written.

Payment period. The length of time from one payment to the next.

Payroll register. A summary of wages earned, payroll deductions, and final take-home pay.

P/E Ratio One measure of a stock's value, this number is found by dividing the current price by the per-share earnings as reported by the company for the four most recent quarters.

Percent. A number equal to a fraction whose denominator is 100.

Percentage (P). A portion of the Base.

Percentage method. One of two primary methods for calculating the amount of income tax to withhold from employee paychecks. After the total withholding allowance is subtracted from an employee's gross earnings, the amount to be withheld is determined by taking a percentage of the balance. The percentage to be used is specified by the IRS.

Percent key. A calculator key that eliminates the need to convert a percentage to a decimal before multiplying or dividing.

Percent of cost. Markup that is based on the cost of the goods for sale.

Percent of selling price. Markup that is based on the selling price.

Period. The unit of time over which compounding interest is computed, typically a day, month, quarter, half-year, or year.

Periodic interest rate. The interest rate used to calculate interest at the end of one compounding period.

Periodic inventory. An inventory system that does not require a continuous physical monitoring of the units and unit costs into and out of the inventory stock. The information about the purchases and sales of items is recorded in the accounting system and compared with the results of the physical inventory; an adjust-

ment to the accounting records is made as needed.

Perpetual inventory. A running count of all inventory units and unit costs based on a physical tracking of every item as it comes into and goes out of inventory.

Personal exemptions. Reductions to taxable income for the primary taxpayer and a spouse.

Physical inventory. An actual counting of the inventory.

Pie chart. A diagram that depicts parts of a whole as percent segments of a circle; circle graph.

Powers of 10. Numbers that are multiples of 10, such as 100 and 1,000. The number 1 followed by one or more zeros.

Preferred provider organization (PPO). Group health insurance coverage with benefits based on use of contracted providers as a means of keeping health insurance costs lower than that of regular group policies.

Preferred stock. A type of stock that has a right to share in earnings before earnings are distributed to common stock.

Premium. Fee for insurance coverage, usually paid every year by the insured person. The difference between a bond's par value and its market value when the market value is more. When bonds are sold at a premium, the yield rate will be lower than the stated (face) rate.

Present value. The estimated worth today of money to be paid or received in the future.

Present value factors. Number factors used to calculate present values.

Present value formula. A formula used to calculate present value using a present value factor.

Present value of an annuity. The dollar amount needed at the beginning in order to be able to take out, or pay out, annuity payments of a certain dollar amount at certain regular intervals in the future.

Price-based markup. The difference between the price and the seller's cost of an item for sale that is calculated as a percent of the selling price.

Prime cost. The price that commission merchants pay for the merchandise when they purchase goods for their principals.

Principal. Amount that is borrowed using credit. The person (client) for whom a service is performed.

Proceeds. The amount that a seller receives from the buyer of a note being discounted; the difference between the maturity value and the discount amount.

Proceeds (from sale of stock). The amount received from the sale of stock; the selling price minus the fee or commission.

Product. The answer to a multiplication problem.

Promissory note. An agreement signed by the borrower that states the conditions of a loan.

Proper fraction. Smaller than one whole unit. The numerator is smaller than the denominator.

Property insurance. Insurance against loss of or damage to property.

Property tax. A tax on real estate or other property owned by the business or an individual.

Purchases (*P*). Those goods for sale that have been acquired during the current time period.

Pure decimal. A number with no whole-number part.

Q

Quantity produced. The total number of units built or manufactured.

Quotient. The answer to a division problem.

R

Rate (R). The stated or calculated percent of interest.

Rate (percent) of decrease. The negative change in two values stated as a percent.

Rate (percent) of increase. The positive change in two values stated as a percent.

Rate of return on investment. A rate that approximates the interest rate that owners are earning on their investment in a company; rate of return on investment = net income ÷ owner's equity.

Rate of yield. The ratio between the dividend and the total cost of stock; one indication of the return on investment in the stock.

Rate of yield to maturity. A more refined estimate of investment yield (income rate) than the average annual yield; annual interest + annual discount (or − annual premium) amortization ÷ average principal invested.

Ratio. The relation of one amount to another.

Ratio of accounts receivable to net sales. Indicates the percentage of sales that have not yet bean paid for by customers; ratio of accounts receivable to net sales = accounts receivable ÷ net sales.

Raw materials. Unused items waiting to become part of the goods manufactured.

Reciprocal. The quotient resulting from dividing 1 by the original number; 1 ÷ original number equals the reciprocal of the original number.

Reconciliation of the bank balance. Comparison of the check stubs or check register with the bank statement to determine the adjusted bank balance.

Recovery amount. The maximum amount that an insurance company will pay on a claim.

Registered bonds. Bonds issued to a person whose name is listed or registered with the corporation.

Relationship of net income to net sales. This ratio indicates the portion of sales that is income; relationship of net income to net sales = net income ÷ net sales.

Remainder. A part of a dividend that is left after even division is complete. The leftover part of division into which the divisor cannot go a whole number of times.

Remittance. Amount that a buyer actually pays after deducting a cash discount.

Resale, scrap, or salvage value (SV). The amount of value that an owner of an item expects to receive from disposing of it at the end of the item's estimated useful life.

Round lot. Usually 100 shares.

Rounding down. The process of dropping unwanted digits to the right of a certain digit.

Rounding off. Rounding up or down.

Rounding up. The process of changing a certain digit to the next larger whole value and dropping unwanted digits to the right.

S

Sales. The revenue realized by the seller for providing a good or service to a buyer in exchange for something else of value, usually money or a promise.

Sales tax. A government charge on retail sales of certain goods and services.

Scrap value (SV). The amount the owner of an asset expects to receive upon disposing of it at the end of its estimated service life.

Secured. Guaranteed by certain assets. A bond that is guaranteed by certain of the issuer's assets.

Securities. Stocks, representing ownership, and bonds, representing debt, that usually can be bought and sold, or traded.

Series of discounts. Two or more trade discount rates available to a buyer for different volume purchases.

Short rates. Insurance premium rates charged for less than a full term of insurance.

Short-term credit. Loans that are 1 year or less in length.

Simple interest. The fundamental interest calculation.

Single-payment interest method. The most direct method of computing interest on loans; the method involves use of the formula I = P x R x T.

Sinking fund. An account set up and increased with regular deposits plus accumulated interest intended to pay a large, anticipated future expenditure, such as principal of a corporate bond issue.

Sole proprietorship. A business that has only one owner and is not incorporated.

Solution. The value or number that makes an equation true.

Standard deductions. Set amounts, provided by the IRS, that may be claimed by those individuals who do not itemize their deductions.

State Unemployment Tax Act (SUTA). Any of various laws passed by states that require the employer to pay a tax, such as 5.4% on the first $7,000 paid to each employee, used to help fund unemployment programs.

Stated interest rate. The interest rate stated on the note or bond, the par or face rate.

Statement of account. A report that shows an opening balance, all deposits and credits, all checks paid, withdrawals recorded, bank service charges, general information about the account, and the balance at the end of the period. A bank statement.

Statistical data. Number facts or information that is organized, summarized, and interpreted.

Statistics. A field of study that deals with the collection, organization, analysis, and presentation of quantitative data. It may also refer simply to the actual quantitative data.

Stockbroker. An agent in the purchase and sale of stock for clients; a broker.

Stock certificate. A special paper containing its own serial number that is evidence of ownership of a stock share.

Stock exchanges. Formal marketplaces set up for the trading of stock.

Stock transactions. The purchase and sale of stocks.

Straight (or ordinary) life insurance. Insurance requiring a certain premium to be paid every year until the death of the insured person. The policy then becomes payable to the beneficiary.

Straight-line (SL) method. A depreciation method that distributes the depreciable cost of an item in equal amounts to designated units or periods covering its useful life; (original cost − scrap value) ÷ estimated total life in units or periods of time = depreciation amount for 1 unit or period.

Substitution. The replacement of one value for a different but equal value.

Subtrahend. Number being subtracted.

Sum. The total of two or more addends.

Sum-of-the-years-digits (SYD) method. A depreciation method based on the assumption that greater use (and greater productivity) occurs in the earlier years of an asset's life; the rate of depreciation is greater than the straight-line method but less than the declining-balance method in the earlier years.

T

Tax rate. The percent used to calculate a tax.

Tax Rate Schedules. Tables formulated by the IRS to compute, depending upon filing status, the tax owed for various levels of taxable income.

Taxable income. The amount of income on which the income tax is determined.

Term insurance. Insurance protection issued for a limited time. A certain premium is paid every year during the specified time period, or *term*. The policy is payable only in case of death of the insured during the term. Otherwise, neither the insured nor the specified beneficiaries receive any payment, and the protection stops at the end of the term.

Term of the loan (or note). The period of time between the loan date and the repayment date.

Terms of payment. A statement on the invoice that informs the buyer of any available discount rate and discount date as well as the due date.

Time (T). Stated in terms of all or part of a year, the length of time used for calculating the interest dollars, the rate, or the principal.

Time line. A line representing time onto which marks are placed to indicate the occurrence of certain activities.

Title. Right of ownership.

Total cost. The sum of the variable costs and the fixed costs.

Total cost equation. The formula for determining the sum of all manufacturing costs: Total cost (TC) equals variable cost (VC) times quantity produced (Q) plus fixed cost (FC).

Total cost (for purchaser of stock). The purchase price of the stock plus a brokerage fee.

Total income. Gross income.

Trade discounts. Discounts given to buyers that generally are based on the quantity purchased.

Treasury bonds. Bonds issued by the United States government.

Truth in Lending Act. A federal law to assist consumers in knowing the total cost of credit.

Turnover at cost. Inventory turnover determined by using an average inventory based on inventory valued at cost.

Turnover at retail. Inventory turnover determined using an average inventory based on inventory valued at retail.

U

Ungrouped data. Numbers listed individually.

Unknown. The letter part of an equation.

Unpaid balance. The remaining loan principal not yet repaid.

Unpaid-balance method. A method of calculating interest on loans in which a new interest amount is determined each payment period; the calculation is based on the unpaid principal balance at the beginning of each payment period.

Unsecured. Not secured. A bond that is not guaranteed by the issuer's assets.

U.S. customary units of weight and measure. The standards commonly used in the United States for expressing amounts of liquids and solids (such as cups and pounds) and of distance, volume, and area (such as feet and cubic or square miles), in contrast to the metric system used in much of Europe and Asia.

V

Variable. The letter part of an equation.

Variable costs. A cost that varies with the number of products manufactured.

W

Wage-bracket method. One of two primary methods for calculating the amount of income tax to withhold from employee paychecks. This method starts by granting a deduction for each withholding allowance claimed. The amount for each withholding allowance is provided by the IRS in a table. This method involves use of a series of wage-bracket tables published by the IRS.

Withholding allowance. An amount claimed on tax Form W-4 by an employee that determines how much income tax the employer will withhold from each paycheck. Each allowance claimed (as for a spouse or dependents) reduces the amount of income tax withheld.

Working capital. The amount of current assets less current liabilities.

Working capital ratio. The amount of current assets that would remain if all a company's current liabilities were paid immediately; total current assets ÷ total current liabilities.

Work-in-process. Partially manufactured goods.

Y

Yield. Income from an investment; generally stated as a percent, or rate.

Index

Progress Report

| Part | Chapter | Assignment | Title | Page | Date Assigned | Date Completed | Score/Grade |
|------|---------|------------|-------|------|---------------|----------------|-------------|
| 1 | 1 | 1.1 | Addition | 21 | | | |
| | | 1.2 | Subtraction | 23 | | | |
| | | 1.3 | Multiplication | 25 | | | |
| | | 1.4 | Division | 27 | | | |
| | | 1.5 | Estimating | 29 | | | |
| | 2 | 2.1 | Word Problems, Equations, and Series | 47 | | | |
| | | 2.2 | Word Problems, Formulas, and Equations | 49 | | | |
| | 3 | 3.1 | Weights and Measures | 65 | | | |
| | | 3.2 | Business Applications | 67 | | | |
| 2 | 4 | 4.1 | Addition and Subtraction of Decimal Numbers | 89 | | | |
| | | 4.2 | Multiplication and Division of Decimal Numbers | 91 | | | |
| | | 4.3 | Decimal Numbers in Business | 93 | | | |
| | 5 | 5.1 | Addition and Subtraction of Fractions | 111 | | | |
| | | 5.2 | Multiplication and Division of Fractions | 113 | | | |
| | 6 | 6.1 | Base, Rate, and Percentage | 131 | | | |
| | | 6.2 | Rate of Increase and Rate of Decrease | 133 | | | |
| | | 6.3 | Business Applications | 135 | | | |
| | | 6.4 | Allocation of Overhead | 137 | | | |
| 3 | 7 | 7.1 | Check Register and Check Stubs | 153 | | | |
| | | 7.2 | Check Register and Bank Statements | 155 | | | |
| | | 7.3 | Bank Balance Reconciliation Statements | 157 | | | |
| | 8 | 8.1 | Payroll Problems | 181 | | | |
| | | 8.2 | Payroll, Earnings Record, Payroll Tax Returns | 183 | | | |
| | 9 | 9.1 | Federal Income Tax | 201 | | | |
| | | 9.2 | Federal Income Tax | 203 | | | |
| | 10 | 10.1 | Sales Tax | 219 | | | |
| | | 10.2 | Property Taxes | 221 | | | |
| 4 | 11 | 11.1 | Commission | 237 | | | |
| | | 11.2 | Applications with Commission | 239 | | | |
| | 12 | 12.1 | Trade Discounts | 257 | | | |
| | | 12.2 | Cash Discounts | 259 | | | |
| | 13 | 13.1 | Markup Based on Cost | 275 | | | |
| | | 13.2 | Markup Based on Selling Price | 277 | | | |
| 5 | 14 | 14.1 | Simple Interest | 291 | | | |
| | | 14.2 | Simple Interest Applications | 295 | | | |
| | 15 | 15.1 | Dates, Times, and Interest | 311 | | | |
| | | 15.2 | Using Interest in Business | 315 | | | |

eport